ALLE ZEIT WACH
1842

Jürgen Kiefer

Biological Radiation Effects

With 273 Illustrations

Springer-Verlag Berlin Heidelberg New York
London Paris Tokyo Hong Kong

Professor Dr. Jürgen Kiefer
Justus-Liebig-Universität
Strahlenzentrum
Leihgesterner Weg 217
6300 Gießen

ISBN 3-540-51089-3 Springer-Verlag Berlin Heidelberg New York
ISBN 0-387-51089-3 Springer-Verlag New York Berlin Heidelberg

Library of Congress Cataloging-in-Publication Data

Kiefer, J. (Jürgen), 1936– [Biologische Strahlenwirkung. English] Biological radiation effects / Jürgen Kiefer. p. cm. Rev. translation of: Biologische Strahlenwirkung. Includes bibliographical references.
ISBN 0-387-51089-3 (U.S.: alk. paper).
1. Radiobiology. 2. Radiation – Physiological effect. I. Title. [DNLM: 1. Radiation Protection. 2. Radiobiology.

Printing: Weihert, Darmstadt. Bookbinding: Lüderitz & Bauer, Berlin
2151/3020-543210 – Printed on acid-free paper.

To the memory of
ALMA HOWARD
and
MICHAEL EBERT
my teachers in science and
friends in life

Preface

The biological action of radiation undoubtedly constitutes an issue of actual concern, particularly after incidences like those in Harrisburg or Chernobyl. These considerations, however, were not the reason for writing this book although it is hoped that it will also be helpful in this respect. The interaction of radiation with biological systems is such an interesting research objective that to my mind no special justification is needed to pursue these problems. The combination of physics, chemistry and biology presents on one hand a fascinating challenge to the student, on the other, it may lead to insights which are not possible if the different subjects remain clearly separated. Special problems of radiation biology have quite often led to new approaches in physics (or *vice versa*), a recent example is "microdosimetry" (chapter 4).

Biological radiation action comprises all levels of biological organization. It starts with the absorption in essential atoms and molecules and ends with the development of cancer and genetic hazards to future generations. The structure of the book reflects this. Beginning with physical and chemical fundamentals, it then turns to a description of chemical and subcellular systems. Cellular effects form a large part since they are the basis for understanding all further responses. Reactions of the whole organism, concentrating on mammals and especially humans, are subsequently treated. The book concludes with a short discussion of problems in radiation protection and the application of radiation in medical therapy. These last points are necessarily short and somewhat superficial. They are only planned to illustrate the practical aspects of the topic.

The questions and problems of biological radiation action are pursued by very many groups all over the world. The field is so vast that it seems rather presumptuous of a single author to cover it in a one-volume book. I am fully aware of these difficulties but the reason for attempting it nevertheless was prompted by the wish of many people – particularly university students – to have a short introduction to the subject. The book has been built on lectures which I have given over the years. This is reflected in the general approach: comprehensiveness is neither intended nor is it possible within the framework given. The manuscript does not pretend to be a "textbook" where everything may be found, important aspects are dealt with in a rather exemplary manner. The selection must be subjective and certainly guided by my one field of interest. It is hoped that nevertheless the number of serious omissions is not too high. It should also be said that the author is a physicist by training – although with a burning affection to biology. This very fact may, at places, have led to more mathematics than felt digestible by some readers. I tried, however, always to demonstrate how a certain result is obtained rather than just giving only the final expression. This caused an enlargement of the number of pages and too many formulae. The non-physicist should not be deterred by this because, even if it does not appear so at first sight, the mathematics is in fact not very complicated. Radiation biology is a special branch of quantitative biology, and this cannot exist without mathematics.

The interaction of radiation and biological systems is as old as life itself, it most certainly played an important role in the evolution of self-organizing structures. This aspect could not be covered here. The role of radiation for further evolution is also not negligible, the close similarity between the absorption spectra of ozone and DNA does not seem merely fortuitous.

I have tried to describe the present state of knowledge in an exemplary manner, sometimes superficially of necessity not just to limit the size of the volume. Historical aspects are more or less completely ignored. Although unavoidable within the given scope, this omission is regrettable since biology as a whole has received from radiation biology a number of important findings which reach far beyond this subdiscipline. Examples are the action spectrum for mutation induction which demonstrated the importance of nucleic acids as carriers of genetic information long before it was proven biochemically. The discovery and elucidation of repair processes and their relevance to human health has also to be mentioned in this context. The history of radiation biology has still to be written – it would make fascinating reading.

The bibliography had to be short, it is restricted mostly to review papers which may serve as starting points for further reading. The original sources are found in the figure legends which thus have an additional purpose. I hope that in this way the readability is retained, the volume limited without neglecting to give credit to the original authors.

There is a long way from the inception of the idea to write a book until its final completion. Quite often the author feels the "loneliness of the long-distance runner" and the pressing desire to give up. I owe a lot of thanks to my family and many colleagues and coworkers for efficient coaching on this long path. This was already the case when the first edition which appeared 1981 in German was prepared but even more so with this English version. It grew out of the German book but it is not just a translation but has been amended and extended in many places. The reader is kindly asked to be merciful to the author whose native language is not English. I thank the publisher SPRINGER for all kinds of support – not just with improving the language. My coworker Michael Kost was very helpful in the final writing of the manuscript and patient in the deciphering of my sometimes obscure handwritten notes. Many students and colleagues gave valuable advice – thanks to them all!

I am sure that in spite of good will and critical proofreading there will still be a number of errors for which I have to take the responsibility. I should be grateful if they were pointed out to me for future corrections.

In closing, I sincerely hope that the book might help to arise interest in our fascinating field and win us new friends within and outside the scientific community.

Giessen, fall 1989 Jürgen Kiefer

Contents

Chapter 1
Types of Radiation: Characterization and Sources

This introductory chapter gives a description of the different types of radiation and briefly outlines the means of their production. Spectral distributions are explained and discussed. A final paragraph deals with the fundamentals of radioactivity.

1.1 Types of Radiation

Radiation is the transport of energy without the necessary intervention of a transporting medium. This may be accomplished either by electromagnetic waves or by particles, e.g. electrons, neutrons or ions.

According to PLANCK and EINSTEIN, the energy transfer by electromagnetic waves can also be described by discrete processes involving elementary units called "photons" or "quanta".

Their energy E is given by:

$$E = h \cdot \nu \tag{1.1}$$

where h is PLANCK's constant and ν the frequency.

Since velocity c, wavelength λ and frequency ν are related as:

$$c = \lambda \cdot \nu$$

this also means that:

$$E = \frac{hc}{\lambda}$$

A sometimes convenient unit for the number of quanta is the "EINSTEIN" (often used in photochemistry) which is the number of quanta in multiples of AVOGADRO's number.

Figure 1.1 gives a survey of relations for parts of the total spectrum of electromagnetic waves. Apart from γ and X rays, ultraviolet radiation will be of particular importance in this book. Its lower wavelength limit is not clearly defined; for most practical purposes it may be set at $\lambda = 200\,\text{nm}$.

Further subdivisions are customary:

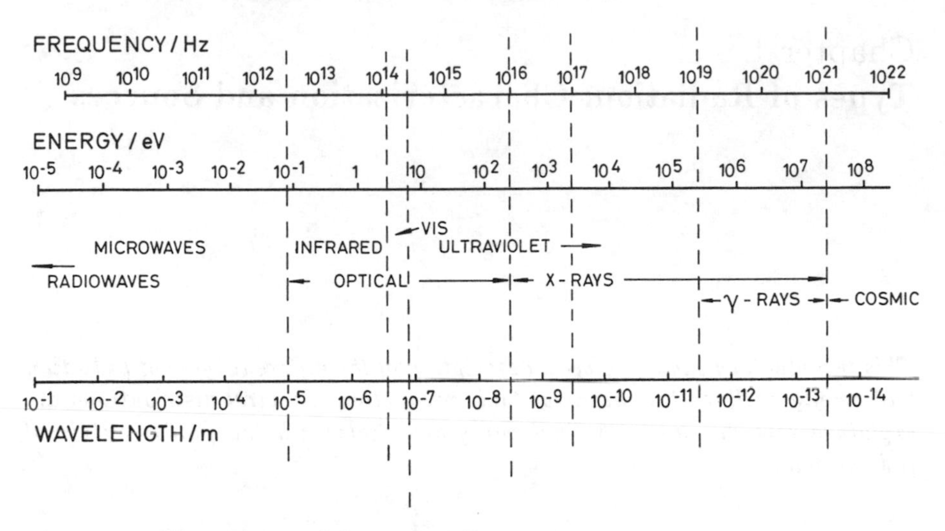

Figure 1.1. The spectrum of electromagnetic waves

far UV: 200–300 nm
near UV: 300–380 nm

or (often used in medicine and biology):

UV A: 315–380 nm
UV B: 280–315 nm
UV C: 200–280 nm

Ionizing corpusculate radiation may consist of charged or uncharged particles. Nuclear physics knows a great variety of such particles, but for radiation biology only a few are important. Table 1.1 lists some examples and their main properties.

Table 1.1. Properties of particles. (After WACHSMANN and DREXLER, 1976)

name	symbol	rest mass m_0 10^{-27} kg	rest mass m_0/m_{e0} [a]	charge	rest energy MeV	half life /s
photon	γ		0	0	0	∞
neutrino, antineutrino	$\nu, \bar{\nu}$		0	0	0	∞
electron	e^-, β^-	$9.1 \cdot 10^{-4}$	1	−1	0.511	∞
positron	e^+, β^+	$9.1 \cdot 10^{-4}$	1	+1	0.511	∞
π-meson	π^-, π^+	0.2489	273.2	−1,+1	139.6	$18 \cdot 10^{-9}$
proton	p^+	1.6725	1836.1	+1	938.26	∞
neutron	n	1.6748	1838.6	0	939.55	700
deuteron	d	3.3443	3675.1	+1	1875.5	∞
α-particle	α	6.6440	7301.1	+2	3727.2	∞

[a] m_{e0}: electron rest mass

The total energy of a particle is the sum of its rest energy E_0 (see below) and its kinetic energy T. The latter is related to the mass m and the velocity v by the well known expression:

$$T = \frac{m}{2}v^2 \tag{1.2}$$

which holds as long as v is very much smaller than c, the velocity of light in vacuo. Therefore, v is often measured in units relative to c, the dimensionless quantity is normally called β:

$$\beta = \frac{v}{c} = \frac{1}{c}\sqrt{\frac{2T}{m}} \quad (v \ll c) \tag{1.3}$$

The following relation is also practical if T is measured in MeV and m in atomic mass units u:

$$\beta = 4.63 \cdot 10^{-2}\sqrt{\frac{T/\mathrm{MeV}}{m/u}} \tag{1.3a}$$

Since the velocity is often the important parameter, which is proportional to the ratio of kinetic energy/mass, sometimes – particularly with accelerated ions – the "specific particle energy", i.e. T/m, is given, usually in MeV/u.

Another quantity is the momentum $\boldsymbol{p}$:

$$\boldsymbol{p} = m \cdot \boldsymbol{v} \tag{1.4}$$

which is related to the kinetic energy by:

$$T = \frac{mv^2}{2} = \frac{p^2}{2m} \quad . \tag{1.5}$$

All expressions given so far are only applicable in "non-relativistic" cases, i.e. where $v \ll c$. If this condition is no longer fulfilled, one has to take into account that the mass is no longer constant but increases with velocity:

$$m = \frac{m_0}{\sqrt{1-\beta^2}} \tag{1.6}$$

where m_0 is the "rest mass". For a free particle, the total energy E is now given by the sum of "rest energy" E_0 and kinetic energy T. According to EINSTEIN's famous equation:

$$E_0 = m_0c^2 \tag{1.7}$$

$$E = T + E_0 \tag{1.8}$$

and

$$E = mc^2 = \frac{m_0c^2}{\sqrt{1-\beta^2}} \tag{1.9}$$

so that

$$T = \frac{m_0c^2}{\sqrt{1-\beta^2}} - m_0c^2 \quad . \tag{1.10}$$

The momentum is then:

$$p = mv = \frac{m_0 v}{\sqrt{1-\beta^2}} \quad . \tag{1.11}$$

Total energy and momentum are related in the following way:

$$E^2 = p^2c^2 + m_0c^4 \tag{1.12}$$

This very useful expression can be verified simply by insertion.

For the relative velocity β, in the general case:

$$\beta = \sqrt{1 - \frac{m_0^2c^4}{E^2}} \tag{1.13}$$

or related to the kinetic energy:

$$\beta = \frac{1 + 2E_0/T}{1 + E_0/T} \quad . \tag{1.14}$$

For practical calculations, when T is measured in MeV and m in atomic mass units (u), the following expression is helpful:

$$\beta = \frac{\sqrt{1 + 1862\frac{m_0/u}{T/MeV}}}{1 + 931\frac{m_0/u}{T/MeV}} \tag{1.14a}$$

and for the special case of electrons:

$$\beta = \frac{\sqrt{1 + \frac{1.022}{T/MeV}}}{1 + \frac{0.511}{T/MeV}} \quad . \tag{1.14b}$$

As an example, one may compare electrons and protons with the same kinetic energy of 1 MeV. While the electron travels with 94% of the speed of light, the respective value for the proton is only about 5%.

Although photons do not possess a rest mass m_0, they must be assigned a momentum which may be derived from Eq. (1.12):

$$p_{photon} = \frac{E}{c} = \frac{h\nu}{c} \quad . \tag{1.15}$$

Consequently, a relativistic particle may also be treated like a "particle wave" with:

$$\nu = \frac{mc^2}{h} \tag{1.16}$$

or

$$\lambda = \frac{h}{mc} \quad . \tag{1.17}$$

Radiation energies are measured either in Joule (J) or electronvolts (eV). Because of the various equivalences given above some numerical relations are summarized in Table 1.2 to facilitate practical calculations.

Table 1.2. Relations between energy units

	eV [a]	J [a]	J mol^{-1} [d]	Hz [b]	kg [a,c]	u [a,c]
eV [a]	1	$1.6 \cdot 10^{-19}$	$9.62 \cdot 10^{4}$	$2.42 \cdot 10^{14}$	$1.78 \cdot 10^{-36}$	$1.07 \cdot 10^{-9}$
J [a]	$6.25 \cdot 10^{18}$	1	$6.02 \cdot 10^{23}$	$1.51 \cdot 10^{33}$	$1.11 \cdot 10^{-17}$	$6.71 \cdot 10^{9}$
J mol^{-1} [d]	$1.04 \cdot 10^{-5}$	$1.66 \cdot 10^{-24}$	1	$2.51 \cdot 10^{9}$	$1.85 \cdot 10^{-41}$	$1.11 \cdot 10^{-14}$
Hz [b]	$4.14 \cdot 10^{-15}$	$6.63 \cdot 10^{-34}$	$3.99 \cdot 10^{-10}$	1	$7.35 \cdot 10^{-51}$	$4.42 \cdot 10^{-24}$
Kg [a,c]	$5.61 \cdot 10^{35}$	$8.99 \cdot 10^{16}$	$5.41 \cdot 10^{40}$	$1.36 \cdot 10^{50}$	1	$6.02 \cdot 10^{26}$
u [a,c]	$9.32 \cdot 10^{8}$	$1.49 \cdot 10^{-10}$	$8.98 \cdot 10^{13}$	$2.26 \cdot 10^{23}$	$1.66 \cdot 10^{-27}$	1

[a] Refers to one particle.
[b] Frequency.
[c] According to $E = m \cdot c^2$.
[d] This is not really an energy unit but signifies the energy content of 1 mole of photons.

1.2 Emission Spectra

Most radiation sources do not emit photons or particles of a single well-defined energy, but cover a whole range. The quantitative description of such a distribution is called the "emission spectrum". This definition, however, is not unambiguous because various quantities may be used which have to be clearly specified. There are the following possibilities:

1. Number of quanta per energy or frequency interval as a function of energy
2. Total energy emitted per energy interval as a function of energy
3. Number of quanta per wavelength interval as a function of wavelength
4. Total energy emitted per wavelength interval as a function of wavelength.

This list is not comprehensive but gives only the most important cases. The actual shapes of spectra may differ drastically depending on the kind of representation; even more so if the abscissa is scaled logarithmically.

This is of no concern, however, when the emission is restricted to few narrow energy intervals. Examples of such "line sources" are low pressure mercury lamps or γ-emitting radionuclides (see below). In most cases there will be a continuous spectrum, perhaps mixed with a few discrete lines. The emission may then be characterized by the number N of quanta emitted over the whole range and a distribution $f_T(E)dE$ which gives the fraction of quanta emitted in the energy interval between E and E + dE. Here the particle spectrum is a function of energy. On the other hand, to describe the fraction of total energy emitted in a certain interval, a different distribution function is used which shall be termed $f_E dE$.

The energy emitted per interval dE is obviously $N \cdot E \cdot f_T(E)dE$. This expression is normalized to yield a distribution function:

$$f_E(E)\,dE = \frac{E\,f_T(E)\,dE}{\int E\,f_T(E)\,dE} \tag{1.18}$$

where the integration has to be extended over the whole range.

Characterization of a given spectrum by "mean" values, e.g. the mean energy is achieved in two ways, depending on the choice of parameters. The mean energy per particle or quantum is given by the denominator in Eq. (1.18). Since it is related to the number of quanta it is termed "number average" E_T. It is also possible, on the other hand, to calculate the "energy average" E_E by using the distribution function $f_E(E)dE$:

$$\overline{E_E} = \int E\,f_E(E)\,dE = \frac{\int E^2\,f_T(E)\,dE}{\overline{E_T}} \quad . \tag{1.19}$$

Both average values are only equal if the spectrum consists of a single narrow line. Considerations of this kind play a role in many aspects of radiation biophysics.

Since energy and frequency are proportional, the above formalism can be immediately applied also to spectra where the frequency ν is the variable. This is not the case, however, if the wavelength is used. It is by no means sufficient to exchange ν and λ; the following has also to take into account:

$$-d\nu = \frac{c}{\lambda^2}\,d\lambda \tag{1.20}$$

yielding:

$$f(\lambda)\,d\lambda = \frac{\nu^2}{c} f(\nu)\,dv \quad . \tag{1.21}$$

This leads, of course, also to new mean values; average frequency ν and average wavelength λ are <u>not</u> related by the usual formula:

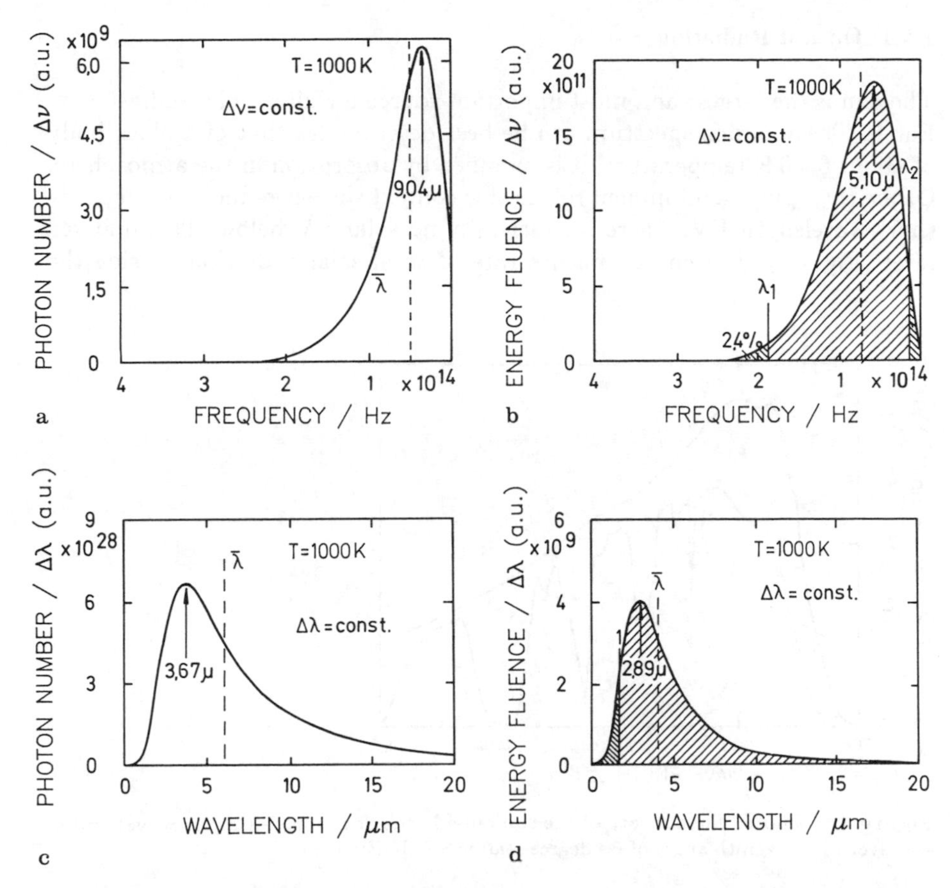

Figure 1.2a-d. The emission of a black body at 1000 K using different representations. (a) photon number as function of frequency, (b) energy fluence as a function of frequency, (c) and (d) as in a and b but as function of wavelength. The mean wavelengths are determined by averaging over the particular spectrum (after SCHULZE and KIEFER 1977)

$$\overline{\lambda} \neq \frac{c}{\nu} \quad !$$

Figure 1.2 illustrates the situation with a black body of 1000 K, temperature as an example. Not only the spectrum shape and the mean values change, but also the maxima are found at different positions.

1.3 Radiation Sources

A comprehensive survey is not intended, only a few representative examples will be described. Also the technical aspects of radiation sources shall not be treated here.

1.3.1 Optical Radiation

The sun is the largest and most important source for ultraviolet radiation on Earth. The emission spectrum can be best described as that of a black body of about 6000 K temperature; it is modified by absorption in the atmosphere. Ozone (O_3) plays a prominent role in this context since it reduces particularly short-wavelength UV. There is practically no solar UV below 300 nm at sea level (Fig. 1.3). The energy fluence rate of total solar radiation outside the

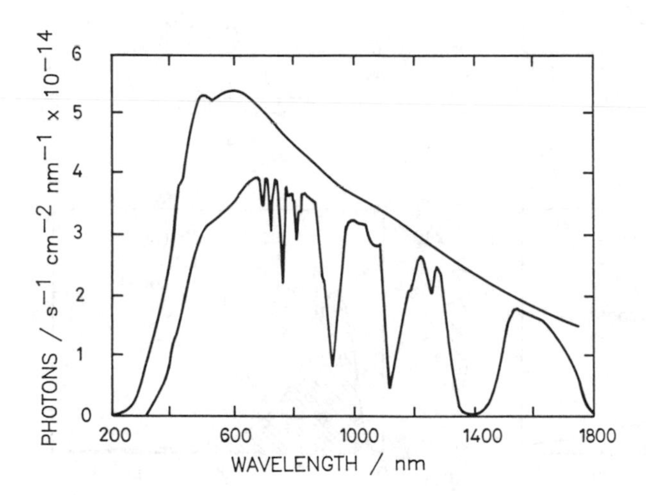

Figure 1.3. Photon fluence rate of the sun outside the atmosphere (upper curve) and at sea level with a zenith angle of 60 degrees (after SELIGER 1977)

atmosphere amounts to 1393 Wm^{-2}, yielding a total power to the Earth of about 10^{17} W. Because of absorption, reflection, and scattering in the air, the fluence rate is much lower at the Earth's surface, in Central Europe the annual average lies around 125 Wm^{-2}. It should be noted, however, that most of the radiation energy is of longer wavelength, as also seen from Fig. 1.3.

Of the many artificial UV sources, gas discharge tubes are probably of greatest practical relevance, particularly mercury tubes. The emission spectra depend on vapour pressure, e.g. low pressure mercury lamps emit about 85% of their energy at 253.7 nm, the so-called "mercury resonance line", which often serves as a monochromatic reference line. At higher vapour pressure, the spectrum shows more lines, especially at longer wavelengths. Their number may be even increased by the addition of other atoms (e.g. cadmium) which leads to a better coverage of the whole spectrum; examples are shown in Fig. 1.4.

Other UV sources are xenon- and deuterium-discharge tubes with a more-or-less continuous emission spectrum (Figs. 1.5 and 1.6). Fluorescence lamps emit radiation in the UV-B and the UV-A region, as shown by the

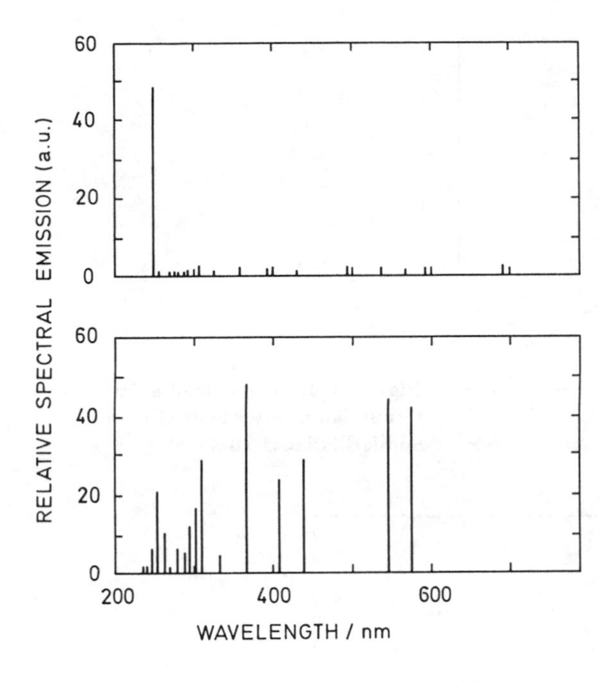

Figure 1.4. Spectra of mercury lamps. Upper panel: low pressure lamp, lower panel: medium pressure lamp (after SCHAEFER and HEINRICH 1977)

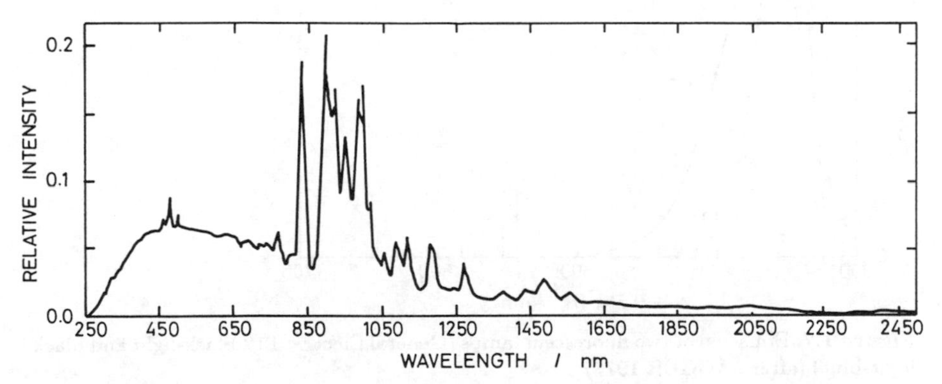

Figure 1.5. Spectrum of a xenon arc lamp (after SCHAEFER and HEINRICH 1977)

examples in Fig. 1.7. If wavelength-dependent phenomena are to be investigated, as often is the case in radiation biology, it is indispensable to produce monochromatic radiation with the aid of grating- or prism monochromators or optical filters.

UV lasers are powerful monochromatic sources; their still very high cost, however, limits their use to rather special scientific applications.

Another source of great potential is "synchrotron radiation" which is produced by circular electron accelerators. It may cover a wide range of energies – from radio waves to very short wavelength UV – but it is, of course, only available at very few places.

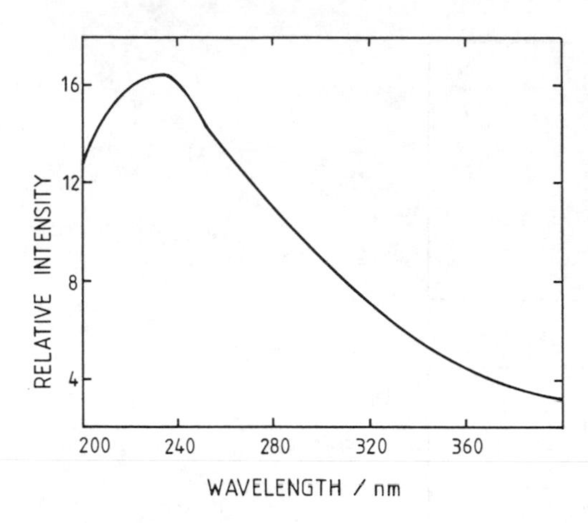

Figure 1.6. Spectrum of a deuterium lamp (after SCHAEFER and HEINRICH 1977)

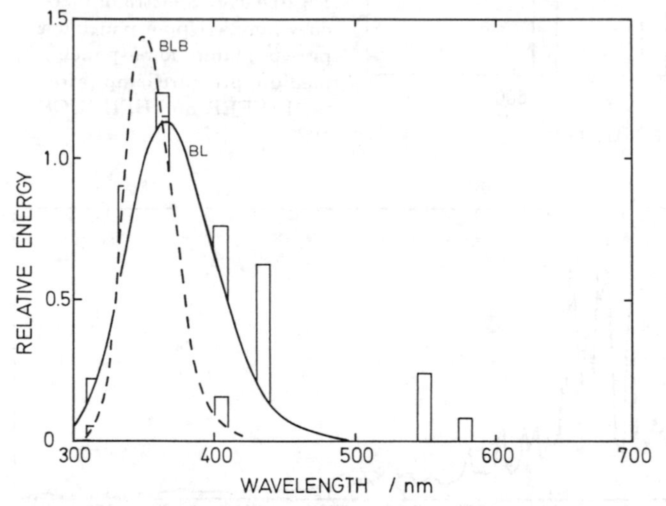

Figure 1.7. Emission of two fluorescent lamps (General Electric T12 black-light and black-light-blue) (after JAGGER 1977)

1.3.2 Ionizing Radiation

There are essentially only two ways to produce ionizing radiations: the acceleration of charged particles, which may then react with suitable targets to yield secondary radiations, and the use of radioactive nuclei (Sect. 1.4).

X rays are generated when accelerated electrons interact with matter. Two processes have to be distinguished: In the first, electrons in the atomic or molecular orbits are excited or ejected. The resultant "holes" are filled by electrons from higher levels; the energy gained is partly emitted as electromagnetic radiation the energy of which depends on the properties of the particular atom. This type is therefore termed "characteristic radiation". The second process is the deceleration of the incoming electrons in the field of the atomic nucleus; the energy loss leads also to the emission of radiation. This "bremsstrahlung" is continuous and covers a wide frequency range whose upper limit is given by the tube voltage. The corresponding spectrum depends on various factors: the material of the anode and of the outlet window, internal and external filtering etc. Low energy "soft" components may be suppressed by the use of metal filters (copper, aluminum) since they are more easily absorbed than the more energetic ones.

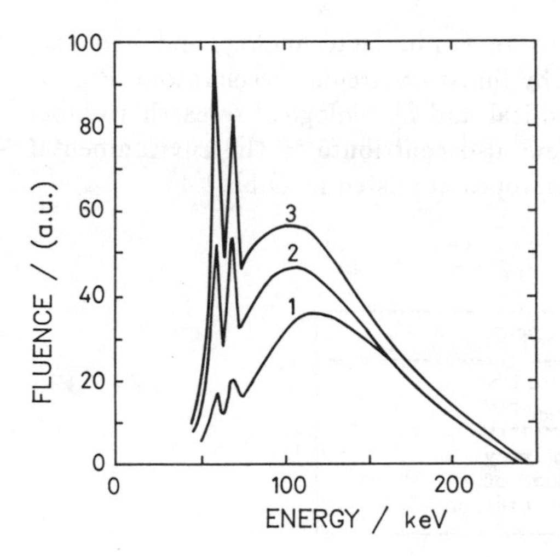

Figure 1.8. Emission of an X-ray tube with tungsten anode operated at 250 kV. 3: without external filtering, 2: 2 mm copper, 1: 2.7 mm copper. The characteristic lines can be clearly seen

Figure 1.8 shows the spectrum of a 250 kV X-ray tube with a tungsten anode and various filters. The lines of the characteristic radiation are clearly seen, as well as the influence of filtering. It is also obvious that the maximum of emission lies below 50% of that corresponding to the tube voltage.

γ *Rays* are emitted as part of nuclear disintegration. In contrast to X rays, they have well-defined energies which are typical for the emitting radionuclide. Otherwise, there is no difference between X and γ rays, especially as far as physical interactions and biological effects are concerned.

Gamma-ray emitters play a role in a number of fields: they may be used as convenient radiation sources in technical and medical applications, as physiological tracers in nuclear medicine, and they also add to the environmental background radiation (Chap. 22). Technical applications generally require long half-lives (apart from a suitable energy), while in nuclear medicine short half-lives are desirable to minimize the patient's radiation burden. Nearly all γ-emitting radionuclides also have a β component which has to be taken into account for medical applications; Table 1.3 lists some examples, more detailed information is found in Table 1.6.

Electrons are also emitted as part of radioactive decay (β radiation), but these nuclides do not play a great role as practical irradiation sources,

Table 1.3. Some important γ-emitting nuclides

name	significance	other decay products
^{60}Co	technical	β: 0.31 MeV
^{137}Cs	technical	β: 0.51 MeV
^{125}I	medical	β: 0.03 MeV
^{131}I	medical	β: 0.061 MeV
^{99m}Tc [a]	medical	-
^{198}Au	medical	β: 0.97 MeV

[a] 'm' indicates an excited metastable nucleus

except for investigating internal exposure. For use in technology and medicine, high-energy electrons are produced by linear or circular accelerators. Pure β emitters are mainly used in biochemical and physiological research to label metabolic key substances. β emitters also contribute to the environmental background radiation. A few such isotopes are listed in Table 1.4.

Table 1.4. Some important β-emitters

name	significance
^{3}H (tritium)	biochemistry
^{14}C (radiocarbon)	biochemistry
^{32}P	biochemistry
^{35}S	biochemistry
^{90}Sr/^{90m}Y	byproduct of nuclear fission

Mesons constitute an important component of secondary cosmic radiation. They are produced in the upper layer of the atmosphere via the interaction of high-energy protons. Recently, the possible application of π^- mesons in radiotherapy has gained some interest, particularly because of their favourable depth-dose distribution (Chap. 4). Their production, however, is rather costly: they are created by the interaction of protons, neutrons or α particles with atomic nuclei, where the kinetic energy has to be larger than the rest energy of the meson (1.4×10^8 eV for the π^- meson). The production cross sections are very small (about 10^{-29} m^2 for protons of 10^8 eV) resulting in very low efficiencies. The acceleration of the projectiles requires large installations, consequently they exist only in few places. Medical investigations are performed in Europe at the SIN (Schweizer Institut für Nuklearforschung) near Zurich, Switzerland, which is a joint venture of a number of European countries. There, 590 MeV protons are used as projectiles. Since uncharged π^0 and π^+ mesons are formed as well as π^- mesons, they have to be separated by magnetic deflection. The available intensities are rather small, dose rates are in the order of 0.1 Gy/min (see also Chap. 4).

α particles are helium nuclei consisting of two protons and two neutrons which are also produced by radioactive decay. The application of the respec-

tive radionuclides is quite limited since the particle energies are comparatively low and hence their range in matter very small. They are useful, however, for fundamental studies in cellular radiation biology. Nearly all α emissions are accompanied by γ components which may not be neglected because of their much larger penetration. Table 1.5 lists a few typical representatives.

Table 1.5. Some important α-emitters

name	significance	other decay products
^{222}Rn	natural	negligible
^{226}Ra	natural	γ: 0.186 MeV
^{239}Pu	nuclear weapons and reactors	negligible
^{241}Am	laboratory application	γ: 0.06 MeV

Ion accelerators are of more practical relevance. It is possible today to accelerate ions of all elements (up to uranium) up to relativistic energies (several GeV per atomic mass unit). This is achieved by the combination of a linear accelerator as injector and a circular accelerator as a second stage (BEVALAC, Berkeley, USA, in the planning state: GSI, Darmstadt, Germany). Since in this case the primary particles are accelerated the intensities can be quite high (up to about 10^{12} particles per second).

Neutrons are produced by a number of nuclear reactions. Historically important is that of α particles with beryllium where either ^{226}Ra or ^{241}Am is used as primary source. Another useful projectile is $^2H^+$ (deuteron) which may be produced in accelerators. If it interacts with a beryllium target, neutrons in the energy range between 3 and 8 MeV are produced. High intensities are thus possible.

Neutron production via the reaction of deuterons with tritium is comparatively simple. The maximum efficiency is reached with deuteron energies around 200 keV, which can be easily achieved. The problem lies here in the technology of the gaseous (and radioactive!) target. It is solved by adsorption onto suitable metallic carriers like tantalum, zirconium or titanium, but the achievable concentrations are low, and the life-time of tritium is rather short. There are also non-negligible radiation protection problems. The resultant neutrons show a rather narrow energy-distribution around 14 MeV. Sources of this kind have recently gained some popularity in radiation therapy.

A "natural" neutron source is the radionuclide ^{252}Cf (californium), a transuranium element. It decays mainly by α emission but there is also a 3% probability of spontaneous nuclear fission yielding 3.8 neutrons on an average. The energy spectrum (Fig. 1.9) covers a broad range typical for fission neutrons. Presently achievable source intensities are quite low (about 2×10^7 neutrons per second). The half-life which is chiefly governed by the α decay is 2.7 years.

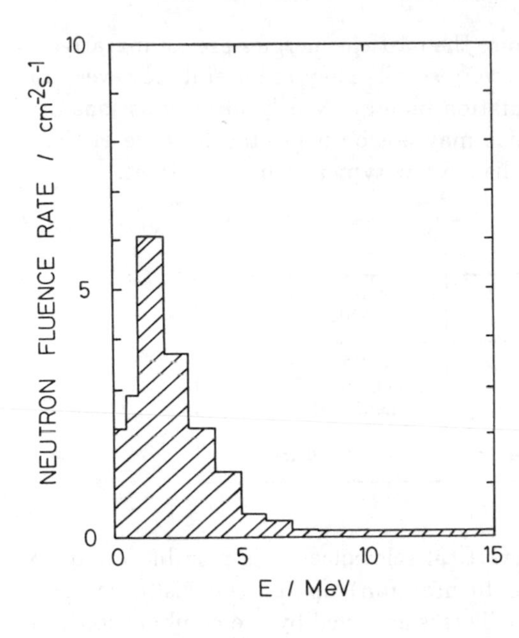

Figure 1.9. Energy distribution of Californium-252-neutrons (after HALL and ROSSI 1974)

The most potent sources of neutrons are, of course, nuclear weapons and nuclear reactors. They are mentioned here only for completeness, their importance as environmental factors is discussed in Chap. 22. Neutrons are also a part of the natural background radiation produced by nuclear reactions in the atmosphere (see also Chap. 22).

1.4 Radioactivity

The term "radioactivity" characterizes the property of certain atomic nuclei to decay spontaneously with the emission of particles and/or electromagnetic radiation. This is a typical example of a stochastic process which can be described by a POISSON distribution (see Appendix I.4). The average number of nuclei disintegrating per unit time – the "activity" A – is directly proportional to the total number N of nuclei:

$$A = -\frac{dN}{dt} = \lambda N \qquad (1.22)$$

λ is the decay constant which is related to the half-life τ, i.e. the time within which 50% of the original nuclei disintegrate:

$$\tau = \frac{\ln 2}{\lambda} \quad . \qquad (1.23)$$

The different types of radioactive decay include:

α decay: A helium nucleus consisting of 2 protons and 2 neutrons is emitted thus reducing the mass number by 4 and the nuclear charge by 2. The emission spectrum is monoenergetic, i.e. consists of a single line characteristic of the given radionuclide.

β decay: In this case an electron is emitted from the nucleus. It is produced by the disintegration of a neutron into a proton, an electron and an antineutrino. The third particle is necessary to conserve the angular momentum. The decay energy is shared between the β particle and the antineutrino so that a continuous β-spectrum results (see Fig. 1.10). Beta decay results in an increase of nuclear charge by 1, leaving the mass number unchanged.

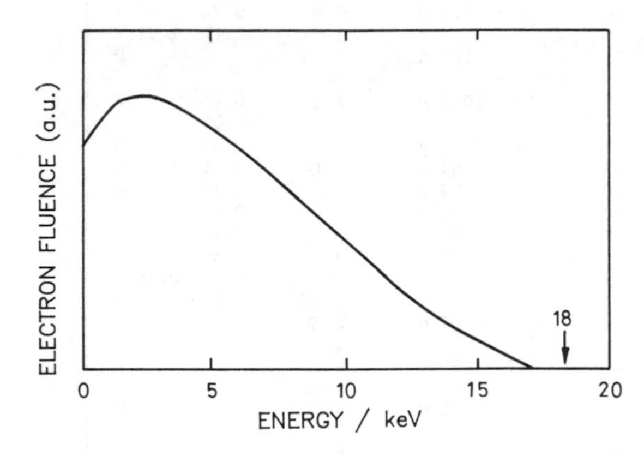

Figure 1.10. Beta spectrum of tritium

In some cases a positron (positive electron) is emitted together with a neutrino leading to a reduction of the nuclear charge. Instead of this process – sometimes also in competition – an electron may be captured from the K shell of the atom ("K capture"; "electron capture", EC). The probability for this to occur increases with higher nuclear charge because of the increased attracting forces between the nucleus and the shell electron. The resultant electron hole in the K shell is filled by electrons from outer orbits leading to the emission of characteristic X rays. Sometimes the energy is used to liberate one or more AUGER electrons from outer shells with X-ray emission (see Sect. 3.2.1.5).

γ emission: α and β decay leave the resultant daughter nuclei generally in an excited state which is deactivated by the emission of γ quanta. They are monoenergetic, reflecting the energy levels of the nucleus. This may be used for the identification of radionuclides (γ spectroscopy). Although γ emission accompanies most decays, it is not always found, e.g. not with a number of light β emitters like ^{3}H, ^{14}C, ^{32}P and ^{35}S.

Radioactive elements are found as natural components in our environment (see below and Chap. 22), but may also be produced artificially from

Table 1.6. Properties of some important radionuclides

charge number	name	symbol	atomic weight	halflife	principal decays	energy MeV	g)
1	hydrogen	H	3	12.3 y	β	0.018	cde
6	carbon	C	14	5730 y	β	0.156	cde
11	sodium	Na	22	2.6 y	β^+,EC	0.55	d
					γ	1.3	
15	phosphorus	P	32	14.3 d	β	1.71	c
16	sulphur	S	35	87 d	β	0.167	c
19	potassium	K	40	1.3×10^8 y	β,EC	1.4	d
					γ	1.46	
24	chromium	Cr	51	27.8 d	γ,EC	0.32	b
26	iron	Fe	59	45 d	β	0.46, 0.27	b
					γ	1.1, 1.3	
27	cobalt	Co	57	270 d	γ,EC	0.14	b
			58	71 d	β,EC	0,47	b
					γ	0.81	
			60	5.3 y	β	0.32	a
					γ	1.17, 1.31	
34	selenium	Se	75	120 d	β		b
					γ		
36	krypton	Kr	85	10.8 y	β	0.47	e
					γ	0.05	
37	rubidium	Rb	87	5×10^{11} y	β	0.27	d
38	strontium	Sr	85	64 d	γ,EC	0.51	b
			89	52 d	β	1.46	e
			90	28 y	β	0.55	e
39	yttrium	Y	90	64 h	β	2.27	e
40	zirkonium	Zr	95	65 d	β	0.4	e
					γ	0.72, 0.76	
43	technetium	Tc	99m	6 h	γ(IT)	0.143	b
53	iodine	I	125	60 d	β,EC		b
					γ		
			129	1.7×10^7 y	β		e
					γ		
			131	8 d	β		be
					γ	0.36	
54	xenon	Xe	133	5.3 d	β	0.35	be
					γ	0.08	
			135	9.2 h	β	0.92	e
					γ	0.25	
55	caesium	Cs	134	2.1 y	β	0.66	e
					γ	0.6, 0.8	
			137	30 y	β	0.51	ae
					γ	0.66	
79	gold	Au	198	2.7 d	β	0.96	b
					γ	0.41	
80	mercury	Hg	203	47 d	β	0.21	b
					γ	0.28	
86	radon	Rn	220	55 s	α	6.3	d
			222	3.8 d	α	5.5	d
88	radium	Ra	226	1600 y	α	4.8	d
					γ	0.19	
90	thorium	Th	232	1.4×10^{10} y	α	4.0	d
					γ	0.06	
94	plutonium	Pu	238	86 y	α	5.5	e
			239	2.4×10^4 y	α	5.1	e
95	americium	Am	241	458 y	α	5.5	ae
					γ	0.06	
98	californium	Cf	252	2.6 y	α,SF	6.2	a

g) Abbreviations for last column (significance):

a technical, b medical, c biochemical (laboratory), d environmental,
e nuclear power reactors, f lab. radiation sources

EC : electron capture
IT : internal conversion
SF : spontaneous fission

virtually every member of the periodic system. Some representative examples are listed in Table 1.6 together with their main properties.

Decay series
Quite often the radioactive decay does not immediately lead to a stable end product but rather to other unstable nuclei so that a whole decay series may result. The most important examples are those starting from very heavy naturally occurring nuclei. Since the mass number changes only with α-decay, all members of a series may be classified according to their mass numbers. There are four of these series, usually named after the "parent" isotope; these are listed in Table 1.7. The total schemes of those which can still be found on Earth are given in Fig. 1.10. The neptunium series is no longer existent since the half-life of the neptunium is so short that it has already completely decayed. The other three series which still do exist are of primordial origin, i.e. they were formed at the birth of our planet. Many members of the natural series have been given historical names which coincide very rarely with their actual chemical nature. They are listed in Table 1.8 for reference.

The parent isotopes starting products of the decay series are not the only radionuclides surviving since the Earth's origin. Other examples of "radiofossiles" are ^{40}K and ^{87}Rb. Their relevance and that of other natural radionuclides will be discussed in Chap. 22.

Several shorter decay series also exist, as, for instance that of ^{90}Sr (Fig. 1.12): This has a half-life of 28 years decaying by β-emission (0.55 MeV) to ^{90}Y, which in turn disintegrates with a half-life of 64 hours to the stable ^{90}Zr emitting an electron of 2.27 MeV maximal energy. Most of the radiation energy is, in fact, contributed by the daughter nucleus.

In hazard assessments, all members of a decay series must have to be taken into account. The relations are described by a system of linear differential equations:

Table 1.7. Natural decay series

name	start nucl.	τ/a [a]	final nucleus	mass formula	range of n
thorium-series	$^{232}_{90}$Th	$1.4\cdot10^{10}$	$^{208}_{82}$Pb	M=4n	58..52
neptunium-series	$^{237}_{93}$Np	$2.2\cdot10^{6}$	$^{209}_{83}$Bi	M=4n+1	60..52
uranium-radium-s.	$^{238}_{92}$U	$4.5\cdot10^{9}$	$^{206}_{82}$Pb	M=4n+2	59..51
uran.-actinium-s.	$^{235}_{92}$U	$7.1\cdot10^{8}$	$^{207}_{82}$Pb	M=4n+3	58..51

[a] half life of the start nucleus

a

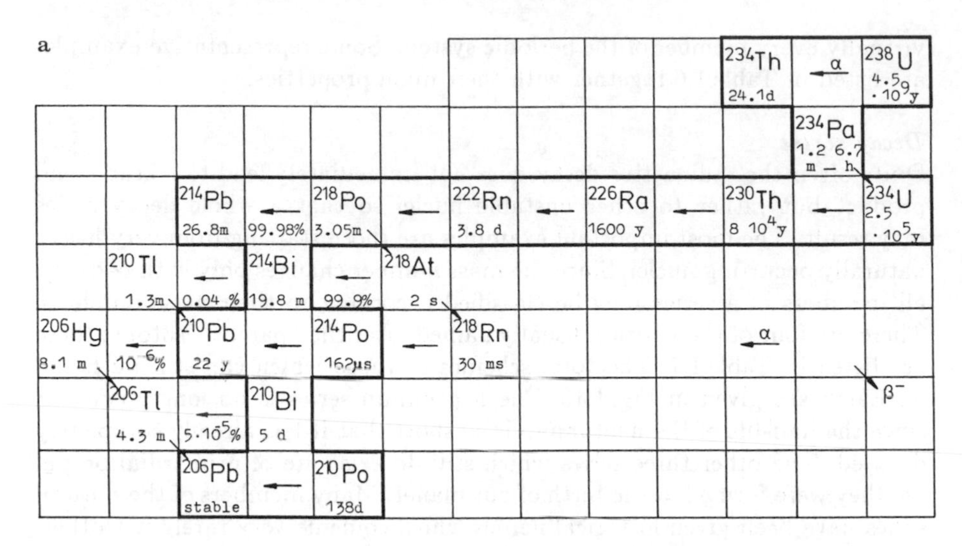

b

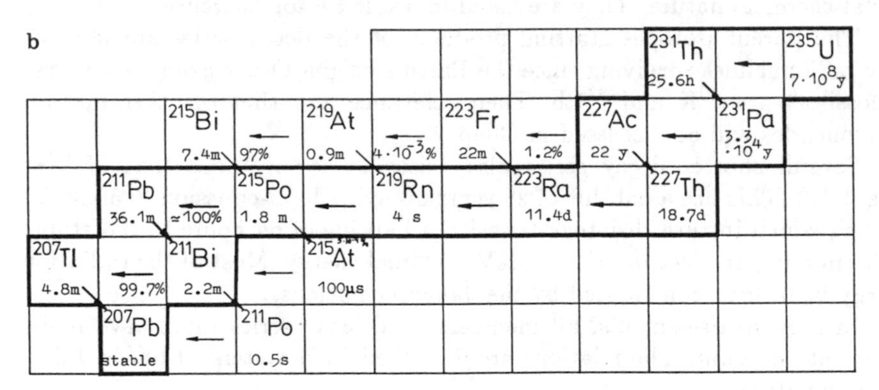

c

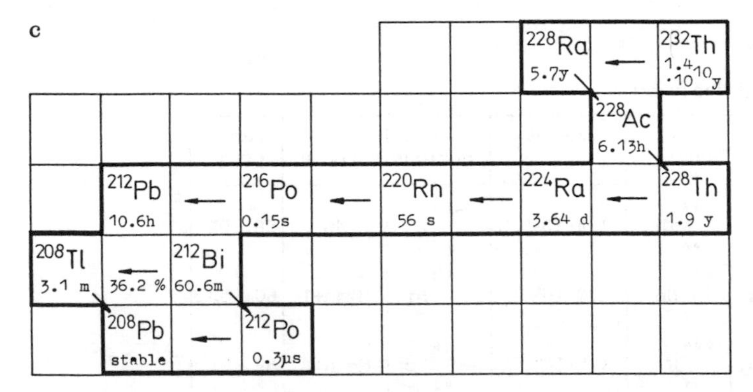

Figure 1.11a-c. The three still existing natural decay series (after KIEFER and KOELZER 1986). (a) uranium-radium, (b) uranium-actinium, (c) thorium

Table 1.8. Historical names of decay products

radium A	^{216}Po	uranium z	^{234}Pa
radium B	^{214}Pb	ionium	^{230}Tl
radium C	^{214}Bi	mesothorium I	^{228}Ra
radium C'	^{214}Po	mesothorium II	^{228}Ac
radium C''	^{210}Tl	radiothorium	^{228}Th
radium D	^{210}Pb	thoron	^{220}Rn
radium E	^{210}Bi	thorium X	^{224}Ra
radium E''	^{206}Tl	thorium A	^{216}Po
radium F	^{210}Po	thorium B	^{212}Pb
uranium I	^{238}U	thorium C	^{212}Bi
uranium II	^{234}U	thorium C'	^{212}Po
uranium X_1	^{234}Tl	thorium C''	^{208}Tl
uranium X_2	^{234m}Pa [a]	thorium D	^{208}Pb

[a] excited metastable nucleus

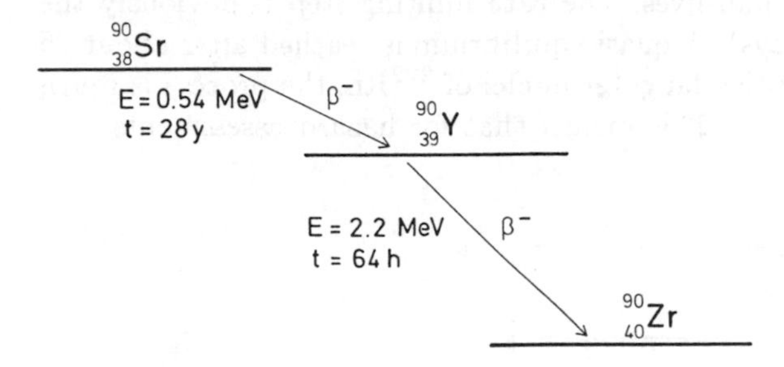

Figure 1.12. Decay scheme of strontium-90

$$
\begin{aligned}
-A_0 &= \frac{dN_0}{dt} = -a_0 N_0 \\
\frac{dN_1}{dt} &= a_0 N_0 - a_1 N_1 \\
\frac{dN_2}{dt} &= a_1 N_1 - a_2 N_2 \\
&\vdots \\
\frac{dN_s}{dt} &= a_{s-1} N_{s-1}
\end{aligned}
\tag{1.24}
$$

where the index "0" designates the start nucleus, "s" the stable end-product. The solution of Eq. (1.24) is given by sum expressions of exponential functions (except for N_0) with negative exponents. Since A_0 never attains zero, a true equilibrium is not possible. If, however, a_0 is much smaller than the others, a "quasi-equilibrium" ("secular equilibrium") is obtained. All time derivatives may then be set zero, so that:

$$a_1 N_i = a_{i+1} N_{i+1} \quad .$$

This means that the activities of all members are equal. The numbers of the nuclei, on the other hand, are related to each other inversely to the decay constants or directly to the half-lives:

$$\frac{N_i}{N_{i+1}} = \frac{a_{i+1}}{a_i} = \frac{\tau_1}{\tau_{a+1}} \quad . \tag{1.25}$$

A secular equilibrium may also be established in a sub-series in case of large differences in decay constants. The velocity of its attainment depends, of course, on the actual conditions. This shall be exemplified with the – also practically important case – of ^{226}Ra and its daughter products, neglecting minor decay pathways (see also Fig. 1.11): The sub-series starts with ^{226}Ra ($\tau = 1680$ years) and ends with ^{210}Pb ($\tau = 21$ years), all intermediate products having much shorter half-lives. The rate limiting step is obviously the decay of ^{222}Rn ($\tau = 3.8$ days). A quasi-equilibrium is reached after about 25 days. If one considers only the daughter nuclei of ^{222}Rn, the process is much faster, as seen from Fig. 1.13. This means that for hazard assessments, not

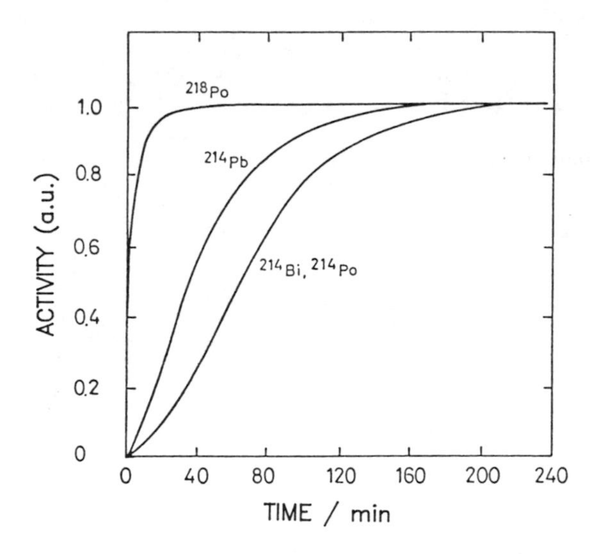

Figure 1.13. Build up of radon daughter products in the course of time

only the α emission of ^{222}Rn has to be taken into account but also those of the short-lived daughter products (2 α decays, 2 β decays).

Quantities and units

The unit of the rate of disintegration, called activity, is the "Becquerel" (Bq) which is 1 decay per second. The older unit was the "Curie" (Ci) originally related to 1 g of ^{226}Ra:

$$\begin{aligned} 1\,\mathrm{Bq} &= 1\,\mathrm{s}^{-1} \\ 1\,\mathrm{Ci} &= 3.7 \times 10^{10}\,\mathrm{s}^{-1} = 37\,\mathrm{GBq} \\ 1\,\mathrm{Bq} &= 2.7 \times 10^{-11}\,\mathrm{Ci} \quad . \end{aligned} \tag{1.26}$$

Dose estimations for incorporated activities are given in Sect. 4.2.5.

The "specific activity", i.e. activity per mass, has the unit $\mathrm{Bq\,kg^{-1}}$. With Eq. (1.22) one obtains

$$\frac{\mathrm{A}}{\mathrm{m}} = \frac{\mathrm{a}}{\mathrm{m_N}} \tag{1.27}$$

where $\mathrm{m_N}$ is the nuclear mass of the respective nuclide.

For practical purposes the following relationship may be useful:

$$\frac{\mathrm{A}}{\mathrm{m}}/\mathrm{Bq\,kg^{-1}} = \frac{\ln 2}{\tau/\mathrm{s} \times \mathrm{m_{Nr}}/\mathrm{u} \times 1.66 \times 10^{-27}/\mathrm{kg\,u^{-1}}} \tag{1.28}$$

where $\mathrm{m_{Nr}}$ is the nuclear mass measured in atomic mass units u.

For measuring the exposure to natural radioactive noble gases (^{222}Rn, ^{220}Rn) a historical unit is still being used, the "working level". One assumes that there is equilibrium within the sub-series between Rn and its daughter products (see above). The numbers of nuclei present are then calculated according to Eq. (1.25). One may then attribute a "potential α energy" to each Rn decay which is obtained by adding the α energies of all decay products weighted by their relative contribution. β-emitting nuclides are to be included since they are also "potential" α emitters because of their daughter products. In the case of ^{222}Rn these are:

^{218}Po ($\mathrm{E}_\alpha = 6\,\mathrm{MeV}$, $\mathrm{a} = 3.73 \times 10^{-3}\,\mathrm{s}^{-1}$)
^{214}Pb (no α, $\mathrm{a} = 4.28 \times 10^{-4}\,\mathrm{s}^{-1}$)
^{214}Bi (no α, $\mathrm{a} = 5.86 \times 10^{-4}\,\mathrm{s}^{-1}$)
^{214}Po ($\mathrm{E}_\alpha = 7.7\,\mathrm{MeV}$, $\mathrm{a} = 4332\,\mathrm{s}^{-1}$)

The potential α energy is calculated per 1 pCi ^{222}Rn activity:

$$\begin{aligned} \mathrm{E}_{\alpha,\mathrm{pot}} &= [1.37 \times 9.77 + 7.7(85.8 + 63.1 + 1 \times 10^{-5})] \\ &= 1.28 \times 10^{3}\,\mathrm{MeV/pCi} \quad . \end{aligned} \tag{1.29}$$

The quantity of 1 "working level" (WL) is defined as the Rn-activity concentration with a potential α energy of 10^5 MeV/l, leading to

$$1\,\mathrm{WL} \approx 100\,\mathrm{pCi/l}\ ^{222}\mathrm{Rn} \quad .$$

For ^{220}Rn ("thoron") one obtains in the same way

$$1\,\mathrm{WL} \approx 7.5\,\mathrm{pCi/l}\ ^{220}\mathrm{Rn} \quad .$$

Further readings

CHARLESBY 1964, GLOCKER and MACHERAUCH 1965, HAXEL 1966, Textbooks of atomic and nuclear physics

Chapter 2
Fundamentals of Radiation Attenuation in Matter

It is the aim of this chapter to introduce the fundamental properties and mechanisms governing the interaction of radiation and matter. This is to provide the basis for more detailed considerations in the following chapters. The important concept of "interaction cross section" is explained and discussed followed by the formalisms of collision processes, both with and without the action of an attracting field. This allows to derive expressions for the amount of energy transferred.

2.1 Interaction Cross Section

If radiation penetrates matter, it is in general attenuated. This may be due to a number of processes, e.g. absorption, scattering etc. In any case they involve an interaction with the constituent entities like atoms or electrons so that a general formal treatment is possible without taking into account the specific type of interaction (Fig. 2.1).

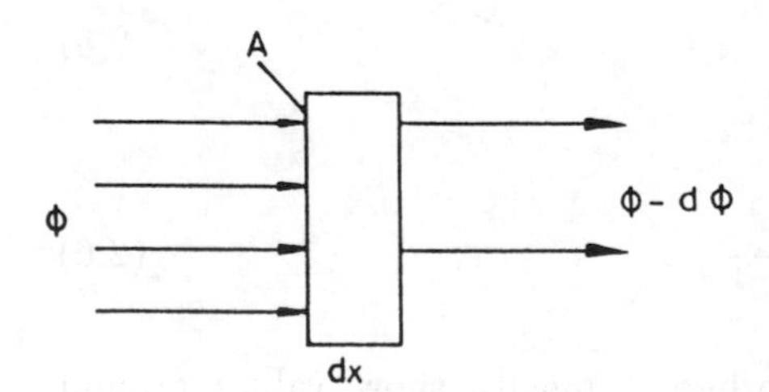

Figure 2.1. Explanation of action cross section

Radiation of particle fluence ϕ (i.e. number of particles per unit area) may hit a mass element dm with area A, thickness dx and density ϱ. Because of the (not specified) interaction, the fluence is reduced by $d\phi$. If dN is the number of interaction centres in dm, then:

$$-d\phi = \sigma \cdot \phi \cdot \frac{dN}{A} \tag{2.1}$$

since the interaction probability is both proportional to the number of particles per area as well as to the number of interacting entities per area. The proportionality factor σ describes the probability of interaction. It has – as seen from Eq. (2.1) – the dimension of an area and is called "interaction cross section". Although this is a strictly formal concept, it may be visualized as the area which has to be hit by the particle in order to cause an interaction.

Equation (2.1) may be written in various ways: The volume of dm is $\mathrm{dV} = \mathrm{Adx}$, dN/dV is the number of centers per volume, i.e. a concentration. To use this in its more usual form as molar concentration, one has to multiply by AVOGADRO's number $\mathrm{N_A}$ ($= 6.02 \times 10^{23}\,\mathrm{mol^{-1}}$). If the molar concentration is abbreviated with c, one obtains:

$$-\mathrm{d}\phi = \sigma \cdot \phi \cdot \mathrm{N_A} \cdot \mathrm{c} \cdot \mathrm{dx} \quad . \tag{2.2}$$

Integration leads to

$$\phi = \phi_0 \mathrm{e}^{-\sigma \mathrm{N_A} \mathrm{cx}} \quad . \tag{2.3}$$

This expression is essentially identical to the so-called "LAMBERT-BEER law" which is used to describe radiation attenuation in homogeneous solutions. For practical purposes, the following form is preferred:

$$\phi = \phi_0 \cdot 10^{-\varepsilon \mathrm{cx}} \tag{2.4}$$

ε – a quantity dependent on radiation wavelength and the nature of the dissolved substance – is termed the "molar decadic extinction coefficient", c is usually given in $\mathrm{mol/dm^3}$ and x in cm.

From a comparison of Eqs. (2.3) and (2.4), it is clear that

$$\varepsilon = \sigma \cdot \mathrm{N_A} \cdot \log_{10} \mathrm{e} \tag{2.5}$$

or numerically

$$\frac{\sigma}{\mathrm{m^2}} = 3.82 \times 10^{-25} \frac{\varepsilon}{\mathrm{mol^{-1}\,dm^3\,cm^{-1}}} \quad . \tag{2.6}$$

Extinction coefficients of strong absorbers typically show values around $10^4\,\mathrm{mol^{-1}\,dm^3\,cm^{-1}}$. According to Eq. (2.6), this corresponds to an interaction cross section of $3.92 \times 10^{-21}\,\mathrm{m^2}$ or to a circle of about $3.5 \cdot 10^{-11}$ m radius which is in the order of magnitude of that for electron orbits in the hydrogen atom ("*BOHR* radius": 5.3×10^{-11} m). This comparison is only meant as an illustration, but it demonstrates qualitatively that light quanta interact with electron orbits.

Optical radiation is selectively absorbed (see next chapter) by "chromophores", while this is not the case for ionizing radiation where all constituents of the mass element contribute to the interaction. In a homogeneous body, the molar concentration is given by:

$$c = \frac{dm}{m_A\,dV} = \frac{\varrho\,dV}{m_A\,dV} = \frac{\varrho}{m_A} \tag{2.7}$$

where mA is the molar mass of the substance in question.

Inserting this into Eq. (2.3) leads to:

$$\phi = \phi_0 e^{-\sigma\varrho N_A/m_A x} \quad . \tag{2.8}$$

This expression is usually written as:

$$\phi = \phi_0 e^{-\mu x} \tag{2.9}$$

where μ denotes the "attenuation coefficient". Its role and modifications will be more extensively discussed in Chap. 4.

By comparison one obtains:

$$\mu = \sigma\varrho\frac{N_A}{m_A} \tag{2.10}$$

where μ is obviously proportional to the density ϱ. It is, therefore, often reasonable to use the ratio μ/ϱ which is called the "mass attenuation coefficient". To give again a typical example: For 1 MeV γ radiation and lead as absorber, μ/ϱ is $7 \times 10^{-3}\,m^2\,kg^{-1}$. The interaction cross section thus is $\sigma = 2.4 \times 10^{-27}\,m^2$ ($m_A = 0.207\,kg\,mol^{-1}$, $\varrho = 11.34\,g\,cm^{-3}$) corresponding to a circle with a radius of 2.8×10^{-14} m.

As will be shown in the next chapter, γ quanta interact mainly with the electrons in matter. Their number in lead is 82, the interaction cross section per electron σ_e then amounts to:

$$\sigma_e = 2.93 \times 10^{-29}\,m^2$$

corresponding to a radius of 3.1×10^{-15} m. This value is nearly equal to the classical electron radius of 2.8×10^{-15} m. One should point out, however, that this surprising agreement is due to the fact that in the energy range chosen, γ absorption is essentially due to COMPTON scattering (Chap. 3) which can be treated as a collision with free electrons. The situation is less transparent in other cases.

2.2 Collision Processes

Collision processes play a central role in the interaction between radiation and matter. Some fundamental aspects shall be described at this point to facilitate the treatment in later chapters.

Consider two particles with masses m_1, m_2 and the vectorial velocities $\boldsymbol{v}_1$, $\boldsymbol{v}_2$. No external forces act on the particles so that any changes of their propagation is only due to interactions between them. In this case the sum of their kinetic energies and of their momenta will remain constant if it is furthermore supposed that there are no processes by which kinetic energy is transformed into other energy forms, e.g. heat. With other words, only elastic collisions are considered. If the velocities are $\boldsymbol{v}_1$, $\boldsymbol{v}_2$ before and $\boldsymbol{u}_1$, $\boldsymbol{u}_2$ after the collision, the following relations hold:

$$m_1 v_1^2 + m_2 v_2^2 = m_1 u_1^2 + m_2 u_2^2 \qquad \text{(preservation of energy)} \tag{2.11}$$

and

$$m_1 \boldsymbol{v}_1 + m_2 \boldsymbol{v}_2 = m_1 \boldsymbol{u}_1 + m_2 \boldsymbol{u}_2 \qquad \text{(preservation of momentum)} \tag{2.12}$$

The choice of the origin of the vectors is arbitrary. For many problems it is very useful to describe the movement of the particles relative to the center of gravity ("cog-system") or relative to each other (relative system).

Center of gravity system (Fig. 2.2)
The space coordinates r_c of the cog are given by:

$$\boldsymbol{r}_c = \frac{m_1 \boldsymbol{r}_1 + m_2 \boldsymbol{r}_2}{m_1 + m_2} \tag{2.13}$$

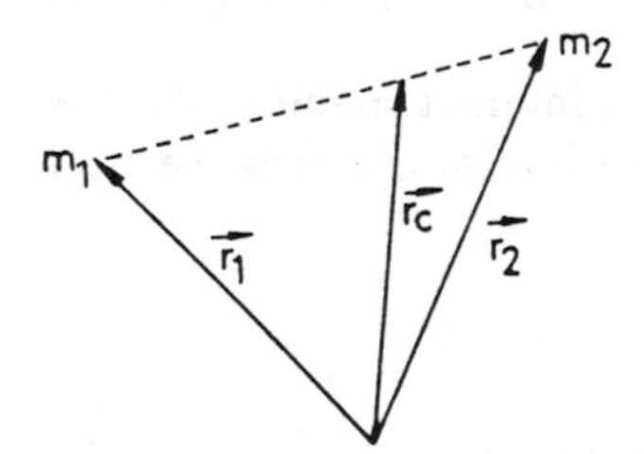

Figure 2.2. Coordinates of the center of gravity

where $\boldsymbol{r}_1$ and $\boldsymbol{r}_2$ denote the space vectors of the two particles. Differentiation with respect to time leads to:

$$\frac{d\boldsymbol{r}_c}{dt} = \boldsymbol{v}_c = \frac{m_1 \boldsymbol{v}_1 + m_2 \boldsymbol{v}_2}{m_1 + m_2} = \frac{m_1 \boldsymbol{u}_1 + m_2 \boldsymbol{u}_2}{m_1 + m_2} \tag{2.14}$$

Equation (2.14) – together with Eq. (2.12) – states that the velocity of the cog is the same before and after the collision. It should be pointed out, however, that Eq. (2.14) in its given form is strictly true only for non-relativistic cases.

The velocities relative to the cog shall be denoted by capital letters:

$$\boldsymbol{v}_1 = \boldsymbol{v}_c + \boldsymbol{V}_1 \quad \text{etc.}$$

If the second particle was initially at rest, the following relations follow:

$$\begin{aligned} V_1 &= v_1 - \frac{m_1 v_1}{m_1+m_2} = \frac{m_2 v_1}{m_1+m_2} \\ &\text{and} \\ V_2 &= -\frac{m_1 v_1}{m_1+m_2} \end{aligned} \tag{2.15}$$

and for the momenta

$$m_1 V_1 = -m_2 V_2 = \frac{m_1 m_2}{m_1+m_2} v_1 = \mu v_1 \quad . \tag{2.16}$$

The quantity $\mu = m_1m_2/(m_1+m_2)$ is termed "reduced mass". Equation (2.16) shows that in the cog system, the two particles have the same momentum but with opposing signs. This means that in the collision, only the direction is altered and that the entry angle equals the exit angle (denoted ψ in the following). This is not true in the "laboratory system" where, upon collision, the two particles are scattered by angles α_1 and α_2 from the original direction of v_1 (the second particle assumed to be initially at rest).

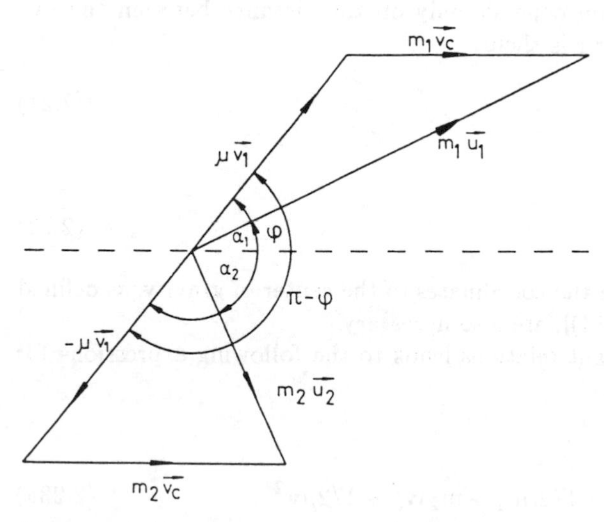

Figure 2.3. Momentum diagramme in the laboratory and the cog-system

The relations are depicted in Fig. 2.3, from which the following relations may be derived:

$$\begin{aligned} m_1 u_1 \cos\alpha_1 &= m_1 v_c + \mu v_1 \cos\psi \\ m_1 u_1 \sin\alpha_1 &= \mu v_1 \sin\psi \end{aligned} \tag{2.17}$$

Division of the two equations leads to:

$$\operatorname{tg}\alpha_1 = \frac{\sin\psi}{m_1/m_2 + \cos\psi} \tag{2.18}$$

The scattering angle of the second particle may be obtained in an analogous way, with the result:

$$\operatorname{tg}\alpha_2 = \frac{\sin\psi}{1 - \cos\psi} \quad . \tag{2.19}$$

This may be simplified to:

$$\cos^2\alpha_2 = 1/2(1 - \cos\psi) = \sin^2\psi/2 \quad . \tag{2.20}$$

It should be reiterated that these simple relations are only valid if the second particle was initially at rest.

Relative system
Here one of the two particles is assumed to be at rest throughout and the movement of the other is described in a coordinate system whose origin is centered at the place of the first. This kind of description is particularly useful where the interaction depends only on the distance between the two particles. The space vector r is then:

$$\mathbf{r} = \mathbf{r}_1 - \mathbf{r}_2 \tag{2.21}$$

and the velocity vector v:

$$\mathbf{v} = \mathbf{v}_1 - \mathbf{v}_2 \quad . \tag{2.22}$$

For a complete description the coordinates of the center of gravity, as defined above [Eqs. (2.13) and (2.14)], are also necessary.

Insertion of the relevant relations leads to the following expressions for the total kinetic energy:

before the collision:

$$1/2 m_1 v_1^2 + 1/2 m_2 v_2^2 = 1/2(m_1 + m_2) v_c^2 + 1/2 \mu v^2 \tag{2.23a}$$

and

after the collision:

$$1/2 m_1 u_1^2 + 1/2 m_2 u_2^2 = 1/2(m_1 + m_2) v_c^2 + 1/2 \mu u^2 \quad . \tag{2.23b}$$

Because of the preservation of energy, one obtains in the absence of external forces:

$$v^2 = u^2 \quad . \tag{2.24}$$

With other words, in the relative system, the velocities change only their direction.

In the following, three important examples are treated in more detail which are fundamental to the understanding of interaction processes with matter:

1. Collision between two point masses (relativistic) (Fig. 2.4)

As usual it is assumed that a particle with mass m_1, velocity v_1 and negligible extensions ("point masses") hits another one (mass m_2) which is initially at

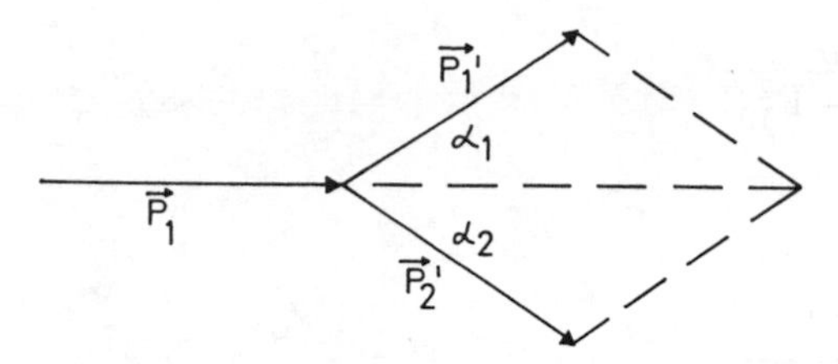

Figure 2.4. Collision between two point masses

rest. The scattering angles are α_1 and α_2, respectively. The momenta after the collision are denoted p_1' and p_2'.

One reads from Fig. 2.4:

$$\begin{aligned} p_1 &= p_1' \cos\alpha_1 + p_2' \cos\alpha_2 \\ 0 &= p_1' \sin\alpha_1 + p_2' \sin\alpha_2 \end{aligned} \tag{2.25}$$

From this follows:

$$p_1^2 + p_2'^2 - 2p_1p_2' \cos\alpha_2 = p_1'^2$$

multiplication by c^2 and reference to Eq. (1.12) yields:

$$E_1^2 + E_2'^2 - E_{20}^2 - 2p_1p_2'c^2 \cos\alpha_2 = E_1'^2 \quad .$$

(E_1 is the energy before the collision; E_1', E_2' are the energies after the collision; E_{10}, E_{20} are the rest energies).

$$E_1 + E_{20} = E_1' + E_2' \quad .$$

With this the expression takes the form

$$E_1E_2' + E_{20}E_2' - E_1E_{20} - E_{20}^2 = p_1p_2'c^2 \cos\alpha$$

or, since $E = E_0 + T$

$$T_2'(E_1 + E_{20}) = p_1 p_2' c^2 \cos\alpha_2$$

and $T_2'^2(E_1 + E_{20})^2 = p_1^2 c^2 p_2' c^2 \cos^2\alpha_2$.

After insertion of Eq. (2.12) and rearrangement, the kinetic energy of the second particle will be:

$$T_2' = \frac{2E_{20}(E_1^2 - E_{10}^2)\cos^2\alpha_2}{(E_1 + E_{20})^2 - (E_1^2 - E_{10}^2)\cos^2\alpha_2} \quad . \tag{2.26}$$

The maximum transferable energy is ($\alpha_2 = 0$):

$$T'_{2\,\mathrm{max}} = \frac{2E_{20}(E_1^2 - E_{10}^2)}{(E_1 + E_{20})^2 - (E_1^2 - E_{10}^2)}$$

which may be simplified to:

$$T'_{2\,\mathrm{max}} = \frac{2T_1\frac{E_{20}}{E_{10}}\left(2 + \frac{T_1}{E_{10}}\right)}{\left(1 + \frac{E_{20}}{E_{10}}\right)^2 + 2\frac{T_1}{E_{10}}\cdot\frac{E_{20}}{T_{10}}}$$

or with the abbreviation $A = \frac{E_{20}}{E_{10}}$:

$$T'_{2\,\mathrm{max}} = \frac{2AT_1\left(2 + \frac{T_1}{E_{10}}\right)}{(1 + A^2) + 2A\frac{T_1}{E_{10}}} \quad . \tag{2.27}$$

With Eq. (2.27) it is possible to rewrite Eq. (2.26) as:

$$T_2' = T'_{2\,\mathrm{max}}\frac{\cos^2\alpha_2}{1 + \frac{T'_{2\,\mathrm{max}}}{2E_{20}}(1 - \cos^2\alpha_2)} \quad . \tag{2.28}$$

This has so far been applied to the laboratory system. The relation of the scattering angles in the cog system can be described via Eq. (2.20) as:

$$T_2' = T'_{2\,\mathrm{max}}\frac{\sin^2\psi/2}{1 + \frac{T'_{2\,\mathrm{max}}}{2E_{20}}\cos^2\psi/2} \quad . \tag{2.29}$$

The question over which system to use depends on the particular problem.

In the non-relativistic case the kinetic energy is much smaller than the rest energy. This then leads to the much simpler relations:

$$\begin{aligned} T_2' &= T'_{2\,\mathrm{max}}\cos^2\alpha_2 \\ &= T'_{2\,\mathrm{max}}\sin^2\psi/2 \end{aligned} \tag{2.30}$$

and for the maximum transferable energy:

$$T'_{2\,max} = \frac{4AT_1}{(1+A^2)} \quad . \tag{2.31}$$

Both in the relativistic as in the non-relativistic case the maximum energy is transferred by a central collision, i.e. if $\alpha_2 = 0$. The amount depends on the value of A, which is the ratio of rest masses. If they are equal (A = 1), then $T'_{2\,max} = T_1$, i.e. the incoming particle loses all its kinetic energy in the collision process.

2. Collision between two spheres (non-relativistic) (Fig. 2.5)

Two spheres with masses m_1, m_2 and radii r_1, r_2 may approach each other at a distance b ("impact parameter"). As usual, m_2 is assumed to be initially at

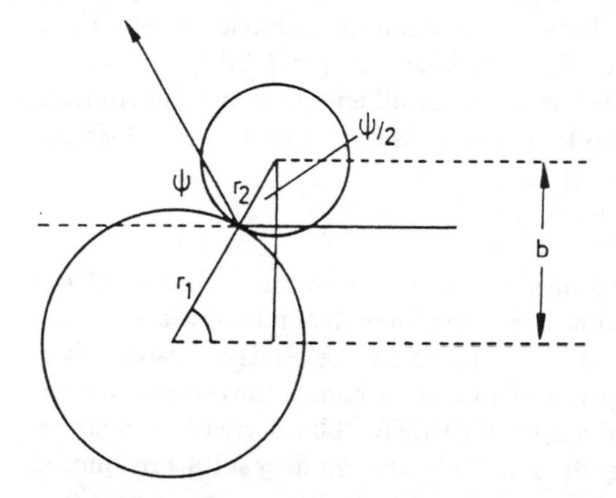

Figure 2.5. Collision between two spheres in the cog-system

rest. In the cog system, the scattering angle ψ is the same for both particles. One reads from the figure:

$$\cos\psi/2 = \frac{b}{r_1 + r_2}$$

and obtains, therefore, with Eq. (2.29) in the non-relativistic case:

$$\frac{T'_2}{T_1} = \frac{4a}{(1+A^2)}\left(1 - \frac{b^2}{(r_1+r_2)^2}\right) \quad . \tag{2.32}$$

As expected, the energy transferred is greatest with a central collision (b = 0), it vanishes if $b \geq (r_1 + r_2)$, i.e. if two spheres miss each other.

To obtain the distribution of transferred energies, the "differential interaction cross section" $d\sigma$ is introduced:

$$d\sigma = 2\pi b\, db = \pi\, d(b^2) \tag{2.33}$$

$d\sigma/[\pi(r_1+r_2)^2]$ describes the relative contribution of collisions with an impact parameter b. With Eq. (2.32) one obtains:

$$\frac{d\sigma}{\pi(r_1+r_2)^2} = \frac{(1+A^2)}{4AT_1}\, dT'_2 \qquad 0 \le T'_2 \le T'_{2\,max}$$
$$\frac{d\sigma}{dT'_2} = \frac{\pi(r_1+r_2)^2}{T'_{2\,max}} \quad . \tag{2.34}$$

Equation (2.34) describes an interesting result, since it shows that within the given limits, the probability for a certain energy transfer – the energy distribution of the hit particle – is equal for all values. The example may be directly applied for collisions between neutrons and nuclei where there is no charge interaction: After the first collision the probability to find an energy T'_2 between 0 and $T'_{2\,max}$ is the same for all energies; the distribution is rectangular. This is, however, no longer the case after multiple collisions.

3. Collision between two charged particles

This situation is more complicated because there is now the influence of the electric field to be taken into account. Since the force depends on the distance (COULOMB's law), the problem is best treated in the relative system. It is formally equivalent to the description of planet or comet movement around the sun; a solution was already given by KEPLER. The derivation is somewhat lengthy and found in Appendix I.3. Only the main results are quoted here:

If one assumes that one of the collision partners has a much larger mass (as in ion-electron interaction), the differential interaction cross section is given by:

$$d\sigma(\varepsilon) = \frac{2\pi k^2}{m_2 v_1^2} \cdot \frac{d\varepsilon}{\varepsilon^2} \tag{2.35}$$

where ε is the energy transferred to the lighter partner of mass m_2 and k the force coefficient (the product of charges in the case of COULOMB's law).

The total cross section per interaction – obtained by integration of Eq. (2.35) – is then:

$$\sigma = \int_{\varepsilon_{min}}^{\varepsilon_{max}} d\sigma(\varepsilon) = \frac{2\pi k^2}{m_2 v_1^2}\left[\frac{1}{\varepsilon_{min}} - \frac{1}{\varepsilon_{max}}\right] \quad . \tag{2.36}$$

Equation (2.35) shows that the probability of an energy transfer is inversely proportional to the square of ε. If – as assumed – $m_1 \gg m_2$, the maximum is:

$$\varepsilon_{max} = 2m_2v_1^2 \quad .$$

The minimum energy ε_{min} is more difficult to evaluate, it is discussed in Sect. 3.2.3.

Further readings
Textbooks of atomic physics and theoretical mechanics

Chapter 3
Interaction Processes

Based on Chap. 2, the most important interaction processes for understanding biological radiation effects are described and treated quantitatively. Charged particles are particularly emphasized since they play an essential role in the effects of ionizing radiation. This also involves fundamental approaches for the determination of ranges and fluence distributions.

3.1 Optical Radiation

Photons of optical radiation can only be absorbed by atoms or molecules if the energy difference dE between two states matches the energy content of the quantum, i.e.:

$$\mathrm{dE} = \mathrm{h}\nu \quad .$$

Electron levels in molecules are superimposed by vibrational and rotational states whose energy differences are considerably smaller. Absorption normally occurs in the electronic ground state, i.e. in the lowest possible one, which does not exclude, however, that vibrational and/or rotational states may be excited as a result of thermal interactions. For photochemical processes, the first excited electronic state is the most important. In principle, excitations to higher states are also possible and do in fact occur but their lifetime is so short that there is no chance for further reactions. They are rapidly deactivated to the first excited state via interactions within the molecule or with the environment. The lifetime of the first excited state is considerably longer so that photochemical processes may take place.

Two kinds of excited states are distinguished: If the spin direction of the excited electron is conserved, the process is called "singlet excitation", which is the normal case. It is possible, however, that the spin direction is reversed so that the electron remaining in the ground state and the excited one possess parallel spins ("intersystem crossing"). This does not contradict the PAULI principle since the two electrons are in different energy states. "Intersystem crossing" is theoretically not allowed in isolated systems, it occurs with a small probability for intra- or intermolecular interactions. Once attained, the lifetime of the triplet state is comparatively long since the probability for reverse intersystem crossing is also low. Triplet excited molecules are, therefore, important candidates for photochemical processes as they have a longer time available to react. Except for rare special cases (e.g. O_2) the ground level is usually a singlet state. Excited states are deactivated either by emission of light quanta (fluorescence with singlet excitation, phosphorescence with triplet excitation), by thermal interactions or by reactions with other molecules. The situation described is diagrammatically depicted in Fig. 3.1 ("JABLONSKI" diagram).

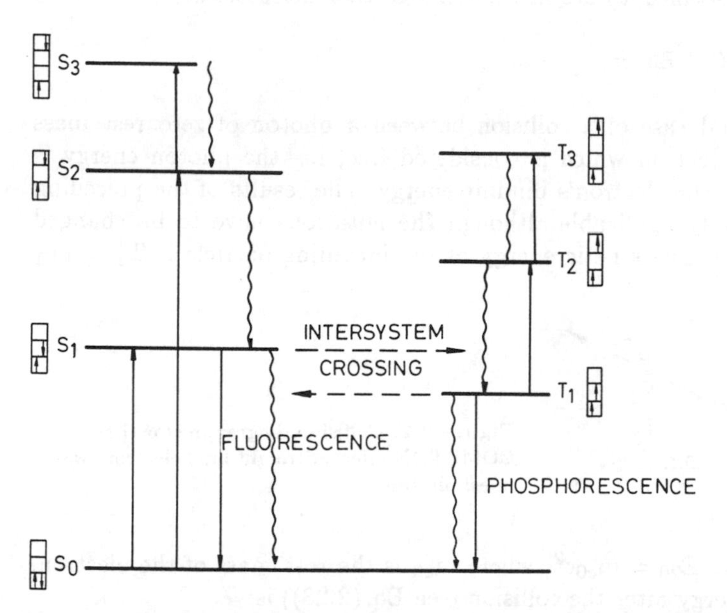

Figure 3.1. JABLONSKI diagram Electron transitions through excitation by light absorption (S: singlet-, T: triplet-states). The spin directions are indicated on the left- and the right-hand sides. Wavy lines indicate radiationless transitions

The probability for an excitation to occur is given by the wavelength-dependent cross section $\sigma(\lambda)$ which is related to the extinction coefficient ε as shown in Eqs. (2.5) and (2.6).

3.2 Ionizing Radiation

3.2.1 Electromagnetic Radiation

3.2.1.1 General

Photons having sufficient energy to cause ionizations in the irradiated material interact – depending on the energy and the atomic composition of the exposed substances – via four processes: elastic scatter without energy transfer, COMPTON scattering, photo absorption and pair formation. The first case will not be treated since it does not contribute to biological radiation effects (although it has to be taken into account in dosimetric measurements). COMPTON and *photoeffects* are essentially collision processes so that the results of Chap. 2 may be applied. Pair formation constitutes a realization of EINSTEIN's mass-energy equivalence. For the sake of completeness, the interaction with nuclei (nuclear *photoeffect*) must be mentioned, but because of its minor importance it does not warrant further discussion.

3.2.1.2 COMPTON Effect

This is the special case of a collision between a photon of zero rest mass with an atomic electron which is considered free, i.e. the photon energy is much larger than the electron's binding energy. The results of the preceding chapter are directly applicable although the notations have to be changed (see also Fig. 3.2): The kinetic energy of the incoming particle is $T_1 = h\nu_1$

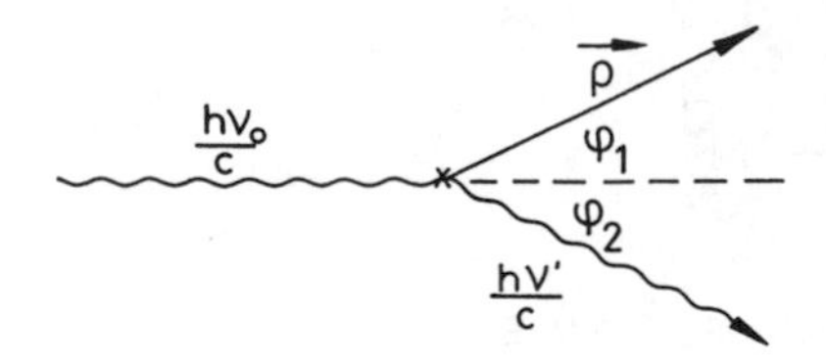

Figure 3.2. Collision diagramme for the COMPTON effect. Straight line: electron, wavy lines: photons

and $E_{10} = 0$, also $E_{20} = m_{e0}c^2$, where m_{e0} is the rest mass of the electron, whose kinetic energy after the collision (see Eq. (2.28)) is:

$$T'_2 = T'_{2\,max} \frac{\cos^2\alpha_2}{1 + \frac{T'_{2\,max}}{2m_{e0}c^2}(1 - \cos^2\alpha_2)} \tag{3.1}$$

with:

$$T'_{2\,max} = \frac{h\nu_1}{1 + \frac{1}{2}\frac{m_{e0}c^2}{h\nu_1}}$$

the final relationship between electron energy and emission angle is:

$$T'_2 = \frac{2\frac{(h\nu_1)}{m_{e0}c^2}}{1 + 2\frac{h\nu_1}{m_{e0}c^2} + \left(1 + \frac{h\nu_1}{m_{e0}c^2}\right)^2 \mathrm{tg}^2\alpha_2} \ . \tag{3.2}$$

It is seen from Eq. (3.1) that the electron can never accept the total energy of the photon, with other words, there is always a scattered photon. The physical reason for this situation is that both energy and momentum have to be conserved. Since the electron is considered free, the remaining momentum cannot be transferred to the atom. This is different for the *photoeffect* (next section) which involves bound electrons.

Equation (3.2) implies a unique relationship between electron energy and emission angle at a given photon energy.

The probability for a certain energy transfer cannot be treated in a similarly simple way since the interaction between the photon and the field of the electron has to be taken into account which requires a deeper insight into the quantum-electrodynamical relations. Since this is beyond the scope of this book, only some pertinent results are given:

The differential cross section $d\sigma_e$ for an energy transfer T_2' per scattering electron is:

$$d\sigma_e(T_2') = \frac{\pi r_0^2 \cdot m_{e0}c^2}{(h\nu_1)^2}\left[2 + \left(\frac{T_2'}{h\nu_1 - T}\right)^2 \left[2 + \left(\frac{T_2'}{h\nu_1 - T_2'}\right)^2 \times \left(\frac{m_{e0}^2 c^4}{(h\nu_1)^2} + \frac{h\nu_1 - T_2'}{h\nu_1} - \frac{2m_{e0}c^2}{h\nu_1} \cdot \frac{(h\nu_1 - T_2')}{T_2'}\right)\right]\right] dT_2' \quad (3.3)$$

$r_0 = 2.818 \times 10^{-15}$ m is the classical electron radius.

$d\sigma_e(T_2')$ is, of course, zero for $T_2' \geq T_{2\,max}'$. Figure 3.3 shows the distribution of recoil electrons for a number of incident photon energies.

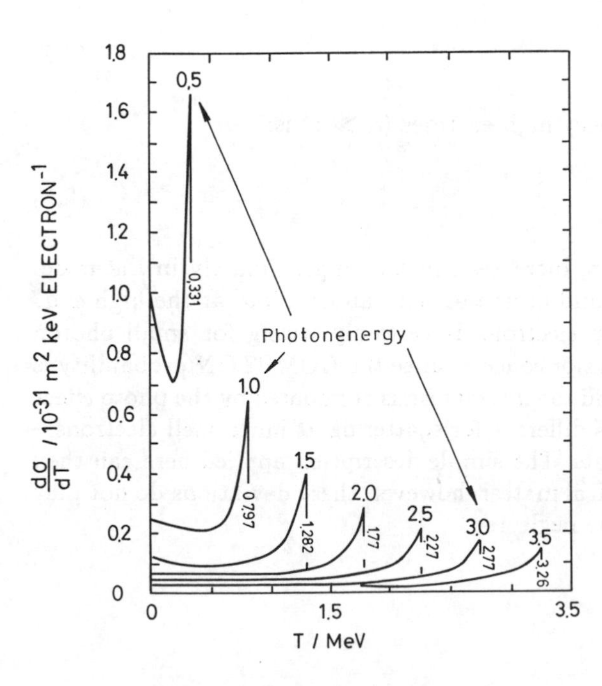

Figure 3.3. Differential cross sections of COMPTON scattering at a single electron with various photon energies (after JOHNS et al. 1952)

The total cross section σ_e for the total number of scattering events is obtained by integration of Eq. (3.3) for $0 \le T_2' \le T'_{2\,max}$ yielding the result:

$$\sigma_e = 2\pi r_0^2 \left[\frac{1+\alpha}{\alpha^2} \left[\frac{2(1+\alpha)}{1+2\alpha} - \frac{\ln(1+2\alpha)}{\alpha} \right] + \frac{\ln(1+2\alpha)}{2\alpha} - \frac{1+3\alpha}{(1+2\alpha)^2} \right] \tag{3.4}$$

where

$$\alpha = \frac{h\nu_1}{m_{e0}c^2} \quad .$$

Equations (3.3) and (3.4) may also be used to calculate the average energy per electron T_2':

$$\overline{T_2}' = \frac{1}{\sigma_e} \int_0^{T'_{2\,max}} T \, d\sigma_e(T) \quad .$$

Compiled data for this and other parameters are given in EVANS (1968).

The probability for a COMPTON interaction depends on the number of electrons in the medium and is, therefore, proportional to Z. The dependence on the incident photon energy is given by Eq. (3.4). For low energies ($\alpha \ll 1$), the following expansion may be used:

$$\sigma_e \approx \frac{8}{3}\pi r_0^2 (1 - 2\alpha \ldots) \quad . \tag{3.5}$$

A good approximation for very high energies ($\alpha \gg 1$) is:

$$\sigma_e \approx \pi r_0^2 \frac{1 + 2\ln 2\alpha}{2\alpha} \quad . \tag{3.6}$$

These relations show that σ_e increases initially approximately in linear dependence of photon energy and decreases with about $1/h\nu'$ at the high end.

The assumption of free electrons is certainly wrong for small photon energies but this is not of major concern since the COMPTON probability is low anyway in this region, and the interaction is dominated by the photo effect (see below). The situation is different for scattering at inner shell electrons – particularly in heavy elements. The simple description applied here can then no longer be used. In biological matter, however, these deviations do not play a significant role and may be neglected.

3.2.1.3 Photoeffect

It was shown in the preceding section that photons can never transfer their energy completely to free electrons. This is not the case with low photon energies where the binding of the electron can no longer be neglected. Conservation of momentum can then be secured by transfer to the atom as a whole. The kinetic energy T'_2 of the electron is then given by the difference of the photon and the binding energy E_B:

$$T'_2 = h\nu_1 - E_B \ . \tag{3.7}$$

The photo effect is strictly speaking a more complicated case of COMPTON interaction. Cross sections, therefore, no longer apply to single electrons but to the atom as a whole. The relations between Z of the medium and the incident photon energy are consequently also more complex. An acceptable approximation is:

$$\sigma_A \approx \frac{Z^4}{(h\nu_1)^3} \tag{3.8}$$

where σ_A is the atomic cross section and Z the nuclear charge of the medium. The probability for the photo effect decreases rapidly with photon energy.

The photo effect is not at all limited to the outer shell valence electrons. On the contrary, inner shell electrons contribute quite significantly, particularly with lighter elements. The resulting electron holes are filled from higher orbitals, the energy gained is emitted either in the form of fluorescence, X rays or used for the liberation of AUGER electrons (see Sect. 3.2.1.5).

3.2.1.4 Pair Formation

Both, the COMPTON as well as the photo effect have a lower probability with higher photon energies although the decrease shows different dependencies. If $h\nu_1 > 2m_{e0}c^2$, i.e., if the photon energy is larger than twice the rest energy of the electron, the photon may "materialize" by the formation of an electron-positron-pair. Conservation of momentum and of charge demands that always two particles of opposite sign are created. Pair formation does not occur *in vacuo*, it requires the participation of an electric field, normally that of the atomic nucleus which receives also part of the momentum. The photon energy is distributed between the rest energy and the kinetic energy of the two particles:

$$hv_1 = T_{electron} + T_{positron} + 2m_{e0}c^2 \quad . \tag{3.9}$$

The calculation of interaction cross sections and the angular dependencies is complex. In a first approximation, the atomic cross section σ_{pair} is proportional to the square of the nuclear charge of the medium and the photon energy:

$$\sigma_{pair} \approx Z^2 \cdot h\nu_1 \quad . \tag{3.10}$$

The particles formed may have considerable energies. They will ionize atoms along their path, but may also emit "Bremsstrahlung" thus giving up again part of the initially absorbed energy. These radiation losses may also play a role with energetic COMPTON electrons. They must be taken into account for dose determinations, particularly with regard to the difference between "dose" and "KERMA" (Chap. 4). Low energy positrons react with electrons in the environment producing "annihilation radiation" where two quanta of $h\nu = 2m_{e0}c^2$ (0.511 MeV) are produced.

3.2.1.5 AUGER Effect

If electrons are liberated from inner atomic shells they leave vacancies which are filled from higher orbitals. This may lead to the emission of fluorescence X rays, but not necessarily so. If the energy gained is greater than the binding energy of the outer electrons, these may be directly ejected. This process is not mediated by the emission of X rays and subsequent photo absorption but directly in a radiationless mode. The phenomenon is called "AUGER effect", the electrons are termed "AUGER electrons". They always occur with inner shell vacancies. Even an avalanche-like formation may occur if the energy is high enough so that several electrons are emitted. The remaining atom will then be multiply ionized.

The simultaneous appearance of several electrons may create high local densities of energy deposition, very similar to high LET-radiation as represented by α particles or other ions. This may have important biological consequences if the emitter atom is a part of a biomolecule. An example is the decay of ^{125}I in ^{125}*Iododesoxyuridine* incorporated in cellular DNA.

It is not possible to give cross sections for the AUGER effect in a general way as this depends critically on the particular atom and the history of preceding energy-absorption events.

3.2.1.6 Summary of Photon Interactions

If matter is exposed to energetic photons, all three interaction processes occur together, although with different contributions depending on the photon energy. The total interaction cross section is the sum of those of the single components. This means that on a per atom basis:

$$\sigma_{total} = Z \cdot \sigma_{compton} + \sigma_{photo} + \sigma_{pair} \quad . \tag{3.11}$$

The COMPTON cross section is multiplied by the electron number (equal to the nuclear charge Z of the atoms in the medium) since it is related to single electrons. While in the photo and pair formation process the photon energy is completely absorbed this is not the case with COMPTON scattering. The absorption cross section σ_a per atom is:

$$\sigma_a = \frac{\sigma_e \cdot T_2'}{h\nu_1} \tag{3.12}$$

so that the total absorption cross section $\sigma_{a\,total}$ per atom becomes:

$$\sigma_{a\,total} = Z \cdot \frac{T_2'}{h\nu_1}\sigma_{e\,compton} + \sigma_{photo} + \sigma_{pair} \quad . \tag{3.13}$$

The dependence on energy, with water as absorbing medium, is shown in Fig. 3.4. The photoeffect dominates at low energies and pair formation at

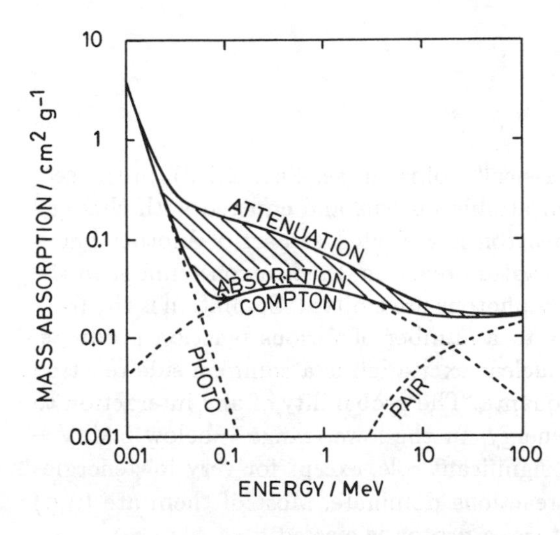

Figure 3.4. The contribution of photo-, COMPTON- and pair-formation effect with photon absorption in water. Total attenuation (including scattering) is also given (after JOHNS and LAUGHLIN 1956)

the far end. The difference between scattering and absorption is most pronounced in the medium range where the COMPTON effect gives the highest contribution.

All three processes result in the production of electrons. They are far more important for the energy deposition than the initial interactions because they ionize much more efficiently. Therefore, electromagnetic radiation is "indirectly ionizing" (similar to neutrons, see next section).

Using an example: 1 MeV of γ quanta liberate – via COMPTON interaction – electrons with an average energy of 440 keV. At 34 eV per ionization (in water) it follows that per initial ionization further 13 are caused by secondary electrons if they are completely stopped in the medium. The actual relations depend, of course, on the photon energy, but it is still true that most of the effect is due to electron interaction.

The atomic cross sections may be transformed into absorption coefficients (Sect. 2.1, Eq. (2.10)):

$$\mu = \sigma_a \cdot \varrho \cdot N_A/M_a$$

For water, this gives us:

$$\mu = \frac{6.02 \times 10^{23}}{18}\sigma_a = 3.34 \times 10^{22}\sigma_a \, cm^{-1} \quad . \tag{3.14}$$

3.2.2 Neutrons

Neutrons may interact with matter via five processes if scattering without energy transfer is excluded:

1. elastic collision
2. inelastic collision
3. non elastic collision
4. capture
5. spallation

The first case represents the "classical" collision (see Sect. 2.1.2); in the second, a neutron is captured by an atomic nucleus and emitted with changed energy; while in the third, the neutron is reejected as part of another particle, e.g. an alpha particle. In the capture reaction, the neutron remains in the nucleus and different particles or γ photons are emitted. Spallation is the fragmentation of the nucleus leading to a number of various reaction products. Except for elastic collision (1), nuclear excitation is a common side-reaction with concomitant emission of γ quanta. The probability of any interaction to occur depends on the neutron energy. In the lower range – below 5 MeV – only the elastic collisions play a significant role, except for very low energies (below 100 keV) where capture reactions dominate. Most of them are (n,p) processes, i.e. upon neutron capture, a proton is ejected.

Inelastic and non-elastic interactions begin to become significant above 2.5 or 5 MeV, respectively, spallation around 20 MeV. In all these cases, part of the energy is also transferred to γ quanta which are emitted from excited nuclei.

For most practical purposes, however, elastic collision is the major energy-transfer process. The fundamental mechanism has been treated in Sect. 2.1.2 where it was shown that the energy loss depends on the mass ratio of the two collision partners and is largest if they possess equal masses. This means that hydrogen nuclei (protons) play the most important role. The differential cross section does – according to Eq. (2.34) – not depend on the transferred energy, with other words, initially monoenergetic neutrons of kinetic energy T_1 cover the whole range between T_1 and $T'_{2\,max}$ already after the first collision. The recoil particles have energies between 0 and $T'_{2\,max}$ if the binding energy is not taken into account. After multiple collisions, the original rectangular distribution function attains a continuous bell-shaped form.

The actual energy deposition in the medium is essentially due to the charged secondary particles, i.e. mainly protons. The situation is, therefore, comparable to that of electromagnetic radiation where the electrons play this role. Neutrons – like photons – are indirectly ionizing. Heavier recoil particles contribute significantly only with higher neutron energies. Fig. 3.5

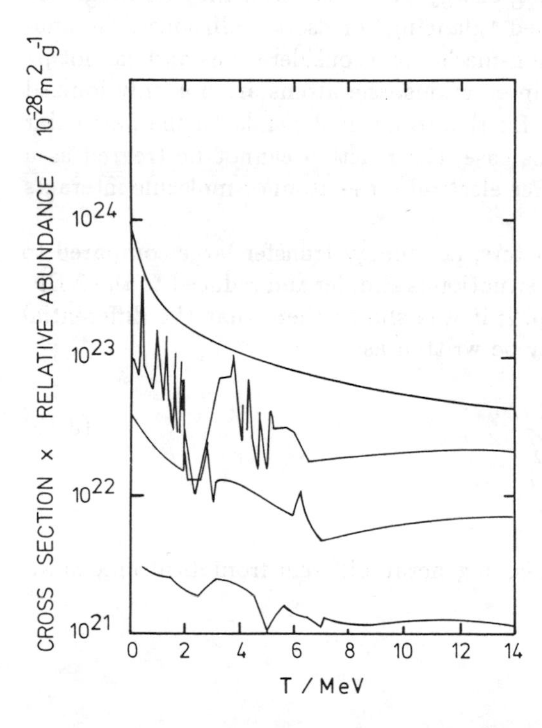

Figure 3.5. Scattering cross sections of neutrons for some elements in tissue. The ordinate values are weighted according to the abundance of the various elements in tissue. (after AUXIER, SNYDER and JONES 1968)

gives interaction cross sections as a function of neutron energy for various nuclei in biological tissue. The values are weighted by the relative abundance of elements in this material.

All what has been said above has to be taken into account if one is interested in the determination of LET distributions (Chap. 4).

3.2.3 Ions

Ions interact – depending on their energy – by three processes with matter:

1. electron capture (low energies)
2. collisions with atomic electrons (medium and high energies)
3. nuclear collisions and nuclear reactions (very high energies)

The last point is of less importance here and will not be detailed; it plays a role, however, with very energetic ions where it leads to fragmentation of the incoming particle so that an originally "pure" beam is contaminated with lighter ions. This effect has to be taken into account in the determination of depth dose curves (Sect. 4.2.2) and LET spectra (Sect. 4.2.1.2).

Most important is the collision with electrons where the interaction is mediated by the electric field of the two partners and depends, therefore, on the impact parameter. With larger distances, the amount of energy transferred is small so that the binding energy of the electron may no longer be neglected. These events are named "glancing" or "soft" collisions. The analytical treatment requires quantum-mechanical considerations and cannot be given here. As a result of these interactions, the atoms are not only ionized but also excited, the probability for this to occur depends on the particular properties of the medium. In this case, the reaction cannot be treated as a collision between the ion and a free electron – the atom or molecule interacts as a whole.

With smaller impact parameters, i.e. energy transfer large compared to the electrons binding energy, the situation is simpler and reduced to the KEPLER problem discussed in Chap. 2; it was shown there that the differential cross section per electron σ_e may be written as:

$$\frac{d\sigma_e}{dT_2'} = -\frac{2\pi Z^{*2} e^4}{(4\pi\varepsilon_0)^2 m_{e0} v_1^2} \cdot \frac{1}{T_2'^2} \quad . \tag{3.15}$$

Here are:

Z^* = effective ion charge which is, in general, different from its atomic number (see below)
m_{e0} = electron rest mass
e = elementary charge
v_1 = ion velocity
ε_0 = dielectric coefficient

To obtain the total differential cross section σ, one multiplies by the electron density in the medium dN_e/dV:

$$\frac{d\sigma}{dT_2'} = -\frac{2\pi Z^{*2} e^4}{m_{e0} v_1^2 \cdot (4\pi\varepsilon_0)^2} \cdot \frac{dn_e}{dV} \cdot \frac{1}{T_2'^2} \tag{3.16}$$

or in modified form

$$\frac{d\sigma}{dT_2'} = -\frac{2\pi e^4}{(4\pi\varepsilon_0)^2 \cdot m_{e0} c^2} \cdot \frac{dN_e}{dV} \cdot \frac{Z^{*2}}{\beta^2} \cdot \frac{1}{T_2'^2} \tag{3.17}$$

where β is the relative ion speed.

The first term consists only of physical constants, its value is 2.55×10^{-23} eV m^2. The electron density in water, as a typical biological medium, is $3.34 \times 10^{29}\,\mathrm{m}^{-3}$. If the first two terms are combined, one obtains for water:

$$\frac{d\sigma}{dT_2'} = C \cdot \frac{Z^{*2}}{\beta^2} \cdot \frac{1}{T_2'^2} \tag{3.18}$$

with $C = 8.5 \times 10^6\,\mathrm{eV/m}$.

Z^* depends on ion energy and the interacting medium. It becomes smaller with low velocities because of electron capture. There are a number of semi-empirical expressions available for its calculation based on different assumptions, but not a generally accepted theory applicable for all media. Because of the lack of experimental data, uncertainties are many, particularly for heavy ions. One useful approximation for water – still widely popular in radiation biology – is the BARKAS formula (BARKAS 1963) which is given as an xample:

$$Z^* = Z[1 - \exp(-125\beta/Z^{2/3})] \tag{3.19}$$

where Z is the ion atomic number and β its relative speed. Figure 3.6 depicts the relationship for a number of higher ions. References to other formalisms can be found elsewhere (KIEFER 1986).

Equations (3.16) to (3.18) can be used to determine the average energy loss of the ion per distance travelled:

$$-\frac{dT_1}{dx} = \int_{T'_{2\,min}}^{T'_{2\,max}} T_2' \cdot d\sigma(T_2') \quad . \tag{3.20}$$

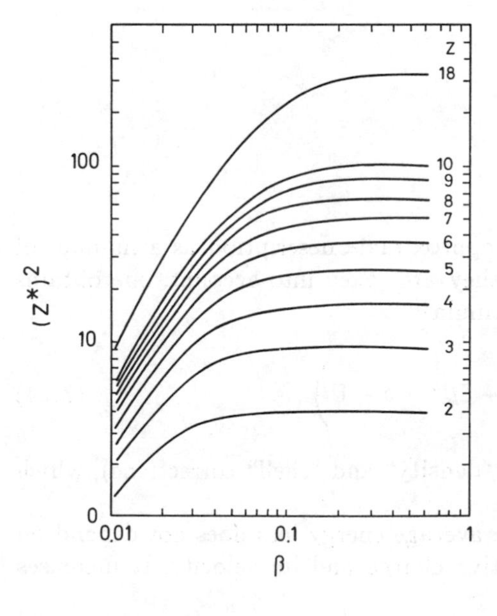

Figure 3.6. Effective charge as a function of energy for a number of ions, calculated according to the BARKAS formula (after BUTTS and KATZ 1967)

All interactions have, of course, to be included in the integrations. "Glancing" collisions transfer only small amounts of energy, but their probability is high. They are important because of their large number. A solution to the problem is only possible by a quantum-mechanical treatment; this yields the following relationship:

$$\frac{T'_{2\,max}}{T'_{2\,min}} = \left(\frac{2m_{e0}v_1^2}{I}\right)^2 \qquad (3.21)$$

where I is the "mean ionization potential" of the medium. With this, Eq. (3.20) takes the form:

$$-\frac{dT_1}{dx} = \frac{4\pi e^4}{(4\pi\varepsilon_0)^2 m_{e0}c^2} \cdot \frac{dN_e}{dV} \cdot \frac{Z^{*2}}{\beta^2} \ln \frac{2m_{e0}v_1^2}{I} \qquad (3.22)$$

or

$$-\frac{dT_1}{dx} = 2C \cdot \frac{Z^{*2}}{\beta^2} \ln \frac{2m_{e0}v_1^2}{I} \quad . \qquad (3.23)$$

The ionization potential can, in principle, be calculated but is usually determined empirically. A few examples are listed in Table 3.1.

Table 3.1. Values of the average ionisation potential for some materials. (After ICRU, 1984)

substance	I / eV
water (liquid)	75.0
methane	41.7
air	85.7
muscle	75.3
bones, cortical	106.4
compact	91.9

Equation (3.23) gives only an approximate description as a number of other influences were neglected. If they are taken into account, one obtains the final form (BETHE-BLOCH formula):

$$-\frac{dT_1}{dx} = 2C\frac{Z^{*2}}{\beta^2}\left(\ln \frac{2m_{e0}v_2^2}{I(1-\beta^2)} - 2\beta^2 - \delta - U\right) \quad . \qquad (3.24)$$

The last two terms are corrections ("density" and "shell" corrections), which play a minor role here.

Equation (3.23) shows that the average energy loss does not depend on the ion mass but only on its effective charge and its velocity. It increases with lower speeds until the effective charge is more and more reduced due to electron capture. It has, therefore, to pass a maximum. It has become customary to use $1/\varrho \cdot dT_1/dx$ (ϱ is the density of the medium), this quantity is called the *mass stopping power*. Figure 3.7 shows its dependence for protons in water.

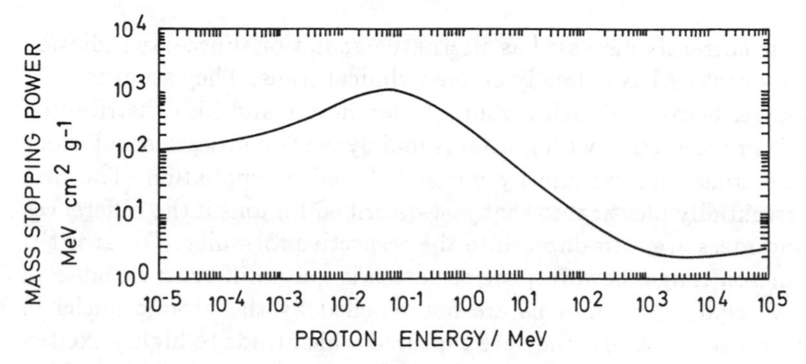

Figure 3.7. Mass stopping power of protons in water (after ICRU 1970)

Equation (3.24) implies that the stopping power of ions with the same speed depends only on Z*. The effective charge represents a useful scaling factor which allows the calculation of the energy loss of any ion if the behaviour of protons is known. This is, however, only a valid statement within

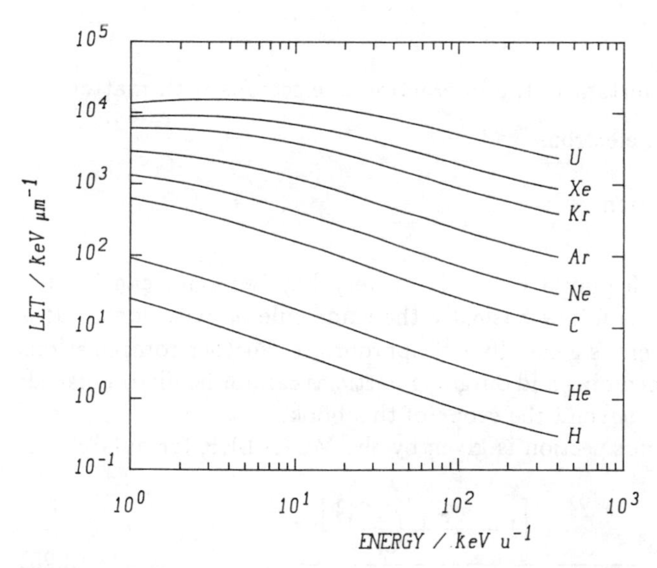

Figure 3.8. Stopping power for various ions in water (after KIEFER and STRAATEN 1987)

the framework of the BETHE formula, i.e. as long as processes other than collisions are not significant, e.g. nuclear reactions.

Figure 3.8 shows calculations of stopping powers for various ions in water in the energy range between 1 and 400 MeV/u.

3.2.4 Mesons

π^- mesons are currently discussed as alternative radiation sources in radiation therapy, a few centers have already entered clinical trials. They seem to offer some advantages because of their favourable depth-dose and LET distribution (Chap. 4). Their interaction with matter is mainly via two processes: collisions with shell electrons and capture by nuclei followed by spallation. The first reaction is essentially identical to that just described for ions if the differences in charge and mass are introduced into the respective formulae. Towards the end of the meson range, however, the situation is quite different: Because of their negative charge, π^- mesons are not repelled by the atomic nuclei in the medium, on the contrary, they are captured which leads to highly excited states. The energy corresponding to the meson's rest energy of 140 MeV is deactivated via disintegration of the nucleus (*spallation*) yielding a number of smaller charged fragments, apart from neutrons and γ quanta. The capture probability varies, in biological matter ^{12}C and ^{16}O are the most important constituents. The latter is responsible for about 75% of all spallations. The fragmentation products deposit their energy very close to the site of their origin since they only have a small range.

3.2.5 Electrons

Four processes are important in the interaction of electrons with matter:

1. collisions with shell electrons
2. Bremsstrahlung
3. CERENKOV radiation
4. nuclear reactions

Nuclear processes which play a role only at very high energies can be disregarded here. Collisions follow basically the same rule as with ions, but a non-relativistic treatment is generally not appropriate. Further complications arise by the fact that incoming and outgoing particles cannot be distinguished; a detailed treatment is beyond the scope of this book.

The differential cross section is given by the MOELLER formula:

$$\frac{d\sigma_e}{dT_2'} = \frac{2\pi e^4}{(4\pi\varepsilon_0)^2 m_{e0} c^2 \cdot \beta^2} \cdot \frac{\left[1 - \frac{T_2'}{T_1} + \left(\frac{T_2'}{T_1}\right)^2\right]^2}{\left(1 - \frac{T_2'}{T_1}\right)^2 \cdot T_2'^2} \quad . \tag{3.25}$$

Integration of Eq. (3.25) leads to the BETHE formula for electrons:

$$-\frac{dT_1}{dx} = \frac{C}{\beta^2}\left[\ln\frac{m_e v_1^2 \cdot T_1}{2I^2(1-\beta^2)} - (2\sqrt{1-\beta^2} - 1 + \beta^2)\ln 2 + (1-\beta^2) + \frac{1}{8}(1-\sqrt{1-\beta^2})^2\right] . \tag{3.26}$$

Equation (3.26) describes only collision losses but includes "glancing" collisions.

With higher energies, *bremsstrahlung* becomes more and more important. Similarly to the generation of X rays (Chap. 1), they are caused by the deceleration of electrons in the field of the nucleus and depend, therefore, on the atomic composition of the medium. A useful rule of thumb for their contribution is given by:

$$r = \frac{T_1 \cdot Z_M}{700} \tag{3.27}$$

where r is the ratio between collision and radiative losses, T_1 the electron energy in MeV, and Z_M the atomic number of the medium.

CERENKOV radiation is generated if the speed of a charged particle passing through matter exceeds that of light in the medium. This criterion sets a lower limit below which this process does not occur (about 500 keV for water). In terms of energy loss, it is usually not important but it may lead to the generation of ultraviolet light which may then be responsible for a certain biological radiation damage.

Figure 3.9 depicts the mass stopping power for electrons in water with and without *bremsstrahlung* losses.

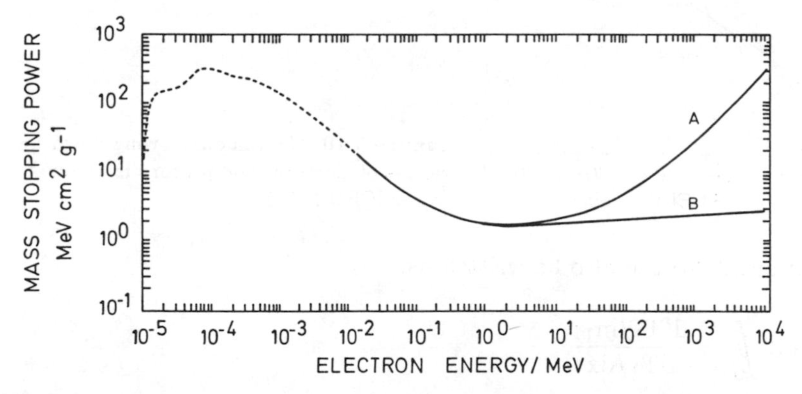

Figure 3.9. Mass stopping power of electrons in water. Curve A includes bremsstrahling losses, curve B is for collision losses only (after ICRU 1970)

3.2.6 Ranges

Contrary to photons, charged particles possess a well-defined range R which could, in principle, be calculated on the basis of the energy loss formula:

$$R = \int_0^{T_{10}} \frac{dT_1}{dT_1/dx} \tag{3.28}$$

where T_{10} is the starting energy.

This simple approach neglects, however, that energy losses are stochastic in nature and that the secondary particles possess an energy and range *distribution*. In addition electrons do not have a straight path but suffer multiple scatter with many changes in direction. Equation (3.28) is therefore only of limited value, it is characterized as "continuous slowing down approximation", csda.

Figure 3.10 gives csda ranges for electrons and protons in water, Fig. 3.11 for some ions. Extensive tabulated data are available in ICRU Report 16 (1970) and 37 (1984).

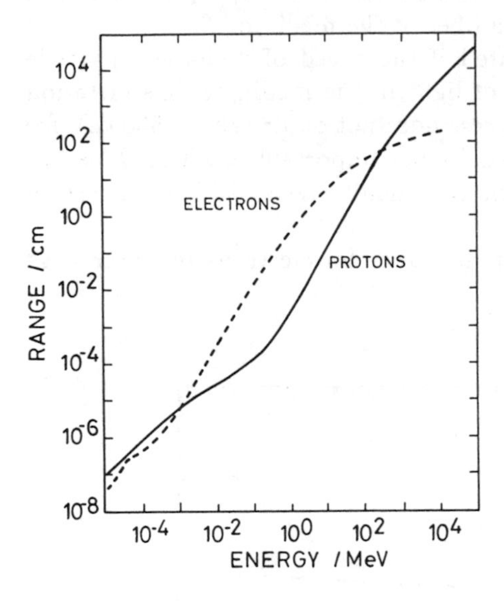

Figure 3.10. Continuous slowing down ranges of electrons and protons in water (after ICRU 1970)

Equation (3.28) can also be written as:

$$R = m_1 \int_0^{T_{10}} \frac{d(T_1/m_1)}{dT_1/dx}$$

and with Eq. (2.23):

$$R = \frac{m_1}{Z^{*2}} \cdot R_{proton} \tag{3.28a}$$

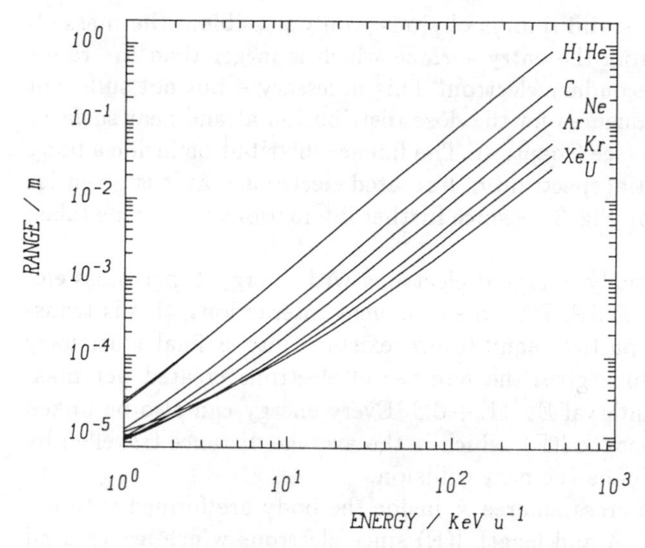

Figure 3.11. Ion ranges in water (after KIEFER and STRAATEN 1987)

where R_p is the range of a proton of equal velocity (or equal T/m). Equation (3.28a) constitutes a practical scaling rule.

All charged particles lose energy when travelling through matter. This leads also to changes in stopping power. The dependence of stopping power on penetration depth is called the "BRAGG curve".

3.2.7 Fluence and Slowing Down Spectrum

In the case of indirectly ionizing radiation, most of the primarily absorbed energy is transferred to secondary particles which in turn cause ionizations and excitations. They are almost exclusively responsible for the energy deposition in matter. Any mass element may be considered a source of these secondary particles but it is also subjected to a particle flow from its environment. If every particle leaving the mass element is compensated by the entry of another one matching the outgoing in type and energy a stationary state is established which is called "secondary particle equilibrium". In the case of photon radiations – only these will be discussed here – one has to deal exclusively with electrons. Their fluence inside an exposed body is isotropic because of multiple scattering. If a secondary particle equilibrium exists, every mass element is penetrated isotropically by a spectrum of electrons which may be characterized by a fluence distribution $\phi(E)dE$. *Fluence* is defined here as the number of electrons entering a spherical volume of radius r divided by the projection area πr^2 perpendicular to the electron direction (r is assumed to be sufficiently small to consider the particle path as a straight line).

Secondary particle equilibrium is obviously only possible if the mass element has a distance from the entry surface which is larger than the range of the most energetic secondary electron. This necessary – but not sufficient – condition bears consequences for the dose distribution at and near surfaces between different media (see Chapt. 4). The fluence distribution inside a body is different from the initial spectrum of liberated electrons – as it is given for the COMPTON effect by Fig. 3.3 – since further interactions have to be taken into account.

The number of initially liberated electrons with energy E per mass element shall be called $n_0(E)dE$. Due to subsequent interactions, this is transformed – if secondary particle equilibrium exists – into a final stationary distribution n(E)dE which gives the number of electrons created per mass element in the energy interval $E \dots E + dE$. Every energy can also be linked with a mean free path length l(E), which is the average distance travelled by the electron before it suffers the next collision.

All electrons which cross an area A inside the body are formed within a cylinder of cross section A and length l(E) since electrons which are created further away than their mean free path length will not reach the reference area. This means

$$\phi(E) \cdot D\,dE = A \cdot l(E) \cdot n(E)\,dE \quad . \tag{3.30}$$

If $n'(E, E_0)$ is the *number* of electrons with energy E which are generated from one initial electron of energy E_0, it also follows that:

$$n(E) = \int_E^{E_{0\,max}} n'(E, E_0) n_0(E_0)\,dE_0 \tag{3.31}$$

and then:

$$\phi(E) = l(E) \int_E^{E_{0\,max}} n'(E, E_0) n_0(E_0)\,dE_0 \quad . \tag{3.32}$$

With the abbreviation

$$y(E, E_0) = l(E) \cdot n'(E, E_0) \tag{3.33}$$

one may write:

$$\phi(E) = \int_E^{E_{0\,max}} y(E, E_0) n_0(E_0)\,dE_0 \quad . \tag{3.34}$$

The function $y(E, E_0)dE$ gives hence the total path length of electrons with energy E generated from one single initial electron with starting energy E_0. It is called the "slowing down spectrum". It may be determined, e.g. by following up reactions and paths of all formed electrons by computer simulation taking into account the different interaction probabilities (*Monte Carlo method*). If $y(E, E_0)$ and $n_0(E_0)$ are known, $\phi(E)$ can be calculated.

To eliminate the influence of different starting energies the slowing down spectrum is usually given in its normalized form $y(E, E_0)/E_0$. Examples are shown in Fig. 3.12. All are identical in the lower energy range which is understandable if one considers that initially very energetic electrons are quickly slowed down, "forgetting" their starting energy.

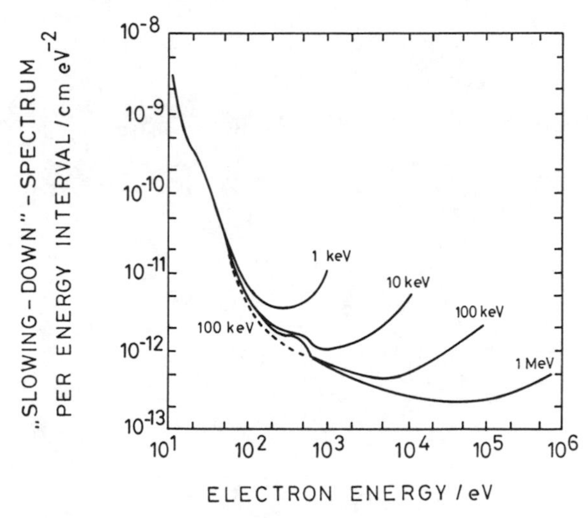

Figure 3.12. Normalized slowing down spectra in water with different starting energies. The heavy lines include AUGER-elctrons which are neglected with the broken lines (after HAMM et al. 1978)

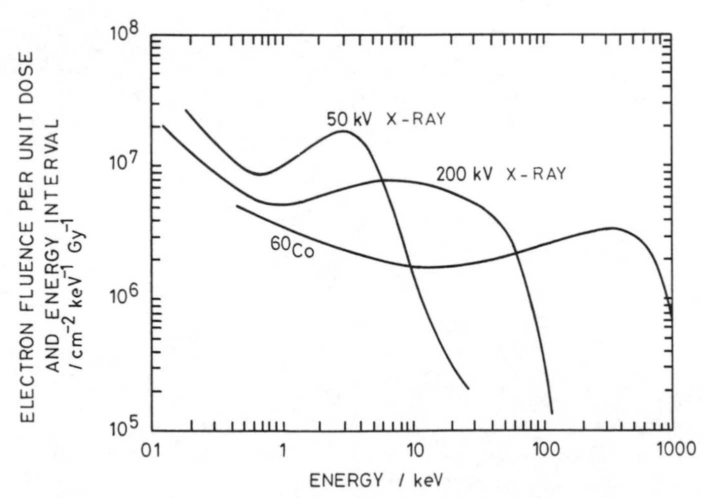

Figure 3.13. Electron fluence spectra for various photon radiations in water (after ICRU 1970)

Fluence spectra computed in the way described are depicted for various photon radiations in Fig. 3.13. It is interesting to compare the shape for ^{60}Co γ rays with the initial spectrum for 1 MeV photons in Fig. 3.3. The equilibrium distribution is markedly shifted to lower energies illustrating impressively the importance of slowing down processes.

Further readings

ATTIX et al. 1968, HINE and BROWNELL 1956, KASE and NELSON 1978, KASE et al. 1985, MORGAN and TURNER 1973, MOTT and MASSEY 1965

Chapter 4
Deposition of Radiation Energy

Energy transfer by radiation to biological systems is both of fundamental and practical importance. The central quantity with ionizing radiation is the dose (absorbed energy per unit mass); However, there is a difference between energy transferred and energy deposited, as will be explained in detail. The spatial distribution of absorption events also plays a significant role which is macroscopically described by the "linear energy transfer" (LET). The applicability of this concept is critically discussed. In microscopical dimensions the stochastic nature of energy absorption events becomes important, an issue which is dealt with under the heading "microdosimetry". This will be followed by a description of the energy deposition in the close vicinity of the track of an ionizing particle. The concept of dose as defined above is generally not meaningful with optical radiation; thus, other relevant quantities such as photon or energy fluence are introduced.

4.1 General Aspects

The first effect of radiation action on biological systems is the transfer of energy to essential cellular components. With optical radiation, this occurs via selective absorption in special molecules, the *chromophores*; with ionizing radiation, all constituents of the exposed medium contribute. This different state of affairs requires the introduction of different concepts and quantities. Dose, in the strict sense of the word, means absorbed energy per unit mass and can meaningfully be used only with ionizing radiation. The energy fluence is often also called "dose" which is inappropriate terminology since it is a property of the radiation field and independent of the object exposed; it may, to a certain extent, be compared with the *exposure* (see next section).

Deviating from the usual sequence, ionizing radiations are discussed first in this chapter because of the higher complexity and the principal importance of the subject.

4.2 Ionizing Radiation

4.2.1 Macroscopic Aspects

4.2.1.1 Dose and Exposure

Ionizing radiations transfer energy to matter through ionizations and excitations as discussed extensively in the foregoing chapter. Most significant is the fact that the interacting entities – particles or quanta – do not deposit energy by a single event and that secondary particles are liberated which are able to transport energy away from the site of primary interaction. A fundamental treatment of the relevant problems is only possible within the framework of general transport theory (CARLSSON 1985); this chapter is restricted to a simplified discussion:

Figure 4.1 summarizes the situation in a small mass element. An ionizing particle or quantum of energy E enters the volume where it loses an

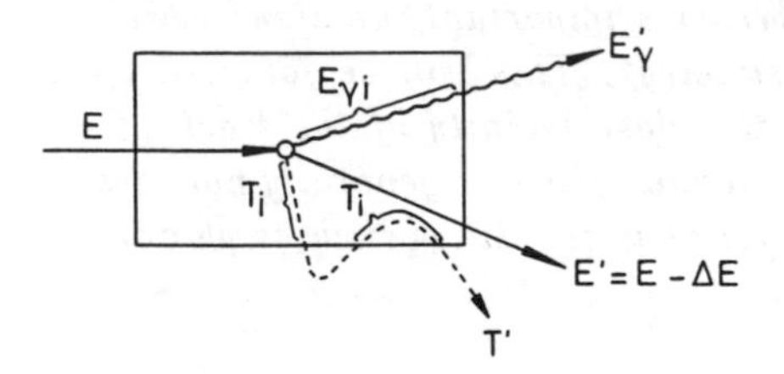

Figure 4.1. Energy deposition by ionizing radiations in small mass elements (see text for explanation)

energy amount dE by either creating electromagnetic radiation (quantum energy $E'_{\gamma i}$) and/or a secondary particle (kinetic energy T'). Only parts of these energies, namely $E_{\gamma i}$ and T_i, remain in the mass element since the secondary radiations may leave the volume carrying away a certain fraction of the originally transferred energy. Here it becomes clear why it is necessary to distinguish between energy *transferred* and *deposited.* Two quantities are relevant in this context: the range of the secondary particles and the relative contribution of radiative losses.

The energy dE_{abs} absorbed in the volume, according to the figure, is:

$$dE_{abs} = E_B + E'_{\gamma i} + T'_i \tag{4.1}$$

where E_B is the binding energy of the secondary particle. The total energy transferred dE is

$$dE = E_B + E' + T' \tag{4.2}$$

i.e.

$$dE \geq E_{abs} \quad .$$

Both are only equal if the path of the secondary particle lies completely inside the mass element and if there are no radiative losses. In the case of secondary particle equilibrium each outgoing particle is exactly compensated by an entering one of the same type and energy. The *dose* D is defined as the expectation value of the absorbed energy divided by the mass dm of the volume:

$$D = \overline{\frac{dE_{abs}}{dm}} \quad . \tag{4.3}$$

The introduction of the expectation value follows from the stochastic nature of energy deposition; this aspect will be extensively discussed in Sect. 4.2.2 (microdosimetry).

The unit of dose is the *Gray* (Gy) which equals 1 J/kg. An older and officially outdated unit is the *rad* (rd):

$$1\,\mathrm{rd} = 100\,\mathrm{erg/g} = 0.01\,\mathrm{Gy} \quad .$$

The remarks made above are particularly relevant with indirectly ionizing radiations, i.e. photons and neutrons. The primary effect here is the liberation of charged particles, electrons and – in the case of neutrons – protons or other ions. With sufficiently high energies they are able to produce *bremsstrahlung* which carries part of the initially transferred energy out of the exposed medium. In order to give a conceptionally clear description a special quantity is defined which comprises the total kinetic energy transferred to secondary particles per mass element. It is called KERMA (*kinetic energy released per mass*) and is measured in $\mathrm{J\,kg^{-1}}$.

Each mass element is, of course, not isolated but part of its environment to which it loses particles and from where others enter. Secondary particle equilibrium is obtained if every particle leaving the element is compensated by an entering one of exactly the same type and energy. A necessary – but not sufficient – condition for this to occur is that the mass element is part of a homogeneous medium at a depth that is larger than the range of the most energetic particle. It is immediately clear that this can never happen at or near surfaces. KERMA and dose, however, are even in the case of secondary particle equilibrium not generally identical; this only if *bremsstrahlung* losses are negligible. Figure 4.2 illustrates this situation:

Close to the entrance surface one finds the largest differences because of the lack of equilibrium. They become smaller with greater depths where they then are only due to bremsstrahlung.

The attenuation of the primary beam is – under neglection of pure scattering – given by the mass absorption coefficient μ/ϱ. Following the above considerations, two further parameters are introduced: the mass energy transfer coefficient μ_K/ϱ and the mass energy absorption coefficient μ_{en}/ϱ. The first is related to KERMA while the second is most relevant for dose determination because *bremsstrahlung* losses are also taken into account.

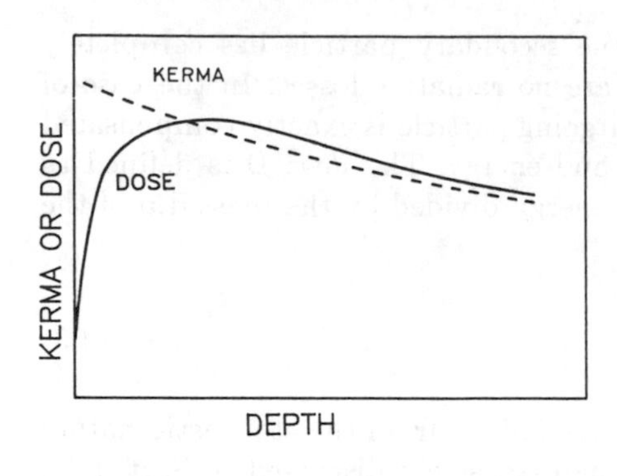

Figure 4.2. Relations between KERMA and dose (after ALPER 1978)

Ionization is important in biological radiation action as it occurs initially in most reactions. Also, it is easily measurable – at least in gases. This is why the number of ionizations in air originally has been used as a measure of dose. This is incorrect, however, since the energy absorbed depends on the nature of the exposed medium. Thus, more correctly, the respective quantity is called *exposure* X. It is defined as the charge (either positive or negative) liberated per mass element of air, its use is by definition restricted to photon irradiation.

$$X = \frac{dQ}{dM} \tag{4.4}$$

(Q: charge).

The unit is $C\,kg^{-1}$; no special name has been proposed. The old unit was the ROENTGEN (R):

$$1\,R = 2.58 \times 10^{-4}\,C\,kg^{-1} \quad . \tag{4.5}$$

As mentioned, X is not a dose quantity but rather a property of the radiation field at the location of exposure. In the case of secondary particle equilibrium and a given medium, conversion factors can be given which depend on the composition of the medium and radiation energy. For this purpose, the energy W necessary to create an ion pair has to be known. It must not be confused with the ionization potential I introduced in Chap. 3. This latter parameter is derived from the binding energies of all electrons in the atom. Since ionizations occur mostly with the weakly bound outer shell electrons, it is clear that $W < I$. W depends on radiation type and energy (apart from the medium); for most sparsely ionizing radiations one may use for air a value of 33.7 eV.

Now the dose D in air can be calculated:

$$D = \frac{W \cdot X}{e} \tag{4.6}$$

(e = elementary charge).

The numerical relation is:

$$D/\mathrm{Gy} = 33.7\,X/\mathrm{C\,kg^{-1}} \tag{4.7}$$

The value for any other medium is obtained by multiplying by the ratio of mass energy absorption coefficients:

$$D_M = \frac{(\mu_{en}/\varrho)_M}{(\mu_{en}/\varrho)_{air}} \cdot D_{air} \tag{4.8}$$

For water and $^{60}Co\,\gamma$ radiation, the following numerical relationship may be used:

$$D/\mathrm{Gy} = 37.5\,X/\mathrm{C\,kg^{-1}} \quad . \tag{4.9}$$

Values for other energies and components are listed in Table 4.1.

Table 4.1. Relation between dose and exposure for various substances and photon energies. (After ICRU 30, 1979)

photon energy keV	water Gy/C kg⁻¹	bones Gy/C kg⁻¹	muscle Gy/C kg⁻¹
10	35.0	140	34
50	34.7	134	34.3
100	36.9	56.1	36.5
200	37.4	37.9	37.0
400	37.5	36.1	37.0
600	37.5	35.9	37.0
1000	37.5	35.8	37.0
2000	37.5	35.6	37.0

These relations are of great practical importance since radiation measurements are very often performed with ionization chambers. It must be stressed again, however, that they are only valid if secondary particle equilibrium is secured.

With indirectly ionizing radiations one is often interested in the relationship between photon fluence ϕ (number of photons per unit area) and dose (with directly ionizing particles it is calculated with the aid of LET, see next section). For photons of quantum energy E and in the case of secondary particle equilibrium:

$$D = \phi \cdot E \cdot \frac{\mu_{en}}{\varrho} \quad \text{(any medium)} \tag{4.10}$$

or for the exposure X in air:

$$X = \phi \cdot E \cdot \frac{e}{W} \cdot \left(\frac{\mu_{en}}{\varrho}\right)_{air} \quad . \tag{4.11}$$

The latter relationship is graphically shown in Fig. 4.3. The minimum is due to the fact that the decrease in energy is overcompensated by an increase of μ_{en} in the lower region, for higher energies, μ_{en} does not change significantly.

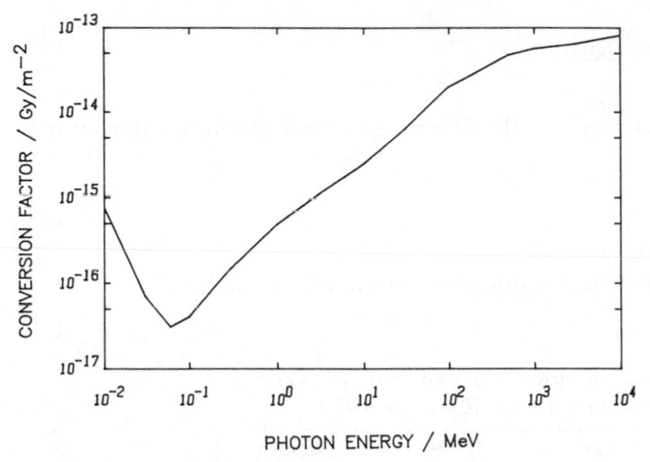

Figure 4.3. Conversion between photon fluence and dose (data taken from ICRP 21, 1971)

Equation (4.11) may also be used to calculate the dose rate of radionuclides. A characteristic figure in this context is the *specific radiation constant* Γ which gives the exposure rate of a point source at a distance of 1 m *in vacuo*. If A is the activity and n(E) the number of photons emitted per disintegration, the exposure X at 1 m distance becomes:

$$X_m = \frac{A}{4\pi} \sum n(E) \cdot E \cdot \frac{e}{W} \left(\frac{\mu_{en}}{\varrho}\right)_{air} \tag{4.12}$$

and hence:

$$\Gamma = \frac{X_m}{A} = \frac{1}{4\pi} \sum n(E) \cdot E \cdot \frac{e}{W} \left(\frac{\mu_{en}}{\varrho}\right)_{air} \tag{4.13}$$

if μ_{en}/ϱ is given in $m^2\,kg^{-1}$.

For example, with $^{60}Co\,\gamma$ rays (two quanta of 1.173 and 1.322 MeV each per decay) this yields:

$$\begin{aligned} \Gamma = {} & \frac{1.601 \cdot 10^{-19}}{4\pi \cdot 33.7}(1.17 \times 10^{-6} \cdot 2.7 \times 10^{-3} \\ & + 1.33 \times 10^{6} \cdot 2.6 \times 10^{-3})\frac{C\,m^2}{kg\,s \cdot Bq} \\ = {} & 2.51 \times 10^{-18} \frac{C}{kg \cdot s}/Bq \quad . \end{aligned}$$

Table 4.2. Specific radiation constants Γ for some γ-emitters. (After ICRU 30, 1979)

nuclide	Γ $\frac{C\ m^2}{kg \cdot s \cdot Bq}$
^{22}Na	$2.3 \cdot 10^{-18}$
^{40}K	$1.56 \cdot 10^{-19}$
^{59}Fe	$1.21 \cdot 10^{-18}$
^{60}Co	$2.51 \cdot 10^{-18}$
^{131}I	$4.1 \cdot 10^{-19}$
^{137}Cs	$6.26 \cdot 10^{-19}$
^{198}Au	$4.47 \cdot 10^{-19}$
^{226}Ra (+ decay products)	$1.59 \cdot 10^{-18}$

Values of μ_{en} are tabulated in ATTIX et al. (1968). Specific radiation constants for a number of radionuclides are listed in Table 4.2.

The situation is less transparent with neutrons as many different kinds of interactions have to be taken into account although collisions with protons dominate. A detailed derivation is, therefore, not given; only the relation between neutron fluence and KERMA is shown by means of Fig. 4.4.

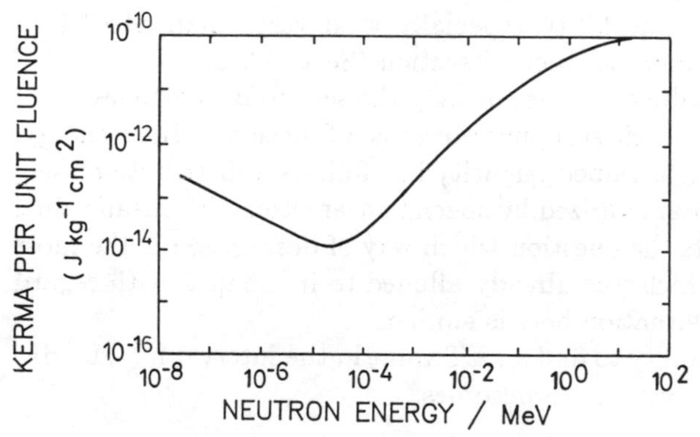

Figure 4.4. Conversion between neutron fluence and KERMA (after AUXIER, SNYDER and JONES 1968)

4.2.1.2 Linear Energy Transfer

As pointed out before, the energy deposition in an exposed body is mediated almost exclusively by charged particles. These cause ionizations on their way loosing parts of their energy in successive steps until they reach the end of their range. Depending on the type of particle, the ionizations are more or less closely spaced, which is, of course, very important if one considers energy deposition onto very small sites. This situation may be described, for instance, by the energy loss of a particle per distance travelled. The corresponding quantity is called *linear energy transfer* (LET) which is defined as the amount of *locally* absorbed energy per unit length. The attribute "locally" is of special importance since it postulates that only that energy fraction is counted which leads to ionizations and/or excitations within the considered site. The remaining kinetic energy of particles leaving the site is excluded. A generally applicable definition is not possible since this would require the specification of site dimensions. This case is particularly relevant with electrons since they may possess considerably long ranges. It has become customary to specify a limit of energy deposition below which the deposition is considered to be local (*energy restriction*); 100 eV has been widely accepted, which corresponds to an electron range of about 5 nm. Electrons of longer ranges are called "δ electrons". Alternatively, a *range restriction* is also possible, but it is seldom used because of greater computational difficulties. The total *transferred* energy per unit length is the *stopping power* $-\mathrm{dT/dx}$, as introduced in Chap. 3. It is numerically equal to LET_∞, i.e. without restriction. There are, however, conceptual differences: stopping power deals with the energy loss of the particle, while LET focuses on the energy deposition in the medium. The energy limits are also called "cut-off energies", their values in eV are indicated by subscripts to LET.

It is inherent in the LET concept that the energy deposition is viewed as a continuous process and that stochastic variations are not taken into account. This limits its applicability, especially with very small sites. This issue is taken up in the "microdosimetry" section (Sect. 4.2.3).

As shown in the preceding chapter, usually the secondary particles cover a wide energy range. Since their stopping power is a function of their energy, LET is normally not a single-valued quantity but follows a distribution. Secondary particles may be characterized by special parameters, like mean value, variance etc. This prompts the question which way of description is the most appropriate, a problem which was already alluded to in Chap. 1 with regard to emission spectra. The situation here is similar.

If $\mathrm{f(L)dL}$ is the probability to find a LET value in the interval $\mathrm{L \ldots L+dL}$, then the *number* or *track average* $\mathrm{L_T}$ becomes:

$$\overline{\mathrm{L_T}} = \int \mathrm{L \cdot f(L)\, dL} \quad . \tag{4.14}$$

The use of this interrelationship is best suited for finding the frequency of energy deposition events, irrespective of their contribution to the total dose. If this is not sufficient, it would be more appropriate to use the *dose* or *energy average* $\mathrm{L_D}$. It is the expectation value of a distribution where every single event is weighted by the amount of energy deposited. The fraction $\mathrm{d(L)dL}$ of the total dose contributed by particles of LET, L, is given as:

$$\mathrm{d(L)\, dL} = \frac{\mathrm{L \cdot f(L)\, dL}}{\int \mathrm{Lf(L)\, dL}} \quad . \tag{4.15}$$

(The integral in the denominator is necessary for normalization) and hence:

$$\overline{L_D} = \frac{\int L^2 f(L)\, dL}{\overline{L_T}} \quad . \tag{4.16}$$

Both averages are only equal if the variance σ^2 of the distribution vanishes:

$$\sigma^2 = \int (\overline{L_T} - L)^2 f(L)\, dL = \overline{L_T} \cdot \overline{L_D} - \overline{L_T^2} \tag{4.17}$$

or

$$\frac{\overline{L_D}}{\overline{L_T}} = 1 + \frac{\sigma^2}{\overline{L_T^2}} \quad . \tag{4.18}$$

From Eq. (4.18) it is seen that $L_D > L_T$ and that both are equal only if $\sigma^2 \ll \overline{L_T^2}$.

The number average L_T allows to establish a correlation between particle fluence ϕ and dose: Consider a sphere of radius r traversed by a particle fluence ϕ with LET distribution f(L)dL. The path length distribution in the sphere is given by s(l)dl.

The dose D is then:

$$\begin{aligned} D &= \phi \cdot \pi r^2 \int_L \int_l \frac{l \cdot s(l)\, dl \cdot L \cdot f(L)\, dL}{4/3\pi \varrho r^3} \\ &= \phi \cdot \frac{\overline{L_T} \cdot \bar{l}}{4/3 \varrho r} \quad . \end{aligned} \tag{4.19}$$

The mean path length l in a sphere is – as shown in Appendix I.2 – $\bar{l} = 4/3r$ so that:

$$D = \frac{\overline{L_T} \cdot \phi}{\varrho} \quad . \tag{4.20}$$

The above given relations may also be used for another considerations which will be useful later in Chap. 16: From Eq. (4.15) it follows that:

$$f(L)\, dL = \overline{L_T} \cdot \frac{d(L)\, dL}{L}$$

and with Eq. (4.20):

$$f(L)\, dL = \frac{D\varrho}{\phi} \cdot \frac{d(L)\, dL}{L} \quad . \tag{4.21}$$

If the fluence of particles with LET values in the interval $L \ldots L + dL$ is denoted $\phi(L)dL$, it is also possible to write:

$$\phi(\mathrm{L})\,\mathrm{dL} = \phi \cdot \mathrm{f(L)}\,\mathrm{dL} = \mathrm{D} \cdot \varrho \cdot \frac{\mathrm{d(L)}\,\mathrm{dL}}{\mathrm{L}} \quad . \tag{4.22}$$

A few examples of actual LET distributions will now follow. These are given as *cumulative* or *sum distributions*, i.e. the ordinate value indicates the fraction of the total dose which is due to energy depositions with linear energy transfers below or equal to the respective value on the abscissa. Figures 4.5 to 4.7 depict these for a number of different radiations. Here photon radiations

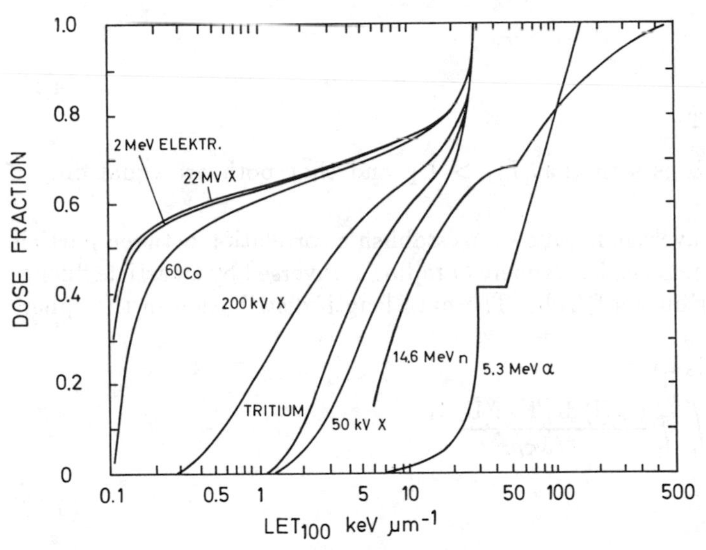

Figure 4.5. LET distributions of several radiation types (after ICRU, 1970)

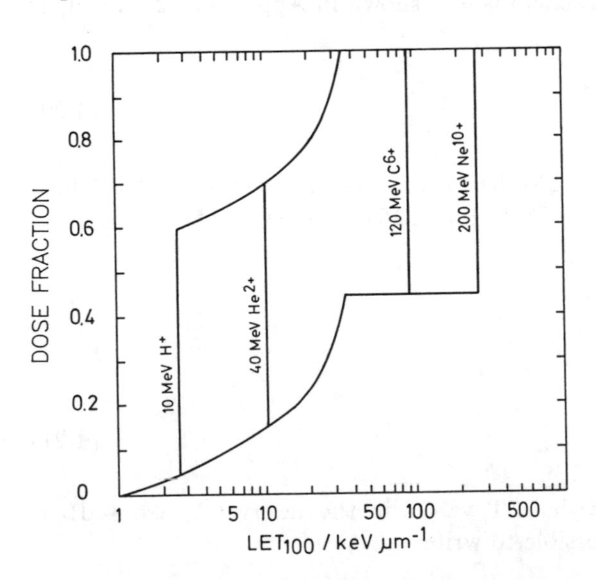

Figure 4.6. LET distributions for various ions but of identical specific energy of 10 MeV/u (after ICRU 1970)

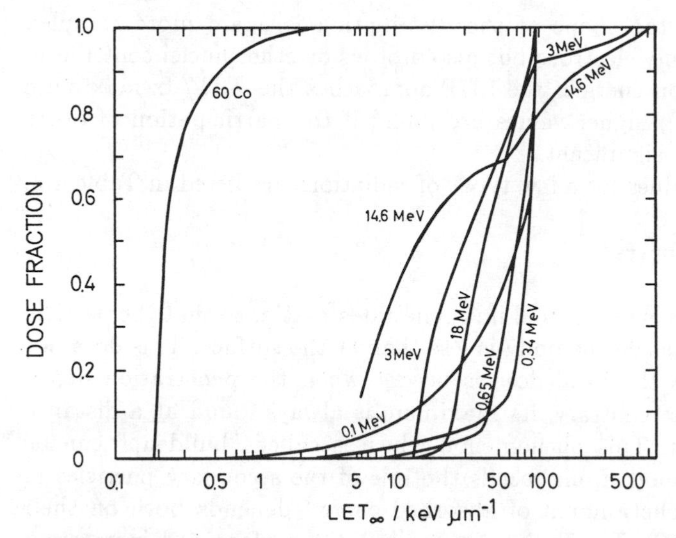

Figure 4.7. LET distributions for neutrons of various energies. ^{60}Co-γ-rays are shown for comparison

are characterized by rather broad distributions as expected from considerations in the foregoing chapter. All radiation types which interact only by the liberation of electrons approach an upper LET limit which is essentially given – neglecting cut-off values – by the maximum of $-dT/dx$ for electrons as shown in Fig. 3.9 (about $30\,\mathrm{keV}/\mu\mathrm{m}$). The α curve in Fig. 4.5 obviously consists of two parts: one is due to δ electrons with the previous maximum; the other results from the primary interaction of the α particles themselves, which leads to higher LET values. This is even more pronounced with heavier ions, as illustrated in Fig. 4.6 which applies to thin layers of material so that the energy loss of the primary particle is negligible (*track segment experiments*). About 50% of the dose is deposited by primary interactions with one well-defined LET, the rest is due to electrons. These relations hold only true for the given kinetic energies but the general picture remains the same as long as the particle energy is not too high.

Table 4.3. Mean LET-values of various radiation types in water. (After ICRU 16, 1970)

radiation	'cut-off'-energy eV	$\overline{L_T}$ keV/µm	$\overline{L_D}$ keV/µm
^{60}Co-γ	∞	0.23	0.31
	100	0.22	6.9
200 kV X-rays	100	1.7	9.4
^{3}H-β	100	4.7	11.5
50 kV-X-rays	100	6.3	13.1
5.3 MeV-α	100	43	63

With neutrons, the shape of the distribution is again more complex (Fig. 4.7) since not only electrons but also protons or other nuclei contribute. With medium neutron energies the LET approaches the $-dT/dx$-maximum for protons (Fig. 3.7); higher values are found if the participation of other recoïl nuclei becomes significant.

Average LET values for a few types of radiations are listed in Table 4.3.

4.2.2 Depth Dose Curves

Any kind of radiation is attenuated if it penetrates matter so that the particle or quantum fluence inside the body is less than at the surface. This does not mean, however, that the local dose is largest when the penetration depth is small, quite on the contrary, its maximum is always found at a distance from the entry point. This phenomenon which is called "build up" can be qualitatively understood if one recalls the role of the secondary particles in energy deposition. The amount of absorbed energy depends both on their fluence and their LET. The fluence is small at the surface and increases – approaching secondary particle equilibrium – at greater depths. The highest energy deposition is near the end of their range so that one would expect the highest dose at a depth corresponding about to the average range of the most energetic secondary particles. This is qualitatively true for photon radiations where only electrons are involved and as long as only narrow beams are used so that additional influences of scattering can be neglected. By using high photon energies, the dose maximum can be placed several centimeters inside the body. This is the physical basis of "high voltage therapy".

In general it has to be taken into account that the properties of the primary beam change with penetration depth with the effect that the interaction processes may be different, both in quantity and in quality. This is not as important with photons, electrons and neutrons but plays a very significant role with π^- mesons and accelerated heavy ions. Mesons interact at high energies like other charged particles only with shell electrons. At the end of their range, however, they are captured by nuclei which causes spallation where fragmentation products of high charge are created depositing their energy in the immediate vicinity. This does not only lead to very high local doses but also to high ionization densities, i.e. to large values of LET.

The situation is similar with accelerated ions although the underlying processes are different. The increase in dose and LET is in this case due to the dependence of $-dT/dx$ on the ion's kinetic energy as essentially described by the BETHE formula (see Chap. 3).

Examples of depth dose curves are shown in Fig. 4.8. One notices the big differences with photon energy and the most impressive depth dose peak with protons. Figure 4.9 illustrates the changes of LET with depth for π^- mesons, nitrogen ions and neutrons for comparison. The advantage of the first two types of particles in terms of depth dose distribution and ionization density is obvious.

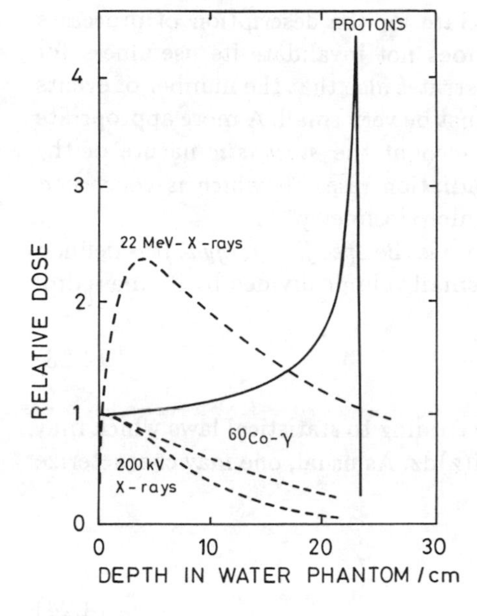

Figure 4.8. Depth dose curves for some radiation types (after RAJU and RICHMAN 1972)

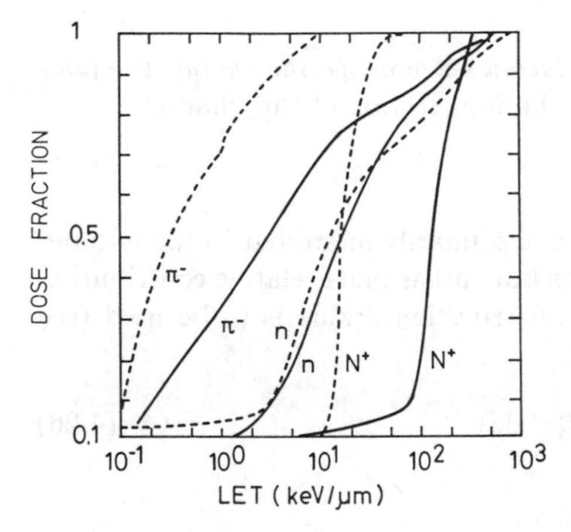

Figure 4.9. LET distributions close to the surface and at the maximum (heavy lines) of the depth dose curve for some particle beams (after RODGERS, DICELLO and GROSS 1973)

4.2.3 Microdosimetry

The specific complexity of biological radiation action lies in the fact that high amounts of energy are locally deposited in very small volumes leading to the alteration or destruction of essential molecular structures. The mean energy transferred to the total body is rather small as illustrated by the following comparison: The mean lethal dose for humans is about 5 Gy, i.e. 5 J/kg. If it were completely given as heat it would cause a temperature increase of only about 1/1000°C! To break a chemical bond, an energy of about 400 kJ/mol is required, i.e. in water 2000 kJ/kg, which corresponds to a *local* temperature increase of 2000°C. This short calculation exemplifies that the macroscopic average "dose" is completely inappropriate for the description of processes in submicroscopical dimensions. This does not invalidate its usefulness for practical purposes. The example demonstrates also that the number of events creating high local energy depositions must be very small. A more appropriate treatment has, therefore, to take into account the *stochastic* nature of the interaction processes. The branch of radiation research which is concerned with these specific problems is called "microdosimetry".

The microscopical counterpart to dose is the *specific energy* z. It is defined as the energy dE deposited locally in a small volume divided by its mass dm:

$$z = \frac{dE}{dm} \tag{4.23}$$

z is a *stochastical* quantity fluctuating according to statistical laws which may be described by a distribution function f(z)dz. As usual, one may characterize it by different mean values:

1. *Frequency mean* $\overline{z_F}$

$$\overline{z_F} = \int z \cdot f(z)\,dz \quad . \tag{4.24}$$

Correctly speaking, z_F is the *expectation value of specific energy*. It equals the dose, as already introduced in the first section of this chapter:

$$\overline{z_F} = D \tag{4.25}$$

2. *Energy* or *dose mean* $\overline{z_D}$: If one is not primarily interested in the number of different energy deposition events but rather in its relative contribution to total dose, an energy weighted distribution d(z)dz is to be used (see also Chap. 1):

$$d(z)\,dz = \frac{zf(z)\,dz}{\int zf(z)\,dz} = \frac{1}{z_F} \cdot (zf(z)\,dz) \tag{4.26}$$

with the respective expectation value:

$$\overline{z_D} = \frac{1}{z_F} \int z^2 f(z)\,dz \quad . \tag{4.27}$$

The total energy deposited in a volume is the sum of a number of single events. For further considerations it is, therefore, necessary to introduce also *single event distributions* which are defined as above but with the subscript $_1$.

The function $f_1(z)$dz does, of course, not depend on dose since it is related to single events. Their mean number n is obviously given by:

$$n = \frac{\overline{z_F}}{\overline{z_{1F}}} = \frac{D}{\overline{z_{1F}}} \quad , \tag{4.28}$$

i.e. it increases with dose. If $D \ll z_{1F}$ the mean number of events is small or, in other words, only few volumina are really affected. This point will be taken up again later.

The next step is to calculate the distribution function for multiple events. For two traversals, it follows that:

$$f_2(z)\,dz = \int_0^z f_1(u) \cdot f_1(z-u)\,du \tag{4.29}$$

where u is an integration variable.

Equation (4.29) expresses the fact that with two events a specific energy z is deposited by all combinations where the sum of the two single transfers is just z. This type of integral is quite common in physics and called *convolution* or *faltungsintegral.* It is usually abbreviated by:

$$f_2(z)\,dz = f_1(z) * f_1(z) \quad .$$

One may obviously generalize Eq. (4.29) in the following way:

$$f_\nu(z) = f_{\nu-1}(z) * f_1(z) \tag{4.30}$$

or – by abbreviating a ν-fold convolution by $f^{*\nu}$:

$$f_\nu(z) = f^{*\nu} \tag{4.30a}$$

To obtain the dose dependent total distribution function f(z,D) it is necessary to multiply each $f_\nu(z)$ with the probability for ν hits and to sum up all contributions. Since hits are "rare" events, one may use the POISSON distribution (see Appendix 1.4), giving:

$$f(z,D) = \sum_{\nu=0}^{\infty} f_\nu(z) \frac{n^\nu}{\nu!} e^{-n} \tag{4.31}$$

It is, in principle, possible to calculate f(z,D) from the knowledge of $f_1(z)$. Special care has to be taken of "zero events". The respective distribution $f_0(z)$ is a degenerate function which comprises only the value "0".

The description at this point, however, shall be restricted to the calculation of some characteristic parameters. It is greatly facilitated by using LAPLACE transforms (Appendix I.5). One finds that:

$$\begin{aligned} \overline{z_F} &= \int z f(z,D)\,dz \\ &= \int \sum_{\nu=0}^{\infty} z f_\nu(z) \frac{n^\nu}{\nu!} e^{-n} \quad . \end{aligned} \tag{4.32}$$

Application of the LAPLACE formalism yields:

$$\int z \cdot f_\nu(z)\,dz = \nu \cdot \overline{z_{1F}}$$

and then after summation:

$$\overline{z_F} = n \cdot \overline{z_{1F}} = D \quad , \tag{4.33}$$

a result which is already known from above (Eq. (4.28)).

The expectation value for the square of specific energy is obtained analogously:

$$\begin{aligned} \overline{z_F^2} &= \int \sum z^2 f_\nu(z) \cdot \frac{n^\nu}{\nu!} e^{-n} \\ &= \sum [\nu(\nu-1)\overline{z_{1F}}^2 + \overline{z_{1F}^2}] \frac{n^\nu}{\nu!} e^{-n} \\ &= n^2 \overline{z_{1F}}^2 + n\overline{z_{1F}^2} \end{aligned} \tag{4.34}$$

and after rearrangement and using Eq. (4.27):

$$\begin{aligned} z^2 &= D^2 + \frac{\overline{z_{1F}^2}}{\overline{z_{1F}}} D \\ &= \overline{z_{1D}} D + D^2 \quad . \end{aligned} \tag{4.35}$$

This is an important result which will be used again in Chap. 16. It states that the expectation value of z^2 depends in a "linear-quadratic" fashion on the dose and that the linear term is governed by the "dose mean" of the single event distribution. It is related to the radiation quality which may be characterized by the LET, as introduced in the preceding section and where the limitations of this concept were already pointed out. They are even more serious if one deals with microscopical dimensions. These problems were, in fact, the starting point of microdosimetry.

In order to remedy the situation to a certain extent, LET is also replaced by a microdosimetric counterpart which is called *lineal energy* (abbreviated y) and defined as the energy deposited in a specified spherical volume divided by the mean pathlength of random traversals. The above descriptions of dose and specific energy apply equally to the relations between LET and lineal energy (distribution functions, mean values, etc.). Since y deals only with single traversals, it is connected with z_1:

$$y = \frac{\varrho \cdot dV}{l} \cdot z_1 \tag{4.36}$$

(dV=volume, ϱ=density, l=mean path length).

For the special case of a sphere (see Appendix I.2 for the mean path length):

$$y = \pi r^2 \cdot \varrho \cdot z_1 \quad . \tag{4.37}$$

Special devices have been developed to determine z_1 and y distributions experimentally. They consist mostly of small proportional counters filled with tissue-equivalent gas at low pressure so that small volumes can be simulated (in the order of micrometers). The subject cannot be further elaborated here, only a few results will be given. In most cases one is interested in the dose-weighted distributions d(y) – similarly as with LET. They are commonly plotted on a logarithmic scale which means that they have to be related to equal *logarithmic* instead of *linear* intervals:

$$d(y)\,dy = y \cdot d(y) \cdot d(\ln y) \quad \text{(see also Eq. (4.15))} \quad .$$

This means that $y \cdot d(y)$ has to be plotted on the ordinate instead of just d(y). A few examples are given in Fig. 4.10.

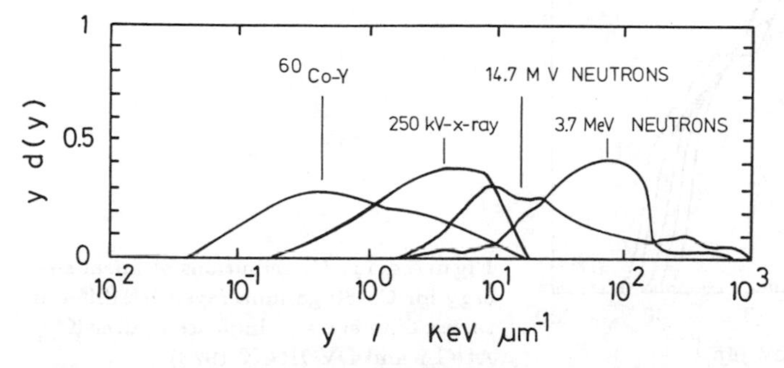

Figure 4.10. Destributions of lineal energy with various radiation types and spherical targets of 1 micrometer diameter (after KELLERER and ROSSI 1971)

Table 4.4. Mean values of lineal energy y for various radiations in a simulated spherical volume of 1 µm diameter with tissue equivalent gas. (After (a) BOOZ, 1976; (b) RODGERS and GROSS, 1974)

radiation	y_F keV/µm	y_D keV/µm	remarks	reference
^{60}Co-γ	0.39	1.86		a
200 kV-X-rays	1.52	4.20	diameter: 0.975 µm	a
65 kV-X-rays	2.11	4.67	diameter: 0.92 µm	a
^{3}H-Tritium β	3.24	5.64	diameter: 0.925 µm	a
4.6 MeV-neutrons	–	47.4		b
15 MeV-neutrons	–	87.7		b

It is seen that the spectra are shifted to higher y values with increasing LET, as was to be expected. Table 4.4 lists a number of experimentally determined mean values.

The y distributions (and hence the mean values) are not independent of the size of the simulated volume, as one might naively expect. The reason for this lies in the fact – as already mentioned in connection with the LET concept – that particle tracks are not infinitely small and that energy transfer occurs discontinuously. Particularly at the microscopical scale there are a number of complicating factors which will be discussed at the end of this section.

Figure 4.11 exemplifies the variation of y spectra with target volume.

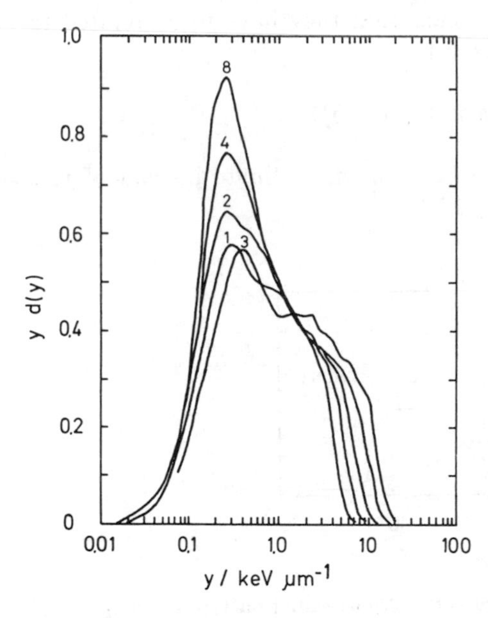

Figure 4.11. Distributions of lineal energy for Co-60-gamma-rays with different target diameters as indicated (after KLIAUGA and DVORAK 1978)

There is, of course, an interrelation between lineal energy and LET (as between specific energy and dose) which can be quantified under simplifying assumptions. A measure of the deviation of specific energy from its frequency mean – the dose – is the variance $\sigma^2(z)$ of the f(z,D) distribution. It follows from Eq. (4.34) that:

$$\sigma^2(z) = \overline{z^2} - \overline{z}^2 = \overline{z_{1D}}D$$

and for the relative variance:

$$\frac{\sigma^2(z)}{D^2} = \frac{\overline{z_{1D}}}{D} \quad . \tag{4.38}$$

The distribution is the narrower the higher the dose is – for a given z_{1D}. On the other hand, the variance is directly proportional to z_{1D}, i.e. it is larger if high amounts of energy are deposited in single traversals, in other words, with densely ionizing radiations.

The relation between LET and lineal energy will be illustrated for the special case of a sphere with radius r. As a first approximation it will be assumed that the particles in question travel in straight lines, cross the volume completely and deposit energy with a common LET. Variations in y are then only due to different path lengths. Their distribution t(l) is given (Appendix I.2) by:

$$t(l)\,dl = \frac{l\,dl}{2r^2} \quad \text{for} \quad 0 \leq l \leq 2r \quad .$$

Furthermore:

$$z_1 = \frac{L \cdot l}{4/3\pi\varrho r^3} \quad \text{and} \quad dz = \frac{L \cdot dl}{4/3\pi\varrho r^3}$$

so that:

$$f_1(z)\,dz = \frac{8\pi^2\varrho^2 r^4}{9l^2} z\,dz \quad \text{for} \quad 0 \leq z \leq \frac{L}{\pi\varrho r^2} \tag{4.39}$$

and for lineal energy:

$$f(y)\,dy = \frac{8}{9L^2} y\,dy \quad \text{for} \quad 0 \leq y \leq \frac{3}{2}L \quad . \tag{4.40}$$

The mean values are then calculated to be:

$$\overline{z_F} = \frac{L}{\pi\varrho r^2} \qquad \overline{z_D} = \frac{9}{8}\frac{L}{\pi\varrho r^2} \tag{4.41}$$

and

$$\overline{y_F} = L \qquad \overline{y_D} = \frac{9}{8}L \tag{4.42}$$

It must be reemphasized that these relations are valid only under the specified assumptions.

It shall now be assumed that the LET is not constant but follows a distribution c(L) – all other assumptions unchanged. In this case variations in y and z are additionally due to LET variations. t(l) and c(L) are independent of each other so that the integrations can be performed consecutively by yielding:

$$\overline{y_F} = \overline{L_F} \quad \text{and} \quad \overline{y_D} = \frac{9}{8}\overline{L_D} \quad . \tag{4.43}$$

Also this is only true under the assumed conditions. They are best fulfilled by ions since straight paths are postulated. Furthermore, all the energy deposited must remain in the volume which means that it must not be too small. If these conditions are met, the frequency mean of y equals that of LET while there is a small difference between the respective dose averages.

Now another case shall be considered: the energy deposition by electrons in a small body. It is again assumed that the particle paths are straight and that there is secondary particle equilibrium. Additionally, it is taken into account that the range is restricted. The mean path length in the volume is then not only governed by geometrical factors but also by particle properties.

The mean range of the electrons is designated x, their mean track length in the spherical target s. As shown in Appendix I.2, the following relation holds true for spheres:

$$\frac{1}{\bar{s}} = \frac{1}{\bar{x}} + \frac{1}{\bar{l}} \tag{4.44}$$

where l is the geometrical mean traversal length, as before. If one approximates LET by:

$$\overline{L_F} = \frac{\bar{E}}{\bar{x}} \tag{4.45}$$

where E is the average particle energy, and by using:

$$\overline{y_F} = \frac{\bar{s} \cdot \overline{L_F}}{\bar{l}} \tag{4.46}$$

the following expression can be given:

$$\frac{1}{\overline{y_F}} = \frac{1}{\bar{\bar{E}}} + \frac{1}{\overline{\overline{L_F}}} \quad . \tag{4.47}$$

For large E, Eq. (4.47) approaches Eq. (4.42). By plotting $1/y_F$ versus $\bar{l}$ (or diameter) a linear relationship should be found. Although the approximations are rather crude for electrons experimental data conform surprisingly well to the theoretical relationship as shown in Fig. 4.12. It allows also to estimate L_T for sparsely ionizing radiation from microdosimetric measurements.

The effects of "low doses" are a matter of great concern in radiation biology, both practically in radiation protection and fundamentally in the understanding of basic mechanism. Microdosimetry may help to delineate this rather loosely defined expression more precisely. Energy deposition in small volumes is caused by essentially two contributions: the number of hits and the specific energy per hit. Because of the discrete nature of energy deposition events there is a lower limit of energy absorbed per mass element which depends both on the dimension of the target and the type of radiation. With low doses any change in mean specific energy is essentially due only to

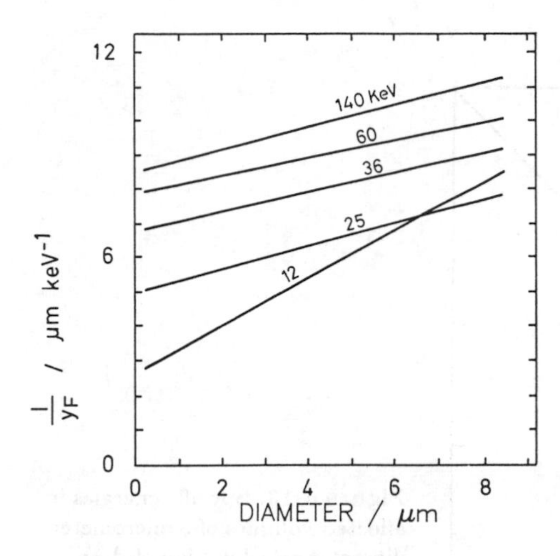

Figure 4.12. Relationship between reciprocal of track average lineal energy and site diameter for some monoergetic photons (after KLIAUGA and DVORAK 1978)

the number of targets affected while the average deposition <u>per</u> <u>hit</u> remains the same. Only with higher doses when many targets are hit there is also an increase in mean specific energy averaged over all affected mass elements. This qualitative statement can be put into more quantitative terms:

Let n again be the mean number of hits. According to the POISSON statistics the fraction of targets without any hit is e^{-n}, hence the fraction of affected entities $1 - e^{-n}$. The average specific energy in *all* targets is $n \cdot z_{1F}$ ($= D$, Eq. (4.33)). The average specific energy per affected site z_{aff} then becomes:

$$\overline{z_{aff}} = \frac{n\overline{z_{1F}}}{1 - e^{-n}} \quad . \tag{4.48}$$

For small values of n, the exponential in the denominator may be expanded:

$$\overline{z_{aff}} \approx \frac{n z_{1F}}{1 - (1 - n \ldots)} = z_{1F} \quad . \tag{4.49}$$

If the mean number of hits is small, all affected sites will receive the same amount of energy. Any increase in dose within that region will only lead to a higher number of affected sites. If $n \gg 1$, then z_{aff} approaches $n \cdot z_{1F} = D$. The border line between "low" and "high" doses is, of course, somewhat arbitrary; it has been suggested to be set at $n = 0.2$ (FEINENDEGEN et al. 1985).

The actual dose value depends both on the size of the sensitive volume and the type of radiation. Taking a sphere of 8 μm diameter as a representative nucleus of a mammalian cell, one finds $z_{1F} \approx 10^{-3}$ Gy with ^{60}Co γ rays and $z_{1F} \approx 3 \times 10^{-2}$ Gy for 9 MeV neutrons. The low-dose region would then be below 2×10^{-4} Gy for γ rays and 6×10^{-3} Gy for neutrons. Equation (4.48) is depicted diagrammatically for the above case in Fig. 4.13.

As shown above, smaller sensitive sites will lead to different delineations.

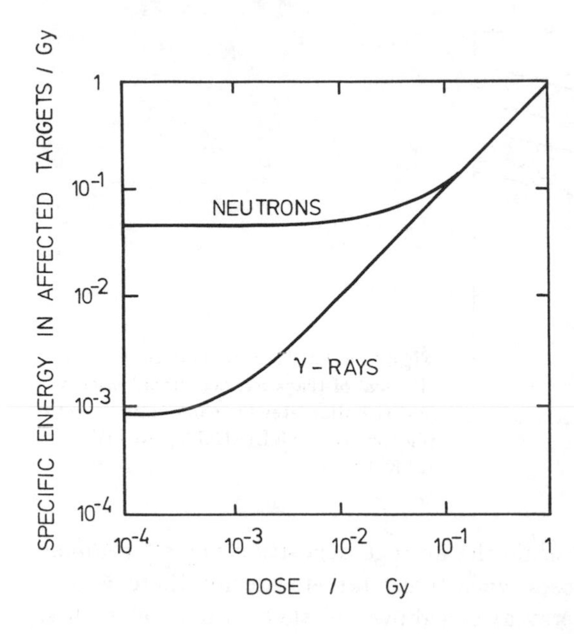

Figure 4.13. Specific energies in affected volumes of 8 micrometer diameter as a function of dose with gamma rays and 9 MeV neutrons (after FEINENDEGEN et al. 1985)

A yeast cell nucleus with a typical diameter of 1 μm would receive $z_{1F} \approx$ 0.05 Gy for γ rays and about 2 Gy with neutrons.

In summary, the major aspects in energy deposition at microscopic scales are the following:

1. *Variations in traversal length*: This does not present a fundamental problem. If the geometry of the target is known, solutions can be found as shown here for spherical volumes.
2. *Range restrictions*: This has also been touched upon, although in a very simplified manner. Most particles do not possess a well-defined range but a range distribution ("range straggling"). Also, the energy loss does change with distance travelled. These points are critical mainly with larger volumes.
3. *Discontinuous energy loss (energy straggling)*: All above calculations are based on the assumption of continuous energy loss. This is far from reality because of the discrete and stochastic nature of interactions. These play an important role with small path lengths, i.e. in small volumes.
4. *δ-electron losses*: If the energy is deposited in small volumes part of it may be carried away by energetic δ electrons. The relative contribution depends on the actual traversal site and its distance from the border of the target. This means essentially that LET becomes a function of space variables and can no longer be unambiguously specified by a single value. The introduction of "cut cff values" cannot completely satisfy of this problem.

The questions raised cannot be treated quantitatively here. Only if all the influences can be neglected, LET constitutes an unambiguously and, therefore, reasonably applicable parameter. Whether this is the case depends on the size of the critical volume and the kind of radiation. The situation is particularly critical with electrons – and therefore also with photons – since range and energy straggling cannot be separated.

Estimates about the applicability of LET have been presented for different radiation types and volume sizes. Some results are given in Figs. 4.14 and 4.15: The energy and site diameter ranges where the above factors play

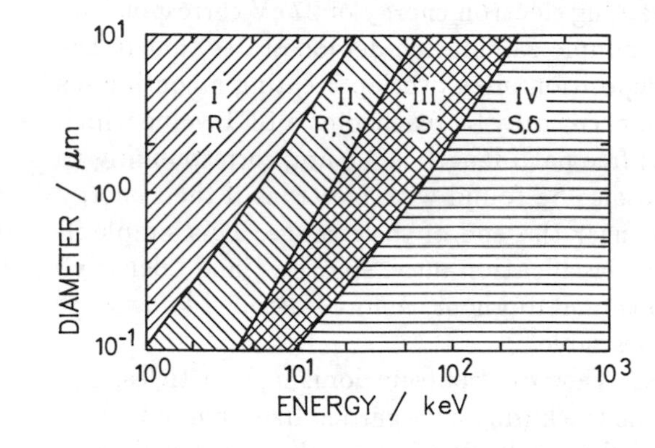

Figure 4.14. On the applicability of the LET concept with electrons (after KELLERER and CHMELEVSKY 1975)

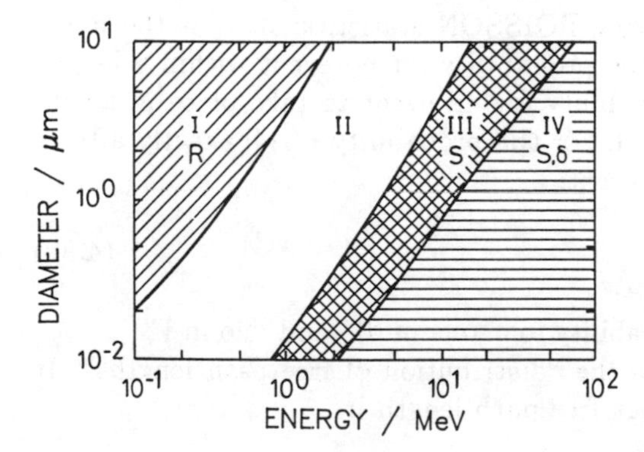

Figure 4.15. On the applicability of the LET concept with protons (after KELLERER and CHMELEVSKY 1975)

a significant role are indicated by letters and different lineatures (R = range straggling, S = energy straggling, δ = δ-electron losses. LET may be used with protons over a certain range (section II), but the picture is entirely different with electrons where all sections overlap. In this case the microdosimetric treatment is mandatory.

4.2.4 Track Structure

Energy transfer by ionizing particles proceeds via discrete events. The track consists, therefore, of a stochastic discontinuous sequence of affected and unaffected domains. An obvious and simple parameter is the average distance between successive events, the *mean free path length*. It depends, on one hand, on the LET or the lineal energy and, on the other hand, on the average amount of energy deposited. This value is not well known; for illustration purposes a reasonable value of 60 eV may be chosen. This leads for ^{60}Co γ rays ($L_T = 0.23\,keV/\mu m$) to a mean distance of 200 nm, for ^{241}Am α particles ($L_T = 43\,keV/\mu m$) to 1.4 nm. The interaction sites, however, are not infinitely small points but cover a certain extension which is mainly due to the generation of secondary electrons. If one assumes the mean ionization energy for water to be 33 eV, there is a remaining electron energy of 27 eV corresponding to a range of about 1.5 nm. This simple calculation demonstrates that in the case of α particles, the energy deposition sites overlap creating a continuous "track core". The limit for this to occur may be reached at a LET value which is small enough so that the mean free path length is 4.5 nm, corresponding to $L_T = 13\,keV/\mu m$. This value can also be found with electrons if their energy is smaller than about 1 keV, e.g. near the end of their range. This simple illustration demonstrates that the classification into "sparsely" and "densely" ionizing radiations is by no means unambiguous. A more detailed knowledge of track structure is, therefore, desirable.

First we should take a closer view of "sparsely ionizing" particles, e.g. energetic electrons. In this case the track may be regarded like a curved string of beads where the single events follow each other randomly. The distribution may be obtained by again applying POISSON statistics: If λ_1 is the mean number of events per distance, the probability for no event within a length l is $\exp(-\lambda_1 \cdot l)$. The mean probability for an event to take place in an infinitesimally small segment is $\lambda_1 dl$. For the probability $p(l)dl$ exactly after a distance l there will be the next event:

$$p(l) = \lambda_1 e^{-\lambda_1 l}\, dl \tag{4.50}$$

which is the product of the probability for "one in dl" and "no in l".

Equation (4.50) is known as the "distribution of free path lengths". It may be used to calculate the *mean* free path length l:

$$l = \int_0^\infty \lambda_1 \cdot l e^{-\lambda_1 l}\, dl = \frac{1}{\lambda_1} \quad . \tag{4.51}$$

As a crude approximation, the track may be seen as a sequence of events distributed according to Eq. (4.50). Even when all the approximations are accepted, this is not true for condensed media, e.g. water. The reason lies in the fact that the liberated electrons can only escape the positive electric field of the ion if they exceed a certain energy. An estimate of this value is obtained by a comparison with the thermal energy KT:

$$\frac{e^2}{4\pi\varepsilon_0\varepsilon_r R} \approx KT \tag{4.52}$$

(ε_0 = absolute dielectricity coefficient, ε_r = relative dielectricity coefficient of the medium, R = electron range).

Only if the thermal energy of the electron is sufficiently high at the end of its range, it is able to escape the retracting field of the ion. ε_r is here considerably smaller than the static value (81 for water) since these are processes which are too fast for the water dipoles to follow. By inserting realistic values into Eq. (4.52), a range of about 20 nm corresponding to an electron energy of about 500 eV can be calculated. Events of this size are characterized by the ability of the liberated electrons to cause secondary ionizations. Therefore, two categories are obtained: isolated primary ionizations (electron energy < 100 eV) and primary ionizations combined with localized secondary ionizations (energy between 100 and 500 eV). The first class is termed "spurs", the second one "blobs". A third group is formed by those events where the spurs overlap because of small mean free path lengths ("short tracks"). The limit for this is about 5000 eV. Secondary electrons possessing energies high enough to cause several events are treated separately as "branch tracks". Since the situation for these is similar as for primary electrons, they have not to be especially considered.

The dose contributions by the different categories may be computed as a function of electron energy; the result is shown in Fig. 4.16. According to

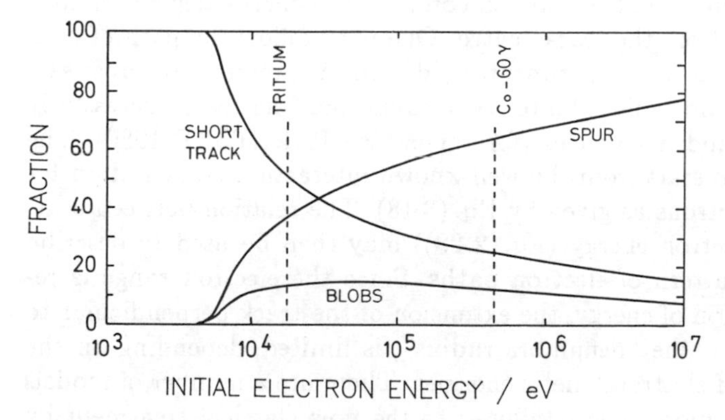

Figure 4.16. Fractions of "spurs", "blobs" and "short tracks" as a function of electon energy (after MOZUMDAR and MAGEE 1966)

the definitions given above, below 5000 eV there are only short tracks which become less and less important with higher energies, while the fraction of spurs increases.

Now the track structure of densely ionizing particles shall be considered. It may be operationally divided into a "core region" where the primary events overlap and a "penumbra" which is made up of secondary δ electrons. The picture of the track structure looks roughly like a test tube brush, as illustrated in Fig. 4.17.

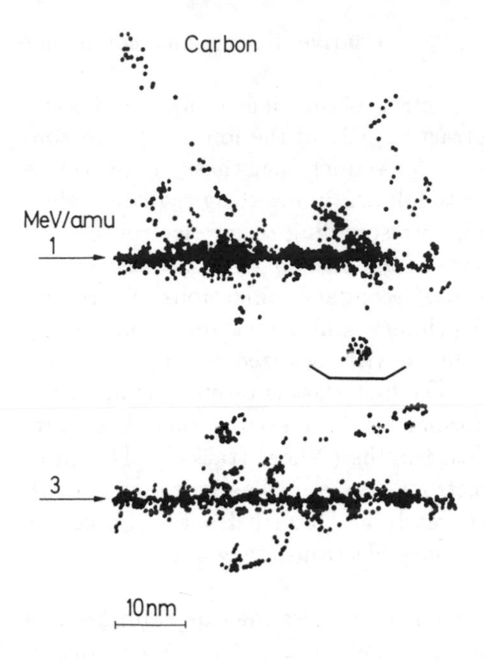

Figure 4.17. Simulated track structures. Each dot represents an energy deposition event (from PARETZKE 1980)

There are, in principle, two ways to compute the energy deposition as a function of distance from the track centre: One is to follow the path of each electron taking into account experimentally determined interaction cross sections in a given medium. This "Monte Carlo simulation" has been successfully applied to protons and α particles (for review see PARETZKE 1986). The other procedure is to start from the well-known interaction cross section for the liberation of electrons as given by Eq. (3.18). The relation between scattering angle and electron energy (Eq. (2.28)) may then be used to describe the initial spatial pattern of electron paths. Since the electron range is restricted and a function of energy, the extension of the track perpendicular to the particle direction, the "penumbra radius", is limited, depending on the initial distribution of electron kinetic energies. There are a number of models in the literature, all more or less tailored to the now classical treatment by BUTTS and KATZ (1967). They all agree in the finding that the local dose varies with the inverse square of the distance from the track centre although the quantitative formulations differ. There are now also a few experimental data available; references are listed in Table 4.5.

Rather wide discrepancies exist concerning the penumbra radius. Since it depends on the energy of the most energetic delta electrons which is in turn related to ion velocity (see Eq. (2.27)), the expressions are generally of the form:

$$r_p = \alpha \cdot \left(\frac{T}{m_i}\right)^{\beta} \tag{4.53}$$

Table 4.5. Available track structure measurements

ion	energy / Mev u^{-1}	references
^{1}H	0.5 - 4	MILLS and ROSSI 1980
	1, 2, 3	WINGATE and BAUM 1976
	0.7 - 2	MENZEL and BOOZ 1976
^{2}H	0.5 - 1	MENZEL and BOOZ 1976
^{4}He	0.25 - 0.75	WINGATE and BAUM 1976
	930	VARMA et al. 1976
	18.3	KANAI and KAWACHI 1987
	0.275 - 1.25	BUDD, KWOK, MARSHALL and LYTHE 1983
^{16}O	2.57	VARMA et al. 1977
^{10}Ne	377	VARMA and BAUM 1980
^{80}Br	0.5	VARMA et al. 1980
^{127}I	0.26, 0.49	BAUM et al. 1974

but with differing values of α and β; some examples are listed in Table 4.6. Fig. 4.18 shows that a reasonable match between experimental and theoretical data may be obtained within certain ion energy ranges. This kind of infor-

Table 4.6 a). Monte Carlo calculations of track structure

ion	energy / MeV u^{-1}	reference
^{1}H	1	PARETZKE 1980
	0.3 - 1.5	ZAIDER, BRENNER and WILSON 1983
	0.25 - 3	WILSON and PARETZKE 1981
	0.3 - 4	CHARLTON et al. 1985
	1 - 20	BERGER 1985, 1988
	1 - 100	WALIGORSKI, HAMM and KATZ 1986
	3	HAMM et al. 1984
^{4}He	0.3 - 5	CHARLTON et al. 1985

Table 4.6 b). Track structure models

penumbra parameters [a] K µm $(MeV/u)^{-\alpha}$	α	reference
0.219	1	BUTTS and KATZ 1967
0.096	1.35	CHATTERJEE et al. 1973
0.226	1.7	FAIN et al. 1974
0.197	1.67	HANSEN and OLSEN 1984
0.0616	1.7	KIEFER and STRAATEN 1986
0.22	1.667	KATZ 1988

a) the penumbra radius r_p is given by: $r_p = K \cdot (T/m)^{\alpha}$

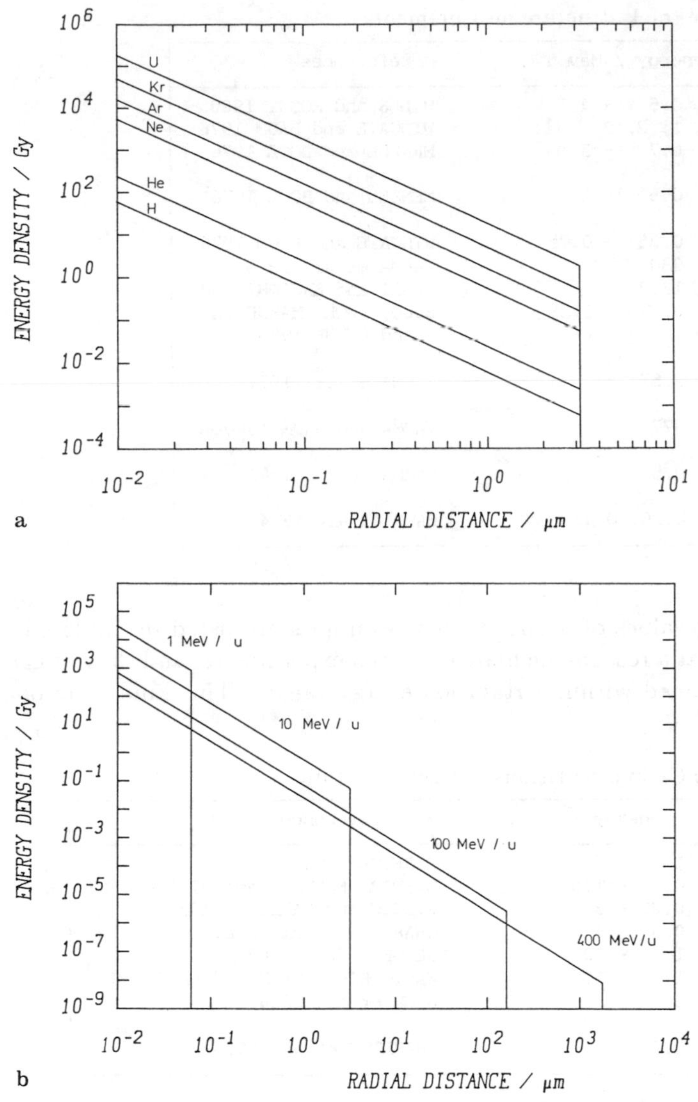

Figure 4.18a,b. Energy density as a function of radial distance. (a) various ions with 10 MeV/u specific energy, (b) neon ions of various energies as indicated (calculated according to KIEFER and STRAATEN 1986)

mation is required if the action of high LET radiation on biological entities shall be quantitatively interpreted.

4.2.5 Dosimetry of Incorporated Radionuclides

Dosimetry of incorporated radionuclides is of great practical importance in radiation protection. Because of the interplay of biological and physical factors, the general situation is rather complex. Some of the more elemental questions will be discussed here, conclusions can be found in Chap. 21.

Throughout this section it will be assumed that the radioactivity is homogeneously distributed in the volume under consideration. This is certainly an oversimplification for most biological systems but it facilitates the treatment and constitutes a useful first approximation in many cases.

The first example is a sphere with radius r which is assumed to be large compared with the range of all primary and secondary particles. In this case all the energy remains in the volume and the dose rate $\dot{D}$ is simply given by:

$$\dot{D} = E_{eff} \cdot A/\varrho V \tag{4.54}$$

where E_{eff} is the effective particle energy (see below), A the activity, ϱ the density and V the volume.

For practical purposes, the following relation is useful:

$$\frac{\dot{D}}{\mathrm{Gy\,s^{-1}}} = 1.6 \times 10^{-19} \frac{E_{eff} \cdot A/(\varrho V)}{\mathrm{eV\,Bq\,kg^{-1}}} \tag{4.55}$$

or in the "old" units

$$\frac{\dot{D}}{\mathrm{rad\,s^{-1}}} = 5.9 \times 10^{-7} \frac{E_{eff} \cdot A/(\varrho V)}{\mathrm{eV\,Ci\,kg^{-1}}} \tag{4.56}$$

These formulae are strictly valid only inside the sphere where the distance from the edge is larger than the particle range. With larger dimensions, the errors, if the edge-effects are neglected, will be small. These conditions apply for α and β emitters of not too high energies.

The effective energy E_{eff} depends on a number of factors: In the case of β emitters, it is the average energy of the continuous emission spectrum which is roughly 1/3 of the maximum energy. If the incorporated radionuclide is part of a decay series, E_{eff} is not only related to the start nucleus but the daughter substances have also to be taken into account. This is particularly important with short half-lives since a radioactive equilibrium is reached rather quickly.

Two examples illustrate the situation:

1. Tritium: The effective energy is 5.7 keV, the maximum 18 keV, corresponding to a range of a few microns in tissue. An activity of 1 μCi/kg, homogeneously distributed, causes a dose rate of $3.4 \times 10^{-11}\,\mathrm{Gy\,s^{-1}}$ or 3×10^{-6} Gy per day.
2. ^{222}Rn: Secular equilibrium with the short-lived daughter substances ^{218}Po (E_α = 6 MeV), ^{214}Pb (β), ^{214}Bi (β) and ^{214}Po (E_α = 5.5 MeV) is assumed. Since most of the dose is due to α particles, the contribution of the electrons will be neglected. Because of the equilibrium, the activities of all members in the series is the same and the effective α energy is the sum of all contributions: $6 + 7.7 + 5.5 = 19.2$ MeV. With Eq. (4.55) or (4.50),

on the basis of 1 μCi/kg, a dose-rate of 1.1×10^{-7} Gy s^{-1} or 9.8 mGy per day is obtained.

If one is not only interested in the actual dose rate but also in long-term irradiation, it is necessary to consider that the activity A is not constant but changes because of radioactive decay or biological elimination processes. Both are usually characterized by a decay constant λ_{eff} (see Chap. 15) assuming an exponential relationship. With an initial activity A_0, the total dose D(t) in a time t becomes:

$$D(t) = \frac{A_0 \cdot E_{eff}}{\varrho V} \int_0^t e^{-\lambda_{eff}} du$$
$$= \frac{A_0 \cdot E_{eff}}{\varrho V \cdot \lambda}(1 - e^{-\lambda_{eff} t}) \quad . \tag{4.57}$$

Calculations of this kind are often required in radiation protection where the total dose accumulated in 50 years is a standard parameter. 1 μCi/kg of tritium is again used as an example: Because of the relatively long physical half-life and the high biological turnover of hydrogen, the effective half-life is comparatively short, namely 11.6 days, corresponding to $\lambda_{eff} = 6 \times 10^{-2}\,d^{-1}$. Then, Eq. (4.57) gives a 50-year dose of 5×10^{-5} Gy, corresponding to 16 days of irradiation with the initial energy. This example demonstrates also that a long physical half-life is meaningless if the compound under consideration is quickly eliminated.

The assumption of short particle ranges compared to the sphere dimensions is not very realistic for energetic electrons and particularly γ rays. Any point in a sphere of radius r is then exposed to the radiation of all sources within this volume.

The dose rate D(x) at a distance x from such a source is given by:

$$\dot{D}(x) = \frac{A}{\varrho V} \int_E F(x, E)\, E\, f(E)\, dE \tag{4.58}$$

where F(x,E) is the fraction of the emission energy E which is deposited in a shell with radius x and thickness dx; f(E)dE is the energy distribution of the sources. If scattering is neglected, the case of photons can be calculated as:

$$F(x, E) = \frac{\mu_{en}}{4\pi \varrho x^2} e^{-\mu_{eff}(E)x} \quad . \tag{4.59}$$

Here μ_{en} is the energy absorption coefficient and μ_{eff} the effective attenuation coefficient which may differ in microscopical dimensions from the macroscopical one. To calculate the actual dose rate, it is necessary to integrate over all possible sources.

An easy example is the centre of the sphere. It is exposed to all point sources in shells with volume $4\pi x^2 dx$, integrated over all distances x from 0 to r:

$$\dot{D} = \frac{A}{V\varrho} \cdot \mu_{en} \int_0^r e^{-\mu_{eff} x} \, dx \cdot \int_E E \, f(E) \, dE$$
$$= \frac{A}{V\varrho} \frac{\mu_{en}}{\mu_{eff}} (1 - e^{-\mu_{eff} r}) \cdot E_{eff} \quad (4.60)$$

with $E_{eff} = \int E \, f(E) \, dE$

If the sphere is small, the exponential in Eq. (4.60) may be expanded yielding:

$$\dot{D} \approx \frac{A}{\varrho V} \cdot E_{eff} \cdot \mu_{en} \cdot r \quad . \quad (4.61)$$

For very large spheres, on the other hand:

$$\dot{D} \approx \frac{A}{\varrho V} \cdot E_{eff} \cdot \frac{\mu_{en}}{\mu_{eff}} \approx \frac{A}{\varrho V} \cdot E_{eff} \quad (4.62)$$

which is identical to Eq. (4.54) which was derived for ranges small compared to the volume dimensions.

The treatment given here excludes the role of secondary electrons. This does not introduce significant errors in gases but may be significant in condensed media.

In the case of homogeneously distributed β emitters, Eq. (4.59) is commonly written in a modified form:

$$F(x, E) \, dx = \frac{F(x/x_0, E)}{4\pi \varrho x^2} \, d(x/x_0) \quad (4.63)$$

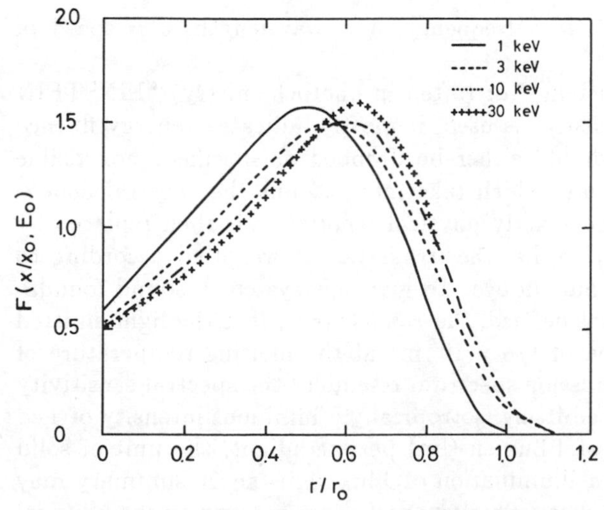

Figure 4.19. Functions for the determination of dose by incorporated beta emitters (after BOOZ and SMITH 1977)

x_0 is the csda range of the electron (Chap. 3). The function $F(x/x_0, E)d(x, x_0)$ gives the fraction of initial energy E which is deposited at the distance x normalized to the range x_0. It varies only slightly with electron energy and may be calculated, e.g., by "Monte Carlo" computation. Some examples are given in Fig. 4.19, numerical data were tabulated by BERGER (1973). An analytical treatment is generally not possible, the doses for different geometrical situations have to be computed numerically.

4.3 Dosimetry of Optical Radiation

The heading of this section gives a misleading impression since it suggests that the same quantities as used with ionizing radiations may also be applied with optical radiations. This is by no means the case and the uncritical common usage of the same terms has often caused unnecessary confusion. The main reason is that optical radiation is *selectively* absorbed by chromophores if the source emission spectrum and their absorption spectra overlap. "Energy absorbed per unit mass" (i.e. dose) does not make much sense here since it depends with a given radiation field on the chemical composition of the exposed body. Nevertheless, the expression "UV dose" is often – and wrongly – used, normally to indicate the energy fluence F_E, the amount of energy arriving perpendicularly at the surface of the exposed sample. Its unit is Jm^{-2}. Sometimes one is more interested in the quantum fluence ϕ which is related to F_E for monochromatic radiation by:

$$F_E = h \cdot \nu \cdot \phi = \frac{hc}{\lambda} \cdot \phi \tag{4.64}$$

(h = PLANCK's constant, ν = frequency, λ = wavelength, c = speed of light).

Its unit is "photons per m^2" or (often in photochemistry) "EINSTEIN per m^2". If the word "intensity" is used, it mostly indicates "energy fluence per time" but this term should either be dropped or specified. For visible radiation, there are also units which take into account the spectral dependence of human vision. The strictly physical *irradiation* is then replaced by the physiological *illumination*, i.e. the irradiation is weighted according to spectral sensitivity of the human eye. To give this system a sound foundation, a special basic unit was defined, the *candela* (cd). It is the light emitted by a black body of an area of $1/6 \times 10^5\,m^2$ at the melting temperature of platinum (1772°C). The emission spectrum resembles the spectral sensitivity of the eye. A point source emitting isotropically a luminous intensity of 1 cd produces a luminous flux of 1 Lumen (lm) per steradiant, the unit of solid angle. 1 lm per m^2 yields a illumination of 1 lux. This short summary may suffice, it is clear that in radiation biology preference is given to the physical units.

When performing experiments it is necessary to realize that the fluence at the surface is generally different from that in the interior because of the absorption of intermittent substances. This is even true for rigorously stirred suspensions where the average fluence is smaller than that at the entry if the medium absorbs. In this case, however, conversion factors can be given: If ε' is the extinction coefficient and x the thickness of the stirred sample, the mean fluence ϕ is:

$$\phi = \phi_0 \frac{\int_0^x e - \varepsilon' u \, du}{x} = \phi_0 \frac{(1 - e^{-\varepsilon' x})}{\varepsilon' x} \tag{4.64}$$

(ϕ_0 = fluence at the entry surface).

If the absorbing species are quantitatively known, it is also possible to calculate the energy absorbed per mass. It has to be distinguished, however, whether this shall be related to *total* or only to chromophore mass. For a homogeneous solution, the relations are simple: If c is the concentration of the chromophore with the decadic extinction coefficient ε, the absorbed energy fluence dF_E is in a volume element dV with area A and thickness dx: (see also Eqs. (2.2) and (2.5)):

$$dF_E = \frac{dE}{A} = F_E \cdot \frac{c \cdot \varepsilon \cdot dx}{\ln 10} \tag{4.65}$$

and hence:

$$\frac{dE}{dV} = F_E \cdot \frac{\varepsilon \cdot c}{\ln 10} \quad . \tag{4.66}$$

The total mass dm_t is $dm_t = \varrho \cdot dV$, so that:

$$\frac{dE}{dm_t} = F_E \cdot \frac{c \cdot \varepsilon}{\ln 10} \quad . \tag{4.67}$$

If the energy absorbed in the chromophore mass dm_c is of interest, one has ($c = 1/M_A \cdot dm_c/dV$; M_A = relative molecular mass):

$$\frac{dE}{dm_c} = F_E \frac{\varepsilon}{M_A \ln 10} \cdot \tag{4.68}$$

Further readings

Ionizing radiation ATTIX et al. 1968, ICRU 1970, 1979, 1983, KASE and NELSON 1978, KASE et al. 1985, MORGAN and TURNER 1973, PARETZKE 1980, ZAIDER and ROSSI 1986

Optical radiation CAMOLA and TURRO 1977, RUPERT 1974

Chapter 5
Elements of Photo- and Radiation Chemistry

This chapter focuses on the basic features of photo- and radiation-chemical reactions as far as they are relevant for the understanding of biological processes. These include the basic mechanisms of photosensitization and of action spectroscopy which allows to identify the primary absorber in a biological effect by studying its wavelength dependence. More specifically, atmospheric photochemical reactions and the light-dependent formation of vitamin D are discussed. The decomposition of water is described and, particular attention is paid to the influence of oxygen and different ionization densities, providing the background for the understanding of cellular phenomena. The chapter concludes with a discussion of direct and indirect radiation effects.

5.1 Photochemistry

5.1.1 Fundamentals

Photochemistry deals with those reactions which are significantly influenced or even made possible at all by the interaction of molecules with optical radiation. There are – at least – two classical fundamental "laws" in this branch of science:

1. The "law" of GROOTHUS-DRAPERT: Photochemical reactions start from those molecules which absorb most strongly.
2. The "law" of BUNSEN-ROSCOE: the photochemical action is proportional to the fluence rate of the radiation applied.

Both rules are derived from the fact that optical radiation is selectively absorbed. They do not really constitute "laws" since they are not always obeyed in complex reaction pathways.

The most significant quantity in photochemistry is the *quantum yield* Q. It is defined as the number of entities altered per quantum absorbed (or number of moles per Einstein absorbed). Usually it is smaller than 1 but larger values may be found in photochemical chain reactions.

Absorption of light causes excitations (Sect. 3.1) which normally terminate in the first excited state, either in the singlet or in the triplet system. Photochemical reactions start from here, they may also be defined as "reactions from excited states". This implies that their course and kinetics are quite different from those in normal (thermal) chemistry.

The primary photochemical processes are in general very fast so that they cannot be followed by chemical methods. It requires the method of "flash photolysis" which is schematically depicted in Fig. 5.1. The principle is simple:

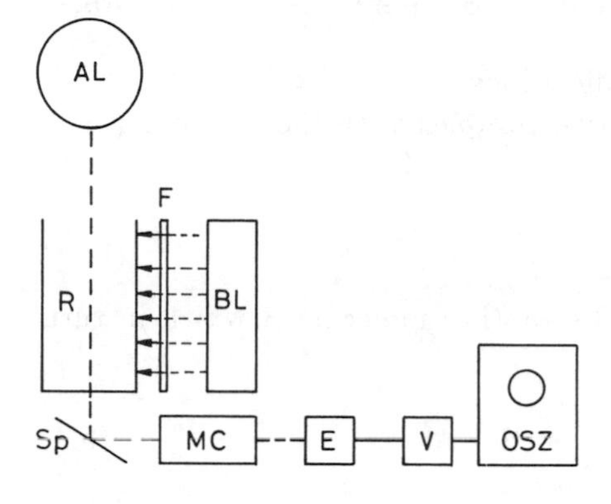

Figure 5.1. Principle of flash photolysis. AL: analysing light, BL: flash lamp, F: filter, R: reaction cuvette, Sp: mirror, MC: monochromator, E: detector, V: amplifier, OSZ: oscilloscope

The reaction is started by a short light pulse and the reaction products are measured as a function of time via their absorption using an analyzing light beam perpendicular to the first one. The time resolution lies in the order of picoseconds (10^{-12} s). In this domain the travelling time of light is already significant (33 ps per cm in vacuo) and has to be taken into account.

Starting from the primary species transient products may be formed (secondary excited molecules, free radicals etc.) until stable end products are reached. Some examples are given below. The reaction may, on the other hand, proceed also in a unimolecular fashion, i.e. restricted to the absorbing molecule. An important example is *photoisomerization* where starting and endproduct differ only in their sterical configuration.

5.1.2 Photosensitization

Photosensitization is a process in which a non-absorbing molecule is photochemically altered via the reaction with an excited chromophore. This phenomenon plays an important role in photobiology, e.g. with the *photodynamical effect* (Sect. 5.3.1). The component responsible for the primary absorption is called the *photosensitizer*. In general a reaction is called photosensitized if it is caused by light and does not occur in the absence of the sensitizer.

There are a number of different types of interaction; a few examples will be given where S denotes the sensitizer, A the acceptor and A′ its ultimate altered form:

1. The binding of S changes the absorption spectrum of A.
2. The excitation energy is directly transferred from the sensitizer to the acceptor (*excitation transfer*):

$$S \xrightarrow{h\nu} S^*, \qquad S^* + A \rightarrow S + A^* \tag{5.1}$$

(* denotes the excitation).

3. The excited molecule transfers or captures an electron (*electron transfer*):

$$S^* + A \rightarrow S^+ + A^- \quad \text{or} \quad S^* + A \rightarrow S^- + A^+ \tag{5.2}$$

A^*, A^- and S^-, S^* may also be radical ions.

4. The excited sensitizer forms a reactive complex with the acceptor (*exciplex*):

$$S^* + A \rightarrow (S^*A) \rightarrow S + A^* \tag{5.3}$$

5. The sensitizer transfers its energy to another molecule B which in turn reacts with the acceptor:

$$S^* + B \rightarrow S + B^*, \qquad B^* + A \rightarrow B + A' \tag{5.4}$$

Sensitized reactions may generally start both from the singlet and the triplet state although the latter case is more frequently found because of the longer life time.

The last example has a special significance in photobiology if the intermediary species is oxygen. It is the only simple molecule whose ground state is a triplet state making it a preferred reaction partner for photochemically generated other triplet-excited molecules. The first excited state of oxygen is a singlet $^1O_2^*$ (singlet oxygen) which is highly reactive and an efficient oxydizer. The excitation energy $^3O_2$?? $^1O_2^*$ is not too high (about 100 kJ/mole $\approx$ 1 eV) but direct excitation by light absorption is rather improbable (only with high intensities) because of the small interaction cross section. The sensitized reaction proceeds via the following scheme:

1. $S \xrightarrow{h\nu} {}^1S^*$
2. ${}^1S^* \rightarrow {}^3S^*$
3. ${}^3S^* + {}^3O_2 \rightarrow S + {}^1O_2^*$
4. ${}^1O_2^* + A \rightarrow AO_2$

This is a special case of *photooxidation* which may also be found with different pathways.

5.1.3 Action Spectroscopy

Since photochemical reactions are functionally related to the spectral absorption efficiency, it is possible to may use the dependence of a particular process on wavelength to determine the primary absorber. This method is called *action spectroscopy*. Although basically simple, it contains a number of pitfalls which warrant a closer inspection. It should be immediately clear from the foregoing that this approach is not able to provide information about the complete chain of reactions but only about the first step. Nevertheless, it is a powerful tool in photobiology if the interpretation of the results is performed critically.

There are a number of necessary conditions which are not always easy to fulfill:

1. The fluence dependence of the effect should be similar for all wavelengths, i.e. the response curves should be superimposable if the fluence is appropriately scaled by a constant factor for each wavelength. If this is not the case, the reaction is most likely caused by different mechanisms.
2. The quantum yield Q should be the same at all wavelengths.
3. The absorption spectrum of the chromophore in the system must be known or it must be secured that it does not differ from that in its pure form.
4. Absorption and scattering of substances not taking part in the primary reaction must be known if they cannot be neglected.
5. The chromophore absorption should not be too high to avoid saturation effects.
6. The effect should be independent of fluence rate since this varies often widely with different wavelengths ("reciprocity").

Response curves may have very different shapes, sometimes well-defined in mathematical terms, but mostly not. In order to exclude any possible bias, action spectroscopy should always be based on the same effect and the *isoeffect fluences* be compared.

If one is dealing with the same mechanism over the whole range (condition 1) and if the quantum yield Q is constant (condition 2), it can be safely assumed that the same effect is caused by the same number of primary alterations. Their number depends on the quantum yield and the number of quanta absorbed; thus, when comparing the same effect at two different wavelengths:

$$\mathrm{Q} \cdot \sigma_{\mathrm{abs}}(\lambda_1) \cdot \phi(\lambda_1) = \mathrm{Q} \cdot \sigma_{\mathrm{abs}}(\lambda_2) \cdot \phi(\lambda_2) \tag{5.6}$$

where σ_{abs} is the absorption cross section and ϕ the quantum fluence. Since $\mathrm{Q} = \mathrm{const}$, one obtains:

$$\frac{\phi(\lambda_1)}{\phi(\lambda_2)} = \frac{\sigma_{\mathrm{abs}}(\lambda_2)}{\sigma_{\mathrm{abs}}(\lambda_1)} \quad . \tag{5.7}$$

The actual task is to determine the quantum fluence necessary to produce a given effect as a function of wavelength. Since σ_{abs} is proportional to the extinction coefficient ε (see Eq. (2.5)) we may also write:

$$\frac{\phi(\lambda_1)}{\phi(\lambda_2)} = \frac{\varepsilon(\lambda_2)}{\varepsilon(\lambda_1)} \quad . \tag{5.8}$$

If the measured quantity is not the quantum fluence but rather the energy fluence F_E, the relation becomes:

$$\frac{F_E(\lambda_1)\cdot\lambda_1}{F_E(\lambda_2)\cdot\lambda_2}=\frac{\varepsilon(\lambda_2)}{\varepsilon(\lambda_1)} \quad . \tag{5.9}$$

In practical action spectroscopy, $1/F(\lambda_1)\cdot\lambda_1$ is usually normalized to a reference wavelength and plotted versus wavelength. If the above conditions are fulfilled, the resultant curve should be equal to the chromophore's extinction spectrum. Examples of action spectra are presented in several of the following chapters.

5.1.4 Special Reactions

5.1.4.1 Atmospheric Photochemistry

Formation of ozone (O_3)
Molecular oxygen (O_2) absorbs UV radiation below 195 nm. The excited molecule may dissociate into atomic oxygen which in turn reacts with O_2 to form triatomic ozone:

$$\begin{aligned} &O_2 \xrightarrow{h\nu} O_2^* \rightarrow 2O^\cdot \\ &O^\cdot + O_2 \xrightarrow{M} O_3 \quad . \end{aligned} \tag{5.10}$$

The second step requires the presence of a third partner M which takes up the remaining momentum. The reaction described proceeds in the outer layers of the Earth's atmosphere and is responsible for the formation of the ozone shield.

Photolysis of NO_2
NO_2, a frequent combustion product (e.g. from car exhaust) may be split by light absorption in the wavelength range 300–400 nm yielding NO and atomic oxygen:

$$NO_2 \xrightarrow{h\nu} NO + O^\cdot \quad . \tag{5.11}$$

This may form ozone (Eq. (5.10)) which is able to oxidize NO to NO_2 so that an equilibrium between NO, NO_2 and O_3 is established under irradiation:

$$\begin{aligned} &NO + O_3 \leftrightarrow NO_2 + O_2 \\ &[O_3] = K\frac{[NO_2]}{[NO]} \quad . \end{aligned} \tag{5.12}$$

K is the equilibrium constant. One sees that an increase of NO leads to a reduction of the steady-state ozone concentration.

Light-induced ozone destruction by chlorofluoromethane
Chlorofluoromethane compounds (general formula F_xCCl_{4-x}) are used worldwide as pressurizing agents (propellants) or as coolants in refrigerators, deep freezers, etc. If they are released into the atmosphere, they may be split by light absorption ($\lambda < 200$ nm) releasing a reactive chlorine atom. This reacts with ozone under formation of molecular oxygen:

$$F_xCCl_{4-x} \overset{h\nu}{\rightarrow} Cl^{\cdot} + F_xCCl_{3-x} \quad (5.13)$$

$$Cl^{\cdot} + O_3 \rightarrow ClO^{\cdot} + O_2 \quad . \quad (5.14)$$

$ClO^{\cdot}$ inhibits the formation of ozone (Eq. (5.10)) by scavenging $O^{\cdot}$:

$$ClO^{\cdot} + O^{\cdot} \rightarrow Cl^{\cdot} + O_2 \quad . \quad (5.15)$$

The overall effect is that Cl atoms catalyse the destruction of ozone:

$$O_3 + O^{\cdot} \overset{Cl^{\cdot}}{\rightarrow} 2O_2 \quad . \quad (5.16)$$

The reactions described are very important for possible UV hazards by changing the solar spectrum on Earth, as discussed in more detail in Sect. 22.2.

5.1.4.2 Photochemical Formation of Vitamin D

Vitamin D regulates the calcium metabolism of the human body. It is particularly important for bone formation in the growing organism. Vitamin-D deficiency leads to a disease known as "rickets". For centuries it has been known that exposure to sunlight has a curative effect, but the underlying mechanism could only be clarified rather recently. There seems to be general agreement that the active component is "vitamin D_3" whose structure is shown in Fig. 5.2. Its formation (Fig. 5.3) involves three light dependent steps where *previtamin* D_3 has a central position. The final product is reached via a comparatively slow reaction which is strongly temperature-dependent but proceeds without light. *Lumisterol* and *tachysterol* as sideproducts are in photochemical equilibrium with previtamin D.

The reactive wavelengths all lie in the UV-B region. The wavelengths at 248, 253, 265, 280 and 302 nm seem to be equally effective if calculated on a *per quantum* basis but 295 nm has a significantly greater effect. The formation of the mentioned sideproducts is also wavelength-dependent with 260 nm being most effective for tachysterol and 310 nm for lumisterol.

Special reference
HOLICK et al. 1982

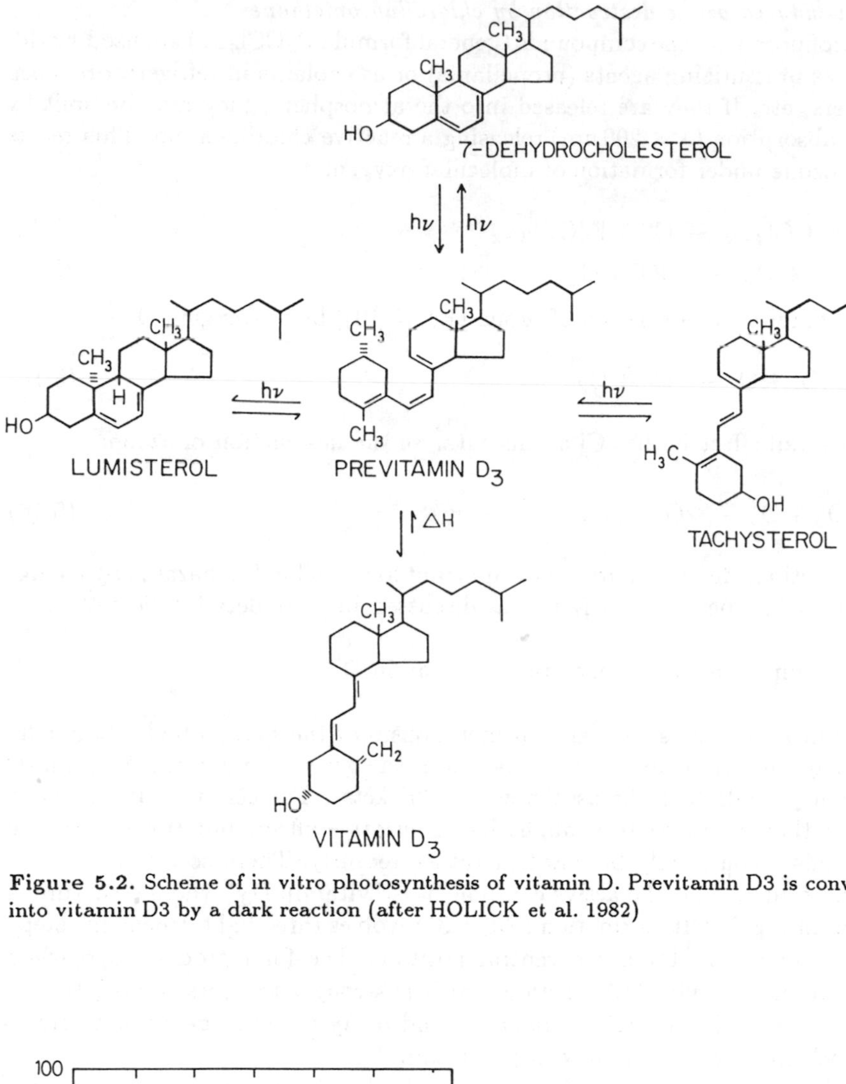

7-DEHYDROCHOLESTEROL

hν hν

LUMISTEROL hν PREVITAMIN D3 hν TACHYSTEROL

ΔH

VITAMIN D3

Figure 5.2. Scheme of in vitro photosynthesis of vitamin D. Previtamin D3 is converted into vitamin D3 by a dark reaction (after HOLICK et al. 1982)

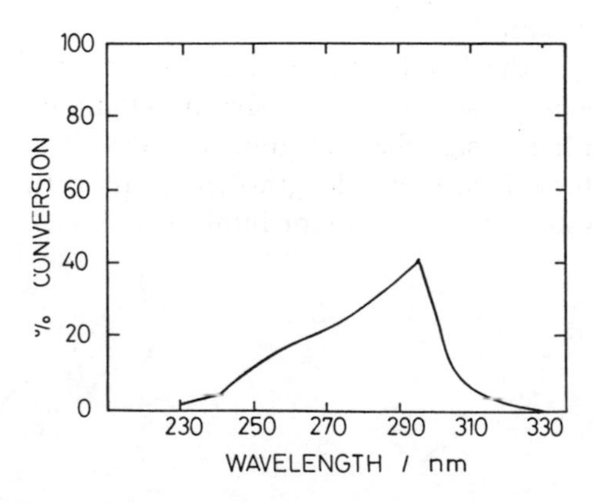

Figure 5.3. In vitro action spectrum for the formation of previtamin D3 from 7-dehydrocholesterol (after HOLICK et al. 1982)

5.2 Radiation Chemistry

5.2.1 Fundamentals

Radiation chemistry deals with chemical reactions caused by *ionizing* radiations. One important difference to photochemistry lies in the fact that absorption is not selectively restricted to certain molecules, the chromophores, but involves the whole reaction mixture. The second point is that secondary particles are formed which can ionize further molecules on their way. Interaction involves not a single type of absorbing species as in photochemistry but rather the system as a whole. Quantum yield would not be a good quantity in this situation, the number of altered molecules is, therefore, related to the total absorbed energy. The respective quantity is termed "G value" and is defined as the number of altered entities (molecules or atoms) per 100 eV absorbed energy:

$$G = \frac{\text{number of altered molecules}}{100\,\text{eV}} \quad . \tag{5.17}$$

It may also be expressed as "molar concentration per unit dose". If c is the concentration, D the dose and ϱ the density of the medium, the following relation holds true:

$$G/(100\,\text{eV})^{-1} = 9.64 \times 10^{6} \frac{c/\text{mol}\,\text{dm}^{-3}}{D/\text{Gy} \cdot \varrho/\text{kg}\,\text{dm}^{-3}} \quad . \tag{5.18}$$

If there are no chain reactions, G values below 10 are usually found. Equation (5.18) shows that doses around 1000 Gy are required to yield millimolar concentrations. This is outside the "physiological" range for most biological systems surpassing, e.g. "mean lethal doses" of mammalian cells by orders of magnitude. Very sensitive detection methods are necessary to investigate the chemical primary processes induced by the radiation which are the starting point for the ultimate biological effect.

Apart from standard chemical procedures, a special method plays a very important role in radiation chemistry, namely *pulse radiolysis.* It is similar to flash photolysis (Fig. 5.1), only the light pulse is replaced by a short electron pulse, usually obtained from a linear accelerator. This technique makes the investigation of very fast primary processes possible and has contributed greatly to our understanding of the basic mechanisms with time resolutions in the nanosecond range (picoseconds in special instances).

As pointed out before, interaction with ionizing radiation does not only lead to localized ionizations and excitations but also to the liberation of secondary particles. The spatial distribution of energy deposition and hence reaction products can be extremely heterogeneous influencing speed and direction of secondary reactions. With sparsely ionizing radiations, the species formed in "spurs" and "blobs" will react with each other before they are able to escape by diffusion. This is even more so in the track of a densely ionizing particle. The product yields will, therefore, strongly depend on ionizing density, both quantitatively as well as qualitatively. The modern development of pulse radiolysis has made it possible to follow even intra-spur and intra-track reactions even if the times involved are extremely short.

5.2.2 Radiation Chemistry of Water

Water is the main constituent of all biological systems. In vegetative cells its fraction lies between 40 and 50% but even in bacterial spores it is still around 20%. Since water does not absorb optical radiation in the range between 200 and 800 nm, this is not directly relevant for photobiology.

The situation is quite different with ionizing radiations. In this case most of the energy is, in fact, absorbed by water. The radiation chemistry of this molecule is, therefore, fundamental to the understanding of radiobiological phenomena.

Since oxygen interferes with many reactions (see below), the situation in its absence is described first.

If a water molecule absorbs energy from ionizing radiation, it is either ionized or excited. Excitation is often followed by a homolytic splitting of the molecule:

$$(1)\quad H_2O \rightarrow H_2O^+ + e^- \quad \text{(ionization)} \tag{5.19}$$

$$(2)\quad H_2O \rightarrow H_2O^* \quad H^{\cdot} + OH^{\cdot} \quad \text{(splitting)} \tag{5.20}$$

The primary products are $H^{\cdot}$ (hydrogen atoms, <u>not</u> ions), $OH^{\cdot}$, H_2O^+ and electrons. All these species possess unpaired electrons, they are *free radicals* and highly reactive. The electron is particularly interesting as detailed below. H atoms may also be formed by secondary reactions, namely:

$$e + H_3O^+ \rightarrow H^{\cdot} + H_2O \tag{5.21}$$

Another important secondary reaction is:

$$H_2O^+ + H_2O \rightarrow H_3O^+ + OH^{\cdot} \quad . \tag{5.22}$$

Free radicals are not the only products, stable molecules may also be formed:

$$e + H_2O^+ \rightarrow H_2O \quad \text{(recombination)} \tag{5.23}$$

or

$$H^{\cdot} + OH^{\cdot} \rightarrow H_2O \quad \text{(recombination)} \tag{5.24}$$

and

$$H^{\cdot} + H^{\cdot} \rightarrow H_2 \quad (5.25)$$
$$e + e + H_2O \rightarrow H_2 + 2OH^- \quad (5.26)$$
$$OH^{\cdot} + OH^{\cdot} \rightarrow H_2O_2 \quad . \quad (5.27)$$

Water radiolysis does not only create the above-mentioned short-lived species but also molecular hydrogen and hydrogen peroxide. The probability of these reactions depends largely on the local concentrations and are favoured within "spurs", "blobs" and "tracks". Interaction with solutes is only possible if the primary species are able to escape these zones.

Table 5.1 lists primary, i.e. before spur reactions take place, and final yields for low LET radiation.

Table 5.1. Initial and final yields of water radiolysis products with low LET radiation. (After BUXTON, 1987)

product	initial yield $(100\ eV)^{-1}$	final yield $(100\ eV)^{-1}$
e_{aq}^-	4.78	2.7
$OH^{\cdot}$	5.6	2.7
$H^{\cdot}$	0.62	0.45
H_2	–	0.45
H_2O_2	–	0.7

The electron warrants further consideration. In water it is normally not free but in a complex with water molecules, i.e. it is "hydrated" which is indicated by the subscript $_{aq}$. Since H_2O is a polar molecule, the electron is able to interact electrostatically with the positive end creating a special spatial structure. This is a highly dynamical process and does by no means imply that the electron is permanently shielded by the same water molecules, quite on the contrary, the partners are rapidly changing. Nevertheless, at any moment of time, the electron is surrounded by water dipoles. This situation

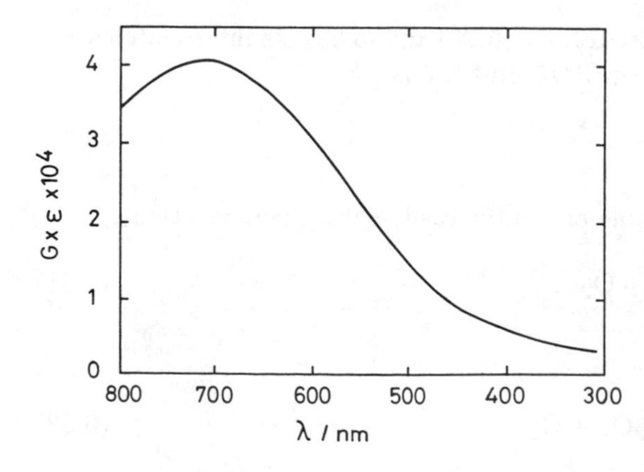

Figure 5.4. Absorption spectrum of the hydrated electron (after KEENE 1963)

leads to interesting consequences: The water shell creates a positive electric potential at the location of the electron. This means quantum-mechanically that it can attain only certain discrete energy states, very similar to the conditions in a hydrogen atom. If this is so, one has to postulate that it can be "excited" and should be able to absorb light. This is, in fact, the case as measured by pulse radiolysis. Figure 5.4 shows the resultant spectrum which is rather broad because of the rapidly changing partners in the hydration shell. Free or (colloquially) "dry" electrons exist only for very short times in aqueous solutions (about picoseconds).

The hydration shell shields the electron and increases its lifetime which is about 600 μs in very pure water. Some properties of e_{aq} are summarized in Table 5.2.

Table 5.2. Properties of the hydrated electron e_{aq}^-. (After BUXTON, 1987)

absorption maximum	715 nm
extinction coefficient at 715 nm	18,500 mol^{-1} dm^3 cm^{-1}
diffusion constant	4.9 x 10^{-9} m^2 s^{-1}
standard reduction potential	- 2.9 V
half-life in neutral water	2.1 x 10^{-4} s
G-value	2.7

So far only reactions in the absence of oxygen were discussed. Since this molecule plays an important role in radiation biology, it has to be included in the further treatment. The following reactions have then to be added:

$$e_{aq} + O_2 \rightarrow O_2^- \tag{5.28}$$

and

$$H^{\cdot} + O_2 \rightarrow HO_2^{\cdot} \quad . \tag{5.29}$$

O_2 does not react with $OH^{\cdot}$. Equations (5.28) and (5.29) are interrelated since there is an equilibrium between HO_2^+ and O_2^-:

$$HO_2^{\cdot} \leftrightarrow O_2^- + H^+ \quad . \tag{5.30}$$

Both are, furthermore, unstable and enter easily subsequent reactions:

$$HO_2^{\cdot} + H_2O^{\cdot} \rightarrow H_2O_2 + O_2 \tag{5.31}$$

and

$$O_2^- + O_2^- + 2H^+ \rightarrow H_2O_2 + O_2 \quad . \tag{5.32}$$

In solutions containing oxygen, there is an increased formation of H_2O_2 while oxygen is partly regenerated. Under irradiation an equilibrium between H_2O_2 and O_2 will be attained. There are, however, additional reactions which have also to be taken into account, the most important in this context being:

$$e_{aq} + H_2O_2 \rightarrow OH^{\cdot} + OH^{-} . \tag{5.33}$$

The decomposition of H_2O_2 thus increases the yield of the hydroxyl radical which may be relevant for the understanding of the role of oxygen in radiation biology.

But even in the absence of oxygen it may be generated radiolytically although only with small yields:

1. $OH^{\cdot} + OH^{\cdot} \rightarrow H_2O_2$
2. $H_2O_2 + OH^{\cdot} \rightarrow HO_2^{\cdot} + H_2O$ (5.34)
3. $HO_2^{\cdot} + HO_2^{\cdot} \rightarrow O_2 + H_2O_2$ or $HO_2^{\cdot} + OH^{\cdot} \quad O_2 + H_2O$.(5.35)

The second step proceeds much slower than the first one or other primary reactions so that Eq. (5.34) is only relevant with high H_2O_2 concentrations, at least with sparsely ionizing radiations.

The situation may be entirely different with high LET particles. With X or γ rays, the yields of primary species is essentially limited by the reactions in "spurs", "blobs" and "short tracks". Because of the comparatively large distances, an interaction between these entities is not very probable.

In the track core of heavy particles, on the other hand, the local concentrations of primary radicals are very high so that many of them will react with each other before they are able to diffuse into the solution. One would expect,

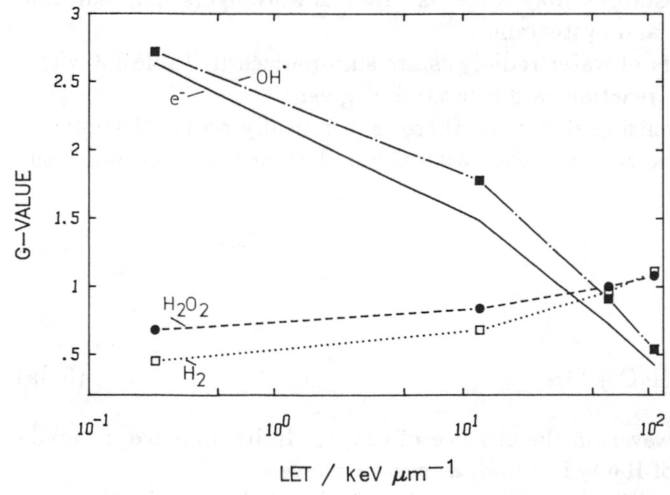

Figure 5.5. Yields of water hydrolysis products as a function of LET (after BUXTON 1987)

therefore, an increase of molecular products at the expense of the radicals. This is in fact found as demonstrated in Fig. 5.5. The values as measured in the solution are given; within the tracks the concentrations are, of course, considerably higher. This means that the probability increases for reactions as shown in Eqs. (5.23) to (5.27). If this is the case, one could visualize high local oxygen concentrations in the track core. This effect is being discussed as a possibility to explain why the extent of the oxygen effect decreases with higher ionization densities (Sect. 9.2.2). It is still unclear at present whether the actual concentrations are really sufficient because reliable measurements are not yet available.

There are a number of other relevant reactions aside from those involving only water radicals and oxygen. Particularly interesting are those where primary species are modified or scavenged. Two examples are given:

Hydrated electrons may be quantitatively transformed into hydroxyl radicals by reaction with N_2O:

$$e_{aq} + N_2O + H_2O \rightarrow OH^{\cdot} + OH^{-} + N_2 \quad . \qquad (5.36)$$

This reaction is not only important for the determination of eaq (it can be measured via the amount of N_2 formed) but even more so because of the presumed biological importance of $OH^{\cdot}$.

$OH^{\cdot}$ radicals may be scavenged by alcohols (e.g. ethyl alcohol):

$$OH^{\cdot} + CH_3CH_2OH \rightarrow CH_3C^{\cdot}HOH + H_2O \qquad (5.37)$$

The resultant alcohol radical is much less reactive than $OH^{\cdot}$ so that the net effect is reduced.

Many other substances may serve as "radical scavengers" in radiation biology, e.g. cysteine and cysteamine.

The main aspects of water radiolysis are summarized in Table 5.3 where also a list of relevant reaction rate constants is given.

With sparsely ionizing radiation, there is practically no net decomposition of water since the radicals react with the molecular products under the formation of water:

$$H_2 + OH^{\cdot} \rightarrow H_2O + H^{\cdot} \qquad (5.38)$$

and

$$H_2O_2 + H^{\cdot} \rightarrow H_2O + OH^{\cdot} \quad . \qquad (5.39)$$

This is only true, however, in the absence of oxygen. In its presence, a steady state concentration of H_2O_2 is found, as shown above.

The situation is different with densely ionizing radiation. In this case there are relatively more molecular products. Thermal decomposition of H_2O_2

Table 5.3. Reactions of water radiolysis products. (After BUXTON, 1987, therein also original references)

reaction	bimolecular rate constant mol^{-1} s^{-1} dm^3
$e_{aq}^- + e_{aq}^- \rightarrow H_2 + 2\,OH^-$	5.4×10^9
$e_{aq}^- + OH^- \rightarrow OH^-$	3×10^{10}
$e_{aq}^- + H_3O^+ \rightarrow H^\cdot + H_2O$	2.3×10^{10}
$e_{aq}^- + H^\cdot \rightarrow H_2 + OH^-$	2.5×10^{10}
$H^\cdot + H^\cdot \rightarrow H_2$	1.3×10^{10}
$OH^\cdot + OH^\cdot \rightarrow H_2O_2$	5.3×10^9
$OH^\cdot + H^\cdot \rightarrow H_2O$	3.2×10^{10}
$H_3O^+ + OH^- \rightarrow 2\,H_2O$	1.4×10^{11}
$e_{aq}^- + H_2O_2 \rightarrow OH^\cdot + OH^-$	1.2×10^{10}
$OH^\cdot + H_2 \rightarrow H^\cdot + H_2O$	4.9×10^7
$OH^\cdot + H_2O_2 \rightarrow HO_2 + H_2O$	2.7×10^7
$HO_2 + HO_2 \rightarrow H_2O_2 + O_2$	2.5×10^6
$O_2^- + HO_2 \rightarrow HO_2^- + O_2$	4.4×10^7
$H^\cdot + O_2 \rightarrow HO_2$	1.9×10^{10}
$e_{aq}^- + O_2 \rightarrow O_2^-$	1.9×10^{10}

yields oxygen so that an explosive mixture of H_2 and O_2 may be formed. This process is quite important for reactor safety where high neutron fluxes deliver a large amount of high LET radiation. To avoid possible dangers, hydrogen is catalytically bound.

5.2.3 Direct and Indirect Effect

It was shown in the foregoing section that highly reactive species are formed by water radiolysis. They are not only able to react with each other but also with solute molecules. One would, therefore, suspect that biological key substances cannot only be inactivated by direct energy absorption but also via interaction with radicals formed in their environment. Because of their high water content this should be especially so in cells. Therefore, it is necessary to distinguish between direct and indirect radiation inactivations. Both cannot easily be differentiated, at least not in closed systems, but it is possible in dilute solutions.

It may be assumed for the sake of illustration that the inactivation dose response curves are exponential. This is not a necessary condition but it facilitates the formal treatment without influencing the general validity.

N is the number of intact entities in a solution of volume V, N^+ the number of inactivated molecules, $\tilde{N}^+$ the respective mean value. The "surviving fraction" y is then:

$$y = \frac{N - N^+}{N} = e^{-\tilde{N}^+/N} \quad . \tag{5.40}$$

For the direct effect the mean number of inactivated molecules is proportional to the dose:

$$\tilde{N}^+ = \alpha \cdot D \cdot N \quad .$$

The rationale for this assumption is that each inactivating event requires a fixed amount of deposited energy.

Thus, for the direct effect one finds:

$$y = e^{-\alpha D} \quad . \tag{5.41}$$

This means that the surviving fraction is only a function of dose and independent of concentration. The number of damaged molecules, on the other hand, is:

$$N^+ = N(1 - e^{-\alpha D})$$

or

$$N^+ = c \cdot V(1 - e^{-\alpha D}) \tag{5.42}$$

(concentration $c = N/V$) with other words, it increases proportionally with concentration. With the indirect effect, the relations are different:

Now the mean number of reactive species R is proportional to the dose:

$$R = \beta D$$

If the concentration is not too high, each radical will react with the solute molecules with a certain probability to inactivate it. This means that not the fraction but the mean number of destroyed entities is linearly related to dose; thus, for the surviving fraction it follows that:

$$y = e^{-\beta' D/N} = e^{-\beta' VD/c} \quad . \tag{5.43}$$

The proportionality coefficient β comprises both radical formation and interaction probability.

If one plots the mean inactivation dose $D_0 = c/\beta V$ versus concentration, a linear relationship is obtained in this case while D_0 remains constant with the direct effect. This is schematically depicted in Fig. 5.6.

The picture just drawn is highly simplified and only applicable in intermediate ranges of concentration. If it is too low, not every radical has the chance to react with the solute and inter-radical reactions will become important reducing the overall yield. With high concentrations, on the other hand, direct effects will more and more predominate.

Indirect effects may even be found in dry systems where they are mediated by diffusible radicals. $H^\cdot$ atoms appear to be important in this respect.

Experiments as just described can, of course, not be performed with cellular systems since intracellular concentrations cannot be easily varied. In this case different approaches have to be used, as discussed in later chapters.

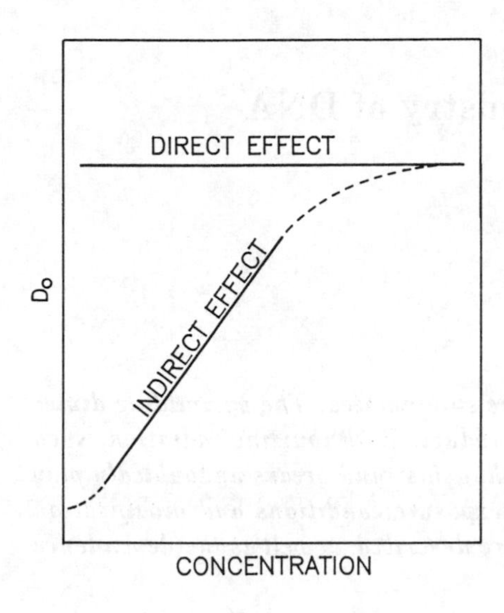

Figure 5.6. Schematic dependence of D_0 values of solute concentration for the direct and the indirect effect

Further readings
FARHATAZIZ and RODGERS 1987, PIKAEV 1967, SWALLOW 1973

Chapter 6
Photo- and Radiation Chemistry of DNA

Radiation-induced alterations of DNA are summarized. The pyrimidine dimer is still considered to be the major photoproduct. With ionizing radiation, such an exclusive statement is not possible although strand-breaks undoubtedly play a very important role. The influence of exposure conditions and modification by additives (photosensitizers, oxygen) are described as well as the dependence on ionization density.

6.1 Photochemical Alterations

6.1.1 General Aspects

Deoxyribonucleic acid (DNA) being the carrier of the genetic information constitutes undoubtedly – but not necessarily exclusively – a principal target for cellular radiation damage. This justifies a somewhat broader discussion of reactions occurring after radiation absorption in this molecule. Its struc-

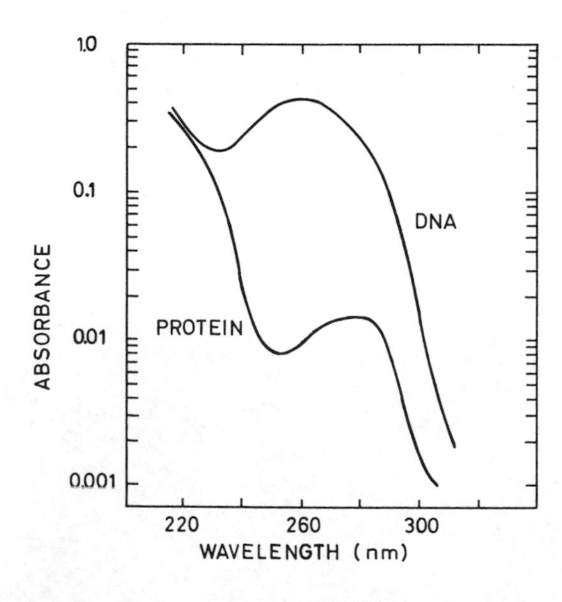

Figure 6.1. Absorption spectra of DNA and a typical protein (bovine serum albumin) (after HARM 1980)

ture and functions, as far as they are helpful for further understanding, are explained in Appendix II.1.

The absorption spectrum of DNA is shown in Fig. 6.1: There is a broad maximum around 260 nm, a trough at 220 nm with again increasing absorption towards shorter wavelengths. Denaturing the molecule, i.e. separation of the strands, leads to a general increase in absorption but without major qualitative changes. The absorption spectrum of proteins (an example is shown for comparison) is distinctively different. This fact allows to delineate the relative importance of these two classes of biomolecules for photobiological effects with the help of *action spectroscopy* (Sect. 5.1.3).

The absorption of DNA is only due to the bases, the other residues – sugar and phosphate – do not contribute in the given wavelength range. Since photochemical reactions start from excited states, it is interesting to compare their energies, as shown in Fig. 6.2. It is seen, that thymine has the lowest

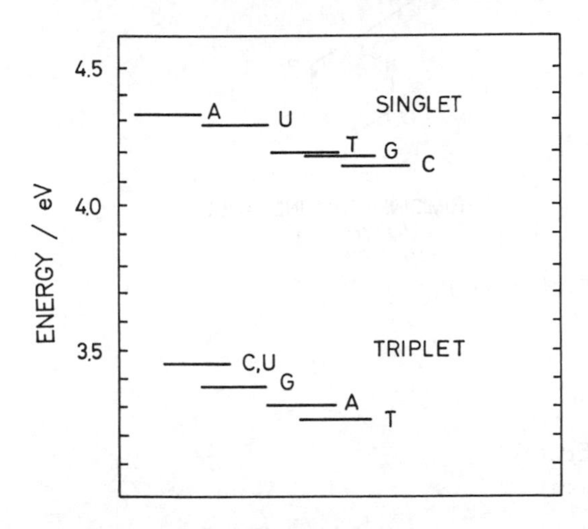

Figure 6.2. Excitation energies of the first excited singlet and triplet states of DNA mononucleotides (after LATARJET 1972)

triplet state making it a primary candidate for photo-induced changes. This is borne out by the experimental findings.

6.1.2 UV-induced Base Alterations

UV exposure in the wavelength range between 200 and 300 nm affects only the bases; sugar and phosphate are not – at least directly – altered. The purine moieties are by about one order of magnitude less sensitive so that essentially only the pyrimidines have to be considered.

The main products found after absorption of UV at 254 nm are summarized in Fig. 6.3:

a) pyrimidine dimers
b) hydrates
c) DNA-protein crosslinks

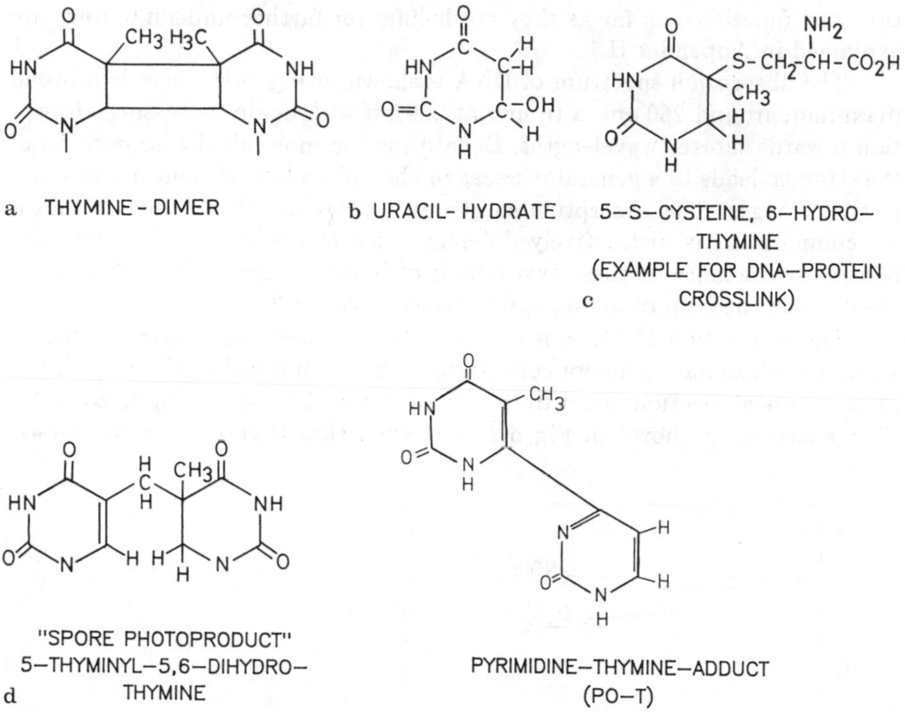

a THYMINE-DIMER

b URACIL-HYDRATE

c 5–S–CYSTEINE, 6–HYDRO-THYMINE (EXAMPLE FOR DNA–PROTEIN CROSSLINK)

d "SPORE PHOTOPRODUCT" 5–THYMINYL–5,6–DIHYDRO-THYMINE

PYRIMIDINE–THYMINE–ADDUCT (PO–T)

Figure 6.3a-d. Photoproducts in DNA (see text)

d) "spore photoproduct"
e) (6–4)pyrimidine adducts
f) thymine glycole (not in the figure)
g) strand breaks (not in the figure)

Pyrimidine dimers represent the most important class of UV photoproducts, at least in terms of quantity. They originate from two adjacent pyrimidines on the same strand via opening of the 5–6-double bonds and formation of a *cyclobutyl* ring structure. It is not clear whether the singlet- or the triplet-state is the precursor in DNA, in solution the reaction proceeds exclusively via the triplet state. Dimer yields are very low in solutions of bases or nucleotides but they are greatly increased in a frozen matrix to provide proper steric fixation of the reactants. Dimers may not only be formed between thymines, in DNA all combinations (T–T, C–T, C–C) and even thymine-uracil and cytosine-uracil may be found. In the latter case, uracil is formed by deamination of cytosine hydrates (see below).

The yields are different – they are highest with T–T – and depend on wavelength: the relative contribution of thymine dimers decreases with longer wavelengths.

Dimer formation is a reversible process, the equilibrium constant de-

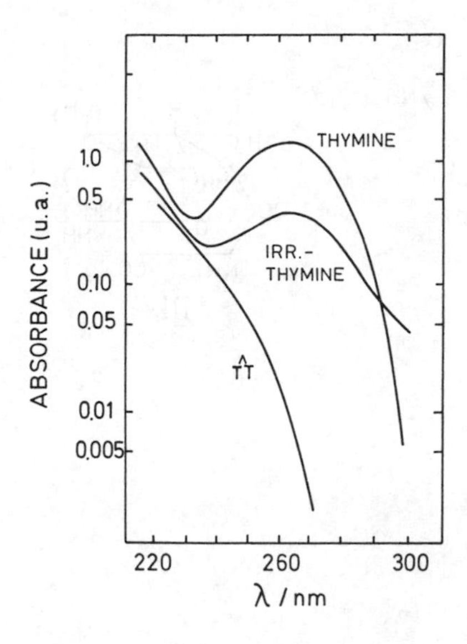

Figure 6.4. Absorption spectra of unirradiated and irradiated (254 nm) thymine and that of the thymine dimer (TT) (after JAGGER 1967)

pends on the wavelength as seen from Fig. 6.4: The absorption band around 265 nm is due to the 5–6-double bond and, consequently, lost in the dimer. Formation of this product is favoured where thymine still has a sizable absorption and the dimer absorbs only weakly. This is the case around 280 nm. Dimer-splitting, on the other hand, occurs predominantly at the absorption minimum of thymine, i.e. around 240 nm. Forward and back reaction can thus be manipulated at will just by changing the UV energy like a photochemical switch. This process being a very useful analytical tool, must not be confused with "photoreactivation" which is a light-dependent enzymatic process occurring at considerably longer wavelengths (see Sect. 13.2.1).

Dimers may have different steric configurations as demonstrated in Fig. 6.5. Two of these (II and III) may furthermore possess different optical activity so that six isomers exist. In cells, essentially only the cys-syn type (I) is formed.

Pyrimidine dimers are chemically extremely stable and survive even acid hydrolysis of DNA so that they are easy to quantitate. This property together with the above mentioned manipulation of the reaction by wavelength shift makes them an ideal probe to follow the course of photobiological events.

Hydrates are formed by water addition at the 5–6-double bond of pyrimidine bases. Contrary to dimers, they are rather unstable reverting easily upon changes in pH or temperature. Cytosine hydrate may lose the amino group yielding uracil. If this happens in a codogenic region of DNA, it may give rise to mutations because of the different base pairing properties. Whether this plays a role in cells is not known.

The (6–4)*pyrimidine adduct* has recently gained some attention. Its yield is smaller than that of dimers (see Table 6.1) but it contributes to lethality and may be even more important for mutations.

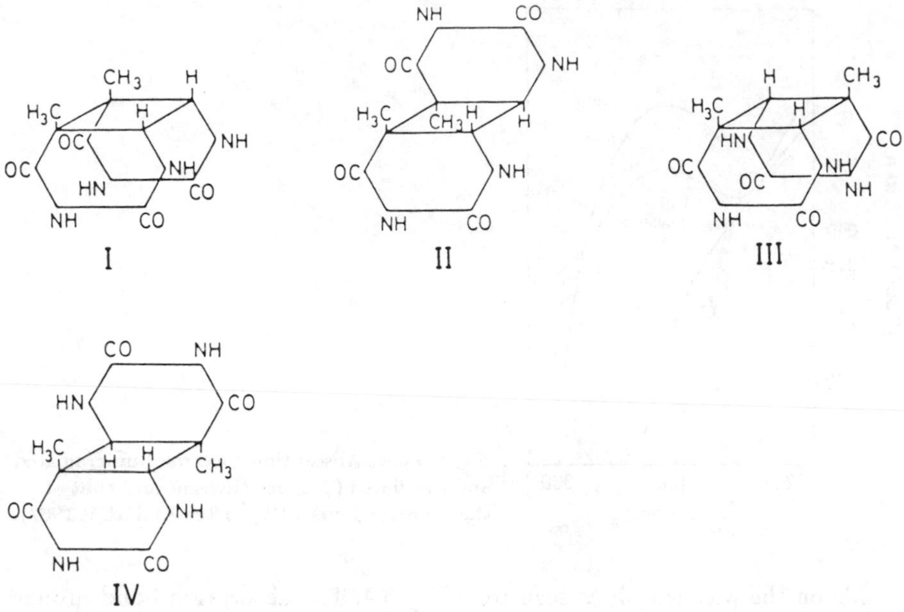

Figure 6.5. Stereoisomers of thymine dimers. II and III have also two isomers of different optical activity

The role of *thymine glycole* remains to be established. Its yield is comparable to that of dimers at longer wavelengths, but since the action spectra for lethality and mutation induction follow essentially that for dimer formation, it appears that glycoles do not play a major role in these processes.

More or less the same may be said about DNA-protein crosslinks and strand breaks. The efficiencies for their formation are much smaller than with dimers, again dependent on wavelength. Table 6.1 gives a summary of relative yields of the products mentioned so far. Examples of action spectra are shown in Fig. 6.6: They all resemble the absorption spectrum of DNA but

Table 6.1. Yield of photoproducts in mammalian cells exposed to 254 nm-UV

product	yield $(10^8\ u)^{-1}\ (J\ m^{-2})^{-1}$	relative to dimers	reference
pyr.-dimers	2.0	1	a
6-4-product	0.4	0.2	a
protein crosslinks	1.3×10^{-4}	6.6×10^{-5}	b
SSB	4.1×10^{-4}	2×10^{-4}	c

References:
- a) ROSENSTEIN and MITCHELL 1987
- b) PEAK et al. 1985
- c) ROSENSTEIN and DUCORE 1983

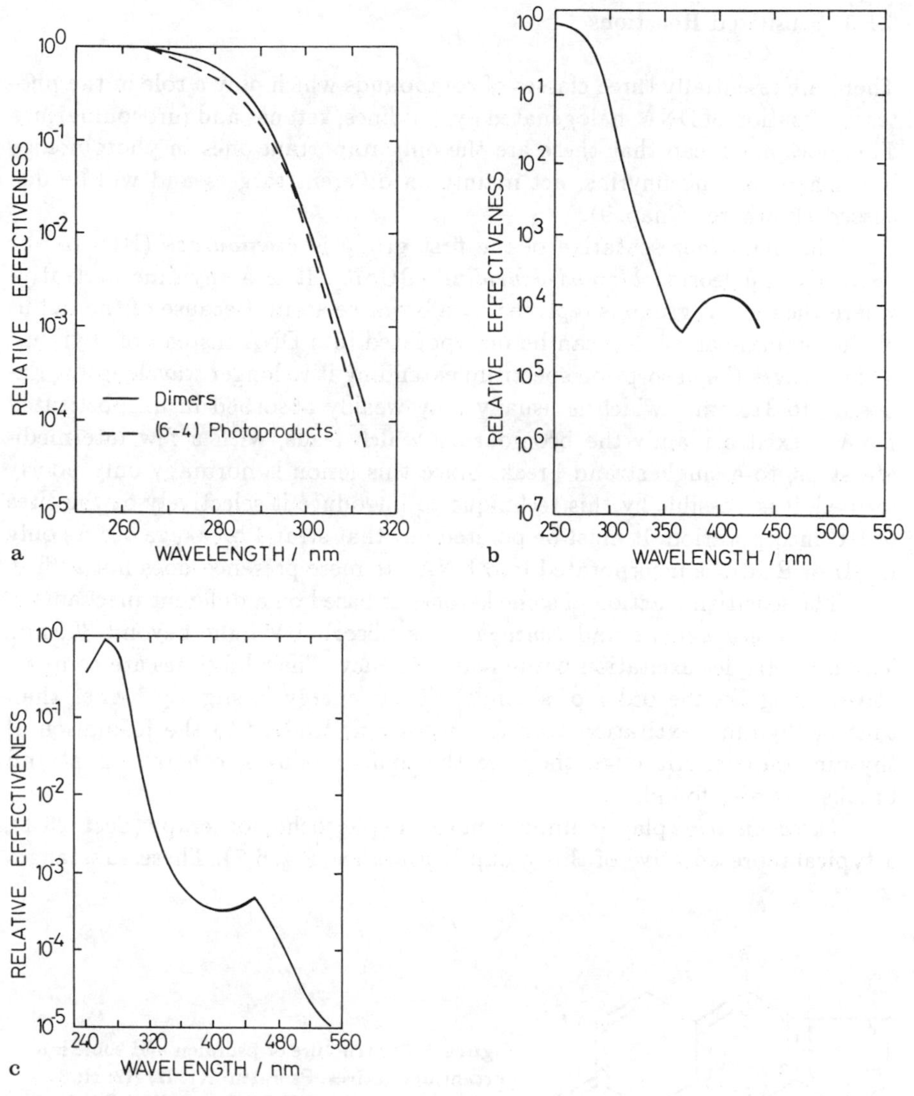

Figure 6.6a-c. Examples of action spectra for photoproduct formation in DNA (a) Dimers and (6–4) adduct (after ROSENSTEIN and MITCHELL 1987) (b) Single strand breaks (after ROSENSTEIN and DUCORE 1983) (c) DNA-protein cross links (after PEAK et al. 1985)

there are noticeable differences with longer wavelengths, particularly in the UV-A region.

The last remaining species constitutes a special case. As the name "spore photoproduct" suggests, it is found in UV-exposed dormant bacterial spores but also in other dehydrated biological systems or dry DNA.

6.1.3 Sensitized Reactions

There are essentially three classes of compounds which play a role in the photosensitization of DNA: halogenated pyrimidines, ketones and furocoumarines. This does not mean that these are the only important ones in photobiology but others, e.g. porphyrins, act mainly on different targets and will be discussed elsewhere (Chap. 9).

The main representative of the first group is *bromouracil* (BU) or the respective nucleoside *bromodeoxuridine* (BUdR). It is a thymine derivative where the methyl group is replaced by a bromine atom. Because of the radius of the bromine atom, BU can be incorporated into DNA instead of thymine. This changes the absorption spectrum extending it to longer wavelengths. Exposure to 313 nm – which is usually only weakly absorbed in unsubstituted DNA – excites mainly the bromouracil which leads, with a few intermediate steps, to a single strand break. Since this lesion is normally only poorly formed, it is possible by this technique to introduce it selectively at the sites of BU incorporation. It must be pointed out that strand breakage occurs only if BU or BUdR is incorporated into DNA, its mere presence does not suffice.

The sensitizing action of some ketones is based on a different mechanism. *Acetone*, *acetophenone* and *benzophenone* absorb UV-light beyond 300 nm, leading to triplet excitation with great efficiency. Their lifetimes are comparatively long (in the order of seconds). Their energy is slightly higher than that of thymine; excitation transfer may occur leading to the formation of thymine dimers. Although they are the main products, others, e.g. strand breaks, are also found.

Furocoumarins play an important role in photochemotherapy (Sect. 23.1), a typical representative of this group is *psoralen* (Fig. 6.7). These substances

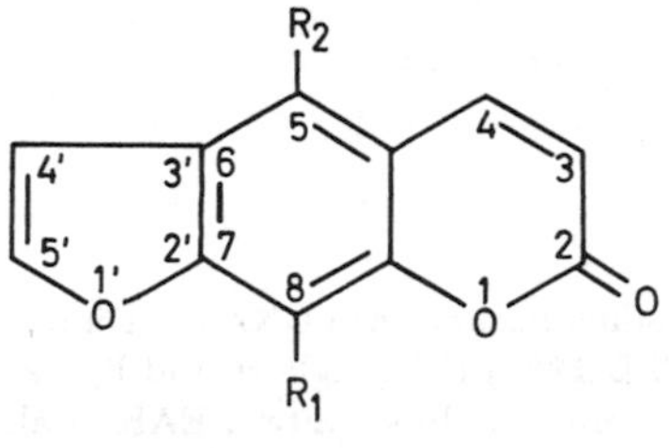

Figure 6.7. Structure of psoralen and some important derivatives. Psoralen: R1: H, R2: H; 8-methoxypsoralen (8-MOP): R1: OCH3, R2: H; 8-methylpsoralen: R1: CH3, R2: H; bergaptene: R1: H, R2: OCH3

intercalate between the stacked bases in the DNA double helix so that excitation transfer can easily occur if they are excited by the absorption of UV at 365 nm. The photochemical reaction proceeds in two steps (requiring, therefore, also two absorption events): the first one is a thymine-psoralen crosslinking, the second an inter-strand crosslink with a pyrimidine base on the opposite strand. The effects on denaturation behaviour or intracellular DNA-replication are obvious.

6.2 Radiation Chemistry of DNA

The radiation chemistry of DNA is far more complex than the field of photochemical alterations. One reason for this is the fact that ionizing radiation is not selectively absorbed by the molecule alone but also by the solvent and other solutes, so that indirect reactions may interfere. This complicates the situation in chemically well defined aqueous solutions, in cells it presents almost unsoluble difficulties. It is, therefore, not surprising that there is presently no general agreement whether cellular biological radiation effects can be traced down to a single radiation-chemical species, it is not even sure whether DNA is the only target. But it is nevertheless important – and, by definition, the origin of at least radiation-induced mutations – so that a consideration of its radiation chemistry is indicated.

There is an extensive literature on DNA constituents – bases, nucleotides and oligonucleotides – a review of which is beyond the scope of this chapter (see "Further readings" for references). In DNA, radiation-induced alterations may be broadly classified as in Fig. 6.8:

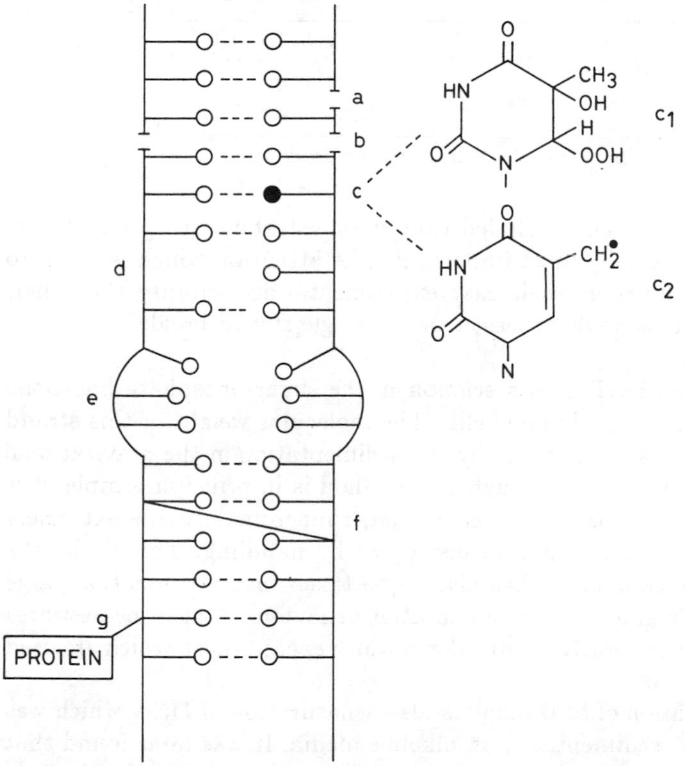

Figure 6.8. Schematic diagramme of DNA alterations after exposure to ionizing radiation (see text). The examples of base damage are: c1: 5-hydroxy-6-hydroperoxythymine, c2: H-abstraction at the mathyl group of thymine (see CERUTTI 1974)

a) single strand breaks (SSB)
b) double strand breaks (DSB)
c) base damage
d) base loss
e) denatured zones
f) intramolecular crosslinks
g) DNA-protein crosslinks

These categories are not very distinct, partly artificial and not necessarily mutually exclusive. Strand breaks, for instance, are often accompanied by base losses, base damage causes local denaturation, etc.

A few G values are listed in Table 6.2. They have more orientating character and should not be overestimated but may serve as a useful guideline.

Table 6.2. G-values of some DNA-alterations caused by exposure with sparsely ionizing radiation in air. (After HUETTERMANN, KOEHNLEIN and TEOULE, 1978, except [b]) ULMER, GOMEZ and SINSKEY, 1979)

type	intracellular	in solution
ESB	0.7 - 1.1	0.3 - 0.5
DSB	0.08 - 0.3[b)]	* [a)]
base damage	-	1.3 - 2.3
denatured zones	-	7.5 - 14

[a)] value cannot be given because of quadratic dose dependence

It should, however, not be concluded that the most abundant products are necessarily the biologically most important. The attention which is given to strand breaks reflects more their easy experimental accessibility than their proven relevance. This holds at least true for single strand breaks.

a) Single strand breaks: This is a scission in the sugar-phosphate backbone in one strand of the DNA double helix. The molecular weight of this strand is thus reduced, as easily measured by the sedimentation in the gravitational field of an ultra-centrifuge. Although this method is in principle simple, it is not without complications: DNA being a large macromolecule is extremely fragile and strand breaks may be caused just by handling. This limits the resolution of this technique. It has also to be taken into account that large molecules exhibit an anomalous sedimentation behaviour. All this necessitates the application of comparatively high doses which exceed those which are used in all biological studies.

The determination of SSB requires also denaturation of DNA which was usually achieved by sedimentation in alkaline media. It was later found that this particular procedure causes the formation of additional breaks that are not detected if the sample was denatured by heat treatment. These "latent" breaks which presumably originate from certain types of base damage are termed "alkali-labile sites" (als). Their exact chemical nature is not known, they may amount up to 30% of the total number of SSB. The chemical pathway of break formation in aqueous solution has been studied in the model compound polyuracil (poly U), as summarized in Fig. 6.9:

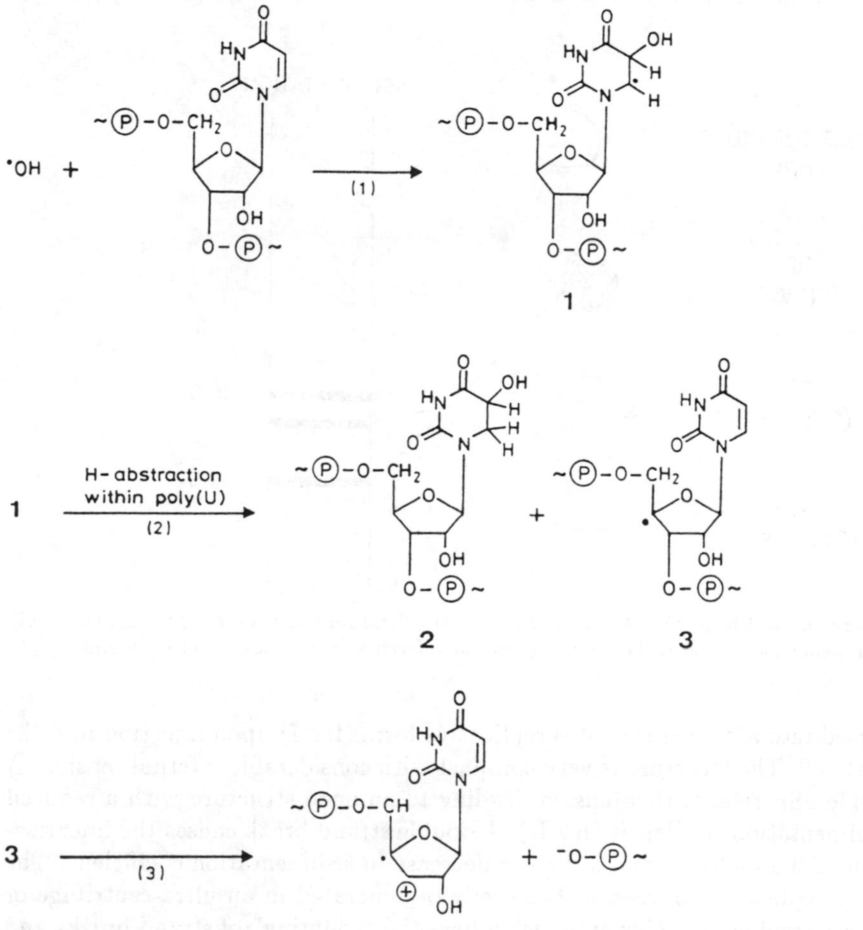

Figure 6.9. Suggested pathway for strand break formation in polyuridylic acid (from VON SONNTAG 1987, with permission)

In a first step, the OH˙ radical adds to the uracil at the C(5) position yielding a base radical. This is followed by a hydrogen abstraction from the C(4)′ position in the sugar (product 3) and subsequent release of the phosphate group from C(3)′. A quantitative evaluation shows that 80% of the OH-reactions occur with the bases and only 20% with the sugar moeity suggesting that the majority of the chain breaks is formed via this or similar pathways. It has still to be evaluated whether the situation is the same in cellular DNA, but nevertheless the reaction shown constitutes a useful guideline for further research. It demonstrates also quite clearly that the molecule acts as a whole and that studies with isolated constituents – useful as they may be – may be rather misleading.

A very useful test object is the bacteriophage ϕX174 (see Fig. 6.10). Its DNA is normally single-stranded (molecular weight 1.7×10^6 g/mole) but converted into a double-stranded replicative form (RF I) upon injection into the host cell. The structure is very compact with considerable internal tension. A single SSB relaxes this tension, leading to an open structure with a reduced sedimentation coefficient (RF II). A double strand break causes the linearization of the molecule and a further decrease in sedimentation coefficient. The three replicative forms can thus easily be separated in an ultra-centrifuge or by electrophoresis. This approach allows the "counting" of strand breaks and does not require the analysis of complex distributions of molecular weights as encountered with irradiated DNA solutions.

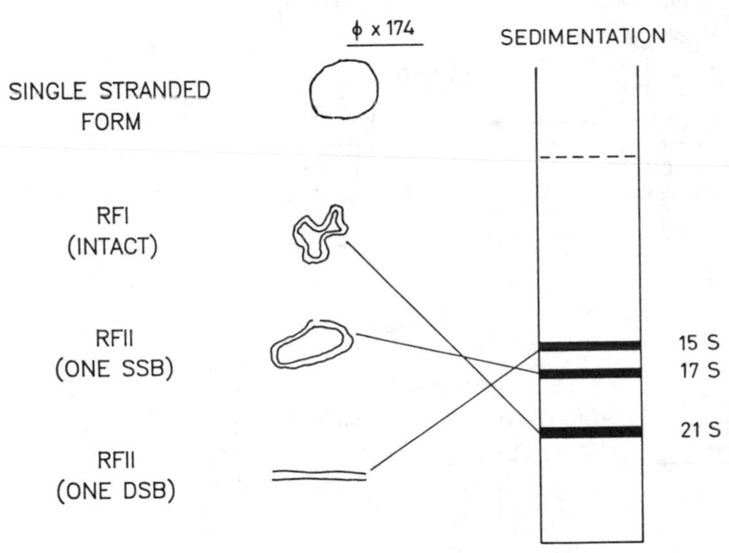

Figure 6.10. Phage ϕX174 as test object for the determination of strand breaks (see text). Sedimentation coeffeicients are also given for reference (see textbooks of biochemistry)

Sedimentation and electrophoretic analysis have in recent years been supplemented by other methods which are less expensive and sometimes more sensitive: hydroxyl apatite chromatography (AHNSTRÖM and EDVARDSSON 1974), fluctuation spectroscopy (WEISSMANN et al. 1976) and the filter elution method (KOHN 1979). Particularly the last one is widely employed, it allows measurements with doses of a few Gray but it requires careful calibration and is – in spite of its simplificity – by no means without pitfalls.

SSB do not form a chemically homogeneous group. They may be formed either in the phosphate or in the sugar moiety, the second case appears to be found only after exposure in the presence of oxygen. Some information may be gained by "endgroup analysis" using enzymes reacting specifically and exclusively with 5′–OH and 3′–OH or free phosphate ends. Some results are listed in Table 6.3. The differences between DNA in solutions and in cells are noticeable reflecting presumably not only the different chemical environment but also the influence of fast intracellular repair processes acting differentially on the endgroups.

Table 6.3. Free endgroups at radiation induced single strand breaks (sparsely ionizing radiation in air). (After [a]) LENNARTZ, COQUERELLE, BOPP and HAGEN, 1975; [b]) MITZEL-LANDBECK and HAGEN, 1976)

type	intracellular (%) [a]	in solution (%) [b]
C5'-OH	10	15
C3'-OH		43
free phosphate ends (nucleotide loss)	40	43
not identified	50	

b) Double strand breaks: They may be either caused by a single energy deposition event or by the interaction of two SSB formed individually in close proximity. The two ways are reflected in the dose-response curve: the relationship should be linear in the first case, quadratic in the second. Figure 6.11 shows that in solution the direct formation of DSB is obviously of minor importance.

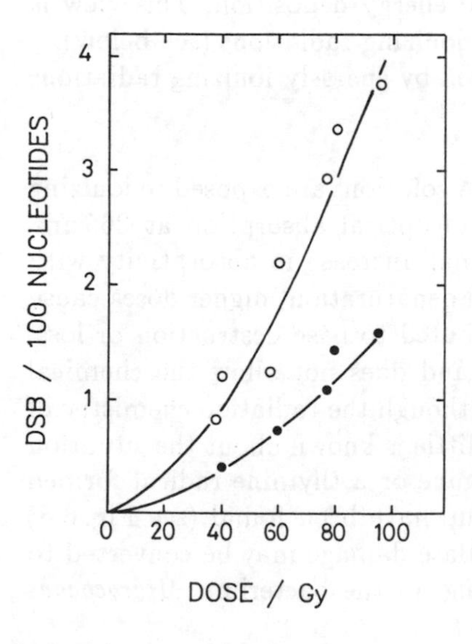

Figure 6.11. DSB yields as function of gamma dose for phage-DNA exposed in solution either in presence (o) or absence of oxygen (•) (after FREY and HAGEN 1974)

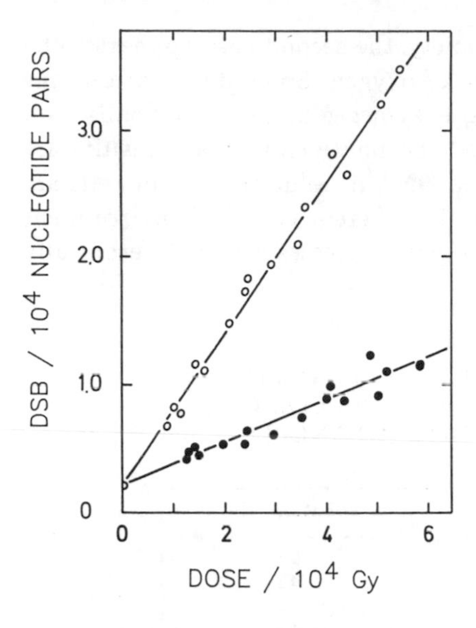

Figure 6.12. DSB-yields for intracellularly irradiated DNA (thymocytes). (o) air, (•) nitrogen (after LENNARTZ et al. 1975)

The situation appears to be quite different for DNA exposed in cells (Fig. 6.12): the relationship is strictly linear. It must be reemphasized, however, that the doses are comparatively high. Recent reports claim that with lower doses and the filter elution technique, non-linear dose-response curves are found (RADFORD 1988) but the debate is not yet closed. It is, despite of these differences, quite clear that the DSB-yield in cells is by about a factor 1000 lower than in solution. There are several reasons for this: the overwhelming contribution of the indirect effect in the solutions, the presence of radioprotectors in cells and differences in the molecular configuration.

Even in solution there is a small linear contribution which is presumably due to the rare events with high local energy deposition. This view is substantiated by experiments with densely ionizing radiations (see below).

Some data about strand break induction by sparsely ionizing radiations are summarized in Table 6.4.

c) and d) Base damage and base loss: If DNA solutions are exposed to ionizing radiation, there is a dose-dependent loss of optical absorption at 265 nm. With double-stranded DNA, one finds a small increase in absorptivity with low doses which is due to radiation-induced denaturation, higher doses cause a progressive decline which has to be attributed to base destruction or loss. This approach is, of course, rather crude and does not allow the chemical characterization of the products formed. Although the radiation chemistry of DNA in aqueous solutions is far advanced, little is known about the situation in cells. 5–6-dihydroxy-dihydro-peroxythymine or a thymine radical formed by the H-abstraction from the methyl group have been found (see Fig. 6.8) but their biological relevance is unknown. Base damage may be converted to SSB by the action of nucleases as found e.g. in the bacterium *Micrococcus*

Table 6.4. SSB- and DSB-yields after exposure to sparsely ionizing radiation. (After ROOTS et al., 1985, original references therein)

system	irradiation source	SSB yield $10^{-10}Gy^{-1}u^{-1}$	SSB energy eV	DSB yield $10^{-10}Gy^{-1}u^{-1}$	DSB energy eV	ratio
λ phage	4 MeV elektr.	2.8	37	0.068	1539	41
SV40	50 kV X	1.5	69	0.2	520	7.5
	Co-γ	2.0	52	0.19	550	10.5
bacterial DNA	Cs-γ			0.09	1150	
	50 kV X	3.1	33	0.41	250	7.6
	Co-γ			0.2	520	
yeast DNA	30 MeV electr.			0.07	1480	
mamm. DNA	4.5 MeV electr.	1.2	87			
	X-rays[a]	2.5	42			
				0.16	650	
				0.12	865	
				0.062	1675	

a) several authors

luteus. This method allows better quantification of the yields but also here the specificity is unknown.

e) Denaturated zones: All changes described so far interfere with the specific base pairing leading to local denaturation. This may be seen in the "melting curves" of irradiated DNA (see Appendix II.1) or with the aid of enzymes which specifically cleave single stranded regions. The G values (Table 6.2) exceed those of the other lesions, which is understandable since denatured zones comprise a number of different damage types.

f) and g) Cross-links: These type of lesions have not been very extensively investigated. Intramolecular interstrand crosslinks lead to drastic changes in the melting curves and cause an increase in sedimentation velocity because of the more compact structure. Cross-links with other compounds – proteins being the most important – make the DNA less extractable from cells.

As pointed out before, DNA may be damaged via the direct and the indirect effect. The relative weight of their contributions differs, of course, between dilute solutions and cells. While the mechanism of the direct effect is still essentially obscure, more can be said about the contribution of the radicals formed by water radiolysis. A first rough estimate may be obtained by considering the reaction rate constants as listed in Table 6.5 together with the G values of the primary products (Chap. 5). One is led to the assumption that the OH-radical plays a major role and this is indeed also shown by experiments with specific scavengers. Contrary to expectations, the hydrated electron does not contribute significantly. This is demonstrated by investigations where DNA solutions saturated with N_2O are irradiated. Under these

Table 6.5. Bimolecular rate constants for reactions of the radical primary products with DNA and its elements in neutral solution. (After SCHOLES, 1978)

	k (10^9 mol^{-1} s^{-1} dm^3) e_{aq}	$OH^{\cdot}$	$H^{\cdot}$
base			
TT	18	5.6	0.38
CC	13.2	4.7	-
UU	15	5.2	-
AA	9	4.4	0.14
sugar			
TdR	-	4.8	0.38
CdR	13.2	4.8	-
UdR	15	4.3	-
AdR	9.2	4.0	0.23
nucleotides			
TMP	1.5	5.3	0.42
dCMP	-	4.4	-
dAMP	-	3.5	-
dGMP	-	5.0	-
DNA	0.6 - 0.14	0.3 - 0.8	0.08

conditions e_{aq} is quantitatively transformed into OH, and the lesion yield increases just by the same amount.

Although yield and rate constants of the H atom are low, its role may not be neglected, as shown by the following experiment (BLOK and LOHMANN 1973): If an anoxic DNA solution is saturated with H_2, there is no change in radiation sensitivity. If, however, a 10:1 mixture of H_2 and O_2 is used, the damage yield is reduced by a factor of 10. This can be explained in the following way: H_2 reacts with OH radicals to form H radicals (Table 5.2). In the absence of oxygen they obviously cause the same degree of inactivation as $OH^{\cdot}$. Oxygen, on the other hand, scavenges H radicals so that the remaining smaller damage is only due to OH.

Water radicals are not the only compounds which cause indirect inactivation of DNA. If other solutes are present, they may also form reactive species which may effectively react with DNA. This emphasizes again the fact that even in dilute DNA solutions, the presence or absence of other solutes may change the lesion yield in a sometimes unpredictable manner.

Influence of oxygen

As discussed in detail later (Sect. 9.2.2), the biological effect of sparsely ionizing radiation is increased in the presence of oxygen. This is also the case in dilute DNA solutions, but the matter is rather complex and there are complicated relationships between the concentration of O_2 and other additives (QUINTILLIANI 1986). In order to compare only the sensitivities in the presence and the absence of oxygen, one may use the "oxygen enhancement ratio" (OER) as the ratio of doses in N_2 and in O_2 to cause the same effect. Table 6.6 lists a few examples. The OER is always around 2, except for phage infectiosity where it is unity, i.e. there is no oxygen effect. This finding has, of course, consequences for the assessment of the biological relevance of the DNA lesions listed. A few data for intracellularly irradiated DNA are given in Table 6.7.

Table 6.6. OER-values for DNA lesions in γ-irradiated phages in solution. (After FREY and HAGEN, 1974)

effect	OER
ESB + alkali labile lesions	2.4
ESB (and heat denaturation)	1.8
end free phosphate groups	1.9
DSB	2.5
base damage	3.5
local denaturing	2.05
phage infectiosity	1

Table 6.7. OER-values for DNA lesions in γ-irradiated thymocytes. (After LENNARTZ, COQUERELLE, BOPP and HAGEN, 1975)

type of lesion	OER
ESB	5.4
alkali labile lesions	2.1
DSB	3.8

O_2 may even act as a protector under special conditions if the damage is caused by reducing agents; an example was given above, others may be found in the literature.

Influence of LET

An analysis of the dependence of damage on radiation quality has so far only been performed for strand breaks. Results obtained with the ϕX174 system are shown in Fig. 6.13: the yield of SSB decreases with increasing LET while that of DSB rises to a maximum. This behaviour may be understood by assuming that SSB are caused by single energy-deposition events, while DSB require multiple events whose frequencies increase with LET. A more detailed analysis will be given in Sect. 7.1.

Summarizing conclusions

The present state of knowledge does not allow to identify unambiguously one

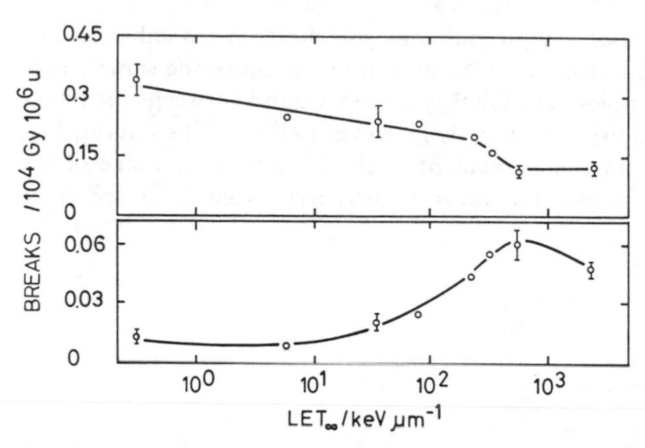

Figure 6.13. SSB- and DSB-yields in the replicative form of phage OX174 as function of LET. The ordinate values are breaks per 10 000 Gy and a molecular weight of 1 megadalton. Upper panel: SSB, lower panel: DSB (after CHRISTENSEN et al. 1972)

particular type of lesion with the cause of biological radiation effects. This is different from UV where the role of pyrimidine dimers is established beyond doubt even if for certain systems and/or wavelength regions, significant contributions of other lesions have to be taken into account. With ionizing radiations, the importance of double strand breaks has certainly be shown but whether they really represent <u>the</u> molecular lesion remains still to be demonstrated. The only acceptable conclusion is that biological radiation effects can most probably not be traced down to a single molecular alteration but rather to a complex interplay of several factors.

Further readings

Optical radiation: KIEFER and WIENHARD 1977, KITTLER and LÖBER 1972, MITCHELL 1988, VARGHESE 1972, WANG 1976

Ionizing radiation: HAGEN 1985, HÜTTERMANN, KÖHNLEIN and TÉOULE 1978, HUTCHINSON 1985, SCHULTE-FROHLINDE and VON SONNTAG 1985, VON SONNTAG 1986, VON SONNTAG and ROSS 1987, WARD 1986

Chapter 7
Radiation Effects on Subcellular Systems

Radiation effects on enzymes and viruses are discussed along with the inhibition of RNA transcription. First, an introduction of "target theory" will be given which describes quantitatively the effects on "small" systems. Investigations using viruses (especially bacteriophages), plasmids or isolated DNA are outlined. A special section deals with repair phenomena at the subcellular level. The chapter includes also a description of "gene mapping", which demonstrates how radiobiological approaches may be successfully applied to problems of general molecular biology.

7.1 Target Theory

Historically, target theory was the first attempt to describe quantitatively in physical and mathematical terms the inactivation of biological entities by radiation. Although it has become quite clear later that this approach is not appropriate for complex systems like cells, it is still useful in all instances where biological secondary processes, e.g. repair and recovery, do not modify the initial damage to any great extent. Furthermore, it is fundamental to most other theories of biological radiation action (see, e.g. Chap. 4 "microdosimetry" or Chap. 16).

Target theory is based on two main postulates: the stochastic nature of energy deposition and the existence of an one-to-one relationship between the number of initial lesions and the ultimate biological effect. It has been pointed out before (Chap. 4) that the average probability for an inactivating energy deposition event is quite small. This means that the basic phenomena can be described by POISSON statistics (see Appendix I.4). It is assumed at first that only the "direct effect" prevails, i.e. that solvent radicals may be neglected. If N_0 is the total number of molecules, N that of those with no hits and λ the average number of hits *per molecule*, the probability p(n) for n hits is given by:

$$p(n) = e^{-\lambda} \cdot \frac{\lambda^n}{n!} \tag{7.1}$$

p(n) equals the fraction of all molecules having received just n hits if N_0 is sufficiently large. Assuming that exactly m hits lead to inactivation, the "surviving fraction" N/N_0 is:

$$N/N_0 = \sum_{n=0}^{m-1} e^{-\lambda} \cdot \frac{\lambda^n}{n!} \quad . \tag{7.2}$$

It is inherent in this formulation that hit numbers smaller than m are completely without effect.

The special case of m = 1 is of particular importance, it is commonly referred to as "single hit process". Quite often it is found in simple systems, examples will be given below. Equation (7.2) takes then the form:

$$N/N_0 = e^{-\lambda} \quad . \tag{7.3}$$

The difficulty of interpretation lies in the problem to equate a "hit" with a concrete physical or chemical alteration. The simple case where its nature is known is rarely realized. One example for this is the inhibition or RNA transcription by pyrimidine dimers (Sect. 7.2): if G is the number of dimers induced per unit fluence and unit mass of the chromophore (the "target"), then:

$$\lambda = G \cdot m \cdot F \tag{7.4}$$

where m is the target mass and F the fluence. The surviving fraction is then (if one dimer suffices):

$$N/N_0 = e^{-G \cdot m \cdot F} \quad . \tag{7.5}$$

This relationship may be used to deduce m from the inactivation curve if G is known.

The situation is more complicated with ionizing radiations. In this case the nature and the yields of inactivating lesions are generally not known. It is sometimes assumed that every primary ionization is sufficient because of the relatively high amount of energy deposited. If ε is the average energy per primary event, the average number per unit mass is D/ε with D being the dose. If the hits are statistically independent (this is, in fact, never the case but to some extent approached with sparsely ionizing radiations), then the following may be written:

$$N/N_0 = e^{-m \cdot D/\varepsilon} \quad . \tag{7.6}$$

Equation (7.6) could, in principle, serve as the basis for a simple determination of molecular masses if the conditions were satisfied and ε known. This approach can be checked by comparing "radiobiological" molecular masses with those known from independent measurements. An example is shown in Fig. 7.1: A linear relationship exists over more than two decades, which is

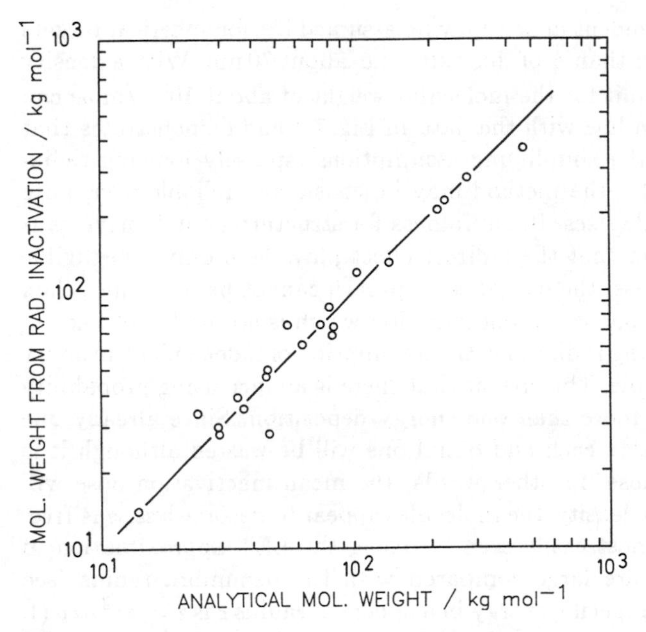

Figure 7.1. Comparison of molecular weights of proteins either determined from radiation inactivation (ordinate) or analytically (abscissa). Data were taken from KEMPNER and SCHLEGEL 1979

rather astonishing considering the simplifying assumptions. From the slope, ε may be calculated as 66 eV, values in the literature vary between 60 and 100 eV.

The method described has been widely used in recent years since it allows molecular weight determinations without chemical identification of the functional unit and avoids cumbersome separation procedures. This is particularly valuable where enzymes are bound to complex structures, e.g. membranes the integrity of which is essential for proper function. It has to be borne in mind, however, that the radiobiological mass is not necessarily equal to its real chemical value because it is limited to those parts responsible for the function. For a further discussion see KEMPNER and SCHLEGEL (1979) and BEAUREGARD and POTIER (1984).

The actual typical situation is not as simple as assumed above. Ionizations are arranged in sequence along particle trajectories and hence not independent. This goes without saying for densely ionizing ions but is also true for electrons, X and γ rays. If the target sizes are significantly smaller than the average distance between primary events along the track, they may be considered independent and reliable results expected. The approximations hold true, therefore, only for small targets and sparsely ionizing radiations. An example may illustrate this: The frequency average LET of ^{60}Co γ rays is 0.23 keV/μm, this means that the mean distance of deposition events of 66 eV is about 280 nm. Independent action may be assumed if – for spherical targets – the radius is not larger than $\frac{1}{4}$ of this value, i.e. about 70 nm. With a density of $1\,g/cm^3$, an upper limit for the molecular weight of about 10^8 g/mole can be calculated. This is in line with the data in Fig. 7.1 and demonstrates that – taking into account the simplifying assumptions (sparsely ionizing radiation, no indirect effect) – the method may be considered reliable over quite a wide range of molecule sizes. Its usefulness for structure-bound enzymes is certainly due to the fact that the indirect effect plays here only a negligible role. If this is not the case, the described approach cannot be used since then the inactivation dose depends on concentration which is normally not known.

With densely ionizing radiations, the assumption of independent primary events is no longer tenable. This means that there is an increasing probability that the target suffers more than one energy deposition. Since already one suffices for its inactivation each additional one will be wasted although it is included in the total dose. In other words, the mean inactivation dose will increase with ionization density, the molecules appear to become less sensitive.

This situation is most easily seen by using the LET approximation: if the target dimensions are large compared with the penumbra radius (see Sect. 4.2.4), the average specific energy in a sphere of radius r is $z = \pi r^2 L/\varrho$ (L = LET, ϱ = density). By neglecting that z follows a statistical distribution and assuming z to be single-valued, the surviving fraction with Eq. (7.6) becomes:

$$N/N_0 = \sum e^{-\lambda} \frac{\lambda^n}{n!} \cdot e^{-\frac{n \cdot m \cdot \pi r^2 L}{\varepsilon \varrho \pi r^2}} = \exp\left[-\lambda\left(1 - e^{-\frac{m \cdot \pi r^2 L}{\varepsilon \varrho \pi r^2}}\right)\right] \tag{7.7}$$

This expression may be simplified: The mean number of hits, λ, equals $\pi r^2 \cdot \phi$ with ϕ being the particle fluence and assuming that the penumbra radius is very much smaller than r. The target mass m equals $4/3\pi\varrho r^3$ so that:

$$N/N_0 = \exp\left[-\pi r^2 \cdot \phi\left(1 - e^{-\frac{4/3 r \cdot L}{\varepsilon}}\right)\right] \tag{7.8}$$

The inactivation cross section σ_{in} is

$$\sigma_{in} = \pi r^2\left(1 - e^{-\frac{4/3 rL}{\varepsilon}}\right) \quad . \tag{7.9}$$

It increases with LET and approaches a limiting value equal to the geometrical target cross section. This is, however, only true within the framework of the assumptions made, especially with regard to the applicability of the LET approximation. It was also tacitly supposed that the mechanism does not change with densely ionizing radiation since ε is taken to be the same as for X and γ rays. For small L, the exponential in Eq. (7.9) may be expanded:

$$\sigma_{in} \approx 4/3\pi r^3 \cdot \frac{L}{\varepsilon} \tag{7.10}$$

i.e. σ_{in} initially increases linearly with LET.

Expression (7.8) may also be written in terms of dose D which is related to LET and fluence by:

$$N/N_0 = \exp\left[-\frac{\pi r^2 \cdot D \cdot \varrho}{L} - \left(1 - e^{-\frac{4/3rL}{\epsilon}}\right)\right] \tag{7.11}$$

with $D = L \cdot \phi/\varrho$ (Eq. (4.20)).

For small LET, this gives, corresponding to Eq. (7.10),

$$N/N_0 \approx e^{-\frac{4/3\pi r^3 D \cdot \varrho}{\epsilon}} = e^{-\frac{mD}{\epsilon}} \tag{7.12}$$

which is identical to Eq. (7.6). With other words, with low LET there is no LET dependence. If the limiting inactivation cross section is approached, however, and particularly with very high LET, the slopes of the exponential survival curves according to Eq. (7.11) become progressively smaller as was already qualitatively stated above. More energy is deposited in the target than actually required for its inactivation. This state of affairs is colloquially nicknamed the "overkill effect".

The treatment just given is obviously too simple for most cases. It can be considerably improved using the concepts of microdosimetry (Chap. 4): The energy deposition by ionizing particles is characterized by the distribution function $f_1(z)dz$ where z signifies the stochastic quantity "energy absorbed per target mass". Without having further information regarding the inactivation mechanism the sensitivity of the system is assumed to be independent of radiation quality and may be derived from X or γ ray data as given by Eq. (7.6). The surviving fraction may then be written with reference to Eq. (4.31) as:

$$N/N_0 = \sum_{n=0}^{\infty} e^{-\lambda} \frac{\lambda^n}{n!} \int f_\nu(z) e^{-\frac{mz}{\epsilon}} dz \quad . \tag{7.13}$$

The integral on the right hand side is just the LAPLACE transform of a convolution integral (see Appendix I.5) so that:

$$\int f_\nu(z) e^{-\frac{mz}{\epsilon}} dz = \left[\int f_\nu(z) e^{-\frac{mz}{\epsilon}} dz\right]^n$$

leading to

$$N/N_0 = \exp\left[-\lambda\left(1 - \int f_1(z) e^{-\frac{mz}{\epsilon}} dz\right)\right] \quad . \tag{7.14}$$

Since in the general case penumbra electrons will contribute to target inactivation, energy may be deposited both by direct hits and by those which pass within a distance given by the penumbra radius r_p. The specific energy depends also on the distance between the target center and the particle's trajectory. Thus, in neglecting straggling, the following relationship can be given:

$$f_1(z(x))\,dz = \frac{2\pi x\,dx}{\pi(r+r_p)^2} \tag{7.15}$$

and

$$\lambda = \pi(r+r_p)^2 \cdot \phi \quad . \tag{7.16}$$

Equation (7.14) then has the form:

$$N/N_0 = \exp\left[-\pi(r+r_p)^2\phi\left(1-\int_0^{r+r_p}\frac{2\pi x}{\pi(r+r_p)^2}e^{-\frac{mz(x)}{\epsilon}}dx\right)\right] \quad . \tag{7.17}$$

The inactivation cross section σ_{in} is:

$$\sigma_{in} = \pi(r+r_p)^2 - \int_0^{r+r_p} 2\pi x e^{-\frac{mz(x)}{\epsilon}}dx \quad . \tag{7.18}$$

Depending on target size and particle, z(x) may attain very high values. A few examples of calculated cross sections for some ions and spherical targets are shown in Fig. 7.2.

A general evaluation of Eq. (7.18) is not possible, it has to be solved for each particular case by numerical computations.

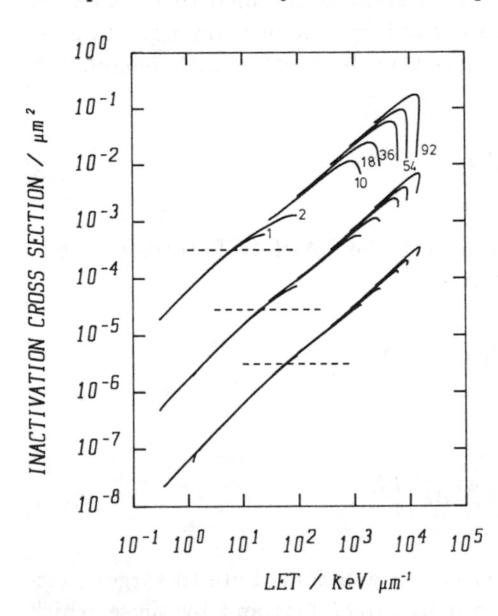

Figure 7.2. Inactivation cross sections as a function of LET with various target sizes and X-ray sensitivies. The geometrical cross sections are indicated by dashed lines. It was assumed that 60 eV has to be deposited in the target volume to cause inactivation. It is clearly seen that there are separate curves for each ion, the atomic numbers are given at the uppermost curve. Calculated according to Eq. (7.18) using the beam model of KIEFER and STRAATEN (1986)

Further reading
HARMON et al. 1985

7.2 Gene Mapping

Pyrimidine dimers in DNA do not only inhibit replication but also transcription. This offers an opportunity to determine functional gene sizes in a simple way. If one dimer suffices to prevent the formation of a functional gene product, its radiation sensitivity should depend directly on the size of the coding DNA segment. Since this is a typical single-hit phenomenon, the relative number of functional products N/N_0 can be expected as:

$$N/N_0 = e^{-\lambda_G} \tag{7.19}$$

where λ_G is the mean number of dimers per gene at a given fluence. If the total number of dimers per genome and unit fluence is known to be α, then:

$$\lambda_G = \alpha \frac{m_G}{m_z} \cdot F \tag{7.20}$$

with m_z as the mass of the genome, m_G that of the gene to be investigated and F the fluence. Using this technique, it is, e.g. possible to assess whether certain proteins are composed of identical subunits coded on the same gene. Also the length of "non-coding" regions may be estimated by comparing the UV sensitivity of transcriptional units and the final translation product. The result, however, is not always unambiguous. If two genes are consecutively transcribed starting from the same promoter, the radiation sensitivity depends on the size of the whole unit except for the starting gene. "UV mapping", however, may then be used to determine the gene sequence within a scripton as shown in Fig. 7.3: It is assumed that the two genes *a* and *b* are transcribed in a tandem fashion from left to right. A dimer anywhere in the total unit will prevent the synthesis of functional *b*-product (B) while a is only inhibited if a dimer is induced in its coding region, damage in region *b* has obviously no effect for its synthesis.

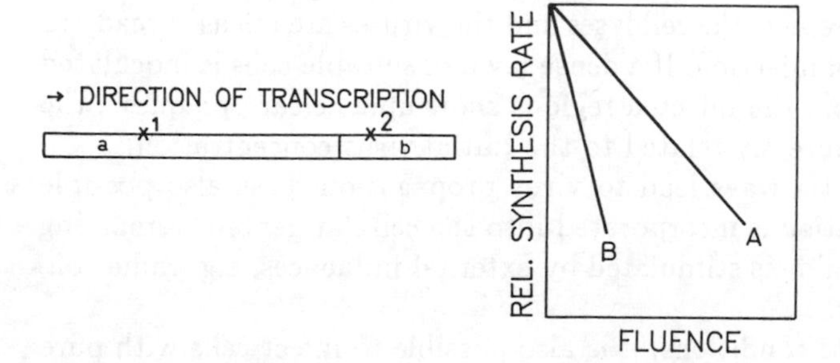

Figure 7.3. The basic principle of gene mapping using UV radiation. Lesion 1 causes inactivation of the complete scripton, lesion 2 inhibits only the expression of b

Applying Eqs. (7.19) and (7.20) appropriately leads to:

$$(N/N_0)_A = e^{-\alpha \frac{m_a}{m_z} \cdot F} \tag{7.21}$$

and

$$(N/N_0)_B = e^{-\alpha \frac{m_a+m_b}{m_z} \cdot F} \quad . \tag{7.22}$$

If the fluences causing a reduction of synthesis to 1/e are termed F_{0A} and F_{0B}, respectively, then:

$$\frac{F_{0A}}{F_{0B}} = \frac{m_a + m_b}{m_z} \quad . \tag{7.23}$$

The fluence ratio is independent of α and m_z making the method very attractive also in less well-characterized systems. An estimate of m_a/m_b is, however, required which may often be obtained from the size of the gene product.

Special reference
SAUERBIER 1976

7.3 Viruses, Plasmids, Transforming DNA, and Vectors

7.3.1 Techniques

Viruses consist of nucleic acid (DNA or RNA, single- or double-stranded) encapsulated in a protein coat. They are not able to replicate autonomously but require the machinery of a suitable host cell. If this is a bacterium, one speaks of "bacteriophages". The virus-host relationship is quite specific, a particular virus propagates only in suitable host cells.

Only the nucleic acid is injected into the host cell for infection, the protein coat is left outside. In the normal case, the cell's metabolism is then geared to replicate the viral nucleic acid and to synthesize the coat protein. If a certain number of viruses is present, the cell lyses and the viruses are released ready to start the next round of infection. If a dense lawn of suitable cells is inoculated with a virus suspension, the infected regions show up as clear "plaques", the number of which are directly related to the initial virus concentration.

Infection does not always lead to virus propagation. It is also possible that the viral information is incorporated into the cellular genome remaining there unexpressed until it is stimulated by external influences, e.g. radiation. ("temperate viruses").

Under appropriate conditions, it is also possible to infect cells with pure viral nucleic acid, i.e. after separation of the protein. If the host cell possesses a cell wall – like bacteria – this has to be removed by enzymatic digestion in order to obtain "protoplasts"; this special technique is called "transfection".

A different, but related, approach is that of "transformation" involving cellular DNA. Few bacterial species only can be used, e.g. *Haemophilus influenzae* and *Bacillus subtilis.* One starts with a mutant (the donor) possessing a genetic property, the *genetic marker*, which is easily testable (e.g. resistance to an antibiotic) and not present in the accepting host.

The donor DNA is carefully isolated and offered to the wild-type host which incorporates in part the foreign DNA and to a certain extent also the genetic marker. Its presence can be proved by assaying for the acquired antibiotic resistance.

The three techniques described so far are schematically summarized in Fig. 7.4. A very useful and potent extension has been made possible with the advancement of DNA-recombinant methods. They allow – at least in principle – to construct viruses which are tailored according to specified properties with the aid of "restriction endonucleases" (see Appendix II.5). If they are able to propagate both in bacteria and higher cells ("shuttle vectors") they may not only be used for transformation studies but also to analyze radiation

VIRUS INFECTION:

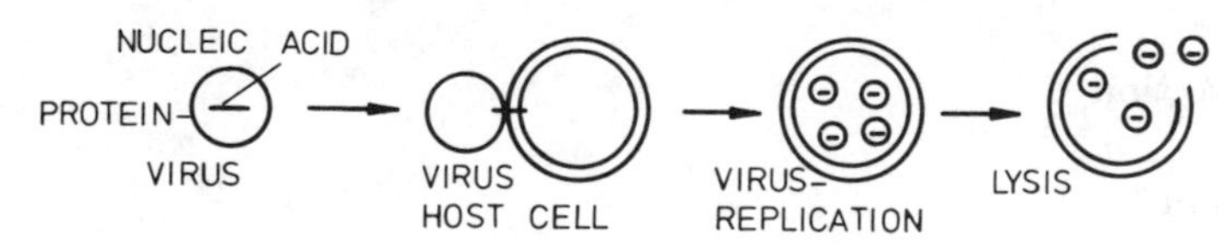

TRANSFECTION :

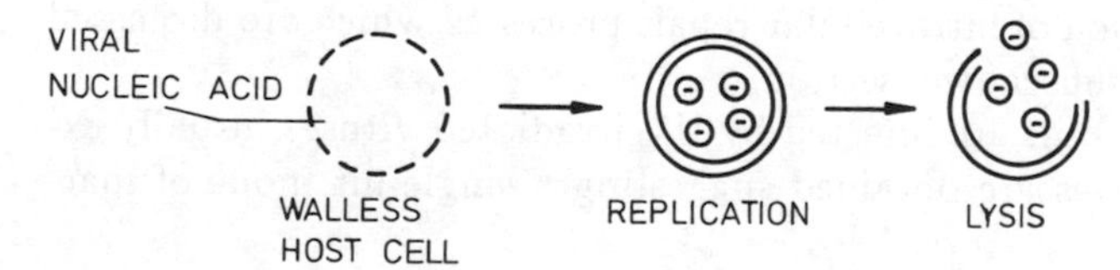

TRANSFORMATION:

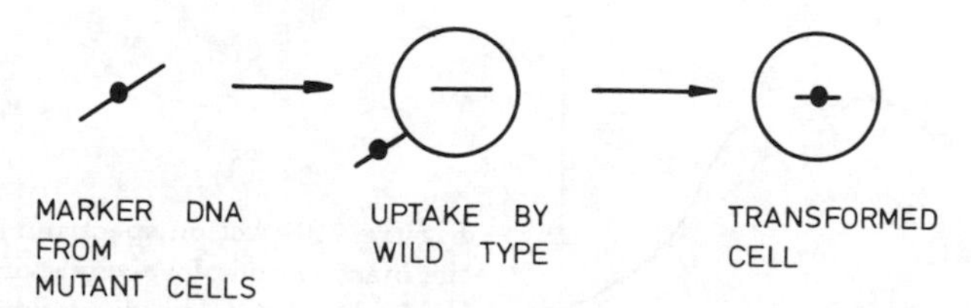

Figure 7.4. Techniques for viral infection, transfection and transformation

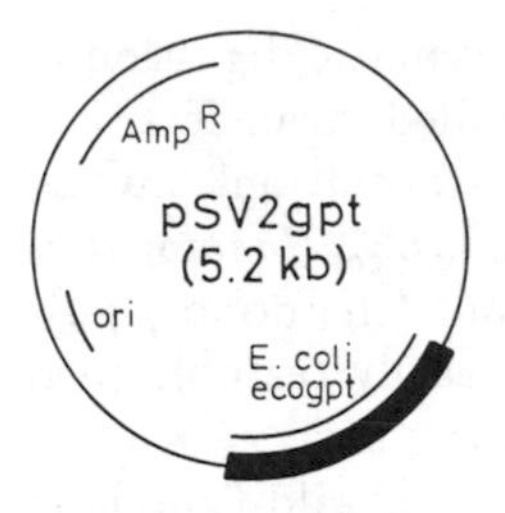

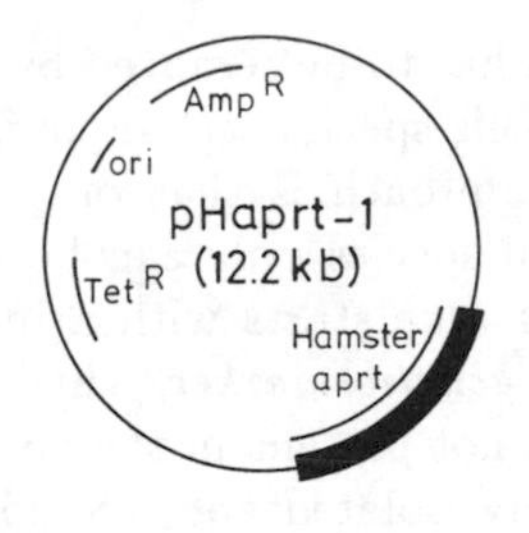

Figure 7.5. Two examples of "shuttle vectors" to investigate tranformation in mammalian cells (see text for further explanation)

induced changes at the molecular level. The altered plasmid may be grown to milligram amounts by multiplication in the bacterial host making the product amenable to biochemical analysis. A typical example is shown in Fig. 7.5.: The recombinant plasmid *p SV2 gpt* contains sequences from the simian virus SV 40 which are necessary for propagation in mammalian cells and also the bacterial plasmid *pBR 322* allowing multiplication in *E. coli*. The "marker genes" *amp* leads to ampicillin resistance and *gpt* codes for xanthine-guanine phosphoribosyl-transferase. This enzyme is not present in mammalian cells, the expression of bacterial *gpt* can hence be used as a selectable marker if the cells are grown in suitable media. Their attractiveness lies in the possibility to clearly separate lesion formation outside the cell and its modification by subsequent processes in cells which have not been irradiated.

7.3.2 Radiation Action

7.3.2.1 Inactivation

The "plaque forming ability" (PFA) for a given dose of irradiated viruses does not only depend on exposure conditions but also on the host cell type. This is a clear indication of intracellular repair processes which are discussed in more detail in the subsequent section.

If repair-deficient cells are infected by UV-irradiated viruses, usually exponential survival curves are obtained suggesting a single-hit mode of inac-

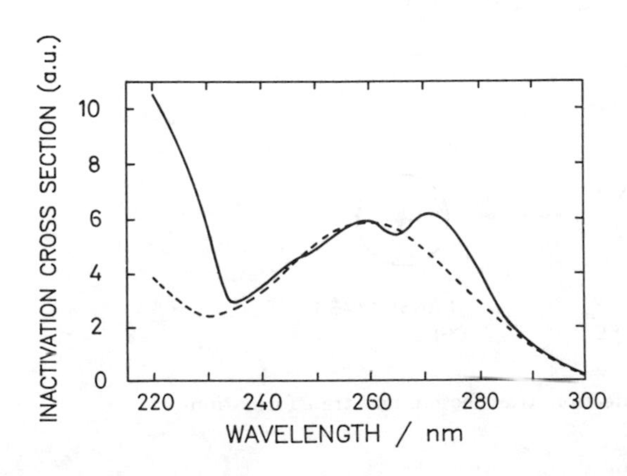

Figure 7.6. Action spectrum for the inactivation of Vaccinia virus. DNA absorption is indicated by a broken line (after SIME and BEDSON 1973)

tivation. The action spectrum (Fig. 7.6) indicates the prominence of damage in the nucleic acid component although the shoulder around 280 nm suggests also a minor contribution of the protein moiety. The chemical nature of the responsible lesion is most elegantly studied by using transforming DNA. The "survival" curve has in this instance an unusual form which may be described by the expression:

$$y = \frac{1}{(1 + kF)^2} \tag{7.24}$$

where y stands for transforming ability, F for fluence and k being a sensitivity parameter. The latter depends on the marker under investigation, its value may be used to estimate the size of the respective gene. Equation (7.24) describes convex curves if y is plotted logarithmically versus fluence on a linear scale (Fig. 7.7). The lesion may be characterized making use of the "photochemical switch" for dimer formation (see Sect. 6.1.1): If the DNA is first exposed to UV at 280 nm, the survival curves shown by bracketed lines in Fig. 7.7 result. A subsequent irradiation at 240 nm increases the transformation ability – i.e. reduces the damage – demonstrating convincingly the importance of dimers being the only lesions which can be reverted by this kind of treatment. Their relevance was later also proven by photoreactivation (Sect. 13.2.1).

Recombinant DNA vectors, as described in Sect. 7.3.1, are now increasingly used to investigate radiation-induced alterations in mammalian cells. The main objective appears to be at present to clarify the molecular nature of mutations and repair. These studies will, therefore, be referred to at the appropriate places.

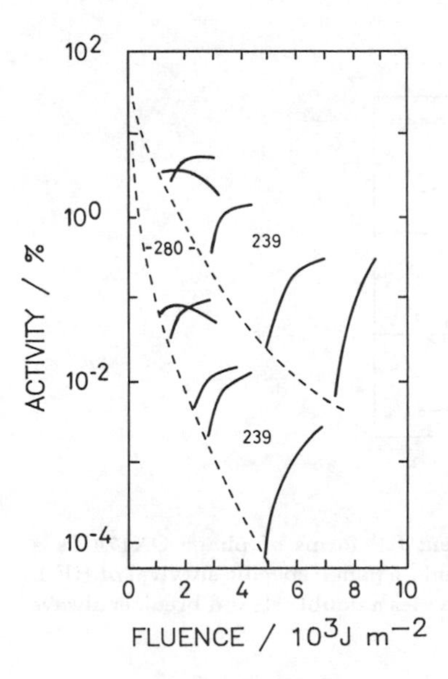

Figure 7.7. "Classical" experiment to demonstrate photochemical reversion of marker damage in transforming bacterial DNA, thus proving the importance of pyrimidine dimers: The DNA was first exposed to 280 nm UV-light and afterwarde to 239 nm. It is seen that the initial inactivation is partly reversed. Upper curve: resistance to cathomycin, lower curve: resistence to streptomycin (after SETLOW and SETLOW 1962)

To gain some insight into the mechanism of ionizing radiation effects, the already mentioned phage ϕX174 offers itself as a particularly useful system. If it is irradiated in its single-stranded form, inactivation is caused by an average absorbed energy of about 60 eV, i.e. one primary ionization. Although one is tempted to identify a single strand break (SSB) with the inactivating lesion, it is not clear whether this is really the case. There is no doubt that one SSB is lethal but the experiments show that other types of damage – presumably base alterations – contribute by up to 50%.

The replicative form of ϕX174 is particularly suitable to investigate double-strand DNA because it makes possible to separate molecules without breaks, with one SSB, and with one DSB and test them for biological activity (see also Chap. 6).

Linear molecules are always inactive, i.e. a DSB is a lethal event in this system. The other fraction loses its activity in a dose-dependent fashion demonstrating that SSB and – in the super-coiled form – other types of damage also contribute significantly. It is estimated that about 9% of total inactivation is due to SSB. This was deduced from the dependence on LET: Suspensions of the RF form containing high concentrations of radical scavengers to suppress the indirect effect were irradiated with accelerated ions (up to ^{40}Ar; with ion specific energies in the order of 10 MeV/u). The sedimentation behaviour of the different RF forms (Sect. 6.2) allowed to analyze the relevance of the various damage types (results are shown in Fig. 7.8):

The overall survival for a fixed dose of the total population (upper panel) remains constant up to LET values around 200 keV/μm, beyond which it decreases. The specific survival ability of RF I and RF II, however, starts to decline much earlier. Comparing this behaviour with that of SSB and DSB formation (Fig. 6.11) allows to conclude that with low LET, base-damage and

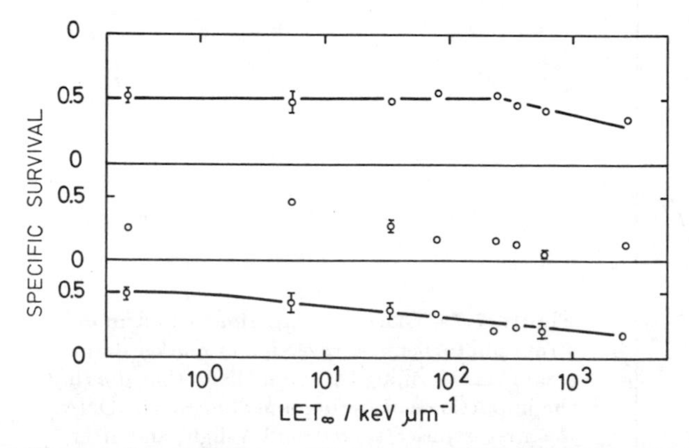

Figure 7.8. Relative specific survival of the different RF forms of phage OX174 as a function of LET. Upper panel: all forms combined, middle panel: specific survival of RF I, lower panel: specific survival of RF II. RF III which carries a double strand break is always inactive (after CHRISTENSEN et al. 1972)

SSB contribute mostly to inactivation. DSB become significantly important only with high LET, but even then their contribution is only about 50%.

The influence of oxygen is very complex. It protects phages exposed in the dry state suggesting that H radicals which are scavenged by O_2 cause inactivation to a certain extent. No oxygen effect is found when irradiation takes place in pure buffer without scavengers – neither with complete phages nor with isolated DNA – although the chemical alterations display OERs around 2, as listed in Table 6.5. The composition of the buffer is critical: phosphate radicals, for example, may be inactivating (see LAFLEUR and LOHMAN 1987 for details). If radical scavengers are added at high concentrations, the overall sensitivity is reduced and a clear oxygen effect is seen (Fig. 7.9). The relevance of these findings is taken up again in connection with the cellular oxygen effect in Sect. 9.2.2.

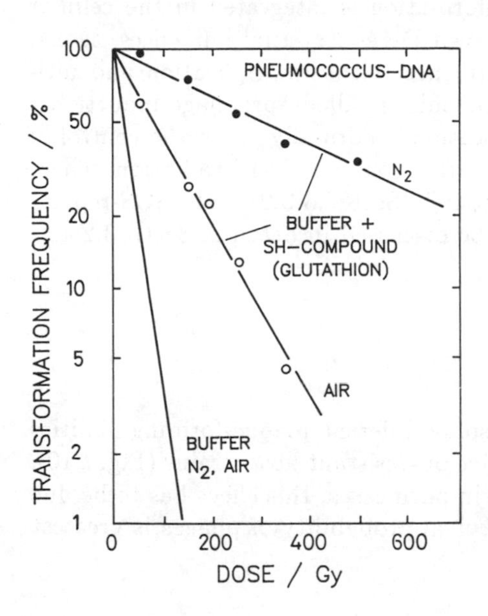

Figure 7.9. Oxygen effect in transforming DNA of Pneumococcus. There is no oxygen effect if exposure is given to DNA in buffer. If glutathion is present the same oxygen effect is found as with intracellular irradiation (after HUTCHINSON 1961)

Phage DNA may also be used to assess the hazard by incorporated radionuclides, an important topic in radiation protection. Single-strand DNA is inactivated by the disintegration of a single incorporated ^{32}P atom because a ^{32}S atom is formed ("transmutation") and the integrity of the sugar-phosphate backbone can no longer be maintained. Also DSB may be caused – although with low probability – since the recoil energy of ^{32}S atoms may damage also the opposing strand. That this is really the case may be seen from a comparison of ^{32}P and ^{33}P which are both β emitters but having different recoil energies. The inactivation probability in double-strand DNA was found to be 0.06 per ^{32}P decay (recoil energy 20 eV) but only 0.02 per ^{33}P decay (recoil energy 1.7 eV), both measured in T4-phage DNA at −196°C.

Incorporated ^{3}H acts only through its β emission but with high local energy densities because of the short electron range.

^{125}I is another interesting radionuclide, also used widely in nuclear medicine. ^{125}Iododeoxyuridine may replace thymidine in DNA like BUdR. ^{125}I decays via electron capture and subsequent emission of AUGER electrons (see Sects. 3.2.1.5 and 4.2.5). This process leads to very high local energy densities, and a highly ionized atom is left behind causing severe instability of the molecular conformation. The probability for DSB formation is, therefore, as high as 50% per disintegration, an effect which has to be taken into account in the practical use of these kinds of radionuclides.

7.3.2.2 Induction

It has been pointed out before that a viral infection does not necessarily lead to cell lysis. In this case the viral information is integrated in the cellular genome and replicated with the cell's own DNA. External influences, as e.g. radiation, heat, etc., may lead to virulence, virus multiplication and ultimately death of the cell. With bacteria this is called "prophage induction". It is known today that phage multiplication is normally under the control of a repressor which may be inactivated after (but not by) irradiation. There are close links to "WEIGLE reactivation" (Sect. 7.3.3.2) and "SOS-repair" (Sect. 13.2.4). The whole process will be discussed in detail in Sect. 13.2.4.

7.3.3 Special Repair Processes

7.3.3.1 Host Cell Reactivation

UV-irradiated bacteriophages may display different plaque forming abilities if they are plated either on radiosensitive or -resistant host strains (Fig. 7.10). Since the primary damage is identical in both cases, this effect has to be due to intracellular modifications. The survival probability of phages is greatest

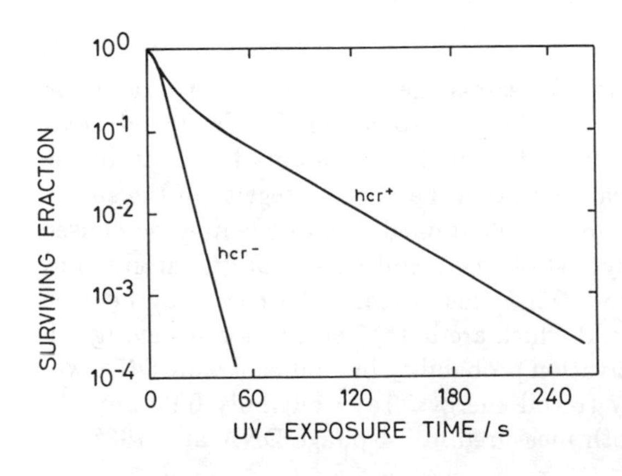

Figure 7.10. Host cell reactivation: UV-irradiated T1-phages were plated either on hcr$^-$- or hcr$^+$-host cells (after HARM 1963)

with resistant bacteria which obviously repair lesions in the phage genome thus reducing their own chance to survive. This phenomenon is called "host cell reactivation", mutants defective in this process are termed *hcr*⁻. Host cell reactivation is not restricted to bacteria but can also be found in mammalian cells and the respective viruses.

More detailed investigations revealed that host cell reactivation is not one single repair process but a combination of several pathways. This is evident from the fact that – depending on the host type – the increase in phage survival is accompanied either by a decrease of or a rise in mutation rate. One has, therefore, to distinguish between error-free and error-prone processes (see next paragraph). Host cell reactivation *sensu strictu* is connected with the proper function of the excision system (see Sect. 13.2.2), with other words, excision-deficient mutants are also *hcr*⁻.

7.3.3.2 WEIGLE Reactivation

If UV-irradiated phages are plated on pre-exposed host cells, an unexpected result is seen: the survival rate is higher than with unirradiated bacteria (Fig. 7.11). This important phenomenon is commonly called "WEIGLE reactivation", named after its discoverer, or "W reactivation" for short. In contrast to normal host cell reactivation, one finds here an increase in phage mutation rate, the process is obviously error prone. It is not only found with UV, but also with X rays. "W reactivation" is intimately connected with cellular "SOS repair" which will be discussed in detail in Sect. 13.2.4.

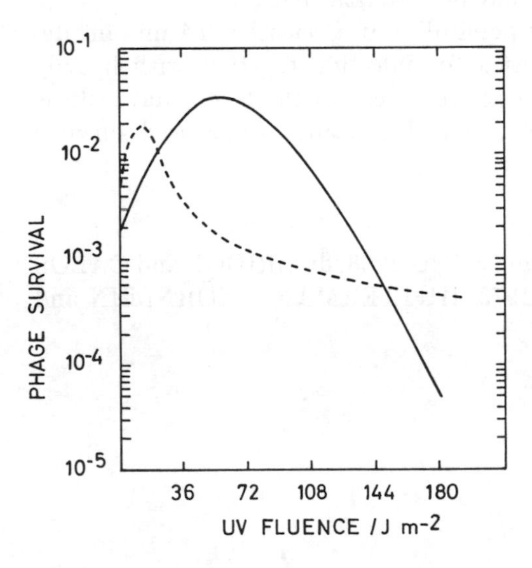

Figure 7.11. WEIGLE reactivation: UV-irradiated lambda phages were plated on excision proficient (heavy line) or excision-deficient E. coli host cells preciously exposed to 254 nm UV-light as indicated on the abscissa (after HARM 1963)

7.3.3.3 Phage-directed Repair

Phage survival does not only depend on the host cell type – as shown above – but also on the kind of phages. Some T phages of *E. coli* have the same survival on hcr^+ and hcr^- bacteria (e.g. T4). A closer analysis reveals that, also in this case, part of the initial damage is repaired but this is directed by phage information as could be demonstrated by the isolation of radiation-sensitive phage mutants. Here two genes play an important role; they are termed "v" and "x". The first one codes for an enzyme which introduces a single strand break near the damage site initiating thus excision repair ("T4 endonuclease"), the action of the other one is less well known.

7.3.3.4 Multiplicity Reactivation

This process is not really a repair phenomenon but since it may lead to increased phage survival, it is mentioned here. Irradiated phages having lost their plaque-forming ability may still be able to enter the host cell. If there is an infection by several phage particles, an exchange of intact pieces of DNA may occur – comparable to intracellular recombination repair (Sect. 13.2.3) – so that a functional phage genome may be reconstituted.

This is, however, not the only possibility. If a sensitive T4 mutant deficient in the T4 endonuclease is used for infection together with heavily irradiated wild type T4 phages, the active v gene of the latter may still be functioning so that v repair may proceed. This leads to increased survival without recombination.

Further reading
DEFAIS et al. 1983, DERTINGER and JUNG 1968, du BRIDGE and CALOS 1988, HANAWALT and SETLOW 1973, HÜTTERMANN, KÖHNLEIN and TÉOULE, 1978, THACKER, 1986

Chapter 8
Loss of Reproductive Ability in Cells

Cell survival, in radiobiological terms, is understood as the ability for indefinite reproduction. After a description of basic techniques, the formalism of survival curves is discussed. The next section shows whether and how the different sensitivities of cell species may be interpreted; it will be shown that it is related – at least with ionizing radiation – to the amount of genetic material. A description of the influence of radiation quality is followed by a consideration of action spectroscopy with UV and the importance of the spatial pattern of energy deposition with ionizing radiation. The concept of "relative biological effectiveness" (RBE) is introduced and critically discussed. The chapter concludes with a discussion of the interaction of different radiation types in combination experiments.

8.1 Survival Curves

The ability of a cell to reproduce itself by division is of central importance for all biological systems. Its impairment by radiation is not only fundamental for radiation therapy but also for the assessment of radiation hazards. In competent cells and the appropriate environment, the division potential is only limited by nutrient deprivation, cell density or by systemic regulatory factors. The loss of reproductive integrity constitutes an important and fundamental parameter in radiation biology.

The principle of the basic technique to measure cell reproduction – as shown in Fig. 8.1 – is very simple: cells are brought on a suitable nutrient medium solidified by agar, in such a density that they are able to divide

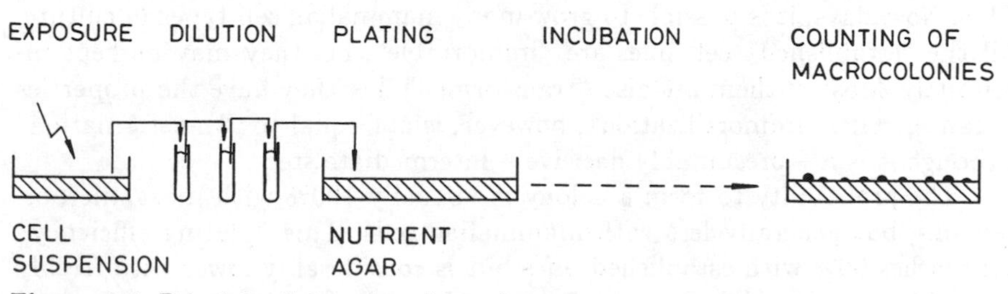

Figure 8.1. Principle of the colony formation assay

without interfering with each other. On incubation, each cell which did not lose its reproductive integrity will grow to a colony visible to the naked eye. Since the initial cell density is usually higher than desired, the suspension has to be diluted. The number of dilution steps has to be appropriately adjusted to the expected survival rate. Thus it is possible to determine the "colony forming ability" (CFA) over many orders of magnitude giving the method a wide dynamical range which is unsurpassed by any other technique.

One has to keep in mind, however, that there is – in biological terms – a long time between the initial insult and the ultimate effect during which many modifying processes may intervene. The irradiated cell does not immediately lose its division ability; even after high doses, a few divisions may still occur. Also not all the cells of a colony are able to divide. Depending on dose, there may be fractions of sterile cells. This phenomenon is called "lethal sectoring".

The division behaviour can be followed by single cell observations ("pedigree analysis"). An example of such an investigation is shown in Fig. 8.2 where

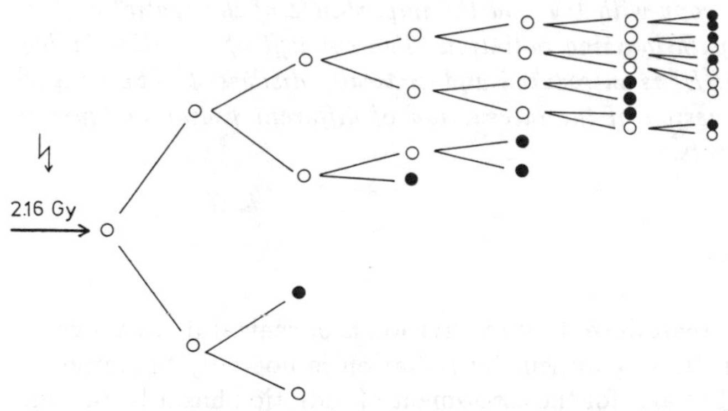

Figure 8.2. "Pedigree analysis" of mammalian cells exposed to X-rays. Open circles: live cells, closed circles: stop of division (after THOMPSON and SUIT 1968)

it may be seen that even after relatively high doses quite a few divisions take place.

The described colony formation assay has long been used for unicellular microorganisms like bacteria, yeast and algae. An important breakthrough is marked by its adaptation to mammalian cells in 1956 by PUCK and MARCUS. Nowadays, it is possible to grow many mammalian cell types in culture. All the "established" cell lines are "immortable", i.e. they may be kept indefinitely. Most of them are also "transformed", i.e. they have the properties of cancer cells. "Immortilization", however, is not equal to "transformation" although it is a – presumably decisive – intermediate step.

The probability to form a colony is virtually 100% with most microorganisms, but generally less with mammalian cells. This "plating efficiency" approaches 90% with established lines but is considerably lower with freshly excised biopsy material. Culture of mammalian cells is not only characterized by experimental difficulties but also hampered by the genetic instability of many lines which is, e.g. exemplified by variation in chromosome number. Experiments with radiation sensitive mutants, being standard with microorganisms, are still restricted to rather few examples.

Colony forming ability is of such central importance that it has become customary to equate it with cell death which is not strictly correct since many other cellular functions may still be working. Nevertheless, colony formers are usually called "survivors". If their fraction is plotted versus dose, one speaks of "survival curves".

Radiation is only one of many agents which inhibit cell proliferation. A comparison of dose-response relationships may be useful to learn more about differences and analogies; Fig. 8.3 shows an example.

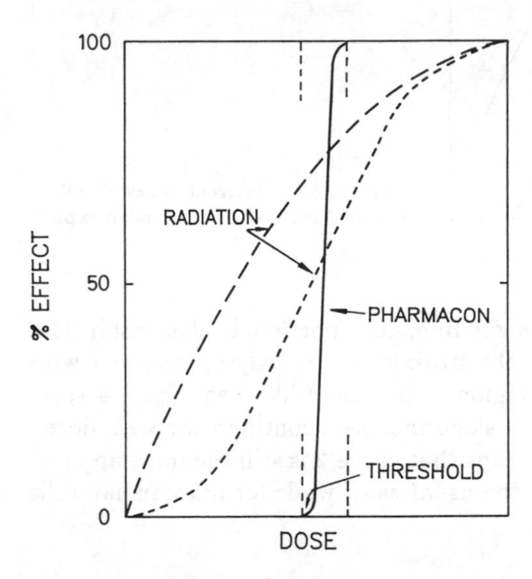

Figure 8.3. Comparison of radiation and "pharmacological" survival curves (after ZIMMER 1961)

One notes immediately that the pharmacological response curve is almost discontinuous and characterized by a clear threshold below which no effect is seen. This is typical for most – but not all – chemical poisons. There are also substances whose reactions resemble those of radiation – "radiation mimetics". Many mutagens and cytostatics belong to this group (see also Sect. 15.4).

Radiation effects start already with small doses, increasing continuously until the final level is asymptotically approached. This behaviour is due – as detailed later – to the stochastic nature of interaction processes.

It is customary in radiation biology to plot the surviving fraction logarithmically on the ordinate versus dose on a linear abscissa scale. Three representatives of such "survival curves" are shown in Fig. 8.4.

Homogeneous cell populations display generally one of these types of survival curves after irradiation. Example 1 is the simplest case: Since the

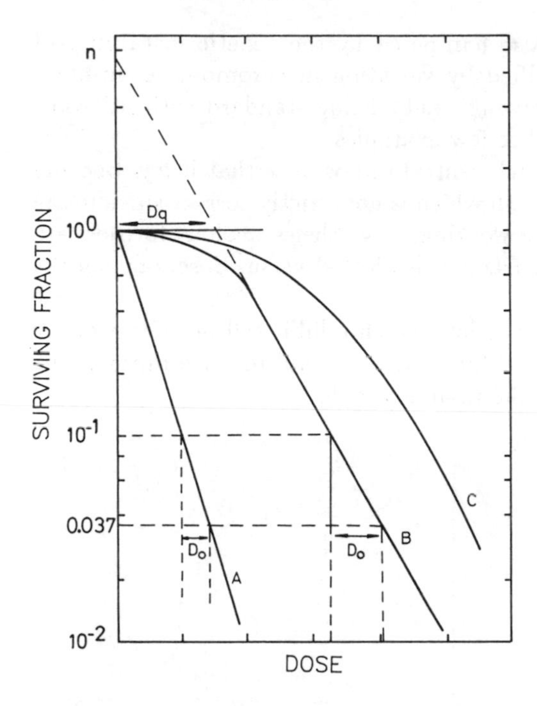

Figure 8.4. Different types of cell survival curves after radiation exposure

semi-logarithmic plot yields a straight line, the functional relationship is a simple exponential. In example 2, the straight line is only approached with higher doses, while in the lower region a threshold-like behaviour is seen. Example 3 is the most complex: the slope increases continuously with dose.

Although widely disputed, it seems that curve 2 is still the most applied. For microorganisms it is certainly the usual case, while for mammalian cells

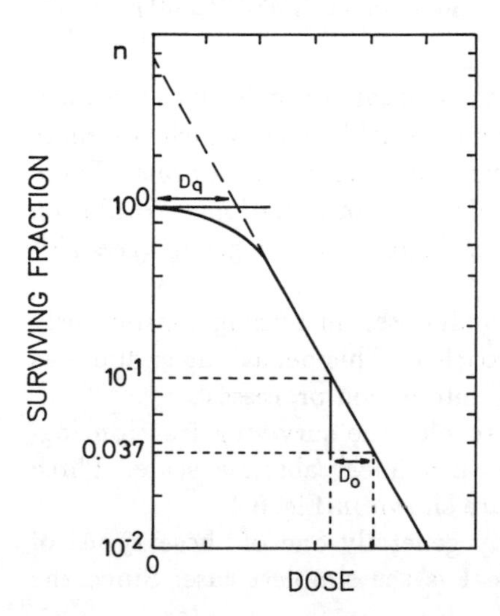

Figure 8.5. Characteristic parameters of shouldered survival curves

its applicability is not generally accepted. It has the advantage that it may be simply characterized by two parameters as seen from Fig. 8.5: One is the slope of the terminal straight part which is commonly indicated by the dose – termed "D_0" – which is necessary to reduce relative survival to e^{-1} (0.37). D_0 is obviously the reciprocal of the slope. There are two ways to deal with the initial part both making use of the backward extrapolation of the straight portion. Its intercept with the ordinate is called "extrapolation number" (n), the dose where it crosses the 100%-survival line is termed the "quasi-threshold dose" D_q. The three parameters are not independent of each other, as also seen from the figure:

$$D_0 = \frac{D_q}{\ln n} \tag{8.1}$$

D_0 is the most important parameter to characterize the sensitivities of different cellular systems. Typical values obtained after UV- and X-ray exposure are listed in Table 8.1. For UV, "energy fluence" is used instead of dose, as

Table 8.1. D_0-values and cell nucleus factors for various systems. (After [a]) UNDERBRINK, SPARROW and POND, 1968; [b]) KIEFER and WIENHARD, 1977)

name	species	DNA/cell [a]) kg	chromosome-number	DNA/chr. [a]) kg	D_0 (UV) [b]) Jm^{-2}	D_0 (X-ray) [a]) Gy
T_1-phage	virus	$8.1 \cdot 10^{-20}$	1	$8.1 \cdot 10^{-20}$	41	2600
E. coli B/r	bact.	$1.9 \cdot 10^{-17}$	1	$1.9 \cdot 10^{-17}$	85	30
Bacillus subtilis	bact.	$7.7 \cdot 10^{-18}$	1	$7.7 \cdot 10^{-18}$	50	33
Saccharomyces cerevisiae	yeast	$2.8 \cdot 10^{-17}$	36	$7.8 \cdot 10^{-19}$	1150	150
Chlamydomonas	alga	$1.4 \cdot 10^{-16}$	16	$8.8 \cdot 10^{-18}$	2500	24
hamster-cells	mammal	$1.5 \cdot 10^{-14}$	22	$6.8 \cdot 10^{-16}$	50	1.4
HeLa-cells	human	$1.2 \cdot 10^{-14}$	78	$1.5 \cdot 10^{-16}$	108	1.4

already discussed in Chap. 4. To facilitate later comparisons (next section), DNA contents and chromosome numbers are also given. There appears to be a general tendency of increasing sensitivity (lower D_0 values) with biological complexity. This will be more extensively analyzed in the following section.

Non-homogeneous populations consisting of cells with different sensitivities display more complex survival curves. A simple example of two components is used for illustration. The two fractions shall be a_1 and a_2 ($a_1+a_2 = 1$); the respective D_0 values, D_{01} and D_{02}. The total surviving fraction y is then:

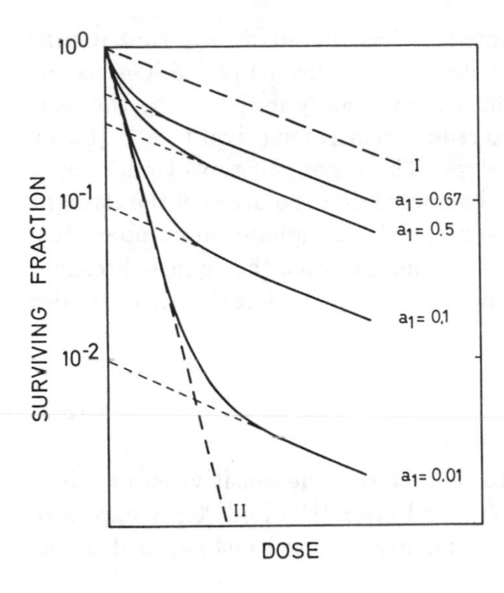

Figure 8.6. Survival curves for inhomogeneous polulations

$$y = a_1 e^{-D/D_{01}} + a_2 e^{-D/D_{02}} \quad . \tag{8.2}$$

The relationship is plotted in Fig. 8.6. The sensitive fraction dominates in the low dose region, while a "tail" is caused by the resistant subpopulation. The intercept of the backward extrapolation with the ordinate allows to estimate the fractional contributions. It must be stressed, however, that this is only possible in case of strictly exponential survival curves. If this not the case, the simple approach fails.

8.2 Radiation Sensitivity and Nuclear Parameters

The data given in Table 8.1 appear to show that X-ray sensitivity increases with cellular DNA content. This would be in line with the following, rather simplistic analysis: Assuming that cell death is caused by one single lethal event caused by the deposition of a critical energy EL anywhere in the DNA, one would postulate:

$$D_0 = \frac{E_L}{m_{DNA}} \tag{8.3}$$

(m_{DNA}: mass of DNA per cell).

Equation (8.3) implies that D_0 and DNA mass are inversely related. It is quite interesting to consider this in more quantitative terms: For mammalian cells, $D_0 = 1.4$ Gy and $m_{DNA} = 1.2 \times 10^{-14}$ kg resulting in $E_L = 1.6 \times 10^{-14}\,\mathrm{J} = 100\,\mathrm{keV}$.

This is a very large value, casting some doubt on the applicability of this very simple picture.

A more detailed experimental analysis shows that the situation is, in fact, more complex. SPARROW and coworkers analyzed a large number of survival curves of many cell types of microbial, plant and animal origin (UNDERBRINK and POND 1976) and found that not only DNA content but also chromosome number plays a decisive role. They concluded that either "mean chromosome volume" (nuclear volume divided by chromosome number) or "average chromosomal DNA-content" (total cellular DNA divided by chromosome number) are suitable parameters to categorize radiation sensitivities. In the first case they all fell into eight, in the second into four groups ("radiotaxa"). In Fig. 8.7 the relationships are shown for the second case.

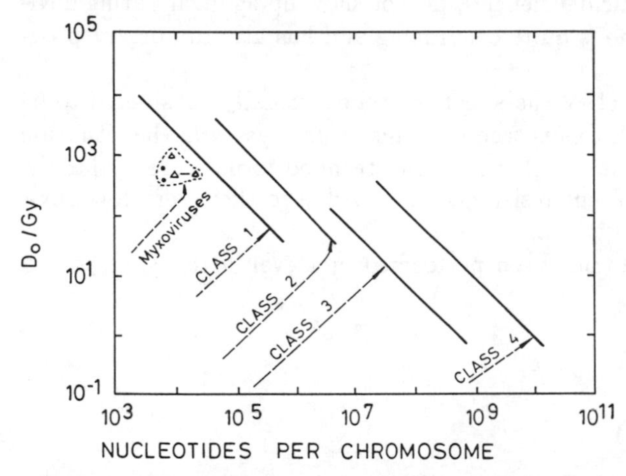

Figure 8.7. The dependence of radiation sensitivity on the average nucleotide content per chromosome (after UNDERBRINK, SPARROW and POND 1968)

The classification found is interesting: the most sensitive radiotaxon (in the given terms) consists exclusively of single-stranded viruses, the second of double-stranded viruses while the third comprises nearly all cellular organisms – from bacteria to humans. The last, not really understood, class has only few members, all of which are higher plants.

It is now possible to calculate by means of Eq. (8.3) the "mean lethal energy per chromosome"; the values obtained are listed in Table 8.2. If they are compared with those necessary for the induction of DNA lesions (Chap. 6), it can be seen that in the first radiotaxon a lethal event may be related to a single strand break, in the second to a double strand break. This comparison does by no means prove that this is really the case but it is biologically plausible.

The value for cellular systems is considerably higher, suggesting that more than one DSB per chromosome may be tolerated. This is in line with

Table 8.2. Absorbed energy per chromosome per 'lethal' event. (After UNDERBRINK, SPARROW and POND, 1968)

class	energy/eV
1	89
2	631
3	2089
4	34666

the more recent discovery that these lesions are repairable (see Sect. 13.2.6). Nothing can be said about radiotaxon 4.

The analysis described spans many orders of magnitude. Small differences, e.g. because of repair deficiencies, do not show up as invalidating deviations. The general scheme is quite convincing and has also important prognostic value.

There is another aspect: Viruses or bacteria are usually considered to be "radioresistant". Although this is true in terms of *dose per cell*, the situation is just opposite at the molecular level: single stranded viruses are by far the most sensitive. Their apparent resistance is only due to their very low DNA content.

A similar analysis has not been performed, not even attempted, for ultraviolet light.

8.3 Radiation Quality

8.3.1 Action Spectra

With optical radiation it is possible to obtain important indications about mechanisms of radiation effects by studying the inactivation efficiency as a function of wavelength. Apart from this fundamental aspect, the obtained relationship is often also of practical interest. The principles of this "action spectroscopy" are outlined in Sect. 5.1.3. Cellular inactivation by UV light provided the first experimental evidence for the importance of nucleic acids as the carrier of the genetic information – long before this fact was biochemically established. This historical action spectrum – depicted in Fig. 8.8 – shows clearly that the wavelength dependence follows closely the absorption of DNA, and not that of proteins as assumed at that time.

Action spectra for the killing of mammalian cells are essentially similar (Fig. 8.9).

The action falls rapidly towards longer wavelengths. It must not be concluded, however, that this region is unimportant, quite on the contrary, just because of the steepness of the action spectrum, small shifts to shorter wave-

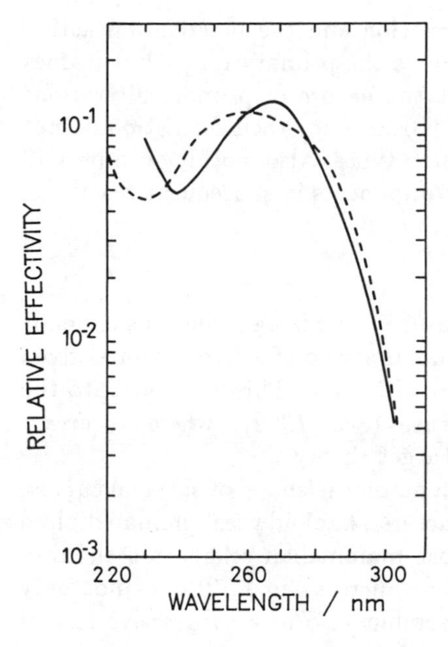

Figure 8.8. A "classical" action spectrum for the inactivation of E. coli. Quantum correction was not performed but the DNA absorption spectrum which is shown comparison (broken line) was accordingly adjusted (after JAGGER 1967 and GATES 1930)

lengths may have dramatic consequences. This is important to realize in connection with the "ozone problem" (Sect. 22.2).

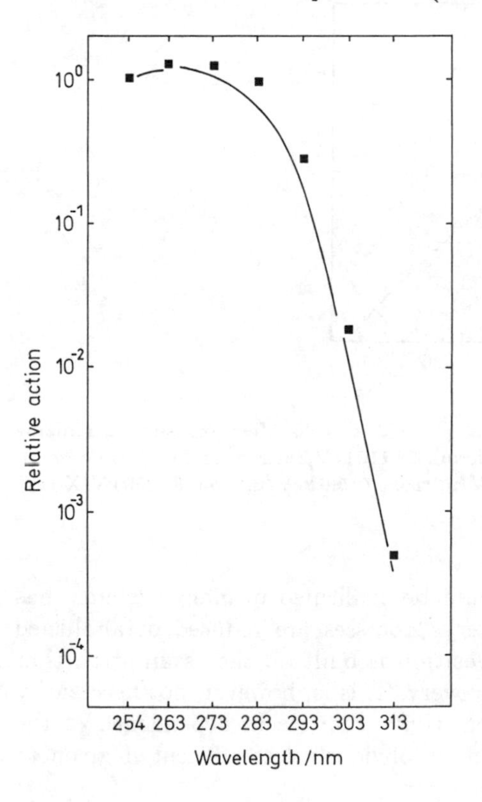

Figure 8.9. Action spectrum for killing of V79 Chinese hamster cells (after ZOELZER and KIEFER 1984)

The close resemblance of DNA absorption and the inactivation action spectrum certainly identifies this molecule as the primary target but it does not allow to draw any conclusions about the nature of primary alterations since the action spectra – at least below 300 nm – for their formation do not allow to differentiate between different lesion types. Also, nothing can be said about the participation of other cellular components in subsequent reactions.

8.3.2 LET Dependence

The survival behaviour of cells after exposure to ionizing radiations depends on radiation quality. It is commonly found that the effectiveness of a given dose increases with ionization density, i.e. with LET. This is different to the behaviour of simple systems, e.g. some viruses (Sect. 7.3.2) – where a *decrease* of effectiveness is found as predicted by target theory.

The rise of effectiveness does not depend on the shape of survival curves, it is found both with exponential (some bacteria, haploid yeast, human diploid fibroblasts) and "shouldered" types (most mammalian cells), as seen from Figs. 8.10 and 8.11. In the second case, an increase in LET does not only lead to steeper slopes of the terminal part but also to a progressive loss of

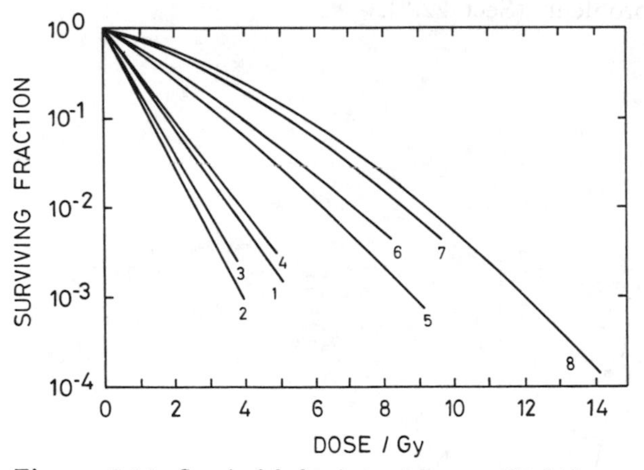

Figure 8.10. Survival behaviour of human T1 kidney cells after exposure to ionizing radiations of different LET. 1: 165 keV/micron, 2: 110 keV/micron, 3: 88 keV/micron, 4: 61 keV/micron, 5: 25 keV/micron, 6: 20 keV/micron, 7: 5.6 keV/micron, 8: 250 kV X-rays (after BARENDSEN 1967)

the shoulder. This finding, which could be confirmed in many systems, has prompted the conclusion that recovery processes are reduced or abolished with high LET radiation. This supposition is built on the assumption that shoulder width is an indication of recovery. This is, however, not necessarily so, it indicates only that cellular sensitivity changes in dependence of the dose applied. A given dose increment is obviously less efficient if given to

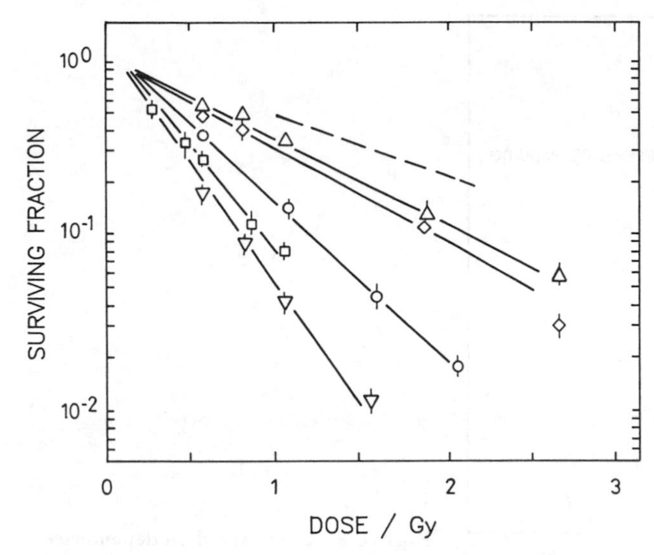

Figure 8.11. Inactivation of human diploid fibroblasts by radiation of different LET. △: 20 keV/micron, ◇: 28 keV/micron, ○: 50 keV/micron, □: 70 keV/micron, ▽: 90 keV/micron, broken line: 250 kV X-rays (after COX and MASSON 1979)

unirradiated cells than applied on top of a substantial pre-dose. Whether the "sublethal" part is, in fact, repaired cannot be concluded just from the shape of the curve; further experiments – like split-dose exposure (Sect. 13.3) – are required to prove that recovery really occurs.

Another point is that with increasing LET, the dose distribution becomes extremely inhomogeneous. A single particle traversing the cell nucleus may deposit more energy than required for inactivation. This case is a real single hit phenomenon with exponential survival curves (see Sect. 7.1). The loss of the shoulder is then simply due to the physical situation, nothing can be inferred about recovery processes.

For a simple quantitative description of the dependence of radiation quality, the concept of "relative biological effectiveness" (RBE) has been introduced. This parameter is defined as the ratio of doses which yield the same effect if one compares the test radiation with 250 kV X rays or $^{60}\mathrm{CO}\,\gamma$ radiation (there is, unfortunately not yet a general agreement):

$$\mathrm{RBE} = \left[\frac{\mathrm{D(250\,kV\ X\ rays)}}{\mathrm{D(test\ radiation)}}\right]_{\text{isoeffect}}$$

RBE has only a unique value if the dose-effect curves are geometrically similar. This is generally not the case as, e.g., seen in Fig. 8.10: The survival curve changes from a shouldered to an exponential type. In this example, RBE becomes larger with small doses and may even approach infinity if the slope at the origin of the shouldered curve is zero. This is illustrated in Fig. 8.12.

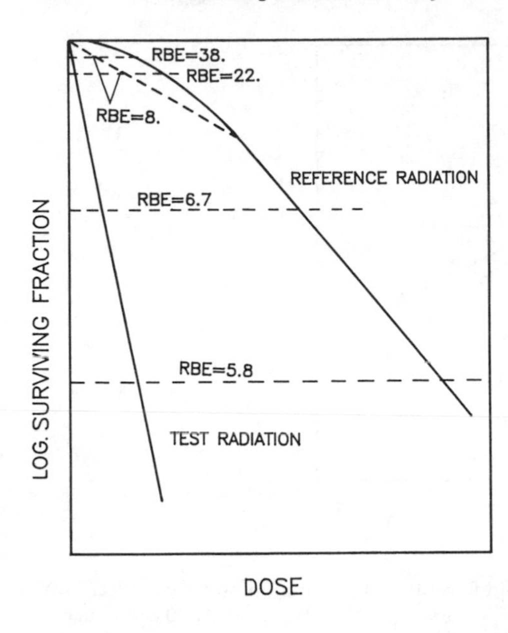

Figure 8.12. On the dose dependence of RBE

The shape of the dose response-curve in its initial part plays a very important role. The RBE will remain limited if the slope does not approach zero.

This problem is essential in radiation protection involving low doses. Experimental data are difficult to obtain and are not sufficiently available. It is, therefore, no surprise that many theoretical models address particularly this question (see Chap. 16).

The RBE concept was originally introduced for the needs of radiation protection. Because of the fundamental difficulties described and because of large variations found in different systems, it is no longer used in the original form and has been replaced by the recommendation of "quality factors" which are, of course, based on RBE measurements.

Since unique RBE values can in most cases not be given as discussed above, one may compare the ratio of slopes of exponential parts of the survival curves to gain a general overview. For a few cell types, this parameter is plotted versus LET_∞ in Fig. 8.13 (the values for human kidney cells were derived from the initial part of the survival curve). The overall behaviour is quite similar: Inactivation effectiveness increases with LET, passes a maximum around 200 keV/μm from where it declines. This last part has been related to the "overkill" situation where more energy is deposited per particle traversal than actually required for inactivation. More detailed measurements of inactivation cross sections (see below) cast substantial doubt on this interpretation so that this point must await further clarification. The initial rise is commonly interpreted to mean that essential lesions, e.g. double strand breaks, are formed with greater efficiency with higher ionization density which is supported by experiments with subcellular systems (see Sect. 7.3).

As already pointed out above, the dose is not a good parameter with

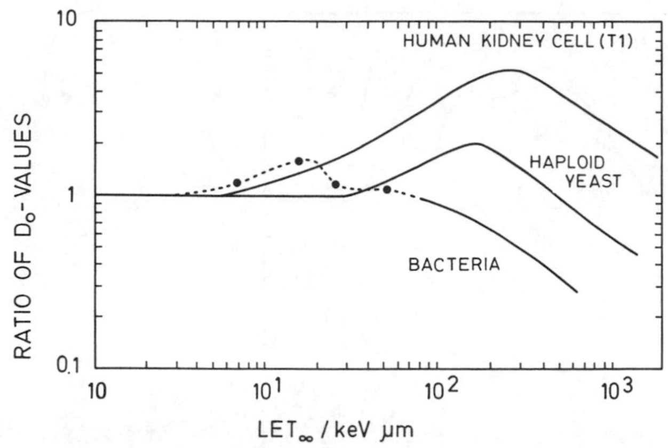

Figure 8.13. Relative radiation sensitivity of various cell types as a function of LET (after ICRU 1970, broken line after ALPER, MOORE and BEWLEY 1967)

densely ionizing particles. Particle fluence, i.e. number of particles per unit area, appears much more suitable. Both quantities are related according to Eq. (4.20). Plotting survival as a function of fluence, the slope in the exponential part of the curve may be termed "inactivation cross section" σ_i. It is interesting to study its dependence on LET and to compare it with the geometrical dimensions of the cell nucleus. Since the development of powerful ion accelerators, many systems and ions have been investigated. A summary of some results is presented in Fig. 8.14: One sees that there is always an increase of σ_i in the beginning which is expected even with constant RBE, as also indicated in the figure. Later on a transient plateau is approached which is, however, substantially smaller than the cross-sectional area of the cell nucleus, at least for yeast and mammalian cells. This suggests that even with very heavy ions, a cell is not killed by the traversal of a single particle. The reasons for this behaviour are not yet clear.

With very high LET values, the uniqueness of the relationship collapses. This is due to the fact that, in this case, cells may be killed by ions which do not directly hit the sensitive area due to the action of penumbra electrons (Sect. 4.2.4). This effect is ion-specific leading to the "hooks" of the curve.

The experiments described so far were performed using accelerated ions. Because of the comparatively narrow range of LET values, they are obviously the best experimental tools to study the fundamental relations between radiation effects and the microscopical pattern of energy deposition. Another approach is the application of very soft X rays, although the maximum LET obtainable is only about 30 keV/μm, as indicated in Fig. 3.7, if one remembers that the action is always due to electrons. An example of this kind of experiment is shown in Fig. 8.15, where results with characteristic radiation of aluminum are depicted (photon energy 1.5 keV, $LET_T = 20$ keV/μm). It is seen that α particles and aluminum X rays of comparable LET have nearly the same effect. This has consequences for the interpretation: The mean range

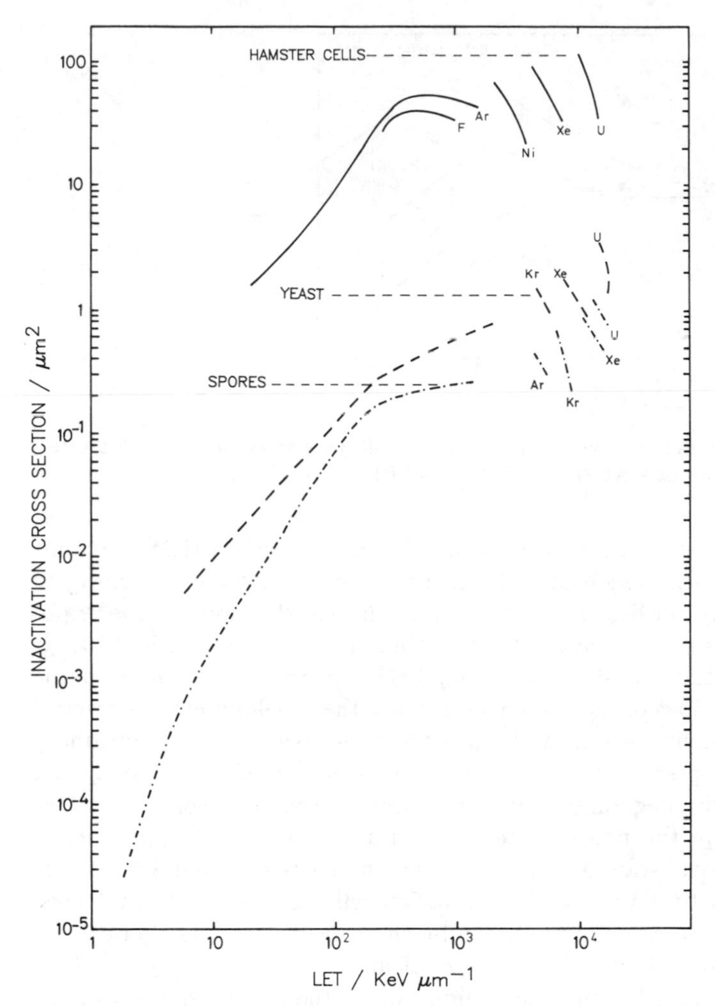

Figure 8.14. Inactivation cross sections of mammalian cells (V79 Chinese hamster cells), diploid yeast and B. subtilis spores as a function of LET with various ions (after KIEFER 1985)

of the photo-electrons is only 0.07 μm, while the δ electrons liberated by the interaction of α particles may travel up to 8 μm. If the increase in effectivity is caused by high local energy depositions, the interaction distances must be very short, indeed.

Most radiation types discussed so far are not suited for practical application because of their small penetration depths. If one is interested to use high LET radiations for the treatment of deep-seated tumours, one has to recur to indirect means. There are three approaches: accelerated heavy ions of very high energy, π^- mesons and neutrons. The main principles of their interaction have been discussed in Chaps. 3 and 4: Both, energetic ions and

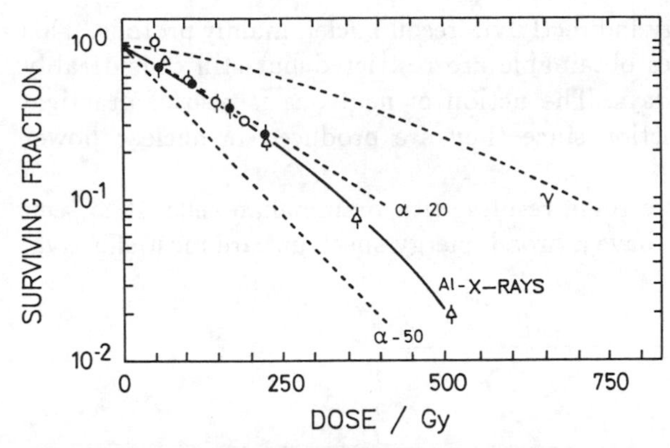

Figure 8.15. Inactivation of Chinese hamster cells by soft X rays (Al-characteristic X rays) compared to that by alpha particles (LET in keV/micron indicated) and gamma rays. The LET of soft X rays is about 20 keV/micron (after COX, THACKER and GOODHEAD 1977)

π mesons, possess comparatively small LET values along the largest part of their path through tissue, only at the end of their range the energy density becomes very high. One would, therefore, expect differences in response with penetration depth. This is, in fact, found experimentally as seen in Fig. 8.16: The effectiveness of a given dose is much higher in the "peak region" than at the entrance of the water phantom. Similar results are obtained with π^- mesons. This issue will be taken up again in Chap. 24.

Ions and mesons appear obviously promising for radiation therapy but the technical effort involved is quite high. A simpler and cheaper way to obtain penetrating high LET radiation lies in the use of neutrons. As detailed in

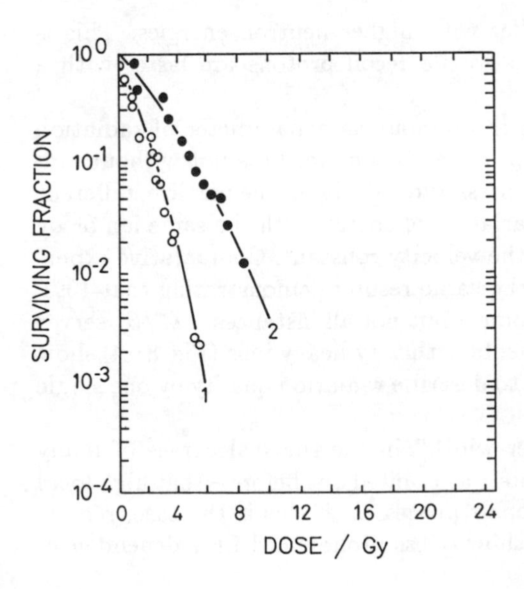

Figure 8.16. Survival of human cells exposed to 400 MeV/u carbon-ions at different depths in a water phantom. 1: at the peak of the depth dose curve, 2: at the entry of the phantom (after BLAKELY et al. 1979)

Chap. 3, they deposit energy indirectly via recoil nuclei, mainly protons. This means that the LET values obtainable are restricted but still considerably higher than with X or γ rays. The action of neutrons is also of practical interest in radiation protection since they are produced in nuclear power plants.

Figure 8.17 summarizes some results with mammalian cells. It is seen that fission neutrons which have a broad energy spectrum are most effective,

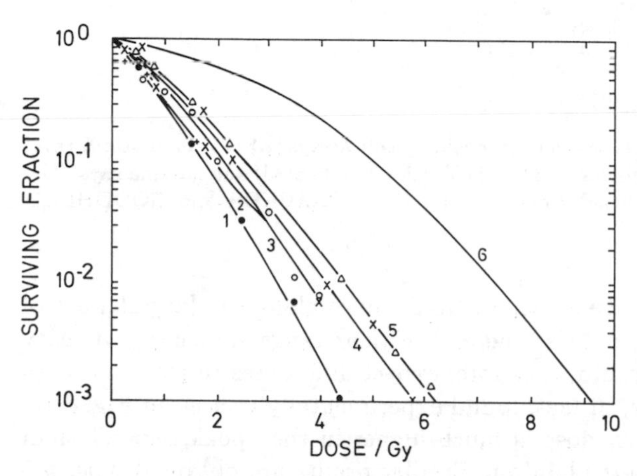

Figure 8.17. Inactivation of human cells by neutrons of different energies. 1: fission neutrons, 2: 3 MeV, 3: neutrons from 20 MeV alpha particles on a Be-target, 4: neutrons from 15 MeV deuterons on a Be-target, 5: 15 MeV, 6: X rays for comparison (after BARENDSEN and BROERSE 1967)

while the effectivity becomes smaller with higher neutron energies. This is due to the fact that, in this case, also the recoil protons are faster with a consequently lower LET.

Although LET has been used throughout as a parameter of radiation quality, one has to recall from Chap. 4 that this quantity is not without ambiguity. Since LET depends both on mass and velocity of the particle, different values may be obtained either by variation of energy with the same ion or by changing the particle type keeping the velocity constant. Comparative experiments of this kind do not lead to the same results demonstrating that LET is not really a good parameter. In some – but not all instances – Z^*/β^2 serves better, but particularly the experiments with very heavy ions (Fig. 8.14) show also that it is generally not possible to describe radiation quality by one single parameter.

The reduction in the "shoulder width" in the survival curves of many systems with increasing LET suggests – as pointed out before – that high local energy depositions interfere with repair processes. If this is the case, repair-deficient, sensitive mutants should show a less pronounced LET dependence.

This has, in fact, been found mainly in bacteria, which seems to support – but not to prove – the assumption.

The role of exposure conditions – particularly the influence of oxygen – has not been discussed here, it is treated in Sect. 9.2.

8.3.3 Interaction Between Different Types of Radiation

8.3.3.1 General Aspects

Interaction experiments, i.e. where two agents are applied concomitantly or successively, may sometimes help to gain a better insight into basic mechanisms. Before going into detail, it is useful to clarify the underlying concepts and to define the relevant terms. In interaction experiments there are several possible outcomes:

a) *Additivity*: One agent is able to replace the other if the dose scales are appropriately adjusted.
b) *Synergism*: One agent sensitizes the system to the other agent.
c) *Antagonism*: One agent reduces the sensitivity to the other agent.
d) *Independence*: Both agents act independently of each other.

These terms may be more stringently defined in mathematical terms: If a and b are doses of the two agents yielding the same effect if given separately, the effect $x(a+b)$ of the combined action may be:

a) additive: $x(a+b) = x(b+a)$
b) synergistic: $x(a+b) > x(2b)$
c) antagonistic: $x(a+b) < x(2b)$
d) independent: $x(a+b) = x(a) + x(b)$

The different types of interaction may be studied by fractionation experiments where a first dose of the first agent is followed by a series of doses of the second one. The four cases introduced above are exemplified for survival curves in Fig. 8.18: With additive action (A), the second survival curve is just a continuation of the first one, synergism (B) leads to an increase, antagonism (C) to a decrease in slope and sometimes – but not necessarily – to a larger shoulder. With independent action (D), the pretreated cells display the same behaviour as unexposed controls. With purely exponential survival curves, a distinction between additive and independent action cannot be made.

8.3.3.2 Interaction Between UV and Ionizing Radiation

The basic mechanisms of cellular UV action, e.g. type of primary lesions, repair processes, etc., are far better understood than for ionizing radiation. Interaction experiments are a way to explore whether and how far common pathways exist for damage expression and cellular recovery. They may also help to answer the question whether the same initial target is involved

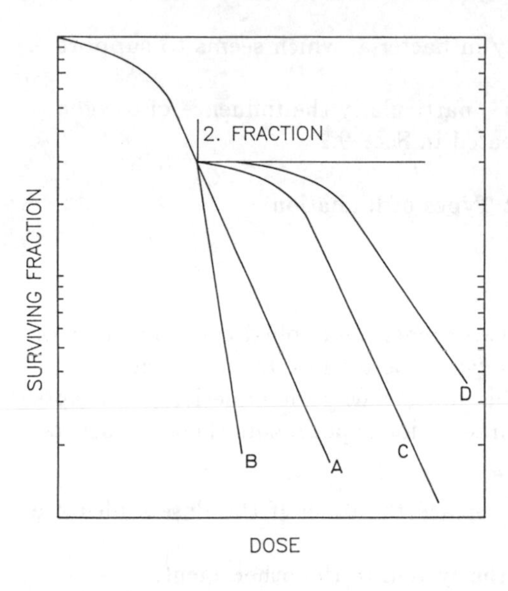

Figure 8.18. Different kinds of interaction between two agents. A: additive, B: synergistic, C: independent, D: antagonistic

which can, for UV, undoubtedly be identified with DNA; results are shown in Figs. 8.19 and 8.20: A UV pre-exposure can obviously abolish the shoulder in the X ray survival curve completely without change in the final slope. This is a clear case of *additivity*.

The situation is slightly more complicated if a first X ray dose is followed by UV exposures: The UV shoulder is reduced but not completely lost, even with quite high doses. It follows that UV can fully replace X rays but X rays only partially UV. This suggests the existence of UV-specific pathways for damage repair and expression.

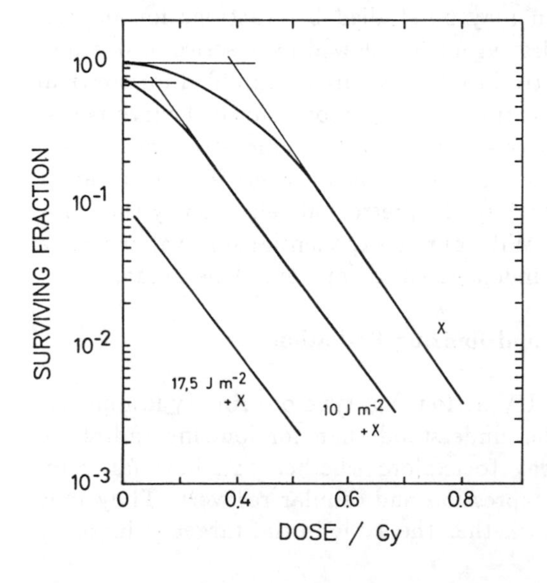

Figure 9.19. Interaction between UV and X rays in Chinese hamster cells. A single UV fluence as indicated was followed by a series of X ray doses (after HAN and ELKIND 1978)

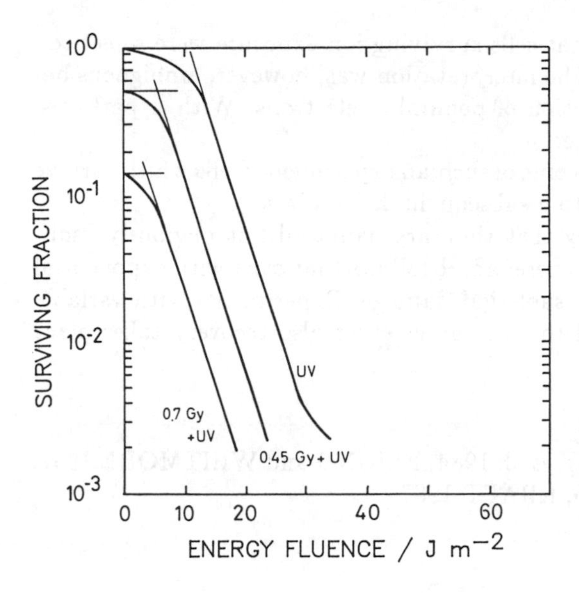

Figure 8.20. As in Fig. 8.19 but in reversed order (after HAN and ELKIND 1978)

8.3.3.3 Interaction Between Low- and High LET Radiation

Ionizing radiations of sufficiently high LET yield exponential survival curves in mammalian cells. Even if there is now agreement that this does not necessarily prove a true single-hit inactivation, it has often been taken as evidence that there is no "sublethal" damage and absence of recovery. The usual experimental protocol – namely split-dose exposure (see Chap. 13) – cannot be used here since it relies on the existence of a shoulder region. Combination experiments with high (exponential curves) and low LET radiations (shouldered curves) may be helpful. In early experiments, very energetic heavy ions

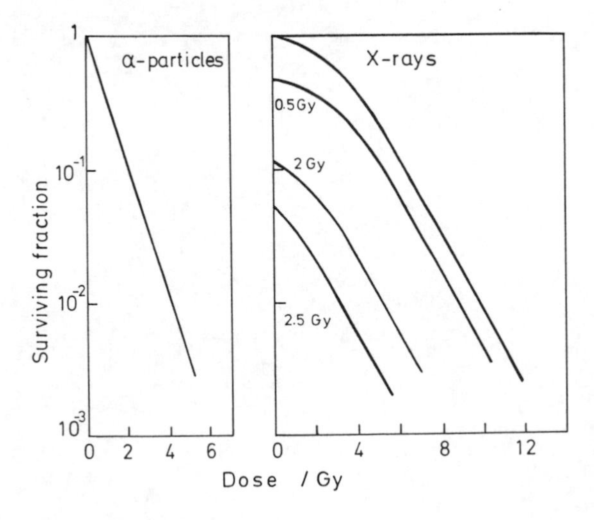

Figure 8.21. Interaction between alpha particles and X rays. Left panel: alpha survival curve, right panel: X ray survival after different predoses of alpha particles (after MCNALLY et al. 1988)

were used and it was found that cells surviving ion exposure were sensitized to subsequent X irradiation. The interpretation was, however, ambiguous because of the possible contribution of penumbra electrons. With α particles, they play only a negligible role.

Figure 8.21 shows the outcome of such an experiment. Cells which survive an α dose are also sensitized to a subsequent X irradiation.

This demonstrates clearly that they are damaged but obviously "sublethally" since they survived. Therefore, it follows that even with exponential survival curves, there may be sublethal damage. Experiments with variable intervals which could be used to find out whether also recovery takes place have not yet been reported.

Further reading

ALPER 1975, 1979, BLAKELY et al. 1984, ELKIND and WHITMORE 1967, ICRU 30 1979, KIEFER 1986, KRAFT 1987

Chapter 9
Radiosensitization and Protection

Recent developments in photosensitization are discussed along with substances having a certain relevance in medical treatment. The mechanisms of protection and sensitization with ionizing radiation are dealt with in greater depth, particularly the "oxygen effect"; hypotheses for its interpretation are discussed. Radiosensitization by chemical agents, especially those with a possible clinical potential, will be discussed by end of the chapter.

9.1 Photosensitization

This is a process in which the presence of a light-absorbing compound increases the cell's sensitivity to the action of optical radiation. The number of known substances is very large, and the mechanisms differ greatly. It is, therefore, only possible to give a few representative and more important examples.

The fundamentals of sensitizer action have already been outlined in Sect. 5.1.2. It is clear from this that the absorption of radiation is a necessary prerequisite but this alone is not sufficient. The excited state thus obtained must either be transferred directly to cellular structures or via reactive intermediate species. An important example of the latter is the formation of singlet oxygen. In this instance, the sensitization is called "photodynamic action". It requires, for obvious reasons, the presence of oxygen and it is modifiable by substances which react with singlet oxygen, e.g. azide or carotenes. Another indirect approach is to expose the cells – if possible – in deuterium oxide (D_2O, "heavy water") in which the life-time of singlet oxygen is considerably prolonged (so that more damage should be registered if singlet oxygen is really involved).

Very many dyes act as photodynamic sensitizers, e.g. riboflavin, acridine orange, acriflavin, bengal red, thiophyramine, hematoporphyrine. Most of them are also fluorescent but this must not necessarily be so. The cellular targets – if known – differ although membrane damage is more frequent than DNA interactions. This may simply reflect the fact that the membrane is easier accessible but also that the compounds are to some extent concentrated in the lipid phase.

Hematoporphyrin has recently gained some practical relevance in the phototherapy of tumours (see also Sect. 24.1). The original interest came from the observation that derivatives of this compound are selectively concentrated in tumour tissue. The preparation method consists essentially of the hydrolysis of hematoporphyrin in a solution of acetic and sulfuric acid. The active compound, commonly abbreviated Hpd, has not yet been unambiguously identified. A dimeric structure as in Fig. 9.1 has been suggested. The

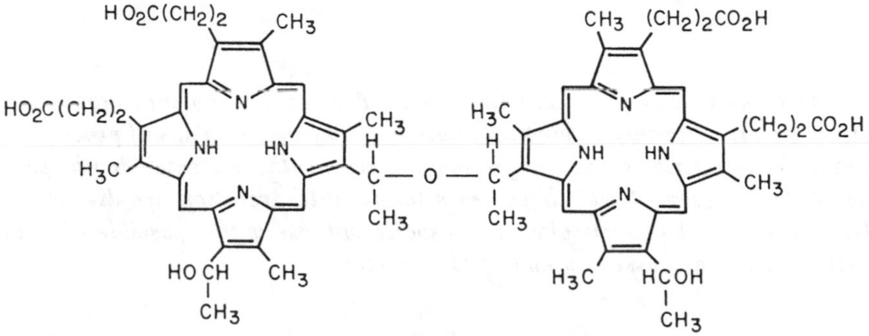

Figure 9.1. Proposed structure of hematoporphyrin derivative (HpD) (after DOUGHERTY, POTTER and WEISHAUPT 1984)

absorption spectrum shows a strong band around 400 nm and a weaker one around 600 nm. The latter region is mostly used for excitation. The action proceeds in cells and tissues via the formation of singlet oxygen. The main target appears to be the cell membrane, although other structures cannot be excluded, particularly after prolonged exposure to the drug.

Another class of sensitizers of practical relevance are the *psoralens*. They interact with DNA, as already discussed in Sect. 6.1.2. When incorporated in cells, they cause cell inactivation and mutations upon exposure to near UV.

The "triplet sensitizers" acetone, acetophenone and benzophenone – although useful in chemical systems (Sect. 6.1.2) – are obviously of limited value in higher cells. They do work well in bacteria but their action in mammalian cells so far could not be unequivocally demonstrated.

Bromouracil – if incorporated in DNA – sensitizes efficiently both lower and higher cells. Its mode of action is primarily via energy transfer to DNA but additionally it interferes with subsequent repair processes. In this respect it represents also a further class of sensitizers, namely repair inhibitors. It is possible that photochemically formed substances inhibit the repair of primary alterations.

Further reading

GALLO and SANTAMARIA 1972, JORI 1987, KESSEL 1983, 1986, POTTIE and TRUSCOTT 1986

9.2 Sensitization and Protection with Ionizing Radiation

9.2.1 Radioprotective Substances

It has long been known that the radiosensitivity of a biological system depends on the medium to which it is exposed. This led to an extensive search for "radioprotectors" in the early 1950s, particularly of such with a non-toxic nature which were to be applicable also in humans, with the hope that they may – among other things – prove useful in nuclear warfare. The results were generally disappointing so that these investigations are no longer carried out with the same intensity. Despite of this, the studies undertaken were quite valuable for the understanding of basic mechanisms.

One possibility for radioprotection is obvious: that involving the "indirect" effect (Sect. 5.2.3) which also contributes to cell inactivation. Thus, substances which react with damaging radicals yielding less dangerous products should reduce the overall effect. In fact, experiments with radioprotectors provide evidence that the indirect effect does play a role, they also give a clue to which radicals are important.

The OH radical and eaq obviously are primary candidates because of their high yield. A number of substances which readily react with the hydroxyl radical proved also to be efficient radioprotectors. This indicates that this radical species is involved in cell inactivation. Representatives of this group are dimethylsulfoxide (DMSO) and many alcohols of which particularly glycerol has to be mentioned because of its low toxicity. Figure 9.2 shows that

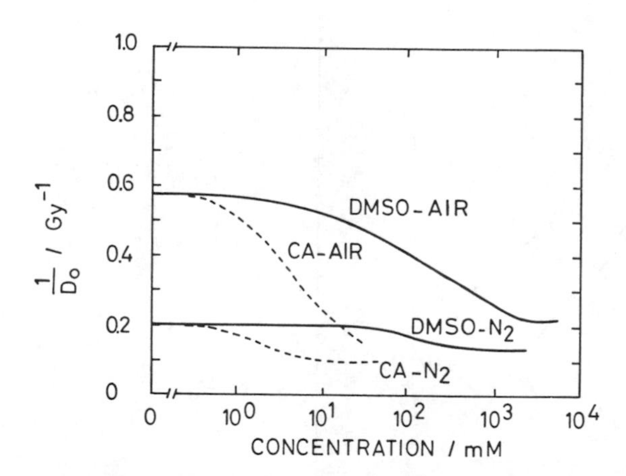

Figure 9.2. Radioprotection by different substances in Chinese hamster cells. The ordinate gives the slope of the exponential part of the survival curve. The different concentration dependencies should be noted (DMSO: dimethylsulfoxide, CA: cysteamine) (after CHAPMAN et al. 1973)

DMSO protects against X ray effects both in the presence and the absence of oxygen but to a considerably lower degree under anoxic conditions. This behaviour is found also with other OH scavengers.

The relative importance of e_{aq} could in principle be tested by using N_2O (Sect. 5.2.2) since it transforms hydrated electrons quantitatively into OH radicals. In subcellular *in vitro* systems it acts as radiosensitizer indicating that e_{aq} is less important than OH. More recently, the same effect has also been shown in mammalian cells.

Another class of protectors are the so-called "SH compounds", i.e. chemicals which have a free SH group, like e.g. cysteamine. They are also OH scavengers and their protecting capacity is, therefore, not surprising. A closer inspection shows, however, (see Fig. 9.2) that this cannot be the sole explanation: They obviously offer protection in addition to that by DMSO and at much lower concentrations. This cannot be due to their higher reaction rates since they differ only by about a factor of 2 (K_{OH}(DMSO) = $6 \times 10^9 \, mol^{-1} s^{-1}$, K_{OH}(cysteamine) = $10^{10} \, mol^{-1} s^{-1}$). It is assumed that they are able to chemically "heal" the primary lesion by a donation of an H atom to a site where it was initially abstracted.

Most cells naturally contain radioprotectors, the most important of which is glutathion (GSH), a tripeptide consisting of glutamic acid, cysteine and glycine. Cells which are genetically deficient in GSH (obtained from 5-oxoprolinuria patients) are markedly more sensitive, as shown in Fig. 9.3.

On the basis of the described experiments, the following scheme of chemical radioprotection against ionizing radiation is suggested (Fig. 9.4):

1. Inactivation is caused both by the direct and the indirect effect, the latter mainly due to OH radicals.
2. OH-scavengers protect by reducing the indirect effect.

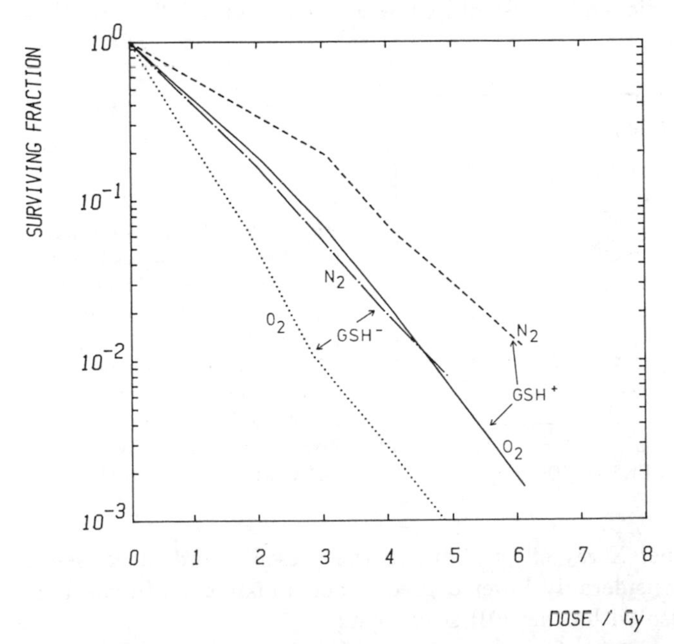

Figure 9.3. Survival of glutathion synthetase proficient (GSH+) and deficient (GSH–) cells (after data from MIDANDER et al. 1986)

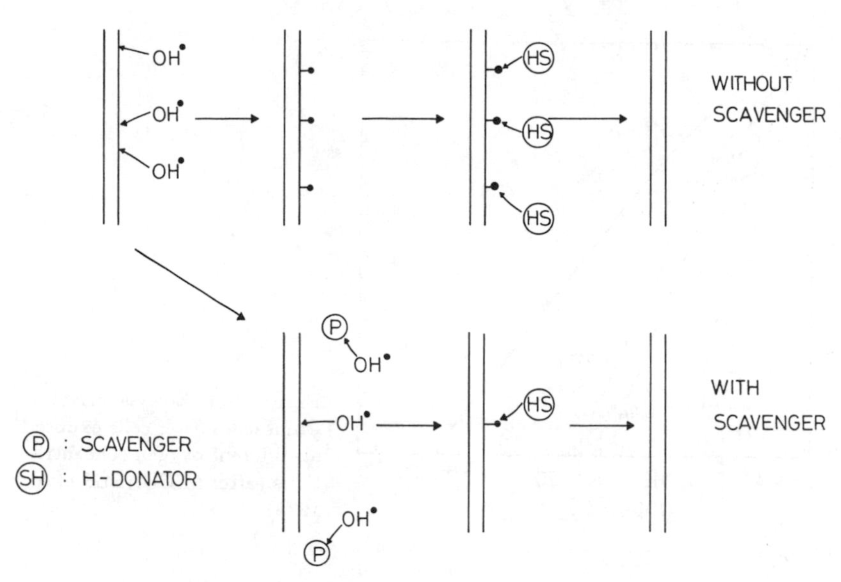

Figure 9.4. Two modes of chemical radioprotection, for explanation see text

3. Reducing, H-donating agents – like SH compounds – react with the primary lesions leading to a chemical restitution of the damaged site.

Chemical radioprotection is hence a two-stage process. The role of oxygen – which does obviously interfere – is discussed in the following section.

9.2.2 The Oxygen Effect

If cells are exposed to X rays in the absence of oxygen, the effect is markedly reduced (Fig. 9.5), in other words, oxygen is a radiosensitizer. In fact, it is the most potent one known. By comparing the survival curves in the figure it can be seen that the extrapolation number and hence the general shape is not changed. This means that all curves can be superimposed on each other just by adjusting the abscissa scale. Agents which act in this way are called "dose-modifying". In the case of oxygen, the dose modification factor is called "oxygen enhancement ratio" (OER). It is defined as:

$$\mathrm{OER} = \left[\frac{\text{dose in the absence of } \mathrm{O_2}}{\text{dose in the presence of } \mathrm{O_2}}\right]_{\text{equal effect}} \quad . \tag{9.1}$$

The OER does not depend on the effect level, at least with higher doses; the situation is less clear for the initial section of the survival curve.

The degree of sensitization is a function of oxygen concentration as also seen from Fig. 9.5. The maximum is already nearly reached with air-saturated suspensions. The OER values (maximum) are found to be 2–3 for sparsely ionizing radiations.

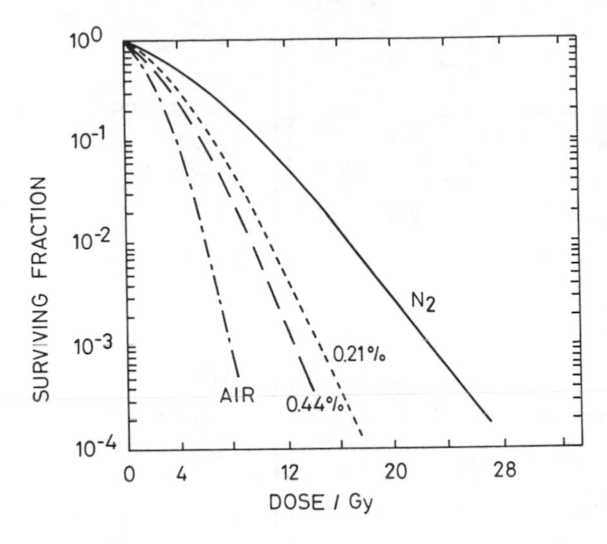

Figure 9.5. X-ray survival of Chinese hamster cells exposed at different oxygen concentrations (after MICHAELS et al. 1978)

The influence of oxygen on radiosensitivity has nothing to do with its physiological role in cellular metabolism as one might expect. If one compares, e.g. respiratory deficient and competent cells which is possible with microorganisms, no difference is seen. Sensitization can only be achieved if oxygen is present during or given very shortly after irradiation. This speaks also against a physiological pathway and points strongly to a radiochemical mechanism.

Further clarification may be gained from the investigation of the time dependence. Much progress has been made in this field by the use of three techniques (Fig. 9.6)

1. *Rapid mix*: An anoxic cell suspension is rapidly mixed with an oxygen-saturated solution either shortly before or after the exposure to a radiation pulse. Time resolutions below 1 millisecond are thus achievable.
2. *Double pulse method*: If cells are exposed on membrane filters to an intense radiation pulse, oxygen is radiochemically depleted but is gradually replaced by diffusion from the outside atmosphere. A second "test" pulse is given at variable intervals after the first one. If the diffusion times are known, the time kinetics of the oxygen effect can be studied. This method is not applicable to fully oxygenated mammalian cells since the depleting doses are so high that the whole population is already killed. It may be used, however, with reduced O_2 concentrations.
3. *Gas explosion technique*: Cells are kept in nitrogen on membrane filters in a thin monolayer. Close to them a vessel with high pressure oxygen may be rapidly opened so that an "oxygen shot" can be delivered shortly before or after a radiation pulse. The time kinetics of the oxygen effect may be studied by varying the interval. The resolution also lies in the submillisecond range.

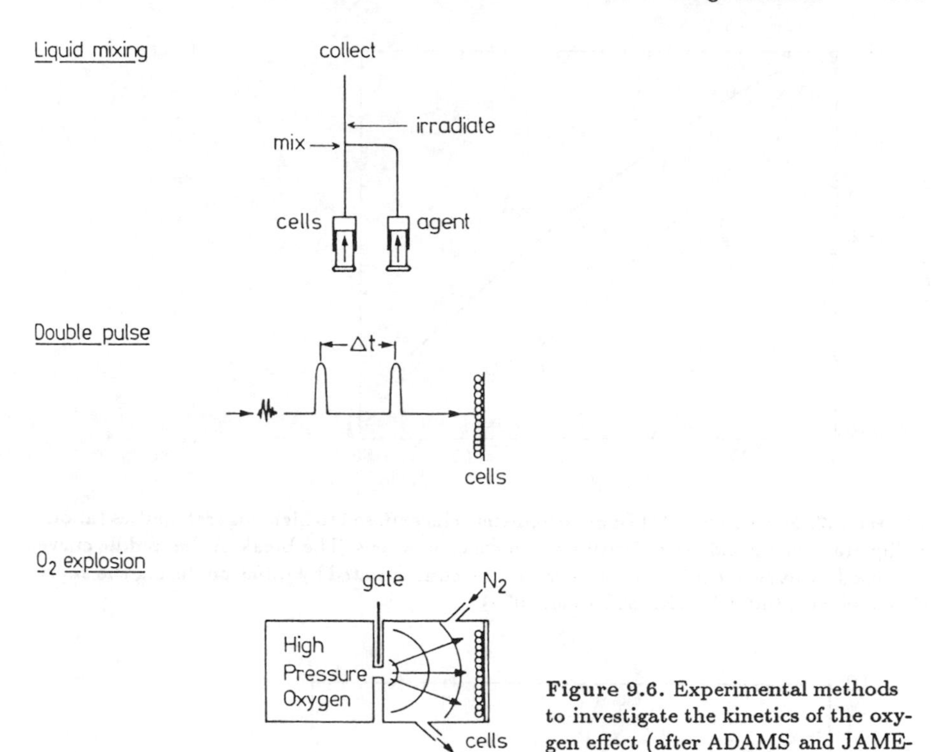

Figure 9.6. Experimental methods to investigate the kinetics of the oxygen effect (after ADAMS and JAMESON 1980)

Radiation-induced oxygen depletion, which is important for the second method but also an interesting phenomenon in its own right, is demonstrated in Fig. 9.7: Cells equilibrated with different concentrations of oxygen are irradiated with intense pulses of short duration compared to oxygen diffusion time. With low doses the expected "normal" survival behaviour is seen, with higher doses the curves bend and approach that for exposure in anoxia. The position of the breakpoint depends on the initial oxygen concentrations. In this way it is possible to "titrate" oxygen radiobiologically and to determine diffusion kinetics. Diffusion sets a natural limit to the time resolution of all three described methods.

Figure 9.8 gives a summary of the time course of the oxygen effect in some cellular systems obtained with the gas explosion technique. Although there are considerable differences – presumably mainly due to different diffusion times – it is obvious that the underlying processes are very fast and essentially completed after a few milliseconds. This, therefore, leads to assume that the basic mechanism is radiochemical.

The dependence of the OER on the oxygen concentration with conventional dose rates is depicted in Fig. 9.9 for some E. coli mutants with different repair capacities. It is seen that maximum sensitization can already

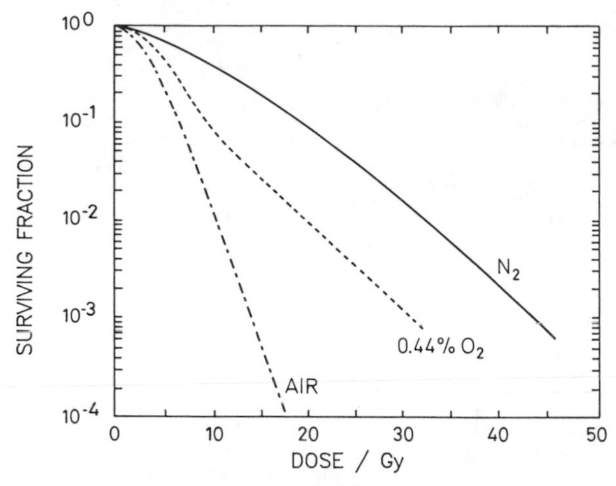

Figure 9.7. X-ray survival of Chinese hamster cells exposed to high dose rate pulses (about 1 Gigagray per second) at different oxygen concentrations. The break in the middle curve is caused by oxygen depletion which cannot be compensated by diffusion during the short time available (after MICHAELS et al. 1978)

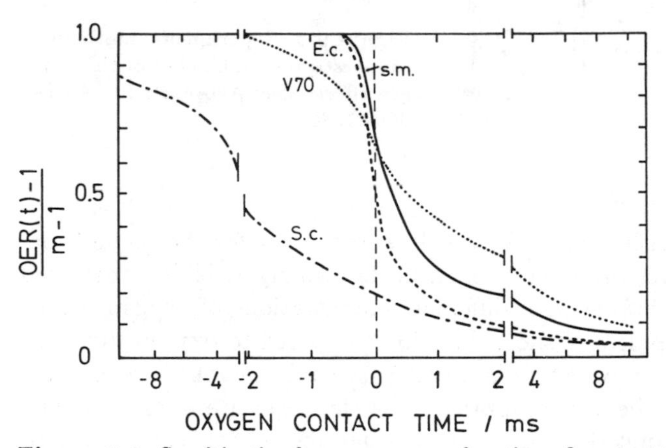

Figure 9.8. Sensitization by oxygen as a function of contact time. The radiation pulse was given at t = 0, negative abscissa values indicate pre-irradiation contact. To facilitate comparison a normalized OER is used for the ordinate. Abbreviations: S.c.: yeast Saccharomyces ceraevisiae, V79: Chinese hamster cells, E.c.: bacterium E. coli, S.m.: bacterium Serratia marcescens (after MICHAEL, HARROP and MAUGHAN 1979)

be achieved with a few percent oxygen. The curve may be described by the so-called ALPER-HOWARD-FLANDERS expression:

$$\mathrm{OER}[\mathrm{O_2}] = \frac{\mathrm{m}[\mathrm{O_2}] + \mathrm{K}}{[\mathrm{O_2}] + \mathrm{K}} \tag{9.2}$$

where $[O_2]$ is the oxygen concentration, m the maximum OER, and K a pa-

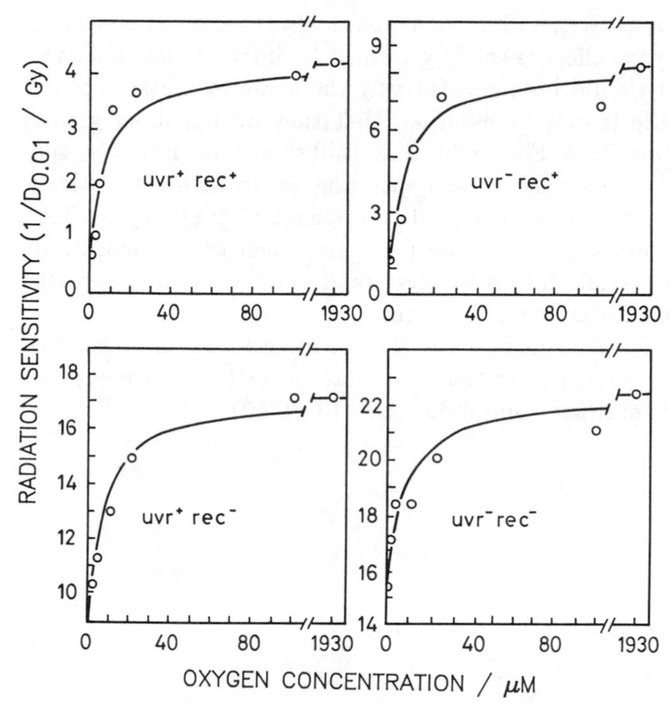

Figure 9.9. Radiation sensitivity as a function of oxygen concentration during exposure with E. coli strains of differing repair capacities. On the ordinate the reciprocal of the 1% survival dose is given (after JOHANSEN et al. 1974)

rameter having the dimensions of a concentration; it is the O_2 percentage where just 50% of the maximum OER is achieved, namely $(m + 1)/2$ (one should note that $m = 1$ in anoxic conditions). Some m and K values as obtained with X rays in a number of organisms are listed in Table 9.1. It is remarkable that the variations are rather small despite the very large dif-

Table 9.1. Parameters of the oxygen effect for various systems. (After KIEFER, 1975)

name	species	m	k 10^{-6} mol dm^{-3}
Shigella	bacterium	2.9	4
E. coli B/r	"	3.1	4.7
E. coli uvr⁺ rec⁺ [a]	"	6.0	8
E. coli uvr⁻ rec⁺ [a]	"	6.7	8
E. coli uvr⁺ rec⁻ [a]	"	1.85	8
E. coli uvr⁻ rec⁻ [a]	"	1.5	8
Saccharomyces (haploid)	yeast	2.4	5.8
Oedogonium	alga	1.9	17.6
Chin. hamster cells	mammal	3.14	8

[a] see also Figure 9.7

ferences in radiation sensitivities. This fact points also to a common general mechanism. If the oxygen effect is solely governed by intracellular radiation-chemical processes, it should be quantitatively the same in mutants of different sensitivities since it may be assumed that they do not differ greatly in their chemical composition. Figure 9.9 and Table 9.1 show, however, that there are considerable variations in m, depending on radiation sensitivity. This means that there is also a biological component of the oxygen effect. It is important to note that the K value for a given cell type is essentially constant and that its variation is generally small, with one exception (the alga *Oedogonium*) where it is unusually high.

To describe the situation more formally, a simple model may be used which is certainly an oversimplification but may nevertheless serve as an illustration of possible pathways and influences (Fig. 9.10):

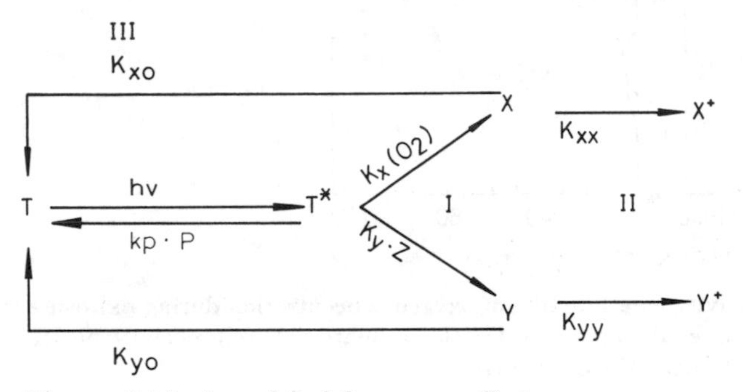

Figure 9.10. A model of the oxygen effect

It is assumed that a target molecule T (which may, but must not, be DNA) is altered by radiation, either directly or indirectly, to a metastable transient damaged form T* (which may be a radical). T* may either react with oxygen (rate constant K_x) or a hypothetical other agent Z (rate constant K_y) to yield intermediary lesions X or Y, respectively, which may be modified by intracellular repair. Chemical reconstitution as described in the preceding section by reaction with intracellular or added radioprotectors P (rate constant K_p) is indicated by the back reaction T* ?? T. X and Y may be permanently fixed (rate constant K_{xx} and K_{yy}, respectively) or biologically repaired (rate constants K_{xo} and K_{yo}), the latter process leading to a biologically intact target molecule T. The permanent lesions are called X^+ and Y^+, their sum determines the survival probability.

If linear kinetics are assumed throughout the following system, differential equations are obtained:

$$\frac{dT^*}{dt} = -(K_x[O_2] + K_p[P] + K_y[Z]) \cdot T^* \tag{9.3a}$$

$$\frac{dX}{dt} = K_x[O_2]T^* - (K_{xx} + K_{xo})X \tag{9.3b}$$

and

$$\frac{dX^+}{dt} = K_{xx}X \qquad \frac{dY^+}{dt} = K_{yy}Y \quad . \tag{9.3c}$$

In the case of a short term exposure, only the final concentration is of interest:

$$X^+ = \frac{K_{xx}}{K_{xx}+K_{x0}} \cdot \frac{K_x[O_2]}{K_x[O_2]+K_p[P]+K_z[Z]} \cdot T_0^* \qquad Y^+ = \frac{K_{yy}}{K_{yy}+K_{y0}} \cdot \frac{K_z[Z]}{K_x[0_2]+K_p[P]+K_z[Z]} \cdot T_0^* \tag{9.4}$$

where T_0^* is the initial concentration of T^*.

In the absence of oxygen, the concentration Y_0^+ of final lesions is:

$$Y_0^+ = \frac{K_{yy}}{K_{yy}+K_{y0}} \cdot \frac{K_y[Z]}{K_p[P]+K_y[Z]} T_0^* \quad . \tag{9.5}$$

The oxygen enhancement ratio is then:

$$\mathrm{OER} = \frac{X^+ + Y^+}{Y_0^+} \quad . \tag{9.6}$$

Introducing the abbreviations:

$$F_x = \frac{K_{xx}}{K_{xx}+K_{x0}} \quad \text{and} \quad F_y = \frac{K_{yy}}{K_{yy}+K_{y0}}$$

after rearrangements, the final result is:

$$\mathrm{OER} = \frac{\frac{F_x}{F_y}\left(1+\frac{K_p[P]}{K_y[Z]}\right)[O_2] + \frac{K_p[P]+K_y[Z]}{K_x}}{[O_2] + \frac{K_p[P]+K_y[Z]}{K_x}} \quad . \tag{9.7}$$

This relationship has the general form of the ALPER-HOWARD-FLANDERS formula (Eq. (9.2)) but the parameters bear now a special meaning. K depends obviously only on radiation-chemical conditions, namely the concentrations of intracellular protectors (P), that of the hypothetical fixing agent Z and the respective rate constants. This is in line with the experimentally found constancy of K (see Table 9.1). Maximum OER, m, on the other hand, is both a function of radiobiological repair (indicated by F_x and F_y) <u>and</u> the concentration of protectors. This agrees with experimental findings that cells with low GSH content show a reduced OER (see below) and repair-deficient mutants are less sensitized by oxygen than their wild-type counterparts (Fig. 9.9).

If the given picture is correct, the K value should be smaller in cells with reduced concentrations of internal protectors. This has, in fact, been found, as shown in Fig. 9.10.

K is also related to the lifetime of the transient species T*. In the absence of oxygen:

$$\frac{dT^*}{dt} = -(K_y[z] + K_p[p])T^* \tag{9.8}$$

and hence for the half-life $\tau_{1/2}$:

$$\tau_{1/2} = \frac{\ln 2}{K_y[z] + K_p[p]} \tag{9.9}$$

implying that $\tau_{1/2}$ increases if [p] is lowered.

An experimental demonstration of this dependence is given in Fig. 9.11.

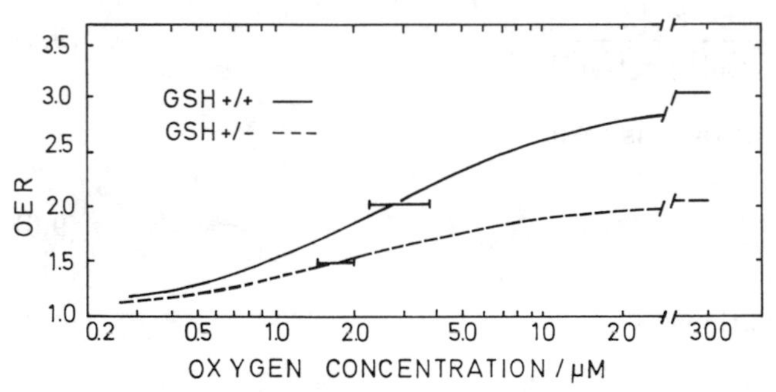

Figure 9.11. Radiation sensitivity as a function of oxygen concentration in glutathion-proficient and glutathion-deficient mammalian cells

The K value of Eq. (9.2) is also related to $\tau_{1/2}$:

$$K = \frac{\ln 2}{\tau_{1/2} \cdot K_x} \quad . \tag{9.10}$$

If the rate constants K_x were known, $\tau_{1/2}$ could be calculated. Assuming $K_x = 10^{-9}\,\mathrm{mol\,s^{-1}}$ (diffusion controlled reactions, see also Appendix I.7) and $K = 10\,\mu\mathrm{m}$ one arrives at a figure of $\tau_{1/2} = 70\,\mu\mathrm{s}$ which is compatible with the experimental measurements.

It must be reiterated that the given model is certainly only a vague approximation to reality. There are a number of indications that the oxygen effect has more than one component. If this is so, the simple scheme has to be modified but it still may serve as a useful illustration of the interplay of chemical and biological factors.

The non-linear relationship of Eq. (9.2) makes the determination of the parameters difficult. It may be facilitated by linearizing transformations which are discussed by POWERS (1983).

The considerations so far are only applicable to sparsely ionizing radiations like X or γ rays. The oxygen effect shows, however, a strong dependence on radiation quality as depicted in Fig. 9.12: Its maximum decreases with LET until it approaches unity around $\mathrm{LET}_\infty = 200\,\mathrm{keV}/\mu\mathrm{m}$. As pointed out before, LET is not an unambiguous parameter as it is a function of both particle effective charge and velocity. This appears to be of particular influence with regard to the oxygen effect, as also seen from Fig. 9.12 where variation of

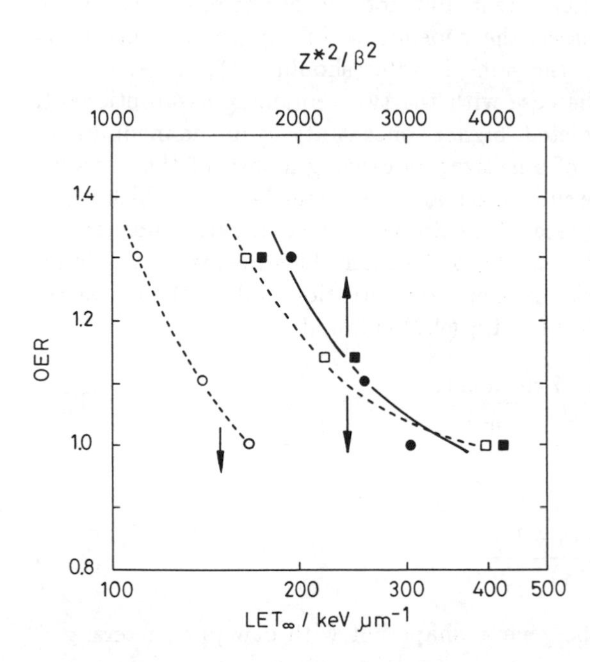

Figure 9.12. The dependence of the OER on radiation quality. Variations in ionization density was either achieved by using different ions of comparable velocity (□, ■) or by using the same ions with differring speeds (∘, •). There are large discrepancies if LET is used as parameter which are substantially smaller with Z^{*2}/β^2 (after CURTIS 1970)

either of these properties leads to two distinct curves. A better agreement is reached if Z^*/β^2 is used instead of LET_∞. Since this parameter characterizes essentially the electron density in the track, it must be concluded that it plays a special role here.

The reduction of the OER with increasing ionization density is not only of great practical relevance in radiotherapy but presents also an interesting theoretical problem which is, however, not yet solved.

There are a number of hypotheses not all mutually exclusive:

1. Lesions caused by high LET radiations are less amenable to repair which – according to the model given – should lead to a reduction in OER.
2. The primary lesion is immediately chemically fixed by intra-track reactions so that there is no chance for oxygen to intervene ("interacting radical hypothesis").

3. The high local energy deposition densities cause a depletion of oxygen so that "oxic" conditions can never be maintained.
4. Oxygen-like sensitizing substances are formed in the track so that real "anoxic" conditions cannot be obtained ("oxygen in the track hypothesis").

Hypothesis no. 1 implies that the increase in LET would mainly affect the "anoxic" survival curve since repair plays a more important role here. Such a behaviour is, in fact, often found.

Hypothesis no. 2 leads to similar conclusions since also here anoxic damage will be changed to a greater extent, but not because of biological but of chemical reactions. In both cases, the dependence of oxygen concentrations should be left unchanged, i.e. the same K value should be found as with X rays. This is, however, not the case with the two remaining assumptions: If oxygen is radiochemically depleted, higher concentrations in the medium are required for the same degree of sensitization causing a shift of the K value. This is also true if oxygen-like substances are formed in the track which is not immediately obvious. In this case, there are never true anoxic conditions as there is always an inherent degree of sensitization. This means that m is reduced. If "a" is the intra-track "oxygen" concentration and m′ the measured maximum OER, a modified form of Eq. (9.2) is found:

$$\mathrm{OER} = \frac{\mathrm{m}([\mathrm{O_2}]+\mathrm{a})+\mathrm{K}}{([\mathrm{O_2}]+\mathrm{a})+\mathrm{K}} \Big/ \frac{\mathrm{m}\cdot\mathrm{a}+\mathrm{K}}{\mathrm{a}+\mathrm{K}} \tag{9.11}$$

and after rearrangement:

$$\mathrm{OER} = \frac{\mathrm{m}[\mathrm{O_2}]\cdot\frac{\mathrm{a}+\mathrm{K}}{\mathrm{ma}+\mathrm{K}}+(\mathrm{a}+\mathrm{K})}{[\mathrm{O_2}]+(\mathrm{a}+\mathrm{K})} \quad . \tag{9.12}$$

This relationship has again the general shape but with new parameters:

$$\mathrm{m}' = \mathrm{m}\cdot\frac{\mathrm{a}+\mathrm{K}}{\mathrm{ma}+\mathrm{K}} \quad \text{and} \quad \mathrm{K}' = \mathrm{a}+\mathrm{K} \quad .$$

It is seen that m is reduced but K increased.

There are only very few measurements of K with high LET radiations, mainly with neutrons, the results being conflicting. A definite conclusion is, therefore, not yet possible. If the amount of O_2 formed in the track is measured, the concentrations found are too low to account fully for the increase in sensitivity. It may be – as indicated – that other products substitute for it. Nevertheless, this cannot be the whole story. If hypothesis no. 4 were the only explanation, there should be no LET-dependent increase of sensitivity with high oxygen concentrations, which is evidently not the case. Also, the curve shape is altered (loss of the shoulder) which cannot be due to oxygen usually being dose-modifying. This points to the involvement of repair processes. Oxygen depletion, on the other hand, can also not fully account for the LET effect since, in this case, the "anoxic" survival curves should remain unchanged, which is also at variance with experiments.

In summary it may be stated that also with the LET dependence of the oxygen-effect, an interplay of chemical (e.g. "oxygen in the track") and biological factors (repair) prevails.

9.2.3 Radiation Sensitizers

As discussed in the previous section, oxygen is a very potent sensitizer or, in other words, its absence leads to an increase in radiation resistance. Since this may be an important factor in radiocurability of certain tumours (Chap. 24), attempts were made to find substances which could replace it and which are not metabolized. Sensitization, however, cannot only be achieved by these "oxygen mimetics" (see below) but also by the inactivation or depletion of natural intracellular protectors, e.g. non-protein sulfhydrils like glutathione (GSH). Examples for this approach are *N-acetyl-maleimide* (NEM) or ***diamide*** the effects of which are illustrated in Fig. 9.13. If the model of the oxygen effect is correct, they should be active in anoxic as well as in oxic conditions as in fact found.

The development of "oxygen-mimetic" radiosensitizers has to start from the particular properties of oxygen. Since it possesses two unpaired electrons in its ground state, it behaves like a stable free radical and reacts easily with

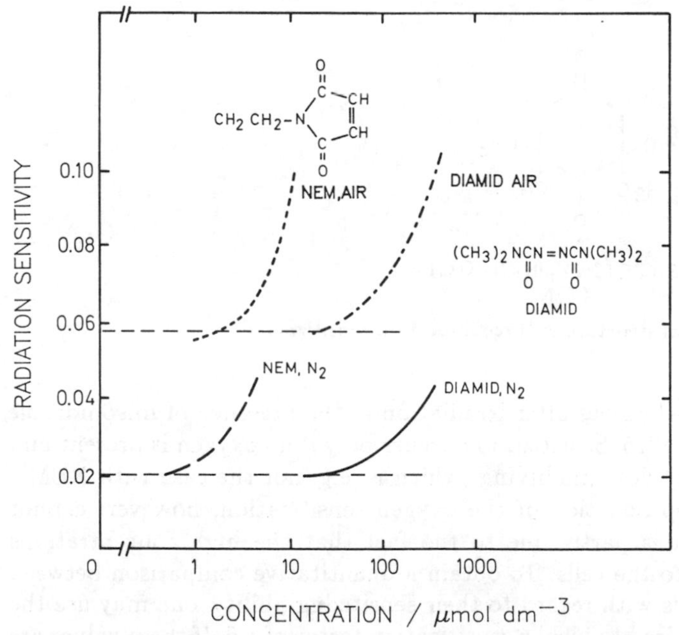

Figure 9.13. Radiation sensitization of Chinese hamster cells by the SH-binding drugs NEM or diamide. The ordinate gives the slope of the terminal part of the survival curve (after CHAPMAN et al. 1973)

other, e.g. radiation-induced, radicals. Furthermore, it is strongly oxidizing, i.e. a potent electron acceptor. This "electron affinity" may be characterized by the "one electron redox potential" which is the energy gained by the transfer of an electron. It is commonly normalized to that of a standard hydrogen electrode. Details which are not essential here may be found in textbooks of physical chemistry.

Redox potentials may be used as a measure of electron affinity. There is, of course, a wealth of oxidizing substances, but for the practical applicability they must fulfill additional criteria, e.g. they should be essentially non-toxic and not enter metabolism. The ideal compound has not yet been found but certain imidazole derivatives have been proven promising. The most important compounds of this kind are *metramidazole (flagyl)* and *misonidazole* whose structural formulae are shown in Fig. 9.14. An example of a free-radical-type sensitizer, namely *triacetonamine-N-oxyl* (TAN) is also given.

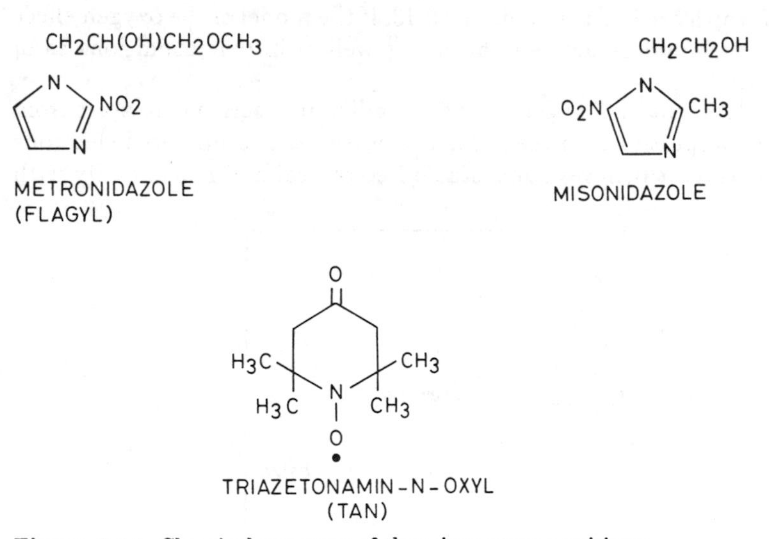

Figure 9.14. Chemical structure of three important sensitizers

Typical survival curves after irradiation in the presence of misonidazole are depicted in Fig. 9.15: Sensitization occurs only if no oxygen is present and the action is strictly dose-modifying (which is, e.g. not the case with NEM).

A complete reproduction of the oxygen sensitization, however, cannot be achieved, which is partly due to the fact that the high concentrations required are toxic to the cells. To obtain a quantitative comparison between different compounds with regard to their sensitizing ability, one may use the concentrations required to yield a sensitization factor of 1.6. If these values are plotted versus the one-electron redox potential, a good correlation is obtained (Fig. 9.16).

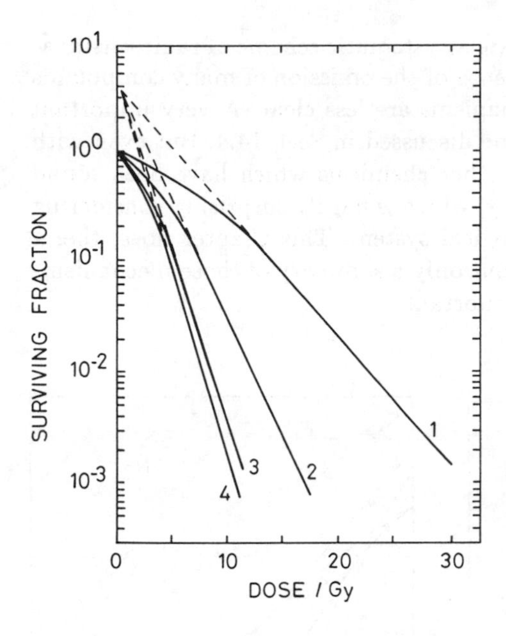

Figure 9.15. Survival of Chinese hamster cells exposed in the presence of various concentrations of misonidazole. 1: nitrogen only, 2: nitrogen plus 1 millimolar misonidazole, 3: nitrogen plus 10 millimolar misonidazol, 4: air with or without sensitizer (after ADAMS et al. 1976)

Stable free radicals – like TAN – represent another group of oxygen-mimetic sensitizers. They have not been studied to the same extent as the afore mentioned electron-affinic substances. They also act only on hypoxic cells, but the basic mechanisms, which cannot be discussed here, appear to be different. This supports the assumption that the oxygen effect may consist of several components.

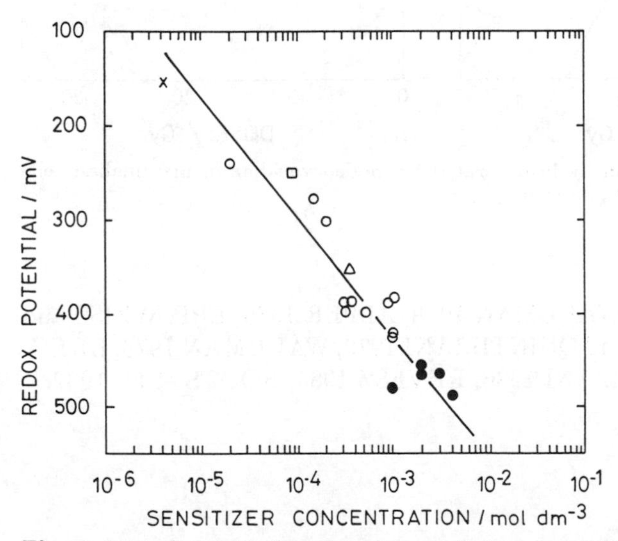

Figure 9.16. Concentration of sensitizers which cause sensitization by factor of 1.6 as function of the one electron redox potential. ×: oxygen, o: 2-nitroimidazole, •: 5-nitroimidazole, □: 5-nitro-2-fuvaldoxime, △: p-nitroacetophenone (after ADAMS and JAMESON 1980)

An attempt was made to delineate a systematic scheme of radiosensitization. This is only possible at the expense of the omission of many compounds which act similarly but whose mechanisms are less clear. A very important class, namely repair inhibitors, will be discussed in Sect. 14.4. But even with their inclusion there are still many other chemicals which have been found to modify cellular radiation sensitivity, which is hardly surprising considering the intricate complexity of the biological system. This chapter does, therefore, not claim to be comprehensive but only a summary of those mechanisms which are thought to be the most important.

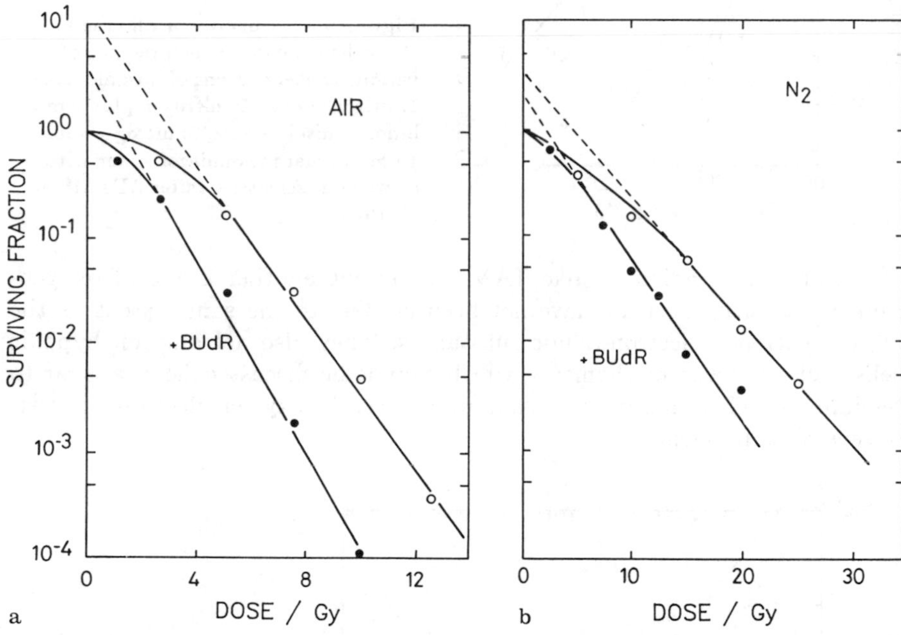

Figure 9.17a,b. Sensitization by incorporated bromodeoxyuridine in mammalian cells (after HUMPHREY, DEWEY and CORK 1963)

Further reading

ADAMS, FOWLER and WARDMAN 1978, ALPER 1979, EPP, WEISS and LING 1976, MOROSON and QUINTILIANI 1970, WARDMAN 1976, BRECCIA et al. 1984, QUINTILIANI 1986, REVESZ 1985, ROOTS et al. 1982

Chapter 10
Radiation and the Cell Cycle

Radiation sensitivity (expressed as loss of colony forming ability or mutability) depends on the cell cycle stage in which the cells are exposed. This is found both with UV and with sparsely ionizing radiations; in many cases, the two types show a mirror-image-like relationship. Radiation also slows down the progression of cells through the cycle; this effect is reversible with low doses. The inhibition of DNA synthesis is particularly relevant in this respect and will, therefore, be discussed in more detail.

10.1 Radiation Sensitivity as a Function of Cell Cycle Stage

If survival curves are determined with synchronized cells (see Appendix II.2), it is found that sensitivity varies with cell cycle stage. The actual relationship is different with X rays and UV, as seen from Fig. 10.1a which is shown as an illustrating example and should not be taken to represent the general behaviour because of considerable differences found between different systems. Nevertheless, it displays the main features: The S phase is most sensitive with UV while it is resistant with X rays, both curves resemble a mirror image. If complete survival curves are determined, the sensitivity changes may be either reflected in different terminal slopes or altered extrapolation numbers (or both), a generalizing statement is not possible.

The pattern shown in Fig. 10.1a is typical for cells with a comparatively short G1 phase. If this is more extended, there is another peak of relative resistance in G1 as illustrated in Fig. 10.1b.

Sensitivity variations as a function of cell cycle stage are not only found with colony-forming ability but also for chromosome aberrations or mutation induction. While the dependence of the former parameters essentially parallels that for survival, it is quite different for the latter, as demonstrated in Fig. 10.2.

The highest mutant yield after X irradiation is found with exposure during G1 with the S phase being quite resistant.

The oxygen effect does not depend on the position in the cell cycle, i.e. the OER remains approximately constant through all stages.

The cyclic variations seen with X or γ rays disappear if the LET of the radiation is increased. With very densely ionizing particles, there is no longer

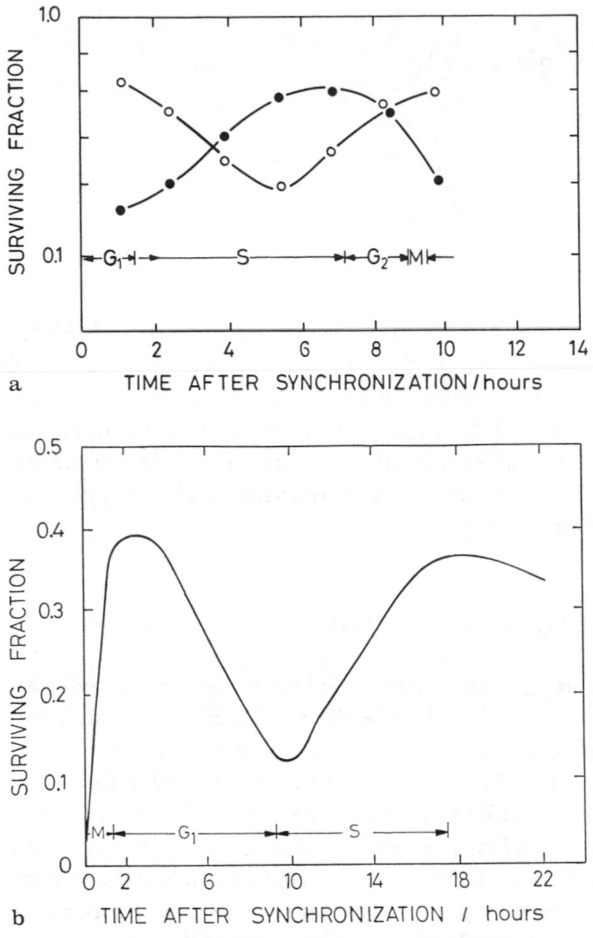

Figure 10.1a.b. Variation of survival after irradiation as a function of cell cycle stage (a) Chinese hamster cells (short G1-period) (after HAN and ELKIND 1977) (o:UV, •: X-rays) (b) human cells (long G1-period, X-rays only) (after TOLMACH et al. 1965)

a dependence of sensitivity on cell cycle stage. This may be taken to indicate that recovery plays an important role with regard to the cyclic behaviour, but there is still no direct proof for this hypothesis.

The cell cycle dependence of radiation sensitivity constitutes an important phenomenon in several respects: Normal cell populations are generally asynchronous. Their survival is hence a superposition of fractions from different parts of the cell cycle, i.e. a weighted average. The situation is complicated by the fact that colony-forming ability depends on dose essentially in an exponential fashion so that theoretical "deconvolution" is very difficult and generally not feasible. This means that the determination of survival curve parameters – if one is interested in their "true" values – requires synchronized populations.

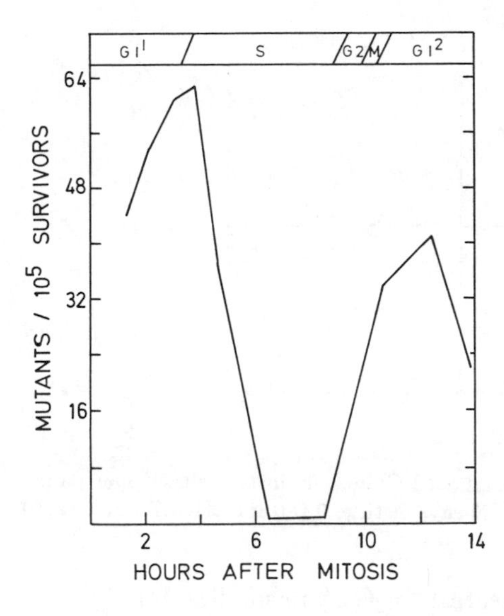

Figure 10.2. Cell cycle dependence of mutation induction by X-rays (after BURKI 1980)

If the exposure is given in two or more fractions, it has to be taken into account that each irradiation changes the population structure since cells in sensitive stages are more readily killed than those in resistant ones. This leads to a partial synchronization which is particularly important in combination experiments.

10.2 Progression and Division Delay

Radiation does not only kill cells but it interferes also with the normal cell cycle progression. Comparatively low doses cause delays in certain phases depending on radiation type. The most obvious effect is a reduction in the mitotic rate which is reversible and – with low doses – not due to immediate cell death. This is a very sensitive parameter and may be detected at exposure levels where survival is hardly affected. The underlying mechanisms are different with UV and X rays: Ultraviolet light leads mainly to a lengthening of the Sphase, while X rays cause an arrest in the G2 period with smaller effects on the S phase. Shifts in cell cycle stage distributions can be most easily shown by the method of cytofluorometry. In Fig. 10.3, the progression of synchronized Chinese hamster cells is shown with and without irradiation. It is clearly seen that irradiation causes a general delay and an accumulation in G2. The quantitative extent of the delay depends on dose and inherent cycle time. There are big differences between strains in terms of "hours/Gy" but a common curve is obtained if the delays are expressed as fractions of cycle time (Fig. 10.4).

Progression delay depends – like survival – also on the stage during which the cells were exposed: With UV it is the larger the earlier in the cycle the cells

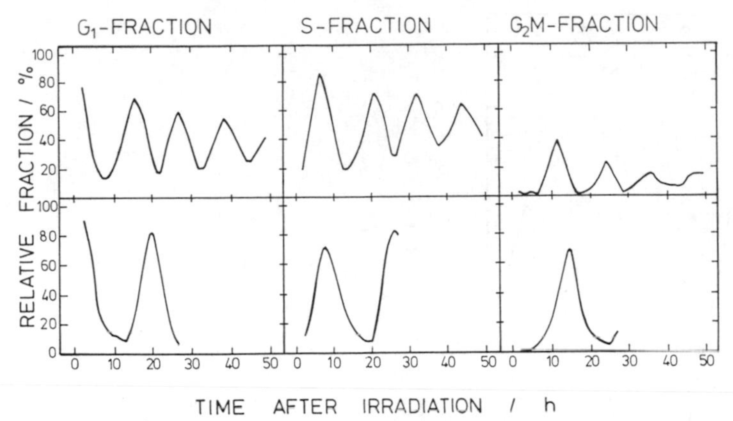

Figure 10.3. Cell cycle progression of X-irradiated Chinese hamster cells. Upper panel: unirradiated controls, lower panel: after 6 Gy X-rays at time 0 (after SCHOLZ et al. 1989)

are exposed, with other words, it is greatest for G1 and smallest for G2 cells. With X rays it appears to be just the opposite: G1 cells are least, G2 cells most affected. There is, however, an important exception to the generality of this statement: If the exposure takes place very shortly before the onset of mitosis, <u>no</u> delay can be seen. This suggests the existence of a critical point in the cell cycle ("point of no return") beyond which mitosis can no longer be stopped. It may be identified with the synthesis time of division proteins; if they are present, even high doses of radiation cannot prevent mitosis.

The different behaviour after UV and X ray exposure may be understood on the basis of the phases which are most delayed by the two radiation types (S with UV, G2 with X rays). If one postulates that lesions are repaired throughout the cycle, the total delay should be largest if the interval between exposure and delay phase is shortest, i.e. G1 with UV and G2 with X rays, as actually found. It fits into this picture that cells which were exposed to UV

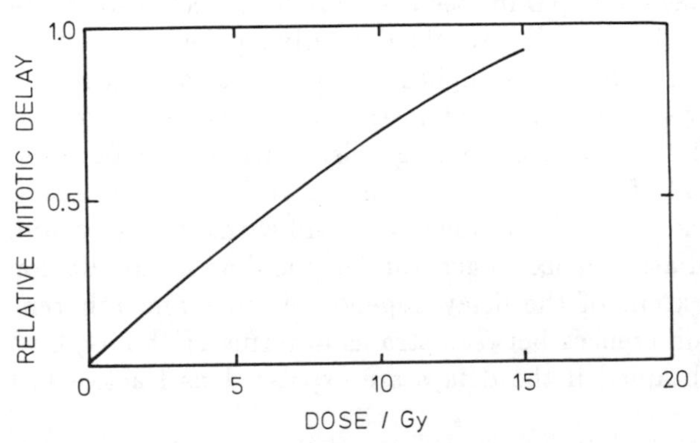

Figure 10.4. Mitotic delay in mammalian cells expressed as a fraction of cell cycle time as a function of x-ray dose (after DEHNEKAMP 1986)

late in the cycle display a delay in the second post-irradiation division. It is largest if the exposure took place in G2 of the preceding cycle. Obviously the lesions are kept until the delay is expressed during the S phase (THOMPSON and HUMPHREY 1970). The situation is less understood with X rays although delays in subsequent cell cycles have also been reported.

It is clear from this description that the cell cycle dependence for the induction of division delay is different from that for survival. This is frequently interpreted to mean that the types of initial damage are also different. But this is not necessarily so and, in fact, unlikely: In cells with photoreactivation capacity (Chap. 13), UV-induced division delay can be reduced by exposure to photoreactivating light, pointing to the involvement of pyrimidine dimers. The general behaviour may be understood by the following hypothetical scheme: Unrepaired lesions in cells reaching the respective "delay" period (S with UV, G2 with X rays) are modified during the delay in such a way that they can no longer be repaired, i.e. they are either repaired or fixed. Only unmodified lesions contribute to the delay. Damage fixation is reflected in loss of colony forming ability. For UV it occurs in S, for X rays in G2 (and perhaps to a smaller extent also in S). Division delay, on the other hand, indicates the amount of unmodified lesions still present in the cell when it reaches the delay period. The different cycle dependence of both parameters can thus easily be accommodated within the postulated scheme. This is, of course, no proof for the correctness of the assumptions but it demonstrates that it is not necessary to invoke different types of lesions.

10.3 DNA Synthesis

DNA synthesis constitutes, without doubt, a key process for cell proliferation and is, therefore, of utmost interest in the study of biological radiation action. It can in principle be easily investigated by measuring the incorporation of specific radioactively labelled precursors, e.g. ^{14}C-thymidine, although there are certain cell types where this is not possible because of the lack of the enzyme thymidine kinase. But even if thymidine can be used, the interpretation of the results is not necessarily unequivocal. Measuring the amount incorporated per cell, i.e. total radioactivity divided by the cell number, any reduction found may be either due to a decreased synthesis rate in *all* cells or to a heterogeneous behaviour of subpopulations. If, e.g. the entry into S phase is delayed, an apparent reduction is seen even if the real synthesis rate is not changed. This problem can only be overcome by determining the radioactivity incorporated in individual cells which is possible with the help of autoradiography. This approach allows to differentiate between cell cycle shifts (i.e. fraction of cells in S) and changes in synthesis rate (radioactivity per cell). But even then misinterpretations are still possible if radiation causes a change in the concentration of endogenous precursors, e.g. by degradation of DNA. Whether this is the case can be checked by a variation of specific activity of the labelled thymidine. Degradation may also be detected by studying whether pre-labelled DNA releases soluble nucleotides into the medium.

In general, radiation reduces overall DNA synthesis in a very sensitive way but one has to differentiate with regard to cellular system and radiation type. In UV-exposed bacteria, the degree of inhibition depends strongly on the functioning of excision repair (see Chap. 13). In its absence the inhibition is much stronger. Photoreactivation studies show clearly that pyrimidine dimers are the responsible lesions. If they are not removed by excision, they block the continuation of replication initially, but it may be later resumed at other sites of the molecule so that the newly synthesized strand contains gaps which may be sealed via recombination, as discussed in Chap. 13. X rays also reduce DNA synthesis in bacteria although the nature of the relevant primary lesions is not known. Oxygen acts – as with survival – as a sensitizer.

Radiation may also cause DNA degradation in bacteria. This process is mediated by enzymes, as discussed later in conjunction with the role of the *recA protein* and with "SOS-repair".

The inhibition of DNA replication in mammalian cells is commonly found to have a biphasic dose-dependence (Fig. 10.5). The initial decrease is com-

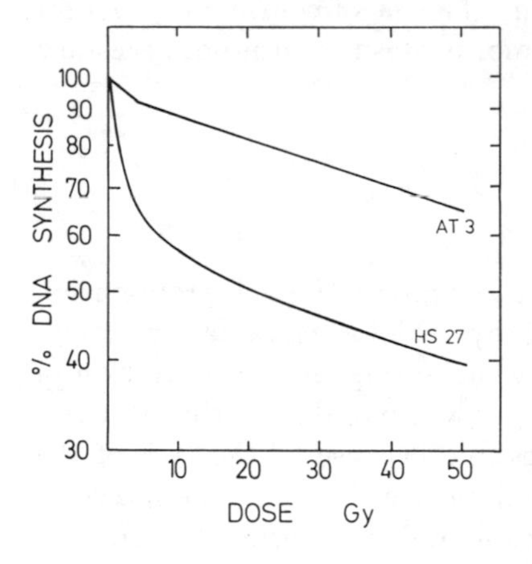

Figure 10.5. Inhibition of DNA synthesis in normal human and Ataxia telangiectasie (AT) cells by X-irradiation (after PAINTER 1986)

paratively steep. Using this to estimate a cross section leads to a figure which corresponds to about 1million base pairs, which is much larger than the normal replicon size of about 50 000 base pairs. It has, therefore, to be assumed that the initiation of replication is inhibited by a single lesion in clusters of about 20 replicons.

The second, less steep part is due to the inhibition of DNA chain elongation. In terms of target size, it corresponds to a single replicon.

Figure 10.5 contains also data for sensitive AT cells (see Chap. 13). This curve lacks the initial descending part seen for normal cells. This has been taken as evidence that in AT cells, initiation is blocked at the single replicon

Table 10.1. DNA-synthesis in mammalian cells after irradiation (CFA: colony forming ability). (After [a]) RAUTH et al., 1974; [b]) PAINTER and YOUNG, 1975; [c]) TOLMACH et al., 1965; [d]) BAKER et al., 1970)

cell type	UV-fluence J/m²	% DNA-S [a]	% CFA [a]	X-ray-dose/Gy	% DNA-S [b]	% CFA
HeLa	10	40	20	5	53	4 [c]
	20	20	1	10	40	< 0.3 [c]
	40	17	< 0.05			
mouse-L	10	45	96	5	73	3 [d]
	20	25	10	10	52	< 0.5 [d]
	40	17	7			

level (and not in a cluster). It is also evident from the figure that DNA replication proceeds – although at a reduced level – even with quite high doses. It has been shown that the reduction seen in AT cells is due to an effect on initiation, not on chain elongation which seems to be highly resistant. This means that DNA lesions do not constitute passive blocks for replication but the inhibition of synthesis may be mediated by certain unknown factors which are absent in AT cells.

UV also inhibits DNA synthesis. This leads to a lengthening of the S phase of the cycle. Pyrimidine dimers seem to be involved although it is not possible to show this directly. They do not constitute complete blocks for replication, similarly as just discussed for X ray lesions. The length of newly synthesized patches are shorter; the distances between replication "gaps" correspond approximately to the number of dimers induced. This suggests that after encountering a dimer, replication stops but is resumed at a new origin.

A few comparative data for mammalian cells are listed in Table 10.1. If equitoxic fluences or doses are compared, i.e. about equal survival levels, UV is certainly more effective in DNA synthesis inhibition, at least with low survival. This reflects the fact discussed above that chain elongation is rather resistant to ionizing radiation.

As pointed out before, the investigation of DNA synthesis requires careful experimentation to avoid erroneous conclusions. A further complicating factor is the so-called "unscheduled DNA synthesis". This is found in cells exposed to radiation either in G1 or G2. It does not represent normal DNA synthesis but repair replication and may be differentiated in the autoradiographic picture because of the much smaller grain density. It is often used as a parameter of general repair ability which is not correct since it does not necessarily represent successful repair. This is exemplified by the fact that its extent is often the same in cells with widely different sensitivities.

Further reading

ALPER 1979, BASERGA 1986, DENEKAMP 1986, EBERT and HOWARD 1972, ELKIND and WHITMORE 1968, PAINTER 1985, PAINTER and YOUNG 1987, WALTERS and ENGER 1976

Chapter 11
Chromosome Aberrations

This chapter deals with cytological alterations. Chromosome aberrations are particularly important because they can be detected already in the first post-exposure cell division and allow, therefore, a closer insight into basic mechanisms. Furthermore, they lead to genetic diseases and constitute also an important tool in practical radiation hazard evaluation.

Colony forming ability as a test parameter can only be assessed at a considerable time after exposure. Visible alterations of the genetic material can already be seen in the first post-exposure mitosis and are, therefore, an important criterion of cellular damage. Furthermore, they allow the investigation of cells which cannot grow to colonies because of their restricted proliferative capacity. Apart from this fundamental aspect, cytogenetic alterations are also of great practical importance in radiation protection for the assessment of radiation hazards.

Generally speaking, chromosome aberrations comprise all alterations of the normal karyotype. They are usually scored at metaphase in which the cells can be arrested by the application of specific spindle poisons like *colchicine* or *colcemide*. The evaluation requires considerable experience, attempts of automation using e.g. pattern recognition devices, have so far not really been successful. Changes in chromosome sizes which constitute a rather crude parameter may, however, be measured using cytofluorometric techniques.

An exact classification of all possible aberrations is rather complex and will not be given in detail, a schematic subdivision suffices here:

The first distinction to be made is between *numerical* and *structural* aberrations. Any change in number from the normal ("modal") value belongs to the first category. If – in the diploid set – one chromosome is lost, this is called "monosomy", if one is duplicated it is called "trisomy". A special case of numerical aberration is "polyploidy" where the whole chromosome set is multiplied. The cause of these alterations is a non-equal distribution of chromosomes during the division process which is referred to as "non-disjunction". Although it is clear that radiation does lead to non-disjunction, quantitative data, especially in mammalian cells, are scarce. This is particularly regrettable since many human hereditary diseases are due to changes in chromosome number. A notable exception to this general statement is "Down's syndrome" (trisomy 21); see Chap. 19.

Table 11.1. Deseases associated with chromosomal anomalies. (After CZEIZEL and SANKARANARAYANAN, 1984)

anomaly	prevalence per 10^4 livebirths	per cent infant and early deaths
Downs syndrome	11.7	34
Patau syndrome	0.1	86
Edwards syndrome	0.2	78
autosomal deletion syndromes	0.1	56
gonadal dysgenesis	0.1	10
Klinefelter syndrome	0.2	28
unspecified	0.2	28

A few more important examples are listed in Table 11.1. This issue will be taken up again in Chap. 19.

Chromosomes at metaphase consist of two chromatids joined together at the centromere. Structural aberrations may either affect the whole chromosome ("chromosome aberrations *sensu strictu*") or single chromatids ("chromatid aberrations"). Within these two classes one has to differentiate between *deletions* (loss of a fragment), and *exchange aberrations.* The latter may occur either in the same chromatid or chromosome ("intrachange") or between different partners ("interchange"). A special example is the exchange between sister chromatids. The final outcome may be quite different as illustrated in Fig. 11.1: All exchange types start with two breakage points which are annealed in variable ways. Information is not always lost, sometimes it is only translocated to a different position. This, however, is by no means without genetic consequences. The new environment may lead to different gene expression and, therefore, profound changes in the phenotype. The best-known example is the so-called "Philadelphia chromosome" (reciprocal translocation between chromosomes 9 and 21), commonly found in leukemia patients, where a normally silent oncogene is expressed.

In detail (Fig. 11.1):

Chromosome aberrations

1. Breaks: They are expressed as deletions of fragments in <u>both</u> chromatids. They are distinguished from *isochromatid breaks* (see below) by the fact that the ends are normally not rejoined. They are also different from *achromatic lesions* which show up in the microscopical picture as discontinuities but without actual breakage.

2. Intrachromosomal exchanges: In contrast to the preceding category, they always require <u>two</u> breakage points. They may be resealed so that the intermediate fragment is lost ("interstitial deletion"), quite often in the form of an *acentric ring* ("acentric" means without a centromere). If the original breakage points lie on different sides of the centromere, *centric rings* are found.

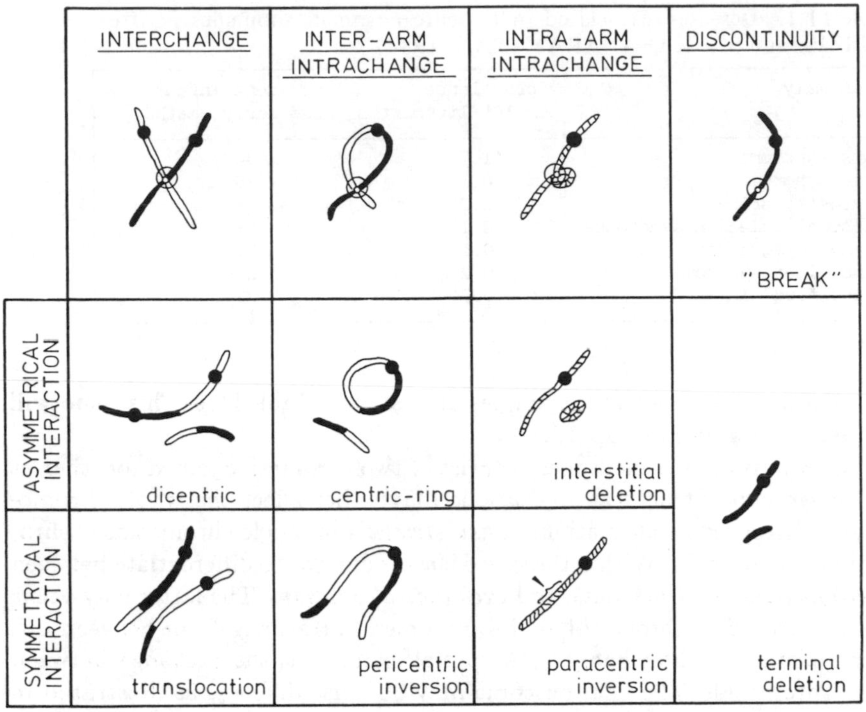

Figure 11.1. Some examples of chromosome aberrations (after SAVAGE 1978)

The fragments are not always lost, they may be reintegrated with different polarity yielding *inversions* or at different positions leading to translocations.

3. Interchromosomal exchanges: They are quite similar to the preceding class with the only difference that the original breakage points are on different chromosomes. Rejoining leads frequently to *dicentric chromosomes* ("dicentrics" for short) and acentric fragments, but also to translocations which are, however, more difficult to detect.

Chromatid aberrations

All what has just been said is also applicable to chromatid aberrations but the situation is more complex because of the interaction between the chromatids. A short summary may suffice, for details the reader is referred to textbooks of cytogenetics.

1. Breaks: They cause the loss of a fragment in one chromatid and appear at metaphase as discontinuities being difficult to distinguish from chromatid lesions (see above). In contrast to the latter, however, they lead to acentric fragments at anaphase. Breaks at homologous sites in both chromatids are called "isochromatid" or "isolocus" breaks. They join usually – distinguishing them from true chromosome breaks – leading to *anaphase bridges* and <u>one</u> acentric fragment.

2. Intrachromatid exchanges: They cause changes only in one chromatid, often together with centric structures as explained above.

3. Interchromatid exchanges: If they occur between homologous chromatids, they are called *sister chromatid exchanges* (SCE). One example is the joining of isolocus breaks. If homologous sections are exchanged, the shape of the chromatids is not changed so that detection is difficult. These exchanges, however, can be investigated by means of the so-called "harlequin technique" (Fig. 11.2):

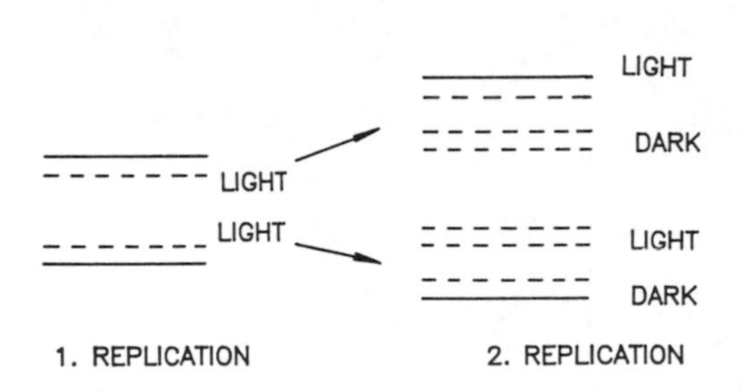

Figure 11.2. Principle of the "harlequin" technique to differentiate between replicated and unreplicated DNA and for the detection of sister chromatid exchanges (see later). During the first round of replication only one strand is labelled by BUdR, the chromatids appear bright. After the second replication the chromatids which were labelled in both strands appear dark (after PERRY and WOLFF 1974, and SCOTT and LYONS 1979)

The cells are grown for two replication cycles in a medium containing bromodeoxyuridine (BUdR) which is incorporated into DNA instead of thymidine. After the first cycle, all cells contain "hybrid" DNA, i.e. one strand only is labelled with BUdR. After the second round 50% of the chromatids has BUdR in both strands, the other half only in one. Since BUdR quenches the fluorescence of DNA-binding fluorescent dyes, the two sister chromatids are easily distinguished by the brightness of their appearance in the fluorescence microscope. Sister chromatid exchanges involving not too small pieces may be spotted by the chequered picture of the respective chromosomes. This gave rise to the name "harlequin technique".

SCE are undoubtedly induced by radiation – more by UV than by X rays – but the quantitative relationship is not well known. This is the more regrettable since they are thought to play an important role in carcinogenesis.

Interchromatid exchanges between non-homologous partners appear as dicentric figures in metaphase, in anaphase as bridges or dicentric figures and fragments. The possible shapes are too numerous to be listed.

Achromatic lesions: These unstained regions are usually only found in one chromatid and persist often for many cell generations. In contrast to true breaks, they do not lead to fragments at anaphase.

A special case of chromosome aberrations is the formation of "micronuclei": If a fragment or even a whole chromosome is not properly segregated during mitosis, it may be separated from the nucleus. It may then form a micronucleus which is – in contrast to normal chromosomal aberrations – most easily seen in stained interphase cells. Micronucleus formation is a comparatively simple and sensitive test for radiation damage.

After this general introduction, the induction of structural aberrations by radiation will be discussed. All types may be found both with UV and X rays, although with differing probability. Exposure before DNA replication leads to *chromosome* aberrations, irradiation after the S phase to *chromatid* aberrations. This kind of behaviour is clearly seen for UV in Fig. 11.3 where

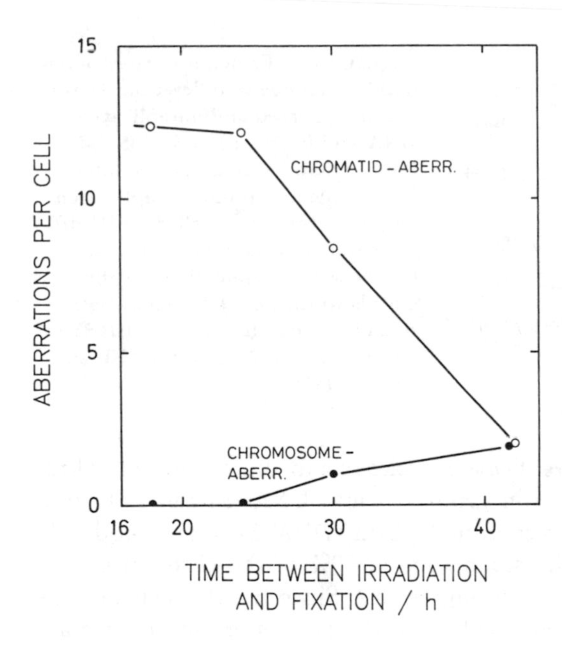

Figure 11.3. Chromosome and chromatid aberrations as function of the time interval between UV-exposure (265 nm) and mitosis (after data of CHU 1964)

the yields are plotted as a function of the time between irradiation and fixation. Since scoring takes always place at a fixed point of the cell cycle, it is not necessary to work with synchronized populations. The results suggest that radiation does not directly induce alterations but rather some kind of latent lesions which are replicated during S phase and expressed at metaphase. An exact investigation, however, is hampered by the fact that cell progression is greatly disturbed by irradiation (see Sect. 10.2).

Figure 11.4 shows an action spectrum. It is comparatively broad with about equal efficiencies at 265 and 280 nm suggesting that both DNA and proteins contribute. There are not very many studies with UV, especially an analysis of the fluence dependence is lacking.

The situation is quite different with ionizing radiation. A good test ob-

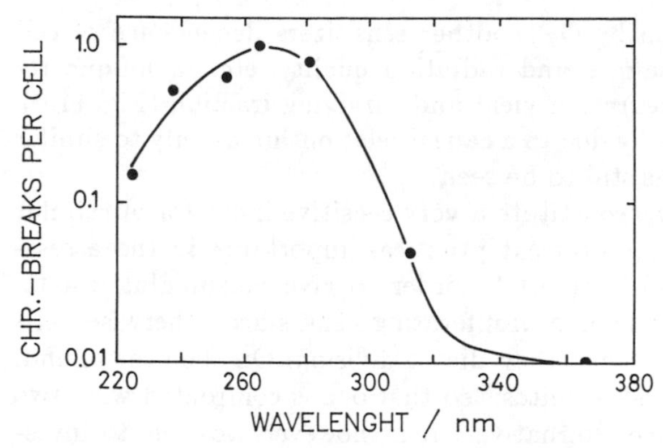

Figure 11.4. Action spectrum for the induction of chromatid aberrations including "achromatic lesions" (after CHU 1964)

ject is the broad bean *Vicia faba* because it possesses only few but large chromosomes. Fig. 11.5 depicts the yields of different aberration types as a function of X ray dose. It is clearly seen that only achromatic lesions show a linear dependence while all other curves are bent. This has consequences for the explanation of aberration formation as detailed below. It suggests that achromatic gaps constitute the true primary lesions which is supported by the fact that they are also formed with the highest yield.

With densely ionizing radiations, the dose-response curves approach straight lines for all aberration types.

Chromosome aberrations are considered by many authors to be the main cause of cell inactivation. It is indeed found that under many different expo-

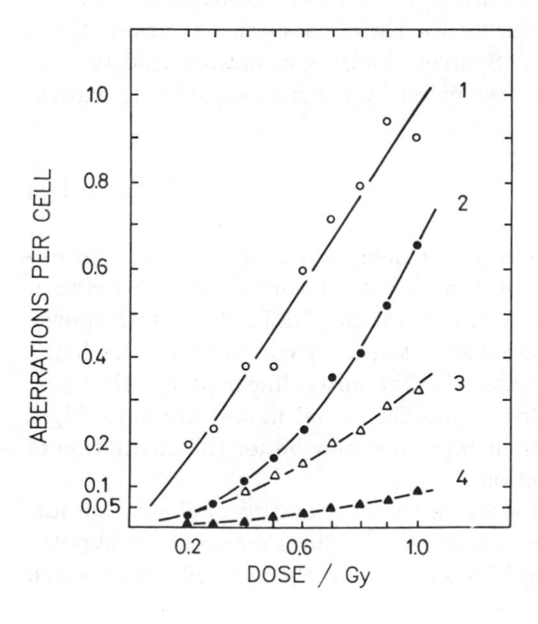

Figure 11.5. Different types of aberrations after X-irradiation in the broad bean Vicia faba. 1: achromatic lesions ("gaps"), 2: chromatid exchanges, 3: isochromatid breaks, 4: chromatid breaks (after REVELL 1966)

sure conditions (sensitization by O_2 or other sensitizers dependence of cell cycle stage, changes of dose-rate and radiation quality, etc.) a unique relationship exists between aberration yield and surviving fraction (Fig. 11.6). Whether this is, however, really due to a causal relationship or only to similar underlying processes remains still to be seen.

Chromosome aberrations constitute a very sensitive indicator of cellular radiation damage and may be of great practical importance in the assessment of radiation hazard to humans. In order to give meaningful results, they should be investigated in non-proliferating cells since otherwise cells with aberrations will be lost when they die at division. On the other hand, aberrations can only be scored in mitosis so that one is confronted with two apparently mutually exclusive alternatives. It is, however, possible to investigate aberrations in a population of cells which do not divide to a great extent in the human body, namely the small lymphocytes in the peripheral blood. They may be stimulated under suitable conditions outside the body to perform a few divisions by the action of ***phytohemagglutinine*** (PHA). This is sufficient to score chromosome aberrations. The time between stimulation and fixation is critical. It should be long enough to yield a high number of metaphases but it must be avoided that some cells are assayed at their second division because of the possible loss of aberration bearing cells. This problem may be solved by an adaptation of the above described "harlequin technique", i.e. the labelling with BUdR and subsequent fluorochrome staining. It is thus possible, on the basis of fluorescence intensity, to distinguish cells with one from those with two rounds of replication. Using this approach, it could be shown that 50% of dicentrics is lost at the first division and that the yield at the <u>first</u> metaphase does not depend on the time of fixation.

Dicentrics are by far the most investigated chromosome aberrations in mammalian cells. Dose-response curves for their induction by various types of radiation are shown in Fig. 11.7. Sparsely ionizing radiations lead to non-linear relationships which can be best fitted by the linear-quadratic expression:

$$X = \alpha D + \beta D^2 \tag{11.1}$$

where X stands for aberration yield, D for dose, and α and β being parameters. Contrary to survival, there is a highly significant difference between 250 KV X rays and ^{60}Co γ rays. With increasing LET, the dose-response curves become linear. The parameters as obtained by computer fits are listed in Table 11.2 together with RBE values for the initial linear part, either normalized to 250 KV X rays or ^{60}Co γ rays. The actual figures are quite high. It remains to be discussed what their relevance may be for the estimation of quality factors in radiation protection.

There are a number of hypotheses on the mechanisms of aberration formation. An obvious and still popular model is the "breakage-reunion hypothesis". It is assumed that the initial lesion is a directly induced break which

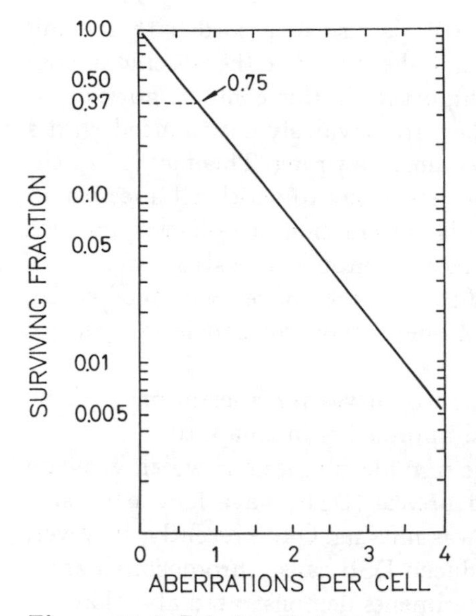

Figure 11.6. Relationship between chromosome aberrations and survival in X-irradiated Chinese hamster cells. Survival reduction to 1/e corresponds to 0.75 aberrations per cell (after DEWEY et al. 1971)

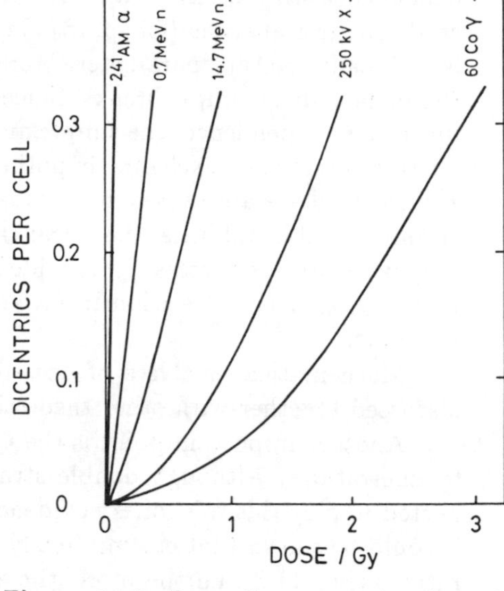

Figure 11.7. Dose effect curves for dicentric chromosome aberrations in human lymphocytes with radiations of different LET (after DU FRAIN et al. 1979)

is either sustained as such or which reacts with others to form exchange-type aberrations. According to this scheme, breaks should follow linear, exchanges quadratic induction kinetics. This is, however, evidently not the case, as seen in Fig. 11.7. A more detailed analysis suggests rather a dose-dependence with the 1.5th power, which led to the so-called "3/2-rule" which is, however, also not generally applicable. The deviation from the purely quadratic dependence is explained by the fact that even with sparsely ionizing radiation, events with high local energy deposition may occur so that a linear-quadratic

Table 11.2. Quantitative dependence for the formation of dicentric aberrations in human lymphocytes by various radiation types. Generally the following relation is used: $x = \alpha D + \beta D^2$ (x: mean number of dicentric aberrations per cell, α, β: coefficients, D: dose). Besides that RBE-values with ^{60}Co-γ- and 250 kV X-rays are given as reference value for the linear part (lower dose-range). (After du FRAIN et al., 1979)

radiation type	α Gy⁻¹	β Gy⁻²	RBE (^{60}Co)	RBE (250kV-X)
^{60}Co-γ	$1.76 \cdot 10^{-2}$	$2.97 \cdot 10^{-2}$	1	0.37
250 kV-X-rays	$4.76 \cdot 10^{-2}$	$6.19 \cdot 10^{-2}$	2.7	1
14.7 MeV neutrons	$2.62 \cdot 10^{-1}$	$8.84 \cdot 10^{-2}$	15	5.5
0.7 MeV neutrons	$8.55 \cdot 10^{-1}$	–	49	18
^{241}Am-α	4.9	–	278	103

dependence may be more appropriate. This is also in accord with the microdosimetric analysis (Chap. 16) since not the dose but the specific energy is the really relevant parameter. More important in this context, however, is the behaviour of simple breaks. Since they are obviously not induced with a linear dose dependence, the simple model must be wrong. This fact led to the postulation of the "exchange hypothesis" according to which all aberrations – even breaks – are caused by the pairwise interaction of radiation-induced chromosomal instabilities, i.e. local changes in chromosome structure, which may be – but not necessarily so – identified with the above-mentioned achromatic gaps. A clear decision in favour of one of the two models can yet not be made.

Mathematical analyses of dose-response curves for aberrations will be discussed together with other theoretical approaches in Chap. 16.

Another important point is the type of molecular lesions which give rise to aberrations. Although double strand breaks (DSB) have long been suspected to play this role, direct evidence was missing. Quite recently, however, it could be shown that enzymatically induced DSB cause chromosomal aberrations (Fig. 11.8). Furthermore, the experiments demonstrated also that the actual form of the breaks is also important: If there are overlapping ends, the aberration yields are much lower – presumably because of spontaneous resealing – than with "blunt" ends.

The necessity to score aberrations in mitosis excludes the possibility to investigate their formation at an early stage so that interfering processes, as e.g. repair, cannot be excluded. The technique of "premature chromosome condensation" (PCC) represents a major breakthrough in this respect.

If interphase cells are fused with mitotic ones the chromosomes condense and fragments and breaks can thus be made visible at any stage of the cycle. The mechanism is not yet clear but it appears likely that certain "mitotic factors" are transferred during fusion.

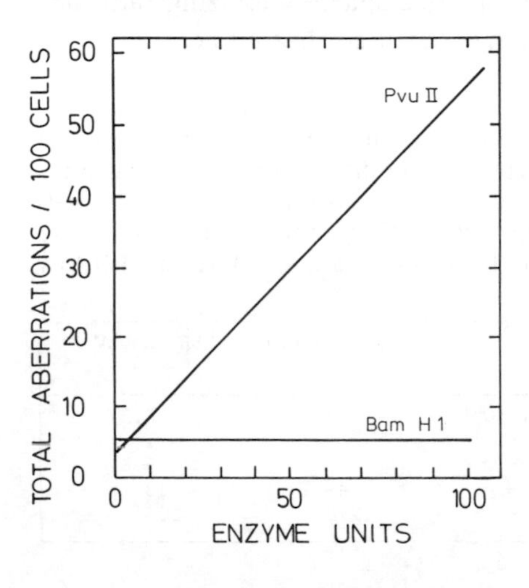

Figure 11.8. Induction of chromosome aberrations by restriction endonucleases in Chinese hamster cells. Pvu II induces "blunt", BamH1 "sticky" double strand breaks (after BRYANT 1984)

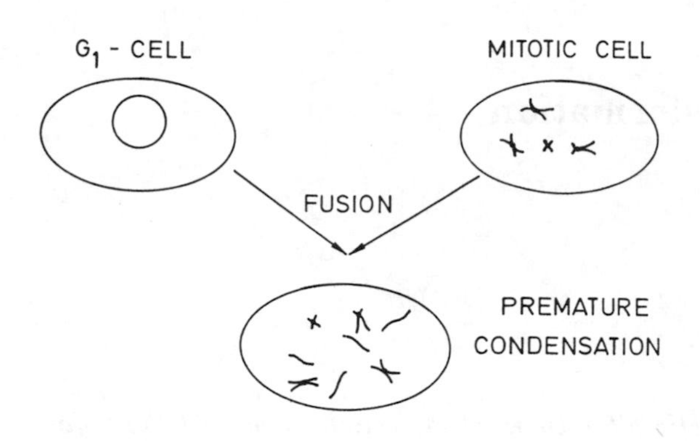

Figure 11.9. Principle of "premature chromosome condensation" (PCC)

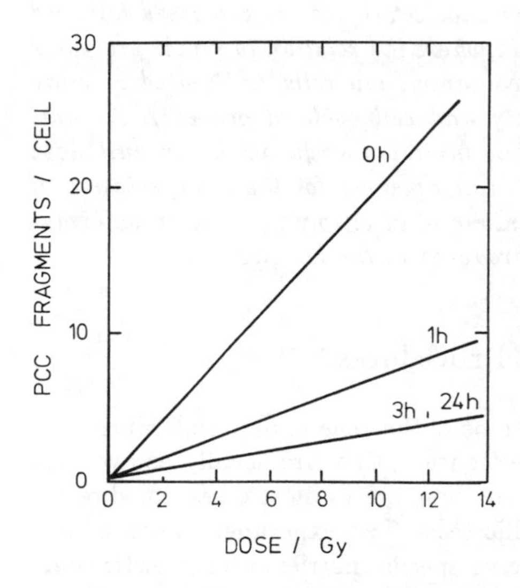

Figure 11.10. Chromosome fragments measured using the PCC-technique immediately and various times as indicated after X-irradiation (after ILIAKIS et al. 1988)

Figure 11.9 shows a dose-effect curve obtained by this technique for chromosome fragmentation in G1. The relationship is clearly linear. After incubation, the yield is reduced suggesting repair. The process seems to be comparatively fast and completed within 4 hours. It is also seen that the linear shape does not change, which is contrary as expected from dose-effect curves measured in mitosis and is not yet understood.

Further reading

ALMASSY 1987, EVANS, BUCKTON, HAMILTON, CAROTHERS 1979, HITTELMAN 1984, IAEA 1986, ISHIHARA and SASAKI 1983, LLOYD et al. 1987, OBE et al. 1982, RIEGER and MICHAELIS 1967, WOLFF 1972

Chapter 12
Mutation and Transformation

This chapter deals with genetic alterations at the cellular level. At the beginning the general fundamental questions and techniques are discussed followed by an outline of experiments in bacteria where the relation to repair processes is well understood. Investigations with mammalian cells are treated in more detail (dependence of radiation quality and cell cycle stages etc). A comparison of different systems shows that mutation frequencies per unit dose increases with DNA content which has consequences for the interpretation of mechanisms. The very important technique of in vitro neoplastic transformation and some results obtained are introduced in the last section.

12.1 Mutation Types and Test Procedures

Mutations are – by definition – alterations of the genetic material. Since they are not easily accessible to direct investigation, they are usually assayed via changes in the cellular phenotype. Even these are in most cases not directly seen but can only been found by specific tests. Two experimental procedures are most widely used: the dependence of specific nutrients for growth (*auxotrophy*) and the resistance to certain toxic agents, very often antibiotics (Fig. 12.1).

If one starts with normal unaltered cells, the *wild type*, the mutants found are due to "forward mutations". If, on the other hand, the reappearance of wild type phenotypes in a mutant population is determined, these are called "reversions". These definitions may be more stringently formulated in molecular terms: Forward mutations are those changes by which a gene product is made inactive or altered. It may – in the case of larger deletions – even be completely lost.

Reversions lead to a total or partial reconstruction of a functional gene product.

Forward mutations are normally not phenotypically expressed in diploid wild type cells since the information is present in two copies on the homologous chromosomes. They have, therefore, to be studied either in haploids or under conditions where the gene on the second chromosome is mutated or not expressed. In very few cases, the mutation may lead to such a change in the gene product that it is toxic to the cells. Under these conditions there

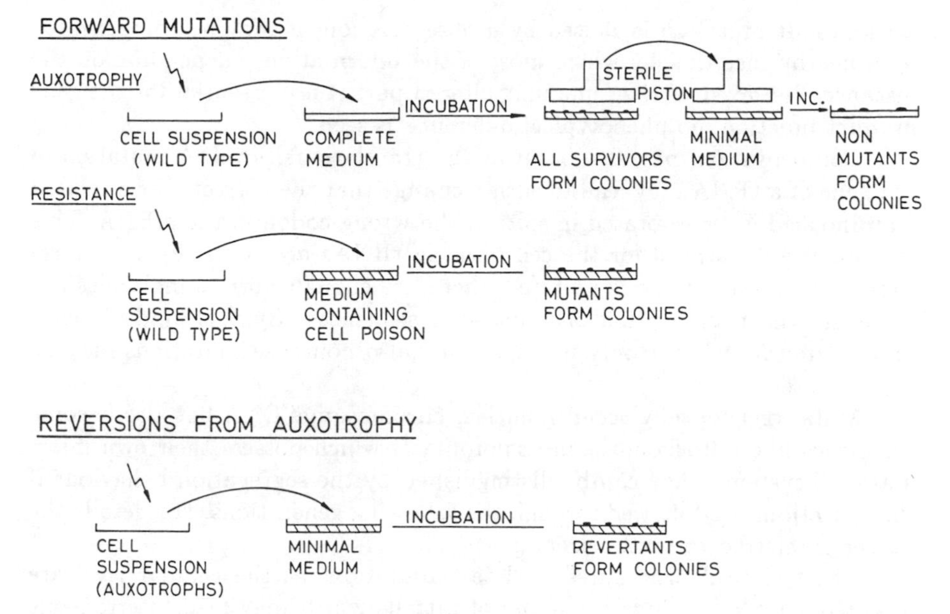

Figure 12.1. Principles of mutation assays

is "dominant lethality". But normally, forward mutations are only expressed in diploids if they are present on both homologous chromosomes, they are "recessive".

Mutations may occur either by the loss of a DNA fragment ("deletion") or a change of the base sequence. A distinction between these two types is possible by testing the revertibility: deletions can obviously not be reverted. Since this is a negative test, it is, however, difficult to arrive at unequivocal conclusions.

Changes which involve only one or few base pairs ("point mutations"), may belong to several categories: If only one base is changed this may either lead to a wrong amino acid in the protein ("missense mutation") or to the stop of translation ("nonsense mutation") if a "stop codon" was formed. It is also possible that an additional base pair was added ("insertion") or that one was lost ("base deletion"). Since the genetic code has no commas the sequence of translation is completely changed, the resulting protein does not bear any resemblance to the original one. This "frameshift mutation" affects the entire rest of the transcriptional unit.

Base substitutions where a purine base is replaced by another purine base or a pyrimidine by another pyrimidine are called *transitions*, the others where the base type is altered have been named *transversions*.

A phenotypically seen reversion does not necessarily represent a complete reconstruction of the original base sequence, this is rather more the exception than the rule. This rare case would be a "true reversion". More frequently, the first error is compensated by a second one at a different site. If, e.g., a frameshift mutation is caused by a base deletion, it may be compensated by a nearby insertion by which most of the information – depending on the distance – is saved. If the remaining altered part is not essential for the gene product function, no phenotypical difference is seen.

Compensation may also occur at the translational level: A mutation in the gene of a tRNA may lead to such a change that the correct – or a related – amino acid is incorporated in spite of the wrong codon in the mRNA. This is usually not harmful for the cell because tRNAs are coded for at several sites so that even if one is mutated, there are enough normal molecules left to secure the transcription of unmutated sequences. By this kind of "suppressor mutation", not only missense, but also nonsense mutations may be compensated.

Mutations not only occur in nuclear chromosomal DNA but also in some organelles like mitochondria and chloroplasts which possess their own informational system. They can be distinguished by the segregation behaviour if the mutations are followed through the following generations. For details the reader is referred to textbooks of genetics.

The technical procedures to detect mutations at the cellular level are basically simple but hide a number of pitfalls which may lead to erroneous interpretations. Since only the principles shall be outlined here, a detailed discussion will not be given (see e.g. AUERBACH 1976). The basic scheme how mutations to auxotrophy or resistance to cytotoxic agents can be scored is sketched in Fig. 12.1: In the first case it is necessary to pick up the mutants by replica plating which makes the method, taking into account the usually very low yield, quite impracticable. This means that forward mutations can be simply found only for the induction of resistance. The colony-forming ability on plates is compared with and without the toxic agent after the necessary concentrations have been determined in preliminary tests. Care must be taken, however: If – as assumed – the resistance is caused by the loss of a certain gene product, this will still be present in the cell immediately after exposure and only slowly degraded. An immediate plating in the selective medium will hence cause also the death of those cells which are mutated but still possess the gene product. It is, therefore, necessary to allow an "expression" period under normal growth conditions. There is also another reason for this: The primary lesions have to be modified by cellular processes to become stable mutations as seen, e.g. from the phenomenon of "mutation frequency decline" (see below). This process is related to repair processes which means that the expression kinetics may depend on the genetic background of the particular cell line. An additionally complicating factor lies in the usually very low mutant yield. This means that high cell densities have to be plated so that the mutants grow on top of a great number of killed cells whose lysis products may influence the result.

Two special tests which are important for mammalian cells are given as examples: The first one is the resistance to 6-thioguanine (6-TG). This purine derivative is a base analogue and may be incorporated into cellular

DNA leading then, however, to cell death. Since incorporation starts from the nucleotide level, both a sugar and a phosphate group have to be attached to the base. This is mediated in the cell by the enzyme *hypoxanthine-guanine-phosphoribosyltransferase* (HGPRT for short). If it is not present, exogenous purines can no longer be used – and also not 6-TG –: the cell has become resistant. Since there are normally other pathways for the synthesis of purine nucleotides, the mutation has normally no effect on growth. If the endogenous synthesis is blocked, however, which is possible by *aminopterine*, the mutants are no longer able to grow. A reversion would then lead to aminopterine-resistance. It is thus, in principle, possible to study both forward and reverse mutations. Another important advantage of this system lies in the fact that the HGPRT gene is located on the X chromosome, the female sex chromosome. Male cells, having only one X chromosome behave, therefore, like haploids. But even in female cells the assay of this recessive mutation is possible since normally only one of the two X chromosomes is expressed. Caution is required in tumour cells which have often more than the normal numbers of chromosomes and an altered mode of gene expression. The particular properties of the HGPRT system make it a rather unique system for mutation studies in mammalian cells. The situation is much easier in microorganisms where one can recur to haploid lines. The other example is resistance to *ouabain* (= g-strophanthine). This compound binds to the enzyme Na-K-ATPase by which the Na- and K-ion transport through the membrane is inhibited, leading to cell death. A mutation in the respective gene may change the enzyme property in such a way that it becomes insensitive to ouabain leaving its normal function intact. This must obviously be a missense mutation and cannot be brought about, e.g. by a deletion. Phenotypically the mutation is semi-dominant, i.e. it is also expressed in diploids.

Auxotrophic mutants are generally only used for reversion studies. In this case, cells are incubated in parallel on media with and without the required component and the mutant yield is determined by comparing the colony forming ability (see Fig. 12.1). It is necessary, however, to supply even on the selective plates a certain small amount of the nutrient to allow the expression of the mutants. Because of the complexity of the growth conditions, these studies are very difficult – and hence rarely used – with mammalian cells.

12.2 Mutation Induction in Microorganisms

Induction and expression of mutations are closely related to repair processes. Some phenomena are thus treated in more detail in the next chapter. The main emphasis in this section lies, however, in the description of experimental results.

UV causes effectively both forward and reverse mutations. In many – but not all – cases the fluence-effect relationship is quadratic which may give a clue for possible interpretations.

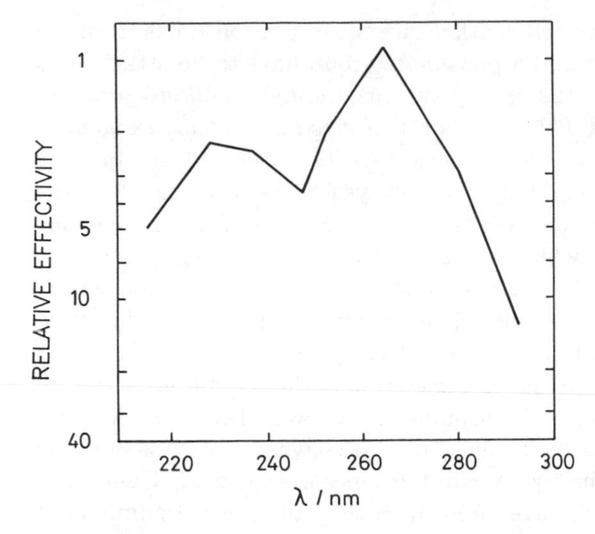

Figure 12.2. Action spectrum for mutation induction in Trichophyton mentagrophytes (after HOLLAENDER and EMMONS 1941)

A "classical" action spectrum is shown in Fig. 12.2: It demonstrates, not surprisingly nowadays, a close resemblance to DNA absorption, but it should be noted that it was published 3 years before the chemical nature of the genetic material was identified by AVERY and coworkers in 1944 (AVERY et al. 1944). Mutations can be photoreactivated like the loss of colony forming ability, but this is not true for all mutation types. Possible reasons for this will be discussed later.

The influence of genetically determined repair processes is complex: Excision-deficient mutants display a drastically enhanced mutation yield com-

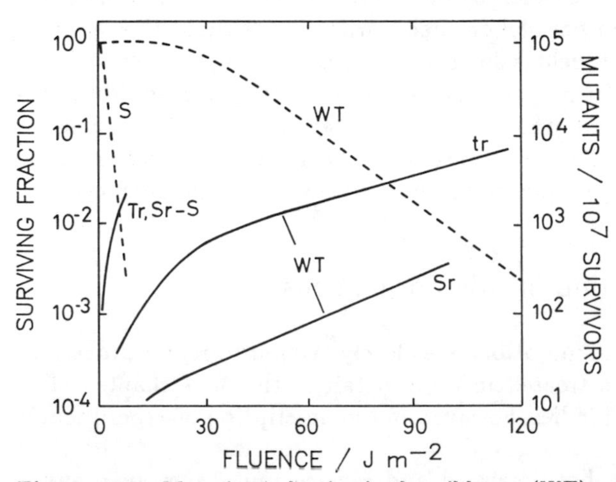

Figure 12.3. Mutation induction in the wild type (WT) and a sensitive mutant of E. coli (S) by UV exposure. Sr: Streptomycin-resistance (forward mutation), tr: reversion from tryptophane auxotrophy. Survival curves are given for comparison (after WITKIN 1966)

pared to the wild type if the same doses are applied (Fig. 12.3). This suggests that excision is "error-proof". But there are interesting differences: Forward and reverse mutations have the same fluence dependence in the sensitive mutant but not in the wild type. By normalizing the fluence scale to equal survival (Fig. 12.4), the differences are even more evident: The rate of rever-

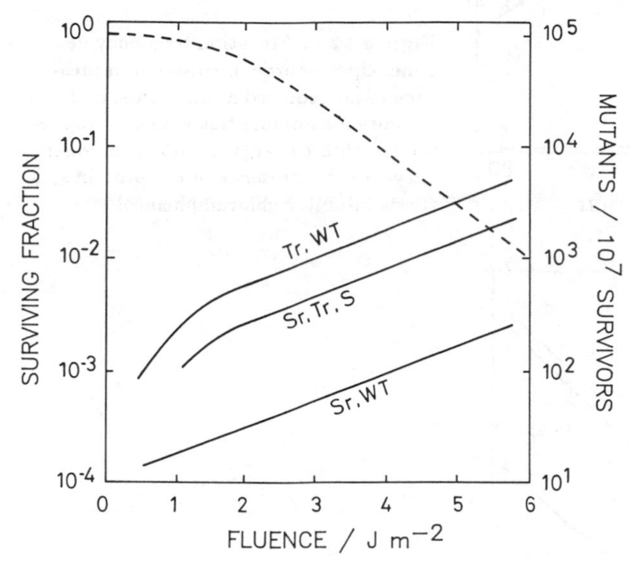

Figure 12.4. As in Fig. 12.3 but with the fluence scale adjusted to give equal survival (factor 16.7) (after WITKIN 1966)

sions is higher in the wild type but that of forward mutations lower. The general significance of this finding is certainly not clear but it casts doubt on the generally made assumption that excision repair is error-free.

Rec$^-$ mutants (see Chap. 13) give an unexpected result: The mutant yields are very low. This was originally taken to indicate that the function of the rec-system is responsible for mutation induction. Today it is known that its involvement is rather more complex. A complete picture in context with repair processes will be given Sect. 13.2.4 (next chapter).

An interesting and important phenomenon is that of "mutation frequency decline" (MFD): The mutant yield is reduced when the cells are kept in a medium in which protein synthesis is not possible before they are finally plated for selection (Fig. 12.5). That a metabolic process is required for mutation "fixation" is evidenced by the fact that inhibition of energy metabolism leaves the mutant yield unchanged with incubation. MFD is thus explained by assuming that in the absence of protein synthesis, repair is favoured over fixation.

MFD is restricted to suppressor mutations suggesting that the mechanisms for forward and reverse mutations are different.

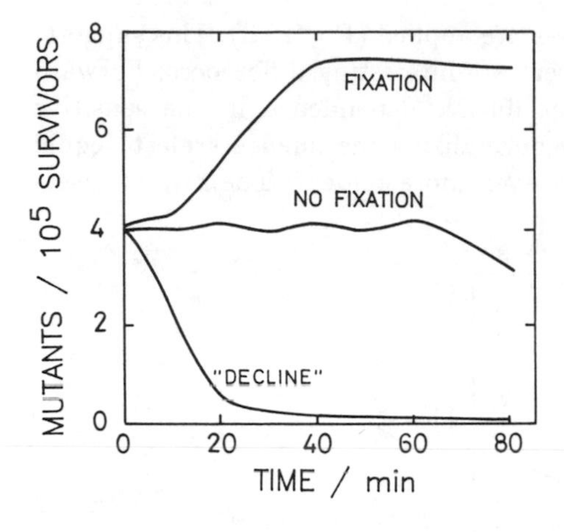

Figure 12.5. Mutation frequency decline. Upper curve: incubation in presence of all required amino acids, middle curve: as before but with dinitrophenol to inhibit energy metabolism, lower curve: in the presence of the protein synthesis inhibitor chloramphenicol

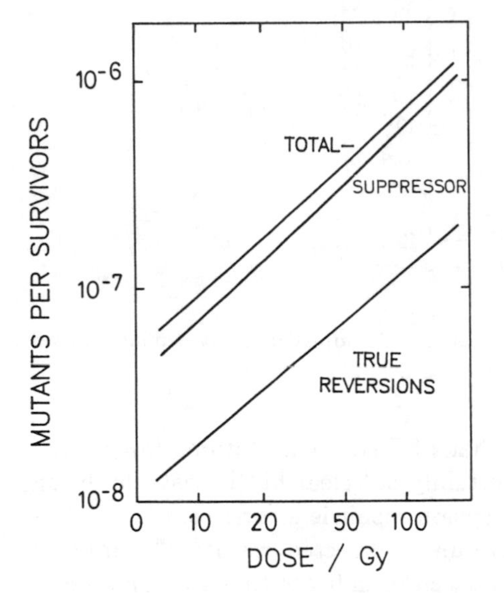

Figure 12.6. Reversions from tryptophane auxotrophy in wild-type E. coli by X-irradiation. The small fraction of true reversions should be noted (after BRIDGES et al. 1968)

Ionizing radiations are much less effective to cause mutations than UV. This may be seen by comparing Figs. 12.4 and 12.6. In the latter, reverse mutations are also subdivided into "true" and suppressor mutations. By far the largest part belongs to the latter group.

12.3 Mutation Induction in Mammalian Cells

The most widely used mutation assay in mammalian cells is the test for resistance to 6-thioguanine (see Sect. 12.1). Figure 12.7 shows a typical dose-response curve for X rays. It has a curved shape which can be fitted by the relationship:

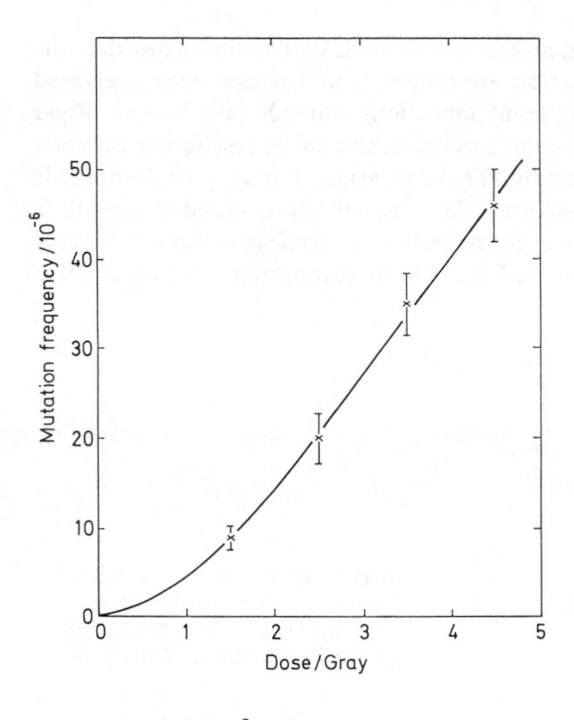

Figure 12.7. Mutation to 6-TG resistance in Chinese hamster cells after X-irradiation (CROMPTON 1987)

$$m = aD + bD^2$$

where m is mutation frequency, i.e. mutants per survivor, D the dose, a and b being parameters.

X ray mutation induction curves are not necessarily curved, in some cell types also purely linear responses may be found (see Fig. 12.8 for an example). Resistance to 6-TG can be induced both by ionizing radiations

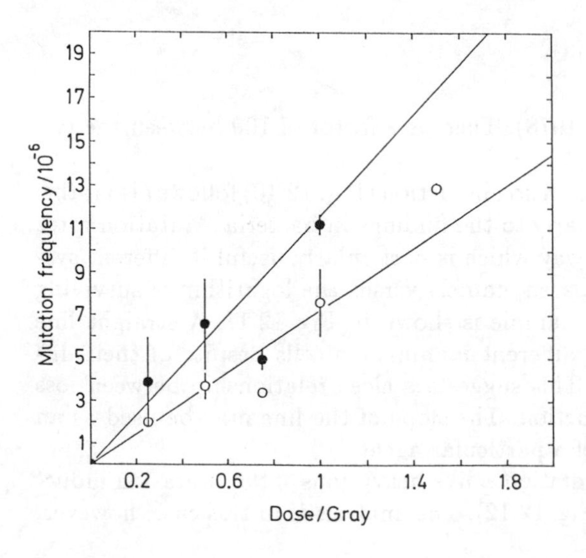

Figure 12.8. Mutation to 6-TG resistance (•) and to fluotothymidine resistance (o) in human TK6 lymphoblast cells (KOENIG and KIEFER 1988)

and UV, resistance to ouabain, however, only by UV. This indicates that the mechanisms of mutation induction are different, and it has been suggested that UV causes predominantly point mutations while X rays lead to larger deletions. Analyses at the molecular level with the aid of restriction enzymes demonstrated, in fact, the predominance of deletions but they could not rule out that point mutations do also occur. More about this is found in Sect. 12.5.

As in microorganisms, UV is considerably more effective than X rays. A typical response curve is shown in Fig. 12.9. In comparing the initial linear

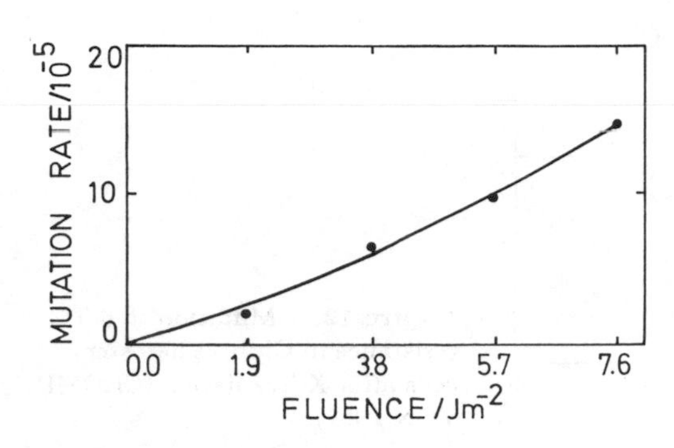

Figure 12.9. Mutation to 6-TG resistance in Chinese hamster cells after exposure to 254 nm-UV (ZOELZER and KIEFER 1984)

parts of the dose-effect curves, the following shapes are found (6-TG resistance in Chinese hamster cells): $1.3 \times 10^{-5}\,\mathrm{Gy}^{-1}$ (X rays) and $3.6 \times 10^{-5}/\mathrm{Jm}^{-2}$ (UV). These values have to be related to the mean lethal doses D_0 as taken from the survival curves. Thus:

$$1.8 \times 10^{-3}/D_0 \quad \text{(UV)}$$

and

$$1.8 \times 10^{-5}/D_0 \quad \text{(X rays)}$$

(Data taken from CLEAVER 1978). There is a factor of 100 between the two radiation types.

The action spectrum of mutation induction (Fig. 12.10) follows also rather closely DNA absorption, similarly to the findings in bacteria. Mutation rates may be plotted in a different way which is particularly useful if different systems or agents are to be compared, namely versus the logarithm of surviving fraction instead of dose. An example is shown in Fig. 12.11. A straight line is obtained fitting data from different mammalian cells despite of their different radiation sensitivities. This suggests a close relationship between loss of survival and mutation induction. The slope of the line may be used as an indicator of "mutagenicity" of a particular agent.

With densely ionizing radiation – like heavy ions – the mutation induction curves become steeper (Fig. 12.12). The analysis is in this case, however,

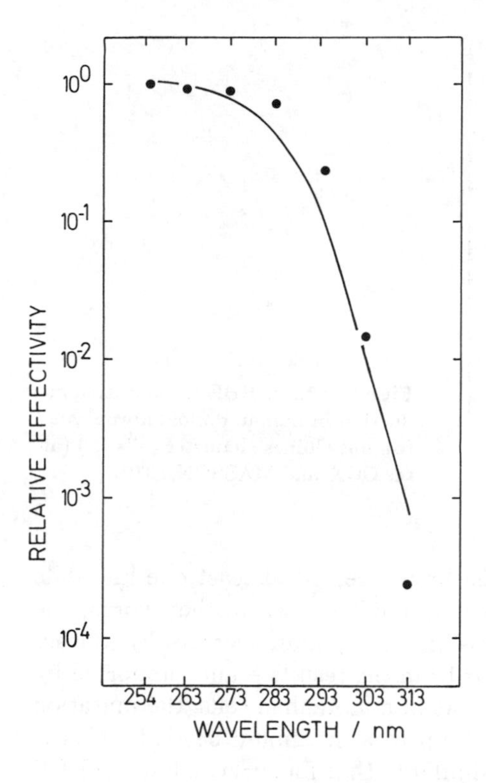

Figure 12.10. Action spectrum for mutation to 6-TG resistance in Chinese hamster cells (points). DNA absorption is indicated by the line (ZOELZER and KIEFER 1984)

hampered by the heterogeneous pattern of energy deposition. Since the cells receive on an average only very few hits, survivors and mutants belong to

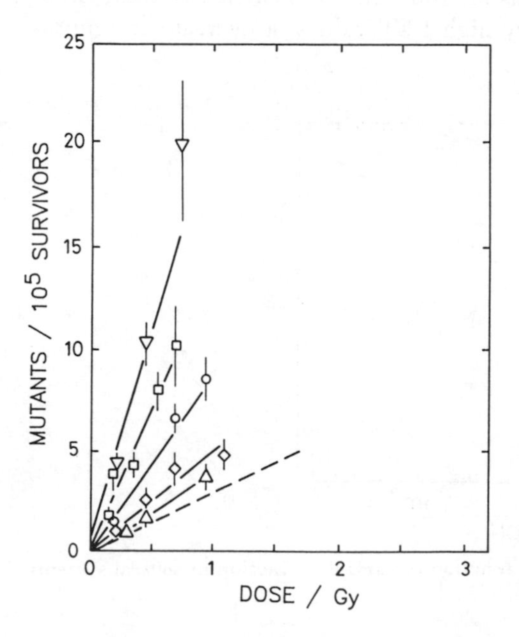

Figure 12.11. Mutation by ion exposure in human diploid fibroblasts with increasing LET (from bottom to top): 20, 28, 50, 70 and 90 keV/micrometer. The broken line gives the gamma-response for comparison (after COX and MASSON 1979)

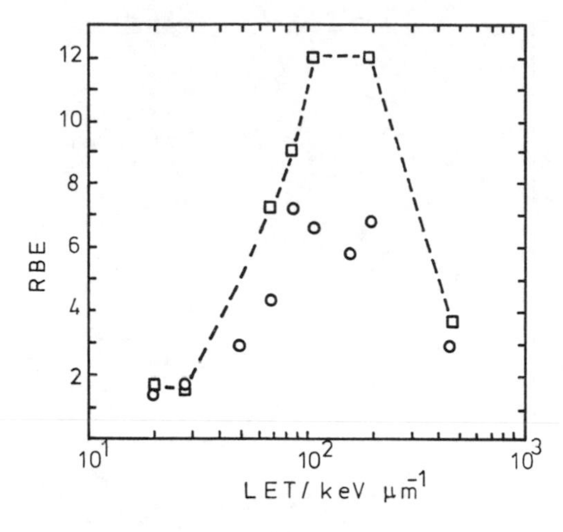

Figure 12.12. RBE for mutation induction in human diploid fibroblasts (o) and Chinese hamster cells (□) (after COX and MASSON 1979)

different subpopulations. Mutants must have received at least one hit while a large part of the survivors is completely unhit or, with other words, the mean absorbed energy in the mutants is higher – (on an average) by at least one particle hit. Nevertheless, it is clear from the results – and supported by a more careful quantitative analysis – that in mammalian cells, the mutation induction efficiency increases with LET up to a maximum beyond which it declines. The qualitative behaviour is similar to that for survival but the RBE values are higher (Fig. 12.13).

The increase seen in Fig. 12.13 is not easily understood. From this result it follows that mutagenic lesions are induced more efficiently than lethal lesions with higher LET. With very high LET values, a decrease in "muta-

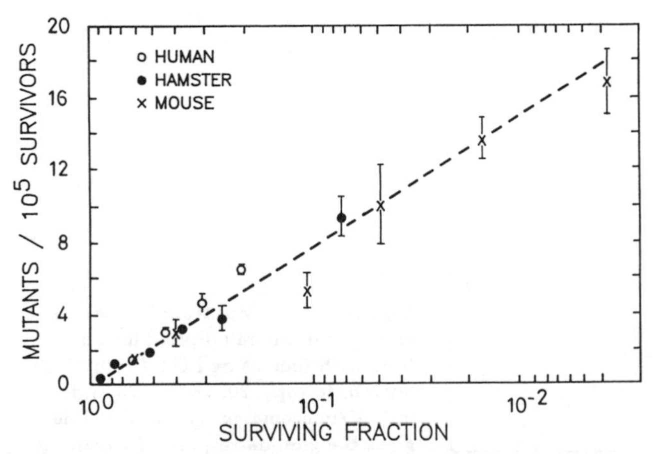

Figure 12.13. Mutant frequencies as a function of surviving fraction in several systems (after THACKER 1979)

genicity" is expected: If the traversal of one particle suffices to kill the cell, radiation-induced mutants cannot be recovered since all survivors left consist only of cells with no hits which, of course, cannot be mutated by radiation. This expectation is borne out by experiments with very heavy ions.

12.4 Comparison of Radiation-induced Mutations in Different Systems

In Sect. 8.2, a systematic relationship of cell sensitivity in terms of survival with nuclear parameters was presented. Similar attempts have also been made for mutation induction. If such a dependence could be unequivocally proved it would open a way to estimate genetic hazards with the help of microbial test systems which are easier to handle and technically more reliable than mammalian cells. An example of such an analysis is shown in Fig. 12.14. It

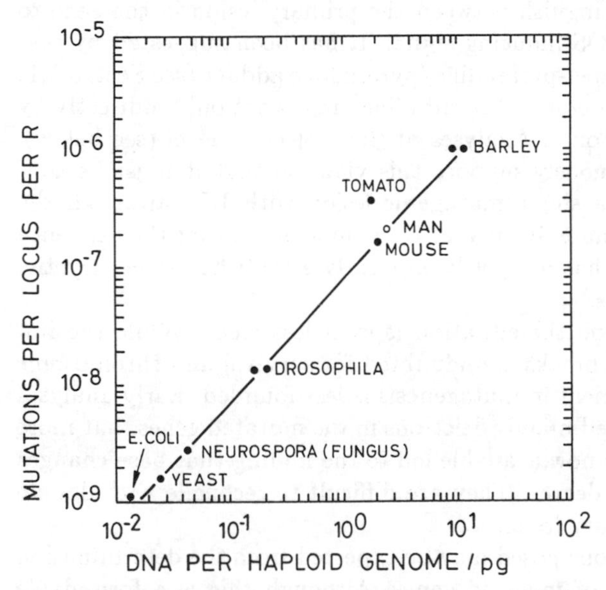

Figure 12.14. Dependence of mutation rate per unit dose on mean haploid DNA content (after ABRAHAMSON et al. 1973)

demonstrates that the mutation frequency per unit dose increases linearly with the DNA content per cell. This analysis has been heavily criticized (see under Further Reading), and it may be doubtful whether the simple relationship really holds. Nevertheless, it is qualitatively correct that the mutagenic potential of a given dose increases with DNA content. Apart from its heuristic value, this statement has important consequences for the understanding of the mutational process: It may be assumed that gene sizes do not differ by orders of magnitude in different organisms. If the gene is the sole target for mutation induction by radiation, one would expect that the sensitivity would not vary to any great extent. This is obviously not the case; thus a mutation is not simply caused by a hit in the respective gene but the total genome (or a significant part of it) is involved in a not yet understood manner.

12.5 Molecular Aspects of Radiation-induced Mutagenesis in Cells

The advent of biotechnological methods has made it possible to study radiation-induced mutations at the molecular level. The first question to be asked is on the nature of the primary lesion. The fact that mutations are photoreactivable in microorganisms points strongly to pyrimidine dimers (see Sect. 13.3.1) but this seemingly obvious interpretation has recently been questioned. At least in *E. coli*, most mutations are created via the SOS system (Sect. 13.3.4), thus we distinguish between the primary lesion in the gene to be investigated and the SOS-inducing signal. It has been suggested by several investigators that the 6–4-pyrimidine/pyrimidone adduct (see Sect. 6.1.1) represents the miscoding lesion and pyrimidine dimers act only indirectly by switching on the SOS response. Analyses at the molecular level (see below), however, do not unambiguously support this view, so that it must be concluded that there is not a single mutagenic lesion with UV, although the 6–4-product may predominate. It must also be emphasized that the existence of error-prone SOS repair has not yet been clearly established in eucaryotic, especially mammalian cells.

With ionizing radiation the situation is even less clear. While the importance of double strand breaks is undoubted for survival and chromosome aberrations, their involvement in mutagenesis is less founded. Early analyses indicated that X rays caused mainly deletions in the mutated genes, but more refined methods which are now available led to the finding that base changes occur also to a significant degree. They are difficult to reconcile with double strand breaks as the primary lesion.

Answers to the questions posed may be expected from the determination of the complete sequence of mutated genes. Although this is a formidable task, it has been achieved in a few cases. The best example is the locus for *adenine phosphoribosyltransferase* (APRT). The gene is autosomal but strains are available which are *hemizygous*, i.e. they possess only a single copy, so that forward mutations may be scored in selective medium containing 8-aza adenine to which the mutants are resistant. The gene is small (3900 base pairs in Chinese hamster cells) which facilitates the analysis greatly.

The basic scheme of the method is briefly summarized in Fig. 12.15: The DNA of established mutants is digested by suitable restriction endonucleases and separated by gel electrophoresis. The fragment containing the aprt gene is then linked to a bacterial vector carrying two suppressible genetic markers.

ESTABLISHMENT OF MUTANTS BY GROWTH
IN 8–AZAADENINE

↓

CUTTING BY RESTRICTION ENDONUCLEASES

↓

SEPARATION BY GEL ELECTROPHORESIS

↓

LIGATION TO A BACTERIAL VECTOR CONTAINING
TWO SUPPRESSIBLE MARKERS

↓

COREPLICATION TOGETHER WITH A VECTOR CONTAINING
supF AND FLANKING SEQUENCES OF THE APRT–GENE.
RECOMBINATION

↓

SELECTION OF RECOMBINANT VECTORS IN A
NON–SUPPRESSOR HOST

↓

AMPLIFICATION

↓

ISOLATION OF THE MUTANT GENE AND SEQUENCING

Figure 12.15. Experimental scheme for the study of mutations at the molecular level (after DROBETSKY et al. 1987)

The plasmid is replicated in a host bacterium together with another vector which carries both a suppressor mutation (supF) and flanking sequences of the aprt gene (but not the gene itself). Co-multiplication leads to a comparatively high frequency of recombinations (10^{-3}) between the two vectors leading to a plasmid containing the mutant aprt gene and the supF gene. The new combination can be selected by plating on a non-suppressor host and amplified for sequencing.

Results are summarized in Table 12.1: No deletions are found after UV treatment while they comprise about one quarter of X-ray-induced mutants. This last figure, however, is quite small in view of earlier assumptions that ionizing radiation does not cause point mutations. The high yield of double mutations at adjacent sites with UV is interesting suggesting some instability of certain DNA-regions. While the X-ray-induced mutations are randomly scattered over the gene, this is not the case with UV where typical "hotspots" are found. It appears, therefore, that either there are sequence specificities or

Table 12.1. Radiation induced alterations in the APRT-gene of CHO-cells (Chinese hamster ovary cells)

dose or fluence	γ-rays [a] 5 Gy	UV [b] 5 J m⁻²	spontaneous [c] 0
transitions	4 (25%)	27 (66%)	22 (71%)
transversions	7 (44%)	12 (29%)	7 (25%)
frame shift	1 (6%)	2 (5%)	1 (3%)
deletions	4 (25%)	0	1 (3%)
double mutants [d]	-	7	1
total	16	34	30

a) GLICKMAN et al. 1987
b) DROBETSKY et al. 1987
c) DE JONG et al. 1988
d) included in the other categories

structural features of DNA which make these parts particularly prone to UV damage.

It is interesting to compare the analysis of the mammalian aprt gene with similar investigations in bacteria, particularly in view of the SOS mechanism whose role in mutagenesis is well established. There are not yet very many studies, and the results have still to be considered as preliminary but they are reported here – deviating from the general philosophy of this book – since they indicate an important line of future research.

Comparative data are summarized in Table 12.2. The procaryotic gene was that for the cI lambda-phage repressor. It is quite small (about 700 bp) and located between two essential genes whose rupture would cause phage inviability. This situation restricts the analysis to some extent since larger deletions would not be recovered. Thus, the results are not without bias. The irradiated phages were either grown in SOS-induced host cells or in those defective in this pathway (recA$^-$ cells). For comparison, they were also

Table 12.2. γ-ray-induced alterations in the lambda-c-1-repressor-gene, analyzed by sequencing of isolated mutants. (After HUTCHINSON, 1987)

	phages in SOS-induced cells	phages in rec⁻-cells	in cell genome	human APRT-gene [b]
dose / Gy	6000	5000	410	5
spontaneous rate	10^{-5}	10^{-5}	3.5×10^{-6} [a]	-
induced rate	1.6×10^{-3}	2×10^{-4}	7×10^{-5}	-
transitions	20 (42%)	16 (76%)	19 (37%)	(25%)
transversions	21 (41%)	2 (10%)	19 (37%)	(44%)
frame shift	7 (15%)	3 (14%)	11 (21%)	(6%)
extended alterations	0	0	3 (6%)	(25%)
total	48	21	52	

a) estimated
b) from Table 12.1

integrated as prophage in the cellular genome. The last example is thought to be closest to the situation of cellular genes. The data from the much larger mammalian aprt gene (taken from Table 12.1) are repeated to facilitate comparison.

There are noticeable differences in the fractions of frameshifts and gross rearrangements (including deletions) while the percentage of transversions is remarkably similar. The absence of SOS repair obviously increases the fraction of transitions considerably.

Sequence analyses as described above are so far restricted to few genes for reasons already mentioned. A method of wider applicability but with less resolution power is to digest mutant genes by restriction nucleases and measure the distribution of fragment lengths.

Results are summarized for a few genes in Table 12.3. They deserve some comments. The data for the aprt gene appear to be compatible with those from the sequencing analysis if one bears in mind the limits of resolution. There are clear differences between different genes with radiation induced mutations but also between species for the same gene. The fraction of deletions is generally larger in the hgprt gene but even more so in the hamster as compared to a human cell line (69% and 37%, respectively). The last two columns refer to a bacterial gene which was inserted into the DNA of Chinese hamster ovary (CHO) cells. In this case, the pattern of mutational events is strikingly different from the other examples with a pronounced overrepresentation of deletions (also for spontaneous mutations) which demonstrates that some caution is indicated when using "exogenous" genes.

Table 12.3. Molecular alterations (%) in spontaneous and radiation-induced mutants of some genes analyzed with the aid of restriction endonucleases. (BREIMER, 1988)

gene	aprt hamster		hgprt hamster		hgprt human		gpt bacterial/hamster	
	spont.	ion. rad.	spont.	ion. rad.	spont.	ion. rad.	spont.	ion. rad.
no detectable alterations	88	81	86	31	82	43	40	0
nuclease cutting site alterations	7	1	–	–	–	–	0	0
partial deletions	4	8	12	27	8	7	43	4
total deletions	–	–	2	41	10	30	17	96
insertions	1	1	0	2	0	20	0	0
others	0	9	–	–			0	0
no. of mut. analyzed	187	80	52	113	28	56	23	25
resolution power/bp	25	25	500	500	500	500	?	?

In summary, the modern methods of "genetic engineering" and sequencing lead to some changes of so-far accepted views of radiation-caused mutagenesis. Most importantly, it can no longer be said that ionizing radiations lead mainly to deletions and do not induce point mutations. The demonstration of mutational "hotspots" with UV is another noticeable result.

Further reading (12.1–12.5)
AUERBACH 1976, BREIMER 1988, BRIDGES 1985, de NETTANCOURT and SANKARANARAYAN 1979, DOUDNEY 1976, KIMBALL 1978, RIEGER and MICHAELIS 1967, ROSSMAN and KLEIN 1988, STRAUSS 1968, THACKER 1986, 1987, WALKER 1984, WOOD 1986

12.6 Neoplastic Transformation *in vitro*

The understanding of radiation-induced carcinogenesis constitutes without doubt one of the central problems in radiation biology, not only because of its importance for hazard assessment but also because of the fundamental interest to unravel the mechanisms of tumour induction. Animal populations are obviously suitable test objects but the experiments are time and cost intensive, and the results not always easily to be interpreted if the question is on the initial cellular effects. The development of techniques to study neoplastic transformation in cell cultures has, therefore, to be considered as an important achievement.

Usual established cell lines cannot be used here since they are already transformed. The starting materials are either fresh explants or special untransformed cell lines which are able to grow in culture maintaining their untransformed state. An example of a widely used approach is depicted in Fig. 12.16: Hamster embryos are dissected and trypsinized to yield a single cell suspension which is plated in the usual way. Untransformed cells grow in a unicellular monolayer. After prolonged incubation, some of them form dense "foci" which may be tentatively identified with transformed cells. That this is really the case is proven by injecting them into isogenic animals where they cause tumours with high probability while non-focal cells are ineffective in this respect. More recently, a special established mouse cell line, termed C3H10T1/2 has been widely used for principally the same assay. These cells are "immortalized", i.e. able to grow in culture for prolonged times but they are not transformed and show contact inhibition. Although simple in principle, the test procedure requires great care and experience to yield meaningful and reproducible results.

Transformation is caused both by ultraviolet and ionizing radiation. The action spectrum resembles that of mutation induction and follows closely the DNA absorption spectrum. Figure 12.17 shows transformation yields (i.e. transformation probability per cell plated, not per survivor) after X ray and neutron exposure. The absolute frequencies are relatively low which demon-

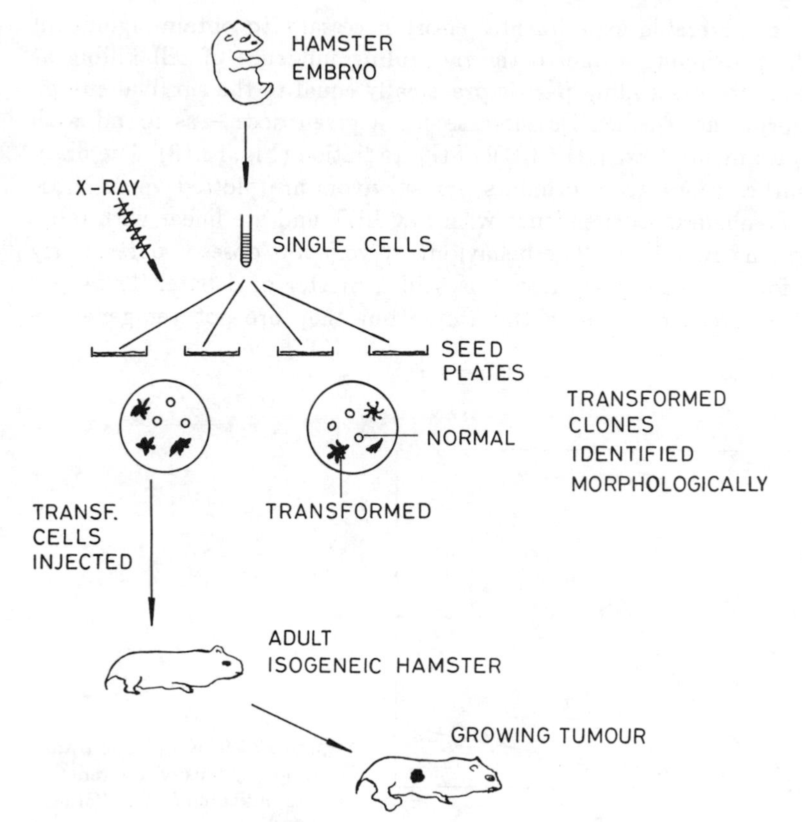

Figure 12.16. A protocol to study neoplastic transformations in vitro (after BOREK 1979)

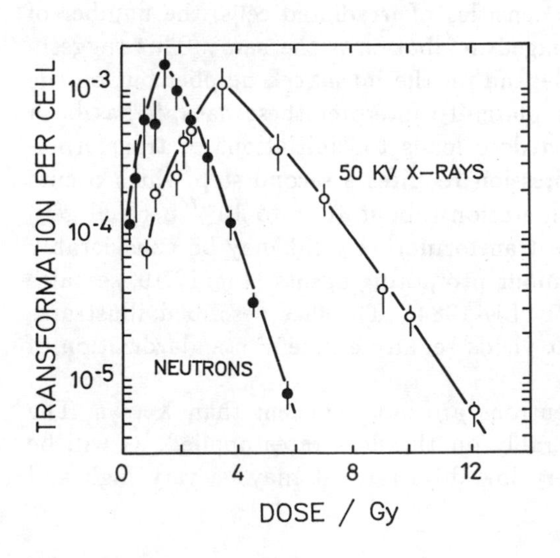

Figure 12.17. Yield of transformants in mammalian cells after X-ray and neutron exposure (after HAN and ELKIND 1979)

strates the considerable experimental effort necessary to obtain significant results. The maximum is due to the overriding influence of cell killing at higher doses. The descending part is practically equal to the survival curve.

Transformation frequencies increase for a given dose – as found with other test parameters – with the LET of the radiation (Fig. 12.18). The dose-response curves where transformants per survivors are plotted versus dose show a linear-quadratic dependence with low LET and are linear with more densely ionizing radiations. The behaviour at very low doses – particularly important for radiation protection – is still a matter of debate. There are suggestions of steeper slopes in this range but they are not yet generally accepted.

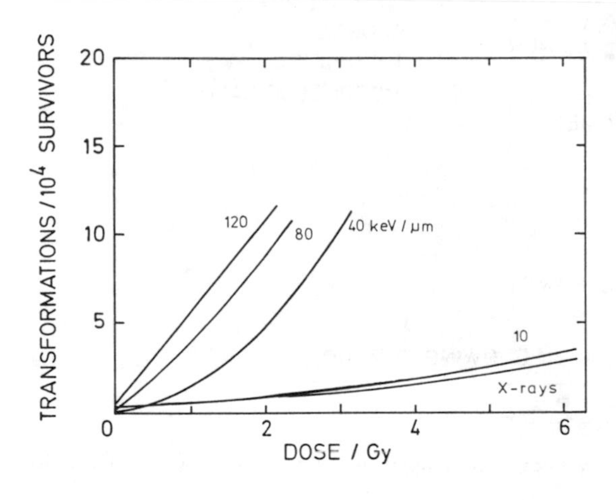

Figure 12.18. Neoplastic transformation by ions of different LET as indicated (after HEI et al. 1988)

In vitro transformation appears to be a two-stage process. If the culture plates are seeded with different densities of irradiated cells, the number of foci which are found after prolonged incubation is the same. This suggests that the foci number does not depend on the initial cell number but on the ***total number of divisions*** on the plate. To interpret these data, it has been suggested that a given radiation dose leads to "initiation" of transformation in all cells but its final expression requires a second step which occurs with low probability during proliferation (about 10^{-6} to 10^{-7} per cell generation). During this period the transformation yield may be considerably increased by treatment with tumour promoting agents (Fig. 12.19, see also Sect. 20.3) (KENNEDY and LITTLE 1984). The fact described illustrates that comparison of transformant yields require a careful standardization of the experimental protocols.

Figure 12.17 shows that neutrons are more efficient than X rays. The RBE, however, depends considerably on the dose rates applied, as will be discussed in Sect. 14.2. With very low dose rates it may be very high and approach values of 70.

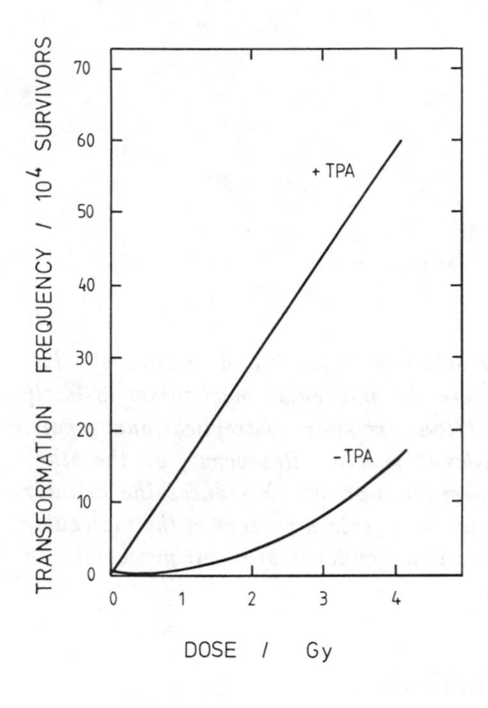

Figure 12.19. The increase of transformant frequency by posttreatment by the tumour promoter TPA (after LITTLE 1981)

Further reading (12.6)
BOREK 1979, 1980, BOREK, ONG and MASON 1987, HALL and HEI 1985, 1986, HAN, HILL and ELKIND 1984, LITTLE 1986, UNSCEAR 1986

Chapter 13
Repair and Recovery

A distinction is made in this chapter between "repair" and "recovery". The first includes only those processes where the molecular mechanism is fairly well known. Examples are photoreactivation, excision, postreplicational repair, mismatch repair, "SOS"- and strand-break repair. "Recovery", on the other hand, is operationally defined as a phenomenon which reduces the cellular radiation effect without specific reference to a certain process at the molecular level. The last section discusses the genetic dependence of repair processes and also their relevance for human health.

13.1 General Aspects and Definitions

The discovery that cells are capable to repair damage in their genome is without doubt one of the most spectacular successes of radiobiological research. Its significance for general biology can hardly be overestimated. The fascination of these phenomena, however, causes often the unproven assumption that every change in radiation sensitivity signifies repair even if there is no evidence at the molecular level. This may lead to unnecessary complications. A clear terminology is, therefore, suggested: "Repair" is only to be used for processes which can be defined at the molecular level and where the mechanism of damage removal can be identified or at least inferred. If only cellular properties or functions are changed, e.g. survival or mutation frequency, the term "recovery" is to be preferred. Examples are given in the following sections. If, e.g., survival is enhanced by dose fractionation, this is clearly "recovery". Whether one or more repair processes are involved is a matter of mere speculation. "Repair" and "recovery" are in this sense in a similar relationship as "genotype" and "phenotype".

13.2 Specific Repair Processes

13.2.1 Photoreactivation

This is a good example for what was said in the introductory section. The process to be described is a light-dependent enzymatic splitting of pyrimidine dimers *in situ*, i.e. in the DNA helix. When it was discovered, the mechanism

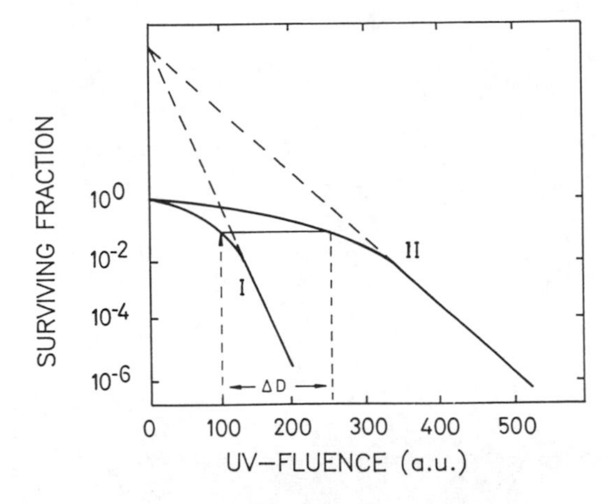

Figure 13.1. Survival after UV exposure with (II) and without (I) photoreactivation. The DMF is here assumed to be 2.5, the photoreactivable sector 0.6. The dose-decrement is also indicated (modified after NOVICK and SZILARD 1949)

was not clear and the rather vague term "reactivation" was used. Although it is now known that it represents a genuine repair process, the name is kept for historical reasons. If UV-irradiated cells are exposed subsequently to near UV or visible light, the original effect is greatly reduced. This holds true, both for survival as well as mutation induction (see Fig. 13.1).

This finding alone, however, is not sufficient to prove photoreactivation proper, i.e. molecular repair. It has been found that the increased resistance may – under certain circumstances – be found in cells which lack the responsible enzyme. In this case, the light exposure may lead to a slowing down of cell progression thus favouring the intervention of other repair processes before the damage is fixed, comparable to "delayed plating recovery" (see Sect. 13.3.2). To distinguish it from true photoreactivation this phenomenon has been called "indirect photoreactivation" (a misnomer) or "photoprotection" (better). Both can be differentiated by the sequence of exposure: Photoprotection is also found if visible light *precedes* UV irradiation. The photochemical splitting of dimers (see Chap. 6) by short wavelength UV must also not be confused with photoreactivation since it proceeds – apart from the different active wavelength region – without the intervention of an enzyme.

Figure 13.1 shows that photoreactivation acts "dose-modifying", i.e. it is characterized by a constant dose-reduction factor, irrespective of the effect level. This "dose modifying factor" (DMF) is the ratio of doses causing the same effect with and without photoreactivation. Another widely used quantity is the "photoreactivable sector" PRS. Both are related in the following way:

$$\mathrm{PRS} = 1 - \frac{1}{\mathrm{DMF}} \quad . \tag{13.1}$$

The substrates for the photoreactivating enzyme, the *photolyase*, are pyrimidine dimers, the process follows the classical scheme (Fig. 13.2):

$$\mathrm{P_yP_y + PRE} \;\; \overset{k_1}{\underset{k_2}{}} \;\; (\mathrm{EP_yP_y}) \;\; \overset{h\nu}{\underset{k_pF}{}} \;\; \mathrm{P_y + P_y + E} \tag{13.2}$$

(f: fluence).

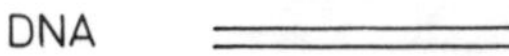

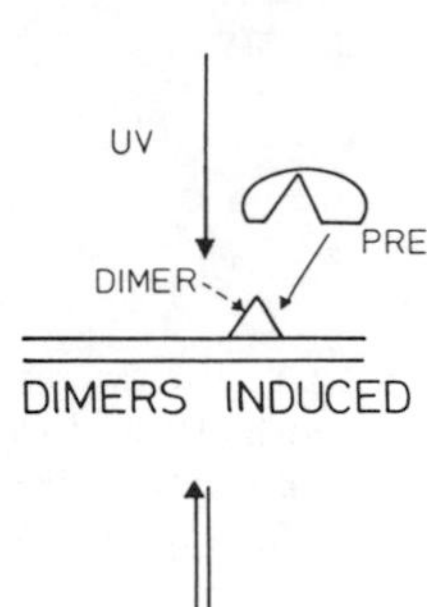

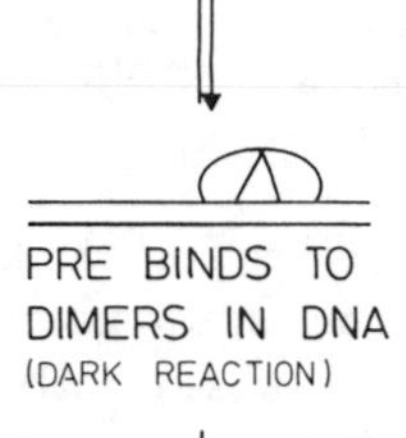

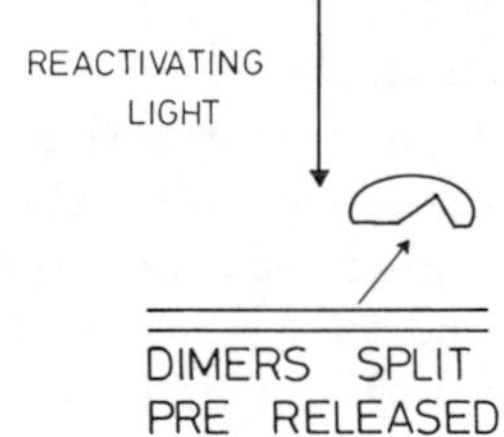

Figure 13.2. The mechanism of enzymatic photoreactivation (PRE = photoreactivating exzyme of photophospholyase)

Pyrimidine dimers P_yP_y form a reversible complex with the enzyme in a dark reaction. It can be split into the monomers and free enzyme upon light absorption.

The photolyase gene from E. coli has been cloned and introduced into a plasmid allowing overproduction of the protein under suitable conditions. Thus, it was possible to examine the properties *in vitro*; they are summarized in Table 13.1. The high relative amount of tryptophane is unusual but its

Table 13.1. Properties of photolyase isolated from E. coli. (SANCAR, 1987)

Molecular weight	54,000 u
No. of amino acids	471
Subunits	1 polypeptide, 1 FAD, 1 second chromophore
Absorption maxima (nm)	280 (1.22 - 15.7 x 10^5)
(extinction coefficients in M^{-1} cm^{-1})	384 (1.81 - 2.13 x 10^4)
	480 (3.6 - 4.7 x 10^3)
	580 (3.6 - 5.0 x 10^3)
	625 (2.8 - 4.4 x 10^3)
Fluorescence excitation maximum	398 nm
Fluorescence emission	470 - 480 nm

significance is unknown. The enzyme contains two chromophores, a flavine-adenine-dinucleotide (FAD) radical and a second one yet unidentified.

The *in vitro* rate constants have been determined; they are listed in Table 13.2.

Table 13.2. *In vitro* and *in vivo* reaction parameters of E. coli photophospholyase. (SANCAR, 1987)

	in vitro	in vivo [a]
Equilibrium constant	$6 - 14 \times 10^7$ M^{-1}	$8.4 - 57 \times 10^7$ M^{-1}
K_1	$1.4 - 4 \times 10^6$ M^{-1} s^{-1}	1.1×10^6 M^{-1} s^{-1}
K_2 slow component	3×10^{-2} s^{-1}	$0.2 - 1.3 \times 10^{-2}$ s^{-1}
fast component	$0.6 - 1.3 \times 10^{-3}$ s^{-1}	$0.6 - 3.2 \times 10^{-3}$ s^{-1}
K_p at λ = 385 nm (reduced form)	1.5×10^{-2} m^2 J^{-1}	1.1×10^{-2} m^2 J^{-1}

a) HARM 1970

The light-dependent step – the splitting of the dimers – depends on the oxidative state of the FAD moiety. It is active only in its reduced form and the measured parameters agree then with *in vivo* data (see below). This led to the proposition of the following action scheme:

$$SC + FADH_2 + P_yP_y \quad SC^* + FADH_2 + P_yP_y$$
$$SC + FADH_2^* + P_yP_y \quad SC + FADH_2^+ + P_yP_y^-$$
$$SC + FADH_2 + P_y + P_y \quad . \qquad (13.3)$$

The second chromophore (SC) is excited and transfers its energy then to the reduced FAD component ($FADH_2$) yielding $FADH_2^*$. This donates an electron to the dimer which ultimately causes the monomerization. It is not yet clear whether the excitation transfer really occurs or whether the photon is directly absorbed by the FAD part. The extinction coefficient at the maximum of absorption is fairly high ($16\,100\,M^{-1}\,cm^{-1}$) and does not change very much if the FAD is oxidized. The quantum yield of splitting, on the other hand, is close to 1 for the reduced form but only 0.07 when it is oxidized. It has thus been concluded that the enzyme is reduced *in vivo*.

The characterization of the in vivo process is possible with the aid of "flash photoreactivation". For this, a very sensitive *E. coli* mutant is used where – because of the lack of other repair functions – already a small number of dimers reduces survival significantly. Upon photoreactivation it is increased, i.e. the cells react as if they had received a smaller dose. The calculated "dose decrement" ΔD (see Fig. 13.1) corresponds to a certain number of split dimers which can be calculated since the number of dimers per unit fluence has been measured. Using this technique, the number of enzyme molecules per cell can be estimated: If the substrate – i.e. pyrimidine dimers – is greatly in excess, virtually all enzyme molecules are bound. If photoreactivation is performed in such a short time (a "flash") that reformation of the complex is not possible, then the number of split dimers (as calculated from the dose decrement) equals that of the available enzyme molecules. The value obtained is surprisingly small – about 20 per *E. coli* cell, similar numbers were also found in yeast –, i.e. only about 10^{-5} of the total protein content. This explains why earlier attempts to purify the enzyme in larger amounts were largely unsuccessful and demonstrates also the power of the new techniques of gene technology.

The equilibrium association constant $K_a = K_1/K_2$ may be estimated by two approaches: The first is to vary the UV fluence and to determine the number of dimers split when there is no substrate saturation. Since the number of enzyme molecules is known, K_a can be calculated (*equilibrium method*). The second – *dynamic* – method consists of a variation of the time between dimer induction and the photoreactivating flash.

The binding coefficient K_1 may be determined by giving a sequence of flashes with fixed intervals whose lengths are varied.

The parameters found are also listed in Table 13.2. There is good agreement between *in vitro* and *in vivo* data.

The third reaction coefficient K_3 depends, of course, on light intensity. If I is the quantum fluence rate (fluence per time interval) the number ΔN of quanta absorbed is given by (see also Sect. 2.1):

$$\Delta N = \varepsilon \cdot c \cdot I \cdot \Delta V \cdot \ln 10 \qquad (13.4)$$

where ε is the extinction coefficient of the absorbing units, c their concentration and ΔV the volume element. The number of monomerized dimers is obtained by multiplying with the quantum yield Q. Since K_3 is the number of split complexes per unit time, it follows that:

$$K_3 = \varepsilon \cdot I \cdot Q \cdot \ln 10 \qquad (13.5)$$

$\varepsilon \cdot Q$ has been measured to be $24\,000\ M^{-1}\,cm^{-1}$ *in vivo* and $16\,100\ M^{-1}\,cm^{-1}$ *in vitro* for the reduced enzyme which shows that the quantum yield is about 1.

The action spectrum is displayed in Fig. 13.3. The wavelength dependence for photoprotection – which extends to considerably longer wavelengths – is also shown for comparison. The spectra of true photoreactivation are similar for all organisms but not identical.

Photoreactivation as measured by increased survival after near UV exposure to far-UV-irradiated cells is not found in all systems. It can readily be demonstrated in all microorganisms – except for photoreactivation-deficient (phr^-) mutants – but not in all higher eucaryotes. It appears that the evolutionary border is with the *marsupialia*, in other words, amphibians, birds and

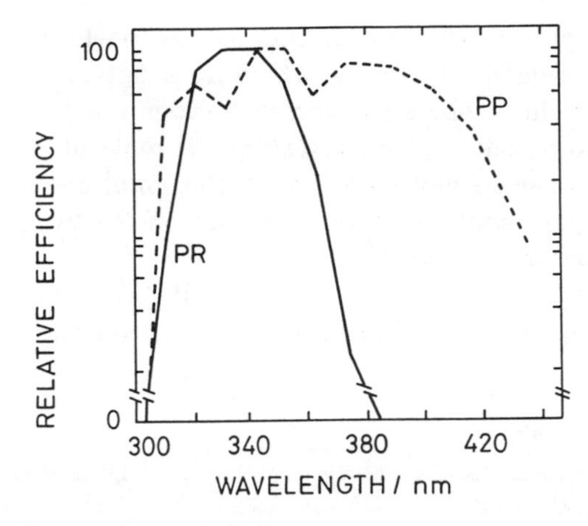

Figure 13.3. Action spectra for photoreactivation (PR) and photoprotection (PP) in E. coli (after SETLOW 1966)

kangaroos have it but not the placental mammals. The last point, however, is still controversial. Enzyme activities which monomerize dimers with the help of light have been frequently found in mammalian cells in culture and also in skin *in situ* but attempts to demonstrate increased survival or reduced mutagenesis have failed so far. The consequences of this discrepancy are not quite clear. It may either be that the results reported are due to unknown artefacts or that pyrimidine dimers play a much minor role for cellular effects in mammals as compared to simple systems.

13.2.2 Excision Repair

The process discussed here is probably the best known and most important one. Its details could be unravelled in recent years, at least in *E. coli.* It was also shown that excision repair does not only play a role for radiation induced damage but removes all kinds of lesions in cellular DNA. It constitutes, thus, a central mechanism for the maintenance of genetic information which is a constitutive factor of life. The first demonstration was the excision repair of pyrimidine dimers in bacteria, it is not yet clear whether similar mechanisms prevail with alterations induced by ionizing radiations.

Schematically the process may be described in the following way: The first step is the creation of a single strand incision close to the lesion which is subsequently removed. The resultant gap is filled by "repair replication" where the other undamaged strand is used as template. The remaining discontinuity is then finally sealed resulting in complete restitution of the originally damaged site.

The sequence described is readily demonstrated by experiments: Incision leads to a reduction of single-strand molecular weight which may be measured by ultracentrifugation after strand separation. Since UV does not induce single strand breaks directly, they have to be produced by enzymatic action. Incision could be demonstrated both for *in-vivo* as well as *in-vitro* cell extracts. It is interesting to note that the number of single-strand gaps is always smaller than the number of pyrimidine dimers. One has to conclude from this that excision repair is a sequential process which does not proceed at random but in a concerted fashion where incision and gap filling are closely linked. This assumption is borne out by the analysis of the molecular mechanism (see below). It makes good sense biologically since the simultaneous presence of many single-strand gaps could lead to an overlap on the two opposite strands thus creating double-strand breaks.

The demonstration of repair replication is not so easy since it has to be ascertained that any measured incorporation of labelled precursors is not due

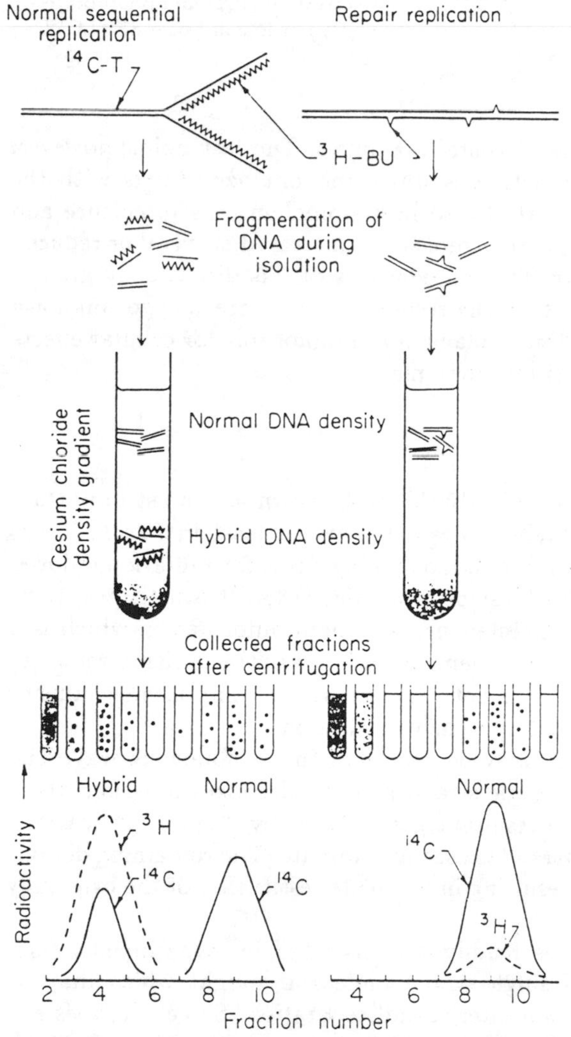

Figure 13.4. Method to distinguish between semi-conservative and repair replication (after HANAWALT and HAYNES 1967, from SMITH 1972 with permission)

to normal semi-conservative DNA replication which may proceed at the same time. The essential difference is that during repair only comparatively small patches are inserted into "old" preformed DNA, while with normal replication complete new strands are synthesized. A distinction between these two alternatives is possible by a modification of the MESELSON-STAHL experiments, which were originally designed to demonstrate semi-conservative replication (Fig. 13.4): The cells are incubated before radiation exposure in a medium containing DNA precursors labelled by stable isotopes of higher nuclear mass (^{15}N, ^{2}H) or BUdR instead of thymidine which also has a higher molecular mass. This leads to a higher density of the DNA which is synthesized during the incubation period thus constituting a "density labelling". If the precursors are additionally radioactively labelled, the "heavy" DNA may be separated by ultracentrifugation in a gradient of increasing density (usually CsCl or Cs_2SO_4). The duration of centrifugation has to be chosen long enough so that equilibrium is reached and the molecules of different densities are distributed according to Archimedes' principle. In order to "isolate" the heavy strains sufficiently, the incubation is usually followed by a short period in "light" medium. After radiation exposure, the cells are grown in the presence of "light" thymidine which is labelled by a different radioactive isotope. After sorting according to density – as described –, repair replication is demonstrated by thymidine incorporation into the preformed "heavy" strand while normal semi-conservative replication occurs only via incorporation in the "light" strand.

Another method to demonstrate repair replication makes use of the fact that BUdR which has been incorporated in DNA leads to single-strand breaks after exposure to 313 nm-UV light. The analogue is added to the incubation medium after radiation exposure. After the repair period, the cells are then irradiated with sufficiently high fluences at 313 nm which is comparatively strongly absorbed by BUdR. Repair replication which is dispersed over the "old" DNA is then visualized by a reduction of single-strand molecular weight, while semi-conservative replication where BUdR incorporation is continuous creates very small DNA fragments. By this method it is also possible to "count" the number of repair patches and to correlate it to the number of dimers induced.

"Unscheduled" DNA synthesis (Sect. 10.3) is presumably also related to repair synthesis but a clear-cut proof is lacking.

The excision step is easy to demonstrate by the appearance of pyrimidine dimers in the extracellular medium. This experimental finding was, by the way, the first indication for the existence of excision repair (SETLOW and CARRIER 1964, BOYCE and HOWARD-FLANDERS 1964).

Ligation – the final step – is the mirror image of incision and readily seen by the increase of single strand molecular weight.

The whole process of excision repair is under tight genetic control (see Sect. 13.4). Recent years brought a complete clarification of all steps involved, at least in *E. coli*, which was only possible because the relevant genes were cloned so that the products could be produced in larger amounts. The present view is sketched in Fig. 13.5: Contrary to previous assumptions, incision and

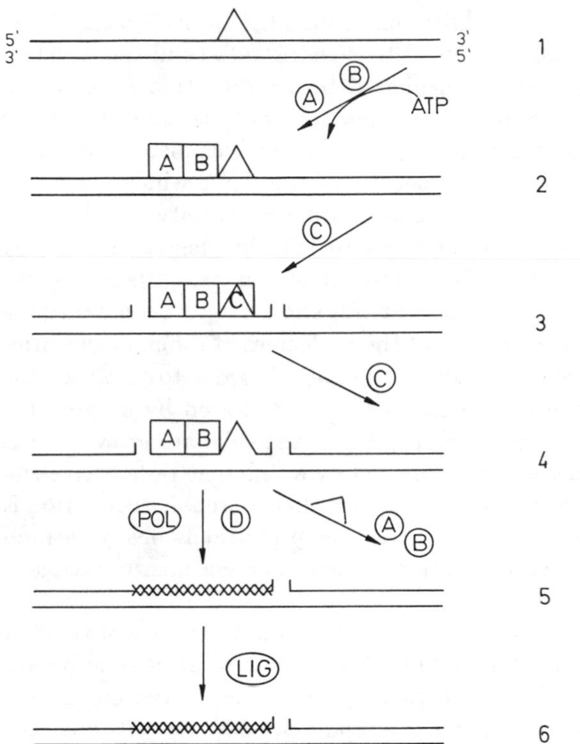

Figure 13.5. The mechanism of excision repair in E. coli (after SANCAR 1987). See text for further explantion

excision are not separated but proceed in a single step. The enzyme responsible which has been termed "ABC excinuclease" consists of three subunits which are coded at different loci. The A subunit binds loosely to damaged DNA and migrates along it until it hits upon a lesion where it is fixed. The energy for migration is provided by ATP hydrolysis which is also catalyzed by the enzyme. The B subunit then binds to the A component to form a very tight complex whose half-life is about half an hour. Incision is only made after completion of the total enzyme by the addition of the C subunit. Two breaks are simultaneously created, one 8 nucleotides apart from the lesion in $5'$ direction and another one 4 or 5 nucleotides away in the $3'$ direction so that a 12 or 13 nucleotides long oligomer is formed. It remains first in place together with the AB complex while C is released. Gap filling and excision is performed through the combined action of polymerase I and helicase II (uvrD product). This nuclease activity of polymerase I is single-strand specific and proceeds in $3'$–$5'$ direction. There is another nuclease activity of this enzyme which acts only on double strand DNA and in the opposite direction. It is responsible for the removal of incorrectly inserted nucleotides ("proof-reading activity") and plays a major role both in normal replication and repair (see also Sect. 13.2.4). If all the enzymes and the necessary cofactors (ATP and NADH for ligation) are present, an *in-vitro* repair velocity of 0.07–0.08 lesions per minute per ABC-complex can be achieved.

It must be emphasized that the mechanism described is only experimentally proven for *E. coli*. Whether it is similar in higher systems, particularly in mammalian cells, is completely unclear at present.

There is no doubt, however, that pyrimidine dimers are excised in mammalian cells although there are species-specific differences. Both the rate and the final extent of dimer removal are much larger in human as compared to rodent cells (Fig. 13.6). This is surprising since the UV sensitivities are com-

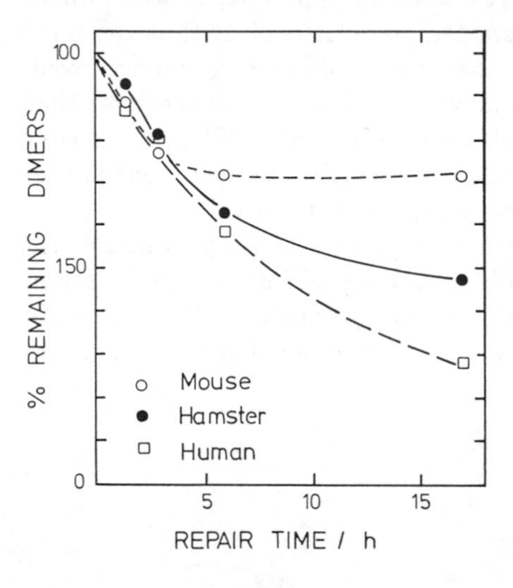

Figure 13.6. Dimer excision as a function of time in different mammalian cells (after MITCHELL 1988)

parable. It has been found recently that in Chinese hamster cells, excision takes place preferentially in active genes where the rate is similar to that in human cells. This fact could explain the mentioned discrepancy but it may also be that other photoproducts like the (6–4) adduct plays a more important role than dimers in mammalian cells. There are indications that their excision kinetics are about the same in human and rodent cells.

13.2.3 Post Replication Repair

The study of mutants defective in excision repair revealed the existence of other repair pathways. But they are also important in normal wild-type cells where they may act on lesions which were not excised. Since excision repair does not stop the normal cell activity, it may happen that DNA replication starts on a damaged template. The remaining lesions may be bypassed by "post-replication repair". As seen below, it involves recombinational activities, it is, therefore, absent in mutants which are unable to perform it ("rec$^-$ mutants"). Double mutants, defective both in excision and recombination, are more sensitive than both single mutants which shows clearly that the two processes are also important in wild-type cells and that they constitute independent pathways.

As usual, the scheme of post-replication repair is best known in *E. coli*; it is depicted in Fig. 13.7: Unexcised lesions which persist until the onset of DNA replication block the progression of the polymerase. This does not lead, however, to a complete halt but replication may be resumed at another starting point later on. In this way gaps are formed in the new strand opposite the lesions. They may be filled by a recombinational exchange from the complementary old strand where the respective information is available. The gaps which are now in the old strand are filled by repair replication where the newly synthesized strand is used as a template. It is clear that the term "repair" is not quite correct here since the original damage is not removed. It is assured, however, that at least the new strands are free of errors so that cell proliferation can proceed but not all progeny is viable. This is reflected in the phenomenon of "lethal sectoring" which describes the fact that not all cells in a surviving colony retain their reproductive integrity.

The mechanism described may be proved experimentally by similar methods as with excision repair, namely double-labelling and ultracentrifugation. Old and new strands are distinguished by density labelling. The gaps are demonstrated by a reduced molecular weight of the new DNA. Their num-

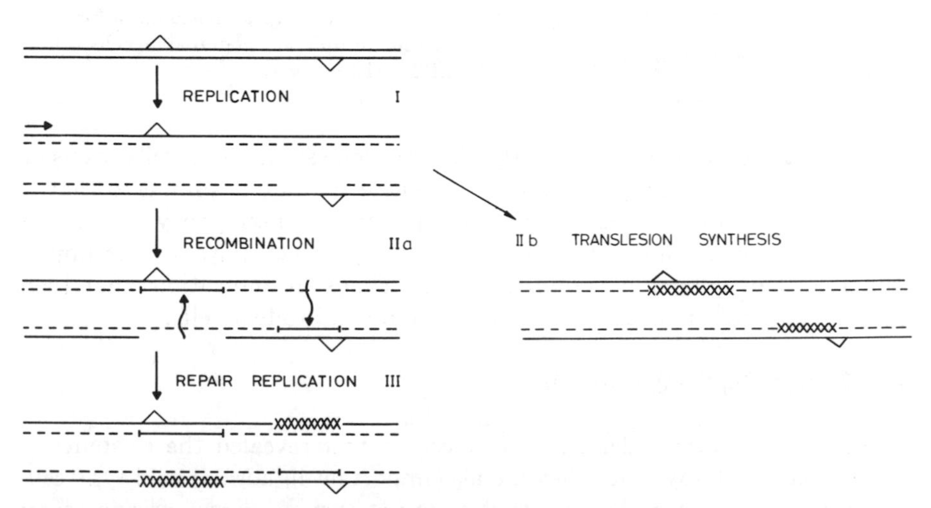

Figure 13.7. The scheme of post-replication repair. Left hand side (after RUPP et al. 1971) for E. coli, right hand side as assumed for mammalian cells

ber equals that of pyrimidine dimers. If the old DNA has been specifically labelled, the recombinational exchange is seen by the appearance of this label in the new DNA. Repair replication is studied as described in the preceding section. The persistence of dimers can be traced by chemical means.

The enzymatic mechanism is less well understood as with excision repair. There are a number of gene products involved. The most important is "recA" (which plays also an essential role in SOS repair; Sect. 13.2.4). It is obviously indispensable for recombination although at least two further genes, recB and recC, are required which code for exonucleolytic activities. They may process the gaps in such a way that they are amenable to recombination. There is, however, no direct proof for this speculation. Rec^- mutants are not only sensitive to UV, but also to ionizing radiations.

Post-replication repair has also been found in mammalian cells but it appears to be different. The gaps are not filled by recombination but by translesion synthesis, the exact mechanism is not yet known (Fig. 13.7).

13.2.4 SOS Repair

The repair processes described so far are the most important – at least for bacteria, where they are best studied. They proceed without errors so that the question arises how mutations may be produced if all lesions can be repaired "error free". The discovery of a further pathway gives at least partly a clue. It has been termed "SOS repair" indicating that it constitutes the cell's last resort if all other attempts to save viability fail. The historical starting point is the "WEIGLE reactivation" which was mentioned in Sect. 7.2.3.2. The interesting feature of it was that preirradiation of the host not only lead to an increase of phage viability but also to mutations. It had to be concluded that the underlying repair process is "error-prone". More detailed investigations showed that not only radiation but also a number of other agents are able to induce it. They all share the property to cause an arrest of DNA replication as do, e.g. also unremoved pyrimidine dimers.

The genetic background of SOS repair in *E. coli* has been clarified to a large extent. A central role is played by a gene called "lexA" (Fig. 13.8): Its product controls the transcription of a number of genes, including recA and also lexA itself. This does not mean, however, that the respective proteins are not synthesized at all but only at a rather low level. If DNA replication is stalled, an inducing factor is produced whose nature is not yet known which reacts with the recA protein. By this the recA protein acquires a protease activity by which the lexA protein is cleaved thus causing the loss of repressor activity. As a consequence, all the genes which are normally shut off are induced, including lexA itself and recA. The proteolytic activity of recA modifies also the DNA polymerases. They possess – at least in bacteria – apart from the already known polymerase and exonuclease activity acting in the same (5′–3′) direction also another exonucleolytic property which con-

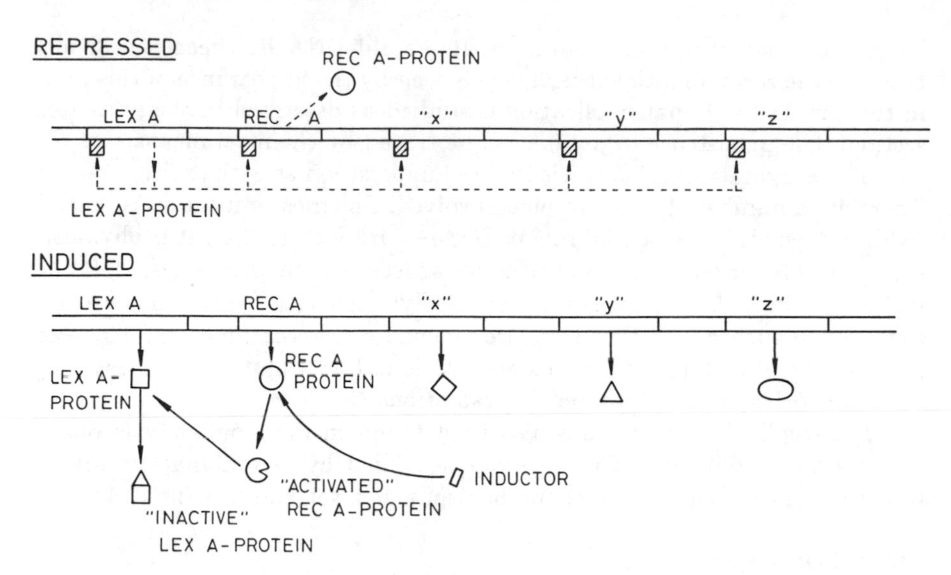

Figure 13.8. Regulation of the SOS response in E. coli

trols the exact insertion of nucleotides. If a wrong one is incorporated, it is immediately removed. This is particularly important if the old DNA contains a lesion where obviously correct base pairing can never be achieved. DNA synthesis stops here in a dynamic manner, every attempt of incorporation is followed by subsequent removal. Under SOS-inducing conditions, the "proof-reading activity" is shut off so that nucleotides can be inserted opposite the lesion although at the expense of possible errors. It appears, however, that this is only a necessary but not a sufficient prerequisite for DNA synthesis to continue.

For polymerization to proceed further, the action of another gene complex, termed umuCD, is required, which is also under the control of lexA.

After translesion synthesis, the inducing factor is no longer produced and the original state is reinstalled with all the genes shut off again. This is very important since the proof-reading activity of the polymerases is essential for the fidelity of DNA replication under normal conditions.

It must be emphasized that the mechanism described and depicted in Fig. 13.9 has only been shown for *E. coli*. It is not even clear whether it is the same in other bacteria, let alone simple eucaryotic or even mammalian cells. Numerous attempts to demonstrate its existence in these systems have so far yielded only inconclusive results. Nevertheless, it is a fascinating example how mutations may be produced as a consequence of "misrepair".

Since SOS repair is an inducible process, it depends on a functionary protein synthesis and can be blocked by suitable metabolic inhibitors. It is obviously important in cells where excision and post-replication repair do not function but it is not restricted to this situation. If the density of lesions is so

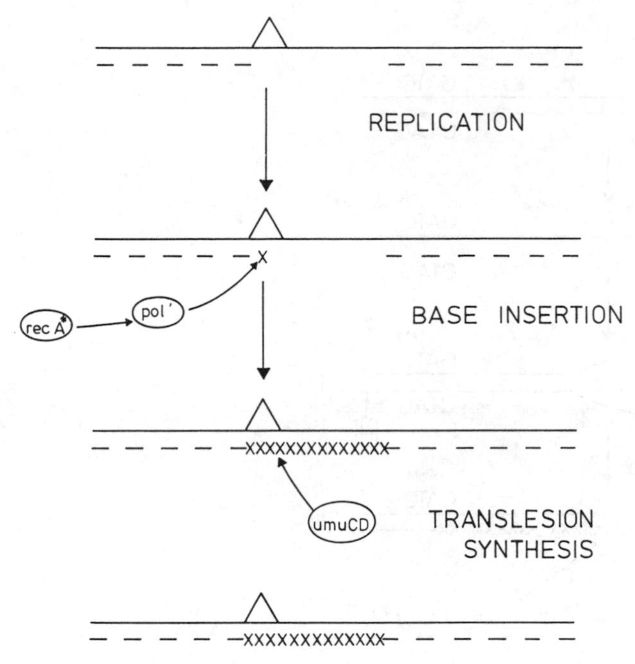

Figure 13.9. The assumed sequence of SOS repair steps

high that the excision or replication gaps may overlap, DNA double-strand breaks might be created. SOS repair prevents this.

As pointed out before, lexA and recA are not the only genes under SOS control. Others include those responsible for prophage induction and – in *E. coli* – filamentous growth. There is even an inducible component of the enzymes involved in excision repair. It appears, therefore, that lexA plays a very central role in all types of repair.

13.2.5 Mismatch Repair

Despite the impressive battery of repair processes, it may still happen that wrong bases are incorporated in newly synthesized DNA strands, causing a "mismatch". This must not necessarily be related to radiation damage but could also occur during normal DNA replication or in connection with recombinational exchanges. Obviously the cell has means to deal also with this situation. As usual, the mechanism is only known for *E. coli* (Fig. 13.10). It is under the control of a number of genes, abbreviated "mut" whose products participate at several stages. The first problem is to discriminate between the "old" and the "new" strand. In *E. coli* this is in principle possible since "old" DNA carries a methyl group at adenine bases which is added to the "new" strand not before considerable time after replication. A certain sequence – CA(met)TC – is important for recognition. If a mismatch occurs between

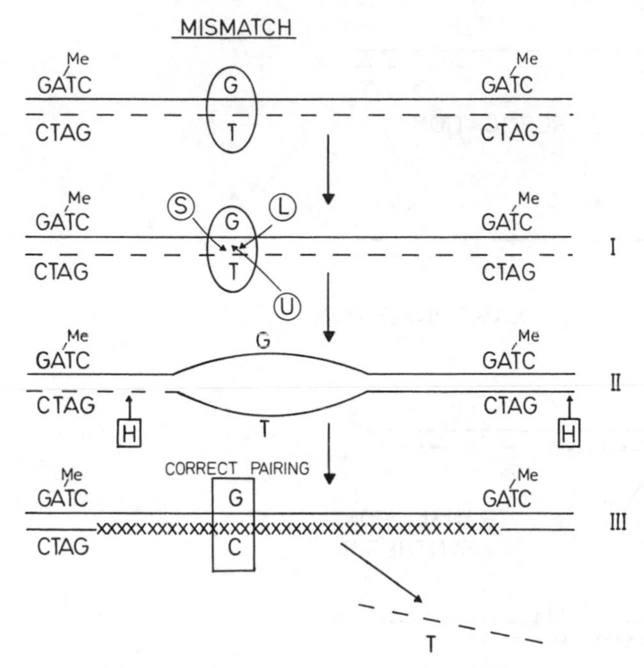

Figure 13.10. Mechanism of mismatch repair in E. coli (after RADMAN and WAGNER 1988)

two of these sequences, the two strands are separated with the aid of the mutU protein. A stretch of new DNA containing the mismatch is excised by the mutH-gene product and the resultant gap filled by repair replication in the usual way. The exact position of the two incisions is still under debate. It seems clear that one is always close to the CA(met)TC sequence while it is uncertain whether the excised part extends always from one of these recognition sites to the next or whether it is, in fact, shorter.

There is indirect evidence that mismatch repair exists also in higher organisms but nothing is known about the mechanism. There is no DNA methylation in yeast, and in mammalian cells the CH_3 group is found at cytosine instead of adenine. It may well be that there are other ways of strand discrimination (e.g. OKAZAKI fragments, see Appendix II.1) but it is not known whether they are used for mismatch repair.

13.2.6 Single-strand Break Repair

DNA single-strand breaks (SSB) are one of the most frequent lesions with ionizing radiations (see Sect. 6.2). In principle, they may be easily measured by ultracentrifugation under DNA-denaturing conditions. In cellular systems, however, care must be taken to avoid that SSB are formed as a consequence of the experimental procedures. This may be achieved by layering the cells whose DNA was previously radioactively labelled on top of a sucrose gradient containing a lysis mixture so that they are decomposed immediately before centrifugation. By chosing an appropriate pH value, it is also ascertained that the strands are separated. Allowing a certain incubation period between radiation exposure and the assay results in the number of SSB being greatly reduced (Fig. 13.11). This indicates that they may be repaired.

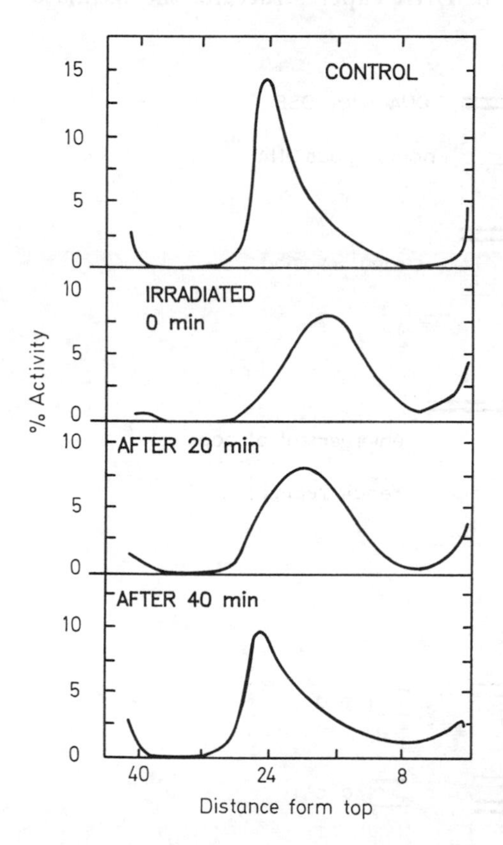

Figure 13.11. Single strand break repair: Cells with radioactively labelled DNA are layered on top of an alkaline gradient, lysed and centrifuged. There are short pieces immediately after irradiation which increase in length upon incubation (after MCGRAWTH and WILLIAMS 1966)

A closer analysis reveals that SSB repair is not a simple process, and that at least three phases may be distinguished, depending presumably on the chemical nature of the lesions. The first part is very fast (in the order of milliseconds) and represents presumably a purely chemical reaction since enzymatic action is considerably slower. The second one which takes a longer time depends on the function of DNA polymerase (as shown from its absence in pol mutants) but does not require nutrient medium. The third one can only be demonstrated in nutrient medium and in cells with a functioning recombination system – it is absent in rec$^-$ mutants.

The behaviour described may be understood by taking into account the different chemical configurations of SSB (Chap. 6). It is possible that they do not necessarily possess the correct structure for the sealing enzymes so that intervening processing may be necessary which may involve "cleaning" by exonucleases.

13.2.7 Double-strand Break Repair

DNA double-strand breaks (DSB) were originally considered to be irreversible since repair of this drastic structural alteration appeared to be quite improbable. The more recent findings of DNA super structure, the intimate

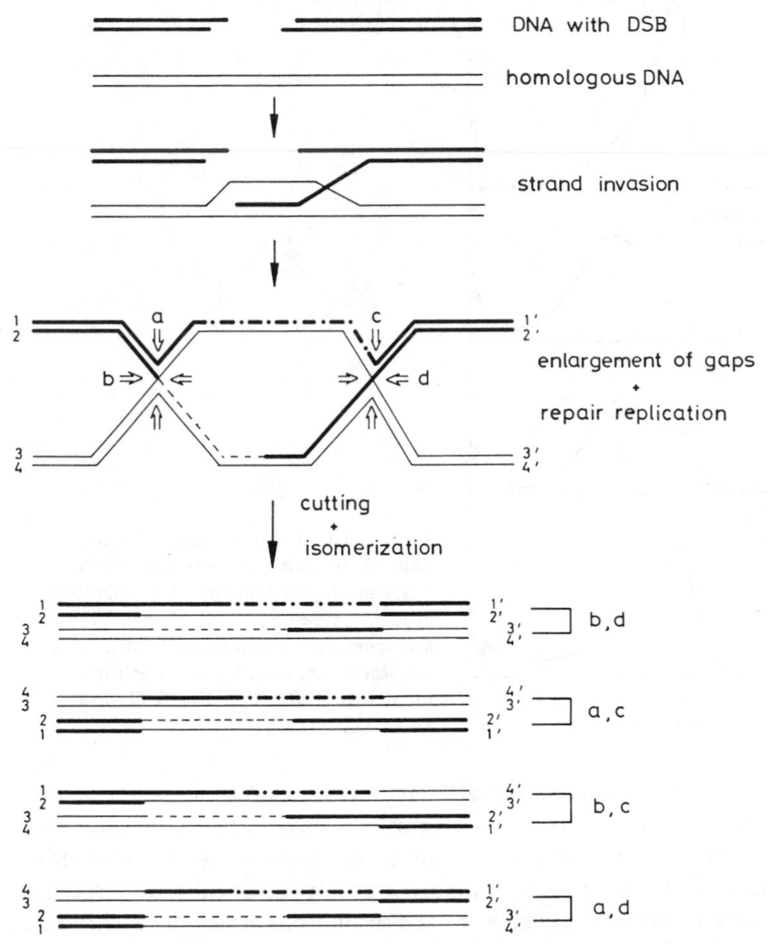

Figure 13.12. A proposed scheme for DSB repair by recombination in yeast: One strand of the DNA containing the DSB (heavy lines) invades the double stand of homologous DNA, the other strand is displaced, creating a so-called D loop. The gaps are enlarged and the filled by repair replication (broken lines). The newly replicated strands are ligated to the end of the DSB. Thus two crosspoints are formed which may be resolved by single strand breaks and ligation. The breaks may be in the inner or the outer strands. Isomerization may then lead to four different structures as shown in the figure. If one inner and one outer strand is cut (b,c and a,d), the larger part of the two DNAs are exchanged (reciprocal recombination) (after ORR-WEAVER and SZOSTAK 1985)

aggregation with proteins in the nucleosomes and the configuration of the nuclear skeleton, have changed this view. It is now quite clear that DSB can be repaired although the fidelity of this process is not known. "Misrepair" is possible but experimental evidence for this is lacking. The procedure to demonstrate DSB-repair is similar to that for SSB repair, only that centrifugation has to be performed under neutral conditions. Another way is by using the "neutral elution technique" (see Chap. 6).

The molecular nature is largely unknown. One model which was suggested for yeast involving recombinational processes is depicted in Fig. 13.12. It is certainly an interesting suggestion but it has still to be proven. The cloning of the genes involved which has already partly been achieved will facilitate this task greatly so that results may be expected shortly.

DSB repair has been found in bacteria, lower eucaryotes like yeast as well as in mammalian cells.

In Chap. 6 it had been pointed out that in cells DSB formation by low LET radiation shows a linear dose-response relationship. This has been questioned for mammalian cells but it is certainly true for yeast. Since DSB are thought to be one of the major lesions responsible for cell killing, "shouldered" survival curves as found in diploid yeast are difficult to explain. It could be shown, however, (Fig. 13.13) that the initially linear relationship is

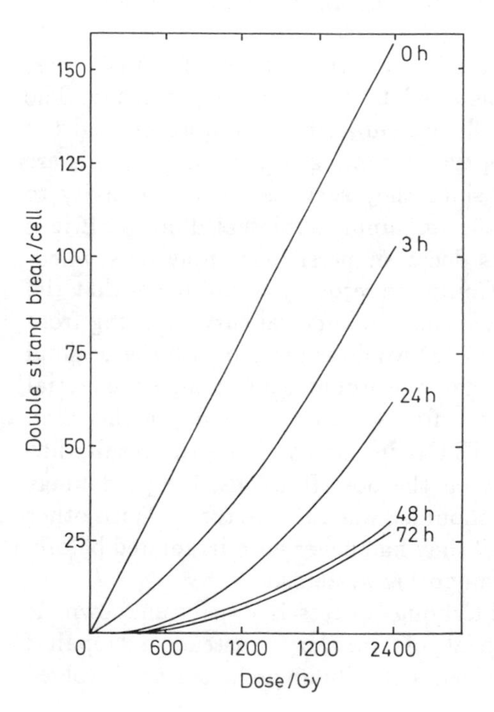

Figure 13.13. Repair of double strand breaks in yeast as a function of incubation time (after FRANKENBERG-SCHWAGER et al. 1980)

changed into a linear-quadratic upon incubation. This suggests that not the initial number of DSB is important but only those which remain after repair has taken place.

DSB repair is not a fast process but takes a few hours for completion in mammalian cells. There are indications that it consists of several components.

13.3 Recovery

13.3.1 "Split-dose Recovery"

The survival curves of most cells display a "shoulder" in the low-dose region. This is an indication of the fact that the effectivity of a certain dose-increment increases with dose or, in other words, that cells react more sensitive to a further irradiation if already exposed before. It means that low doses do not contribute to the same extend to the loss of viability as higher ones but that they cause a sensitization to further exposures. The initial damage is obviously partly "sublethal", and the shape of the survival curves may also be described by stating that the cells are able to accumulate a certain amount of sublethal damage. This does not, however, imply anything about the molecular nature of the lesions. It is quite possible that "sublethal" and "lethal" alterations are identical and that only their number determines the fate of the cell. If, e.g. m hits are necessary for killing, all numbers smaller then m will be sublethal.

It is, of course, an important question whether sublethal damage persists in surviving cells. It may be answered by split-dose experiments. The principle is shown in Fig. 13.13: If cells are subjected to a dose beyond the shoulder region of the survival curve, the remaining colony-forming cells are obviously only sublethally damaged since they survived. The sensitivity to a second dose is then a measure of the remaining sublethal damage. If it is investigated as a function of time its decay or persistence may be studied. There are two extreme alternatives: Complete recovery would mean that the survivors behave like unirradiated cells, the new survival curve starting from the lower level (second frame in Fig. 13.13) would exactly match the original one. Complete persistence of the predamage would lead to an exponential curve following the straight part of the first one. Experiments of this kind were performed by ELKIND and SUTTON in 1960 yielding the result that after an interval of a few hours between the dose fractions, the predamage had completely disappeared, i.e. the shoulder was fully restored. With other words, the survivors behaved then as if they had never been irradiated before. This result was later confirmed in many other systems.

The molecular nature of the underlying process is largely unknown. It is, therefore, quite appropriate to speak of "recovery" instead of "repair". There is no doubt, however, that biochemical cellular processes are involved since a functioning energy metabolism is required, and the recovery can be prevented by RNA- but not DNA- or protein-synthesis inhibitors. Only in yeast it was possible to show that split-dose recovery is linked to the repair of double-strand breaks (Fig. 13.14). It may be presumed that the situation is similar in mammalian cells but direct evidence is still lacking.

"Recovery from sublethal damage" may also proceed under irradiation: If the dose rate is sufficiently lowered and if exposure takes place in a medium favourable for recovery, the sensitivity in terms of cell killing is greatly reduced as more fully discussed in Sect. 14.1.

The time dependence of the process shows a biphasic behaviour (Fig. 13.15): The surviving fraction does not increase monotonically but passes a maximum and a minimum before reaching its final value. This may be understood by taking into account the cell-cycle dependence of radiation sensitivity (Chap. 10). The first exposure leads to a certain selection, the survivors are mostly from the most resistant subpopulation. They progress on further incubation necessarily into a more sensitive state, i.e. their survival probability is reduced. This time course is superimposed on the recovery curve which explains the resulting minimum ("recovery-progression model").

Shouldered survival curves are also found with UV. Split-dose recovery in this case appears to be restricted to the S phase of the cell cycle and it takes considerably longer time than with X rays.

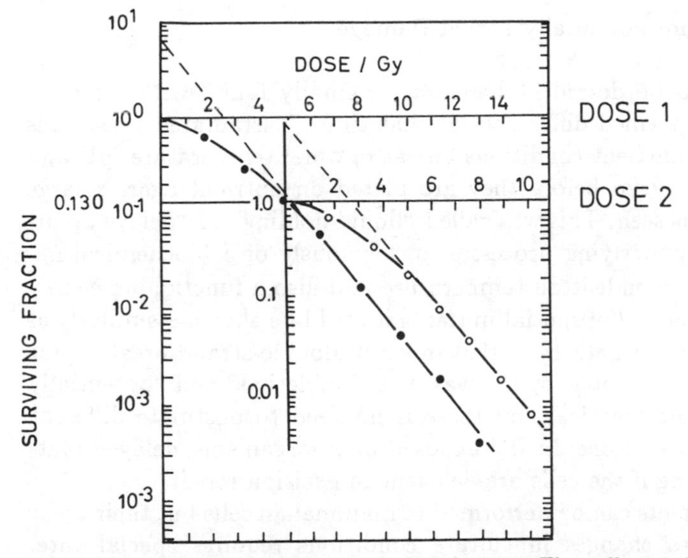

Figure 13.14. Recovery from sublethal damage: The cells are first exposed to a dose resulting in a surviving fraction of about 0.1 and then subjected to a series of second doses. With no recovery the survival curve indicated by full circles is expected which is the continuation of the single dose curve. Full recovery is indicated by open circles. The second case is experimentally found (Chinese hamster cells, after ELKIND and SUTTON 1960)

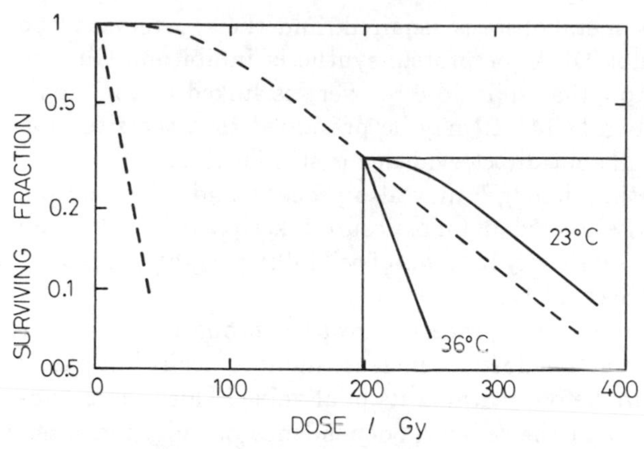

Figure 13.15. Split-dose recovery in a yeast strain with temperature-conditional DSB-repair. The second frame gives the result after fractionated exposure. If the cells are held at the permissive temperature the reappearance of the shoulder signifies recovery which is completely absent after incubation at the restrictive temperature (after FRANKENBERG et al. 1984)

Since radiotherapy is commonly given in a number of fractions, the experiments described are not only of fundamental importance but also practically relevant.

13.3.2 Recovery from Potentially Lethal Damage

The phenomenon to be described here was originally found with microorganisms – although termed differently at that time. If irradiated yeast cells are kept under non-nutrient conditions but at optimal temperature, pH and in the presence of oxygen before they are plated on nutrient agar, a large increase in survival is seen. This was called "liquid holding" or "delayed plating recovery". The underlying processes are obviously of a biochemical nature since they are dependent on temperature and also a functioning energy metabolism. With the aid of special mutants it could be shown – similarly as for split-dose recovery (Fig. 13.14) – that in yeast, double-strand break repair is involved. This demonstrates, by the way, that "sublethal" and "potentially lethal" lesions may be identical and there is no need to postulate different types of molecular alterations. In UV-exposed microorganisms, delayed plating recovery is missing if the cells are deficient in excision repair.

Similar experiments can be performed in mammalian cells but their great general sensitivity to changes in culture conditions requires special care. Recovery can be demonstrated if protein synthesis is prevented by specific metabolic inhibitors, or if the cells are kept in so-called "conditioned" medium in which they were grown to stationary phase or if the incubation buffer is carefully adjusted. Examples are shown in Figs. 13.16 and 13.17, from which also the extent and the time course can be seen.

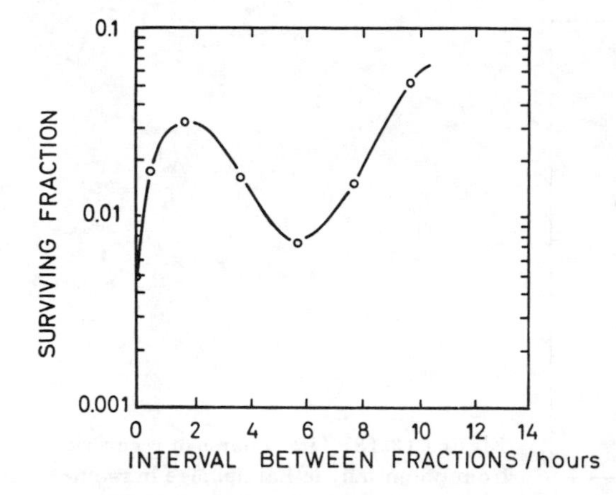

Figure 13.16. The time course of split-dose recovery in Chinese hamster cells (after ELKIND and SUTTON 1960)

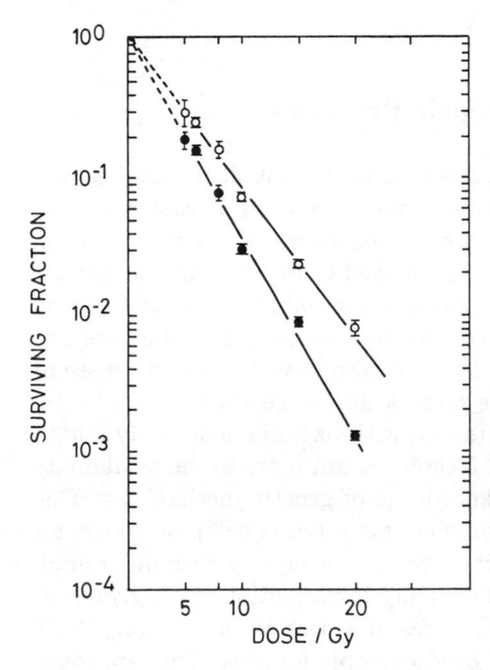

Figure 13.17. Recovery from potentially lethal damage in mammalian cells. Upper curve: immediate plating, lower curve: delayed plating after holding in conditioned medium (after HAHN and LITTLE 1972)

It has become customary to term the described phenomenon as "recovery from potentially lethal damage". The general interpretation is that incubation under non-optimal conditions prevents DNA replication so that lesions may be more readily repaired before they are ultimately fixed.

There is a certain resemblance to "mutation frequency decline" in bacteria (Chap. 12), but the two processes are not identical.

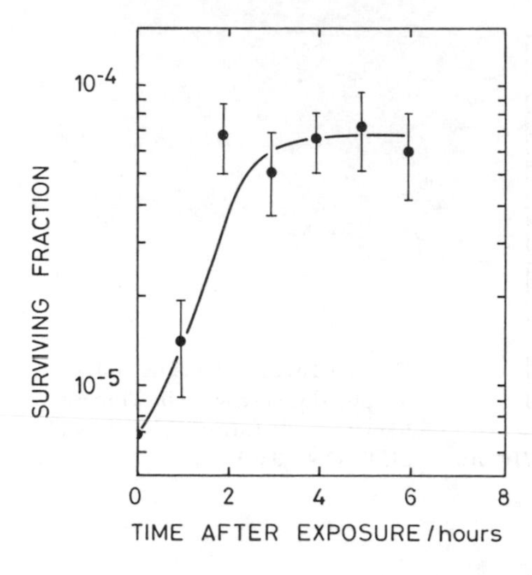

Figure 13.18. Time course of recovery from potentially lethal damage in mammalian cells (after HAHN and LITTLE 1972)

13.4 Genetic Dependence of Repair Processes

The detailed knowledge of repair processes is to a great deal based on experiments with microorganisms, especially the bacterium *Escherichia coli*. More recently, investigations have been extended to more complex systems, even mammalian cells. They have not only helped to understand the cellular phenomena but gave also a better insight into the nature of certain human hereditary diseases. This demonstrates that radiobiological studies are not only important in this field of science but may also have an impact on seemingly unrelated disciplines of medicine such as clinical genetics.

The impressive success of investigations with bacteria is not only due to the relative simplicity of experimental techniques but more to the availability of stable mutants and the advanced knowledge of genetic mechanisms. This explains why the genetic dependence of repair processes is still best known for *E. coli* although other systems, especially yeast, are rapidly gaining ground. Nevertheless, it is still true to say that many investigations in higher cells were tailored to studies in bacteria. The present search for mutagenic, "SOS like" repair in mammalian cells is a good example for this. This approach may be misleading since it has to be kept in mind that the structure of the genetic material and its replication cycle is significantly different in higher eucaryots. Yeasts being eucaryotic microorganisms are hoped to bridge the gap.

There are more than 60 gene loci in *E. coli* which influence radiation sensitivity although not all of them code for enzymes directly involved in repair. An exhaustive compilation shall not be given but only the main pathways discussed. Table 13.3 lists an abridged summary together with the position on the *E. coli* chromosome.

1. uvr mutations: This group regulates UV sensitivity via the excision system. As explained in more detail in Sect. 13.2.2, uvrA, uvrB and uvrC code for the components of the excision complex while uvrD carries the information of a DNA helicase which is necessary for further processing.

2. rec mutations: This group has rather pleiotropic effects. Although the name suggests an involvement in recombinational repair – which is true –, its involvement goes far beyond, and particularly recA plays also a central role in SOS repair. The recA protein (formerly called "protein X") is inducible by DNA alterations but it is also synthesized constitutively (see Sect. 13.2.4). RecB and recC code for subunits of exonuclease V which "cleans" DNA gaps in preparation for recombinational exchanges. Its activity is controlled by the recA protein, its lack leads to extensive DNA degradation ("reckless mutants"). Without exonuclease V recombinations cannot proceed.

Table 13.3. Human hereditary diseases, linked to increased radiation sensitivity. (After HANAWALT and SARASIN, 1986)

syndrome	clinical features	likelihood of cancer	cellular characteristics
xeroderma pigmentosum (classical XP)	sunlight hypersensitivity exposed skin epithelioma often neurological abnormalities ocular defects	+++	hypersensitive to and hypermutable by UV and chemical carcinogens defective in early step of excision repair nine complementation groups
xeroderma pigmentosum (XP variant)	similar to classical XP but no neurological problems	++	hypermutable by UV deficient in DNA replication on UV damaged templates
ataxia telangiectasia	cerebellar ataxia telangiectasia neurological deterioration partial immunodeficiency hypersensitive to ionizing high incidence of lymphomas	+++	hypersensitive to ionizing radiations and certain alkylating agents no DNA synthesis inhibition after X-ray treatment hypomutable by ionizing radiations high frequency of chromosome aberrations chromosome 14 rearrangement
fanconi's anemia	growth retardation pancytopenia bone marrow deficiency anatomical defects increased incidence of leukemia	+	hypersensitive to some cross-linking compounds high frequency of spontaneous chromosome aberrations (triradials)
bloom's syndrome	sunlight hypersensitivity growth retardation high incidence of malignancies	+++	high frequency of spontaneous sister chromatid exchanges and spontaneous chromosome aberrations (quadriradials)
cockayne's syndrome	cachectic dwarfism mental retardation (microcephaly) sunlight hypersensitivity	(+)	hypersensitive to UV and chemical carcinogens premature aging
hereditary retinoblastoma	malignancies of both eyes	+++	partial hypersensitivity to ionizing radiation homozygosity in tumours at *Rb-1* locus (13q13.1-q14.5)

3. lex mutations: As shown in Sect. 13.2.4, lex codes for the repressor of radiation-inducible genes, e.g. recA, prophage induction and also lex itself. A mutation in this gene would lead to constitutive expression of all these functions.

4. pol and lig mutations: This class is not only important for repair but also for normal DNA replication. Pol mutations cause a loss or a change of DNA polymerases, three of which are known in *E. coli*. PolA codes for polymerase I, the "Kornberg enzyme" (see Appendix II), polB for polymerase II and polC for polymerase III which also is involved in repair. Special mutations are also known which interfere only with certain activities of multifunctional enzymes, e.g. polAex where the 5′ exonuclease of polymerase I is inactive or polA1 where only the polymerase proper is not functioning.

The sealing of the gaps which are finally left after completion is mediated by a "ligase". Since this enzyme is also essential in normal DNA replication for the joining of OKAZAKI fragments (see Appendix II.1), its loss is lethal. There are, however, mutations where ligase activity is temperature-conditional so that its action may be studied by changing the incubation temperature.

5. umu mutations: These loci are important for induced mutagenesis as part of the SOS process. If their gene products are missing no mutations can be found suggesting that they play an important role in mutagenic translesion synthesis.

6. phr mutations: Their relevance is obvious, the respective loci code for the photoreactivating enzyme or enzymes (there may be two in *E. coli*).

The yeast *Saccharomyces cerevisiae*, a unicellular eucaryote, may be considered as a link between bacteria and mammalian systems. Studies of repair processes in this cell type may, therefore, provide some clues for the understanding of mechanisms operative in animals and humans. The ease by which genetic manipulations may be performed makes it particularly suitable for analysis. Three independent pathways of repair, termed *epistasis groups* have been established, all comprising several gene loci which in yeast are characterized as "rad mutants". The most important examples are listed in Table 13.4. Many of them have recently been cloned but the analysis of molecular mechanisms is not yet as far advanced as with *E. coli*.

The *rad3-epistasis group* contains those genes which are responsible for excision repair. Mutations in any of them leads to increased UV sensitivity with little effect on X ray response. The mechanism may be similar to that in bacteria but the exact nature could not yet be established.

The *rad6 group* is involved both in UV as well as X ray sensitivity. It plays also a major role in mutagenesis since rad6 mutants are virtually non-mutable by radiation. This is, however, not true for other members of the group sug-

Table 13.4. Radiation sensitivity gene loci in the yeast Saccharomyces cerevisiae. (Modified after HAYNES and KUNZ, 1981)

group	RAD3	RAD6	RAD52
mainly sensitive to	UV	UV, ionizing rad.	ionizing rad.
assumed mechanism affected	excision repair	mutagenic repair (?)	DSB-repair, recombination
locus designation	rad1 rad2 rad3 rad4 rad7 rad10 rad14 rad16 rad23 mms19	rad5 rad6 rad8 rad9 rad15 rad18 rev1 rev3 umr1 umr2 mms3 spo1	rad50 rad51 rad52 rad53 rad54 rad55 rad56 rad57

gesting a multiple branching of the repair pathway. There is presently no convincing explanation for the underlying molecular mechanisms.

The *rad52 group* is rather unique since the respective genes govern the repair of double-strand breaks, a situation which is not as clearly established in other organism. One special mutant, *rad54-3*, is particularly useful for experimental analysis since it is temperature-conditional: double strand repair may be switched on or off just by incubation at the permissive (23°C) or the restrictive temperature (36°C). Mutants of the rad52 group are completely deficient in genetic recombination indicating that double-strand repair involves recombinational processes (see also Sect. 13.3.6).

The genetic analysis in mammalian cells is far less advanced as with simpler systems. Only recently, a few stable radiation-sensitive cell lines could be isolated *in vitro* (see below) but most of the present knowledge comes still from the analysis of certain human hereditary diseases. The most important examples are listed in Table 13.5 together with some of the clinically found symptoms. All are recessively inherited with the exception of hereditary retinoblastoma which shows a dominant trait. Fortunately they are quite rare in the human population.

The most studied and best characterized example is *Xeroderma pigmentosum* (XP). Patients with this disease exhibit extreme sunlight sensitivity and are prone to skin cancer early in life. The cells when grown in culture reflect this behaviour, they are hypersensitive to UV and some chemicals, but not to ionizing radiations. The molecular analysis revealed a deficiency in an early step in excision repair. Nine complementations, presumably related to as many genes, have been found. Apart from this "classical" XP, a number

Table 13.5. The most important mutations with influence upon radiation sensitivity in E. coli and the position of the respective genes on the bacterial chromosome (After HANAWALT, COOPER, GANESAN and SMITH, 1979)

name	position	gene products
uvrA	91	UV-endonuclease (part)
uvrB	17	UV-endonuclease (part)
uvrC	42	UV-endonuclease (probably part of uvrB)
recA	58	'recA-product', protein X
recB	60	exonuclease V (subunit)
recC	60	exonuclease V (subunit)
polA	85	DNA-polymerase I
polA1	85	polymerase-activity of polymerase I
polAex	85	5'-exonucleaseactivity of polymerase I
polB	2	DNA-polymerase II
polC	4	DNA-polymerase III
lig	51	DNA-ligase
phr	16	photoreactivating enzyme

of "variants" exist which display similar overall characteristics but where a later step in repair, namely DNA replication on UV-damaged templates is impaired.

Another prominent example is *Ataxia telangiectasia* (AT). Individuals with this disease are extremely sensitive to ionizing radiation with little change in UV sensitivity. At the cellular level, it is interesting to find a certain degree of hypomutability to X rays and nearly no impairment of double-strand break repair. The most important feature may be that AT cells do not show X ray induced inhibition of DNA synthesis (see Chap. 10). Also there is a high frequency of chromosome aberrations.

The other examples listed are less well characterized. A common feature is proneness to cancer and chromosome fragility. It is also interesting to note that most of these diseases are linked with neurological disorders. The reason is not yet understood.

This short subsection demonstrates that cellular repair processes are not only interesting as general biological phenomena but are obviously of great importance for human health. They figure prominently in the evolution and the conservation of life on Earth. Radiation biology has not only led to their discovery but has also greatly contributed to the understanding of the underlying mechanisms.

Quite recently a number of radiation sensitive mutants of Chinese hamster cells could be established, which are listed in Table 13.6. If known, their molecular deficiency is also indicated. They will be of great help in the future to clarify the mechanisms of repair in mammalian systems.

Table 13.6. Radiosensitive Chinese hamster cell lines (HICKSON and HARRIS, 1988, original references therein). Cell lines only sensitive for special chemicals are not listed. (CHO: Chinese hamster ovary cells; V79: V79 lung fibroblasts)

signature	sensitivity	defect	origin
UV5, UV20	UV	incision	CHO
UV24, UV41, UV135	UV	incision	CHO
UV1	UV	post-repl.-rep.	CHO
XR1	X-ray	DSB-rep.	CHO
xrs-1	X-ray	DSB-rep.	CHO
Apr^r-4	UV	DNA-polymerase	CHO
BLM-2	X-ray	ESB- and DSB-rep.	CHO
XR 2	γ	?	CHO
CHO 12 RO	X-ray, UV	?	CHO
CHO 33 RO	X-ray, UV	?	CHO
CHO 43 RO	X-ray, UV	?	CHO
UV61	UV	?	CHO
UV^s-7, CH^s-1	UV	?	V79
UV^s-40, UV^s-44	X-ray, UV	?	V79
irs 1	X-ray, UV	?	V79
irs 2, irs 3, irs 4	X-ray	?	V79
V-H1, V-B11	UV	?	V79
V-B7, V-E5	UV	?	V79
V-C4, V-C8	X-ray, UV	?	V79
V-G8	X-ray	?	V79

A first success can already be reported: Two human genes could be isolated and sequenced which are able to complement repair deficiencies in sensitive rodent cells. They are called ERRC-1 and -2, standing for "excision repair complementing in Chinese hamster cell". ERRC-1 shows widespread homologies with the yeast rad10-gene product. This indicates that repair genes may be quite stable during evolution.

Further reading

BOOTSMA et al. 1987, CLEAVER 1974, CLEAVER and KARENTZ 1987, COLLINS, JOHNSON and BOYLE 1987, ELESPURU 1987, FRIEDBERG 1985 a, b, HANAWALT 1987, HANAWALT et al. 1979, HANAWALT and SARASIN 1980, HANAWALT and SETLOW 1975, HOGG and SMITH 1987, HICKSON and HARRIS 1988, HOEIJMAKERS 1987, ILIAKIS 1988, KIMBALL 1987, LASKOWSKI 1981, MOUSTACCHI 1987, OSAWA et al. 1986, PATERSON 1979, SANCAR 1987, SARASIN et al. 1985, SEDGWICK 1986, 1987, SIMIC et al. 1986, SMITH 1987, TIMME and MOSES 1988, TOWN, SMITH and KAPLAN 1973, WITKIN 1976

Chapter 14
Modifications of Radiation Effects by External Influences

It is the aim of this chapter to demonstrate with a few important and typical examples how the biological radiation effect may be modulated by external parameters, apart from those which were already treated in more detail. The time pattern of exposure plays an important role, both fundamentally and practically. With respect to the recent interest in hyperthermia in radiation therapy, the combination of heat and radiation at the cellular level is discussed. Modifications by pharmaceutical agents can only be described by a few examples, mainly repair inhibitors. The last section deals with changes in tonicity to demonstrate the complexity of biological radiation action and may serve as a safeguard to avoid the premature acceptance of too simple mechanistic models.

14.1 General Aspects

Even with one single biological test system and well-defined radiation conditions, sensitivity cannot be considered as a constant since it may be modified by many external parameters. Examples were already given, particularly in Chaps. 9 and 13. In many cases repair phenomena are inferred although direct experimental evidence is often lacking, a problem which has already been alluded to in the previous chapter. It is the purpose of this chapter to introduce a few of these parameters without claiming comprehensiveness. The selection is necessarily subjective, it was guided by the presumed relevance for the understanding of basic processes and by the role they may play in radiation therapy. The last section is also thought as a demonstration how important "minor" factors may be.

14.2 The Time Factor in Radiation Biology

Although the time course of chemical and biological reactions has been mentioned several times before, it is worthwhile to discuss this aspect in a more comprehensive manner. Table 14.1 gives a simplifying, but illustrative summary of the time scale of processes which play a role in the expression of biological radiation actions at all levels.

Table 14.1. Time scale of biological radiation action (modified after ADAMS and JAMESON, 1980)

time/s	process
	PHYSICAL PROCESSES
10^{-18}	passage of fast particles through small atoms
10^{-17} - 10^{-16}	ionisation: $H_2O_2 \rightarrow H_2O^+ + e^-$
10^{-15}	electron excitation: $H_2O \rightarrow H_2O^*$
10^{-14}	ion. molecule-reactions: $H_2O^+ + H_2O \rightarrow OH^\cdot + H_3O^+$
10^{-12}	hydration: $e^- \rightarrow e^-_{aq}$
	CHEMICAL PROCESSES
< 10^{-12}	reactions of non-hydrated electrons
10^{-10}	radical reactions at high reactant concentrations (≈ 1 mol dm^{-3})
< 10^{-7}	reactions in 'spurs'
10^{-7}	homogeneous distribution of radicals
10^{-3}	radical reactions at low reactant concentration (≈ 10^{-7} mol dm^{-3}) oxygen effect
1	end of radical reactions
1 - 10^3	biochemical reactions
	BIOLOGICAL PROCESSES
10^3 - 10^4	repair processes, cell division inhibition
10^2 - 10^5	'CNS-death', 'intestine-death'
10^5 - 10^6	bone marrow death
10^7 - 10^8	late effects

One notes that 26 orders of magnitude are spanned if whole body effects, treated in later chapters, are included. The importance of the time factor thus needs no further justification.

The total duration of exposure constitutes an essential parameter. Since certain effects, e.g. survival, mutation, etc., can only be measured within limited dose ranges, time factor investigations imply a variation in dose rate. While the practical lower limit is set by the environmental background (and by the necessity to obtain results in acceptable times), the use of high dose rates depends on the technical development of very powerful sources.

Field emission devices ("Febetron") permit nowadays dose rates up to $2 \times 10^{11}\,\mathrm{Gy\,s^{-1}}$, corresponding to 10^3 Gy in 5 nanoseconds. The influence of high dose rates on cell survival has not been sufficiently investigated to allow general conclusions. An important, but in this context trivial, effect is radiation-induced oxygen depletion which leads to breaks in survival curves (see Sect. 9.2.2). The oxygen effect offers ways and means to study the kinetics and relevance of fast processes at the cellular level, as discussed in Sect. 9.2.2.

In some systems it was found that in anoxia, but not in the presence of oxygen, increasing the dose rate to very high values leads to greater effects. An example is shown in Fig. 14.1. This line of research has not been followed up; how far the results may be generalized is still an open question.

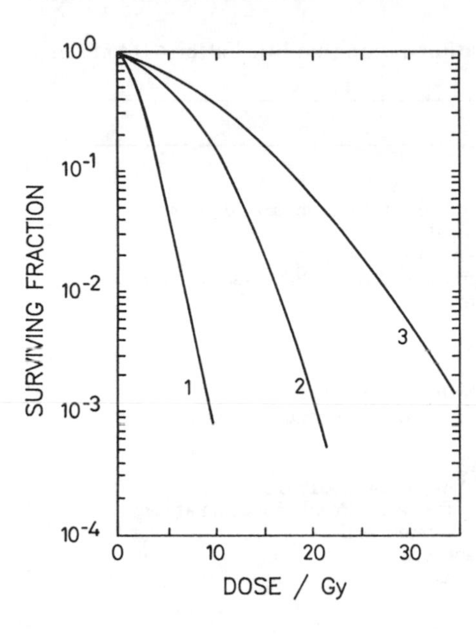

Figure 14.1. Dependence of survival of human kidney cells on dose rate with sparsely ionizing radiation. 1: exposure in presence of oxygen, low and high dose rate, 2: exposure in nitrogen, single 40-ns pulses, 3: exposure in nitrogen with convential dose rates (after PURDIE, INHABER and KLASSEN 1980)

The influence of lower dose rates is far better documented. There are, however, principal difficulties: One would expect from split-dose experiments (Chap. 13) that with low dose-rate exposure recovery from sublethal damage proceeds during irradiation leading to an overall reduction of the effect. Generally speaking, this is in fact found but it must be realized that the biological system changes during the treatment time. Radiation-induced progression delay (Sect. 10.2) leads to a redistribution of cell cycle stages. As

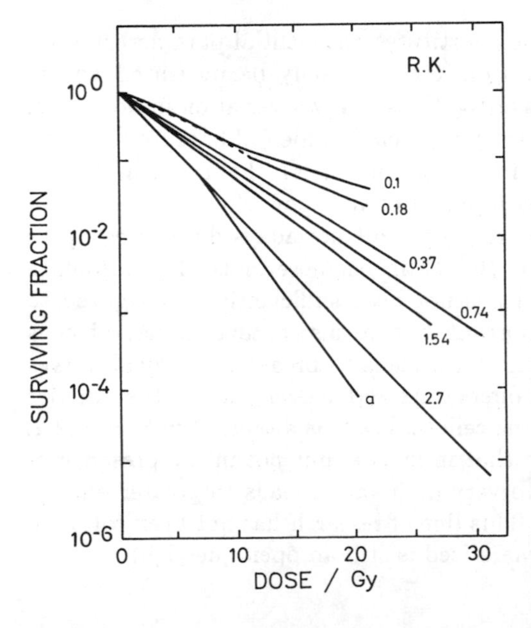

Figure 14.2. Survival of growing pig cells after exposure with different dose rates as indicated (Gy/hour). a: acute exposure with 1.43 Gy/minute (after MITCHELL, BEDFORD and BAILEY 1979)

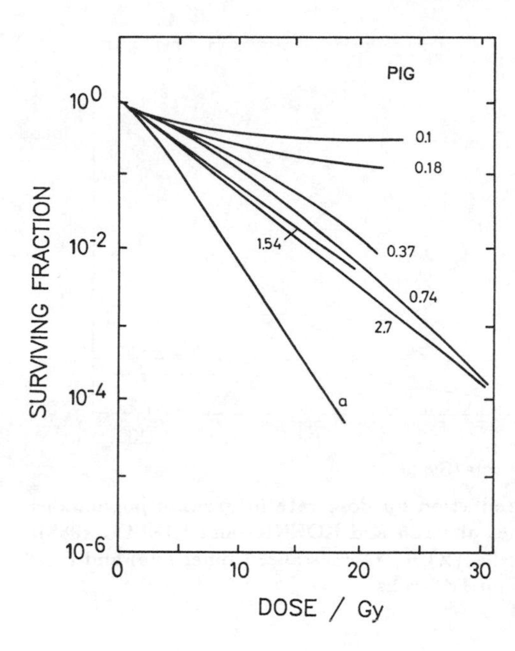

Figure 14.3. As in Fig. 14.2. but for rat Kangaroo cells. The differences with long term and the similarity with acute exposure are to be noted (after MITCHELL, BEDFORD and BAILEY 1979)

shown before, there is an arrest in G2, a radiation-sensitive phase, so that two opposing effects are superimposed: reduction by recovery and increase due to a larger fraction of G2 cells. This is the reason why the influence of dose rate is in general less than expected. There are additional interesting consequences: As progression delay may be different even in cells which have the same survival after acute doses, reducing the dose rate may then lead to large differences in survival, as demonstrated in Figs. 14.2 and 14.3. But it is still true that higher survival is found with lower dose rates. The situation is much "cleaner" if cell progression and division are prevented by holding

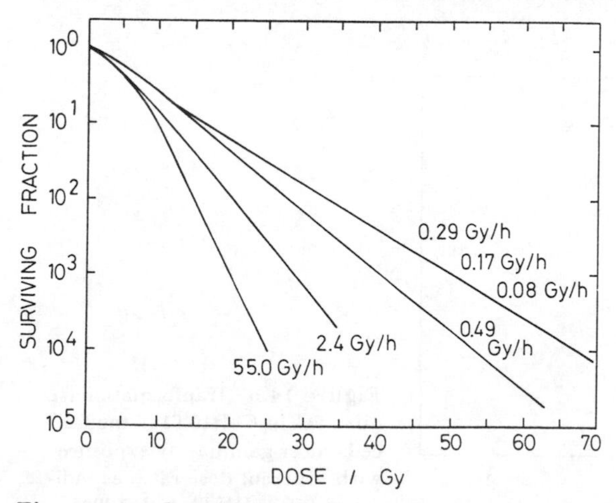

Figure 14.4. Low dose rate survival of mammalian cells held under non-growth conditions (after WELLS and BEDFORD 1983)

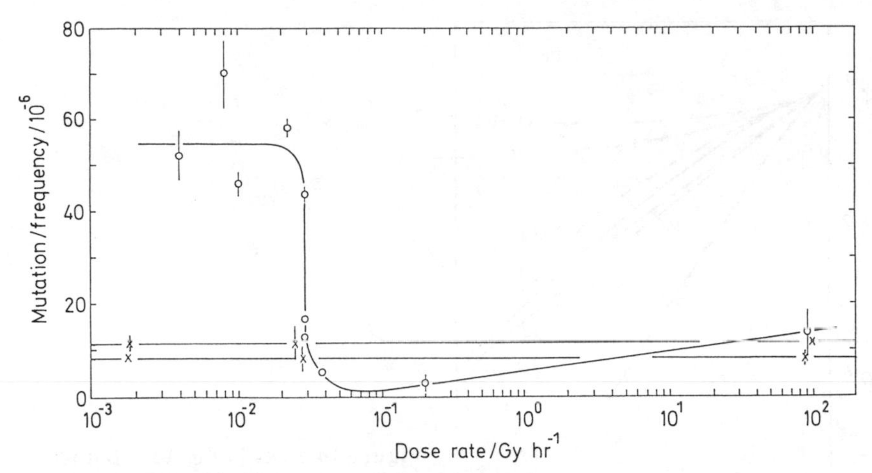

Figure 14.5. Dependence of mutation induction on dose rate in growing populations of mammalian cells (after CROMPTON et al. 1985 and KOENIG and KIEFER 1988). (o) 6-TG-resistance in Chinese hamster cells, (x) 6-TG-resistance (upper line) and fluorothymidine resistance (lower line) in human TK6-cells

the cells in a confluent state where they do not progress, as demonstrated in Fig. 14.4.

The extent of the oxygen effect appears also to be lower with reduced dose rates. This result is not yet generally accepted; it fits, however, in the model of the oxygen effect (Sect. 9.2.2) where repair and recovery phenomena were incorporated.

The influence of dose rate on mutation induction is even less well understood. If, as shown for *E. coli*, mutation is due to a radiation-inducible error-prone repair system, one would expect an *increase* with lower dose rate. Experiments with bacteria are not available. Investigations with mammalian

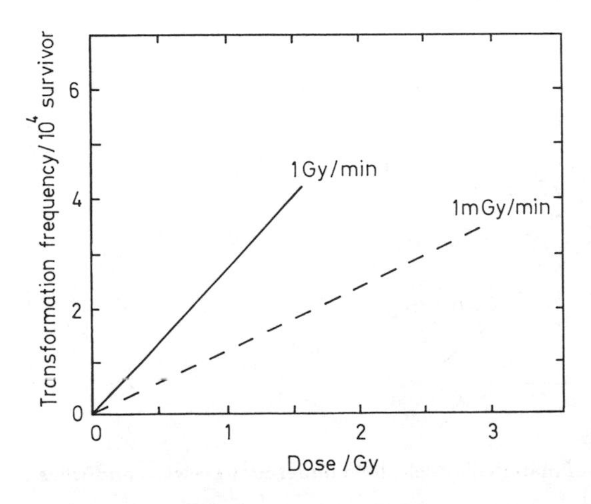

Figure 14.6. Tranformation frequencies in C3H10T1/2 mouse cells after gamma-ray exposure with different dose rates as indicated (after HILL et al. 1984)

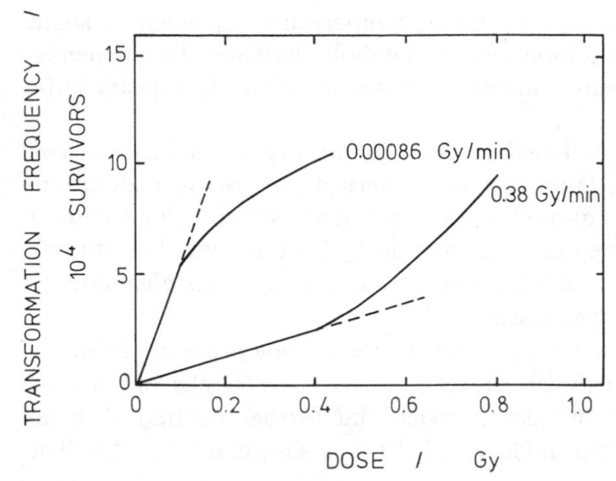

Figure 14.7. As in Fig. 14.6 but with neutrons (after HILL et al. 1984)

cells show that moderate dose rate reduction leads to a decrease in mutation frequency per unit dose. Long term exposure (up to several weeks) of growing cell populations, however, has been found to yield *higher* mutation rates (Fig. 14.5) but not in all systems.

With regard to neoplastic transformation, there is an important influence of radiation quality: Low dose-rate γ rays yield lower transformation frequencies per unit dose if compared to acute exposures (Fig. 14.6). The opposite is true for neutrons, as shown in Fig. 14.7. The reasons for this behaviour are not yet understood.

Further reading
ADAMS and JAMESON 1980, BEDFORD 1987

14.3 Temperature

Temperature is not only an important and interesting experimental parameter by which cellular radiosensitivity may be modified, but there is also great practical relevance since the combination of heat and radiation is considered to be a promising possibility to improve the treatment of tumours (see Sect. 24.2.3). Notwithstanding the special situation within the body in terms of heat effects (influence of vascularization, immune responses, etc.), one may hope to gain some knowledge about fundamental mechanisms from cellular *in-vitro* studies.

One should be aware not to draw premature conclusions since there are wide variations between different systems and even within a single species. Standardization of the experimental protocols is indispensable. Temperature changes affect the physiology of a cell in many different ways without actually damaging it. Enzymatic processes are highly temperature-dependent, a slight change may lead to drastic alterations in metabolic balance. This depends, furthermore, also on other environmental factors of which pH appears to be the most important.

Mammalian cells may be killed by heat. In contrast to radiation where DNA is the principal target, damage to membranes plays here the most prominent role. Heat sensitivity is also cell-cycle-dependent, with S being the most sensitive phase. If cells are treated for some time by heat at sublethal "doses", they develop some resistance to later acute lethal heating. This phenomenon of "thermotolerance" is not yet understood.

Heat inactivation does not seem to constitute a single process, as demonstrated by plotting ARRHENIUS diagrammes (see textbooks of physical chemistry for the rationale of this approach and further details). A summary of results obtained with different cell lines is shown in Fig. 14.8. The

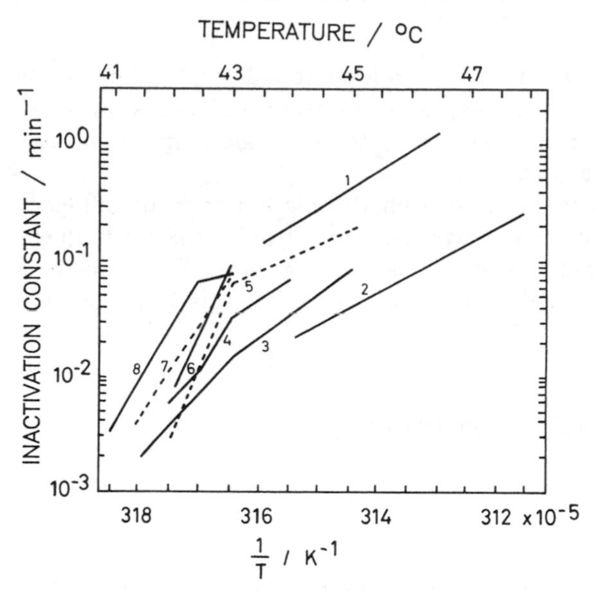

Figure 14.8. ARRHENIUS diagrammes for heat-inactivation of mammalian cells. Data from several laboratories. Chinese hamster cells: 1, 4, 6, 7. Human HeLa cells: 3, 8. Rat tumour cells: 5. Pig kidney cells: 2. The ordinate gives tge inverse of the slope of the exponential part of the survival curves (after ROSS-RIVEROS and LEITH 1979)

breaks in most of the curves are indicative of more than one mechanism, also there are wide differences between different cell types.

If heat and ionizing radiation are given concomitantly or closely spaced in time, sensitization occurs as shown in Fig. 14.9 which is an early example. Similar responses are found in most mammalian cells but the quantitative relations differ. Radiation sensitization by elevated temperatures depends on

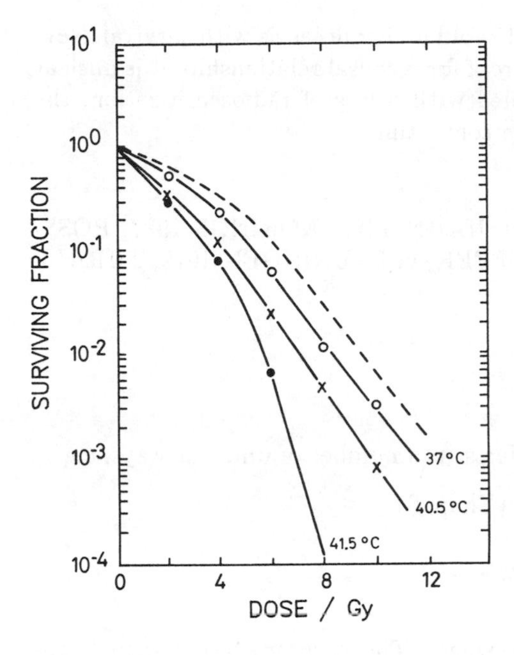

Figure 14.9. Sensitization of Chinese hamster cells to radiation by two hours post-treatment at the indicated temperature. Broken line: no heat (after BEN-HUR, ELKIND and BRONK 1974)

the time interval between the two treatments (Fig. 14.10). If it is too long, the interaction of the two agents is merely additive.

Although the mechanism is by no means understood, there appears to be the common view that impairment of repair is the most important parameter. This is substantiated by the fact that split-dose as well as potentially lethal damage recovery are impaired at elevated temperatures and by measurements of strand break repair. Not only that the rate of repair is diminished, but also the fraction of unrepaired breaks is increased. Their number also depends on the timing of the two treatments, as also shown in Fig. 14.10. It has to be stressed when comparing the two diagrammes that strand-break analysis had

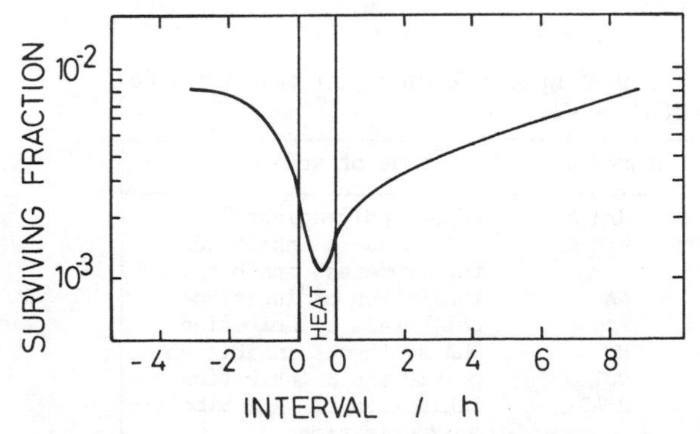

Figure 14.10. Dependence of radiosensitization by heat treatment on the time interval. The abscissa gives the time between radiation and heat exposure (after KONINGS 1987)

to be performed by applying about 10-fold higher doses as with survival. Nevertheless, it may not be taken as a proof for a causal relationship. It is unclear whether "thermotolerance" is coupled with a loss of radiosensitization, the results reported in the literature are conflicting.

Further reading
ALPER 1979, DEWEY et al. 1980, HAHN 1982, KONINGS 1987, ROSS-RIVEROS and LEITH 1979, STREFFER, v. BEUNINGEN 1979, THRALL et al. 1976

14.4 Chemicals

Chemicals may modify radiation effects in a number of different ways, e.g.

- as radioprotectors or sensitizers (Chap. 8)
- as radiomimetics (Sect. 15.4)
- as metabolic and/or repair inhibitors
- unspecifically.

The first two points were treated elsewhere. The third one has attracted considerable interest recently and warrants a slightly more extended discussion.

Repair inhibitors are, on one hand, expected to improve radiation therapy if they can be selectively applied to the tumour, on the other hand, they may help to clear up the mechanisms of the underlying processes.

Table 14.2 gives a list of some representative substances (together with commonly used abbreviations) which is certainly not comprehensive but contains most of the compounds presently under discussion together with their presumed mode of action. A useful guide is the scheme in Fig. 14.11.

It is necessary to distinguish between strand breaks and base alterations as primary lesions. In the first case, it is thought that the ultimate ligation proceeds by only a few "cleaning" steps involving a limited amount of polymerization. Its initiation may be blocked by topoisomerase inhibitors

Table 14.2. Compounds with assumed influence on repair processes (see also COLLINS, DOWNES and JOHNSON, 1984)

name	abbreviation	mode of action
9-β-D-arabinofuranosyladenine	ara A	repair polymerization
1-β-D-arabinofuranosylcytosine	ara C	polymerase α inhibition
novobiocin	–	topoisomerase inhibition
nalidixic acid	NA	inhibition of incision
aphidicolin	Apc	polymerase α inhibition
hydroxyurea	HU	DNA synthesis inhibition
dideoxythymidine	ddT	polymerase β inhibition
3-aminobenzamide	3-AB	inhibition of (ADP-ribose)-polymerization
caffeine	–	prevention of G2-arrest

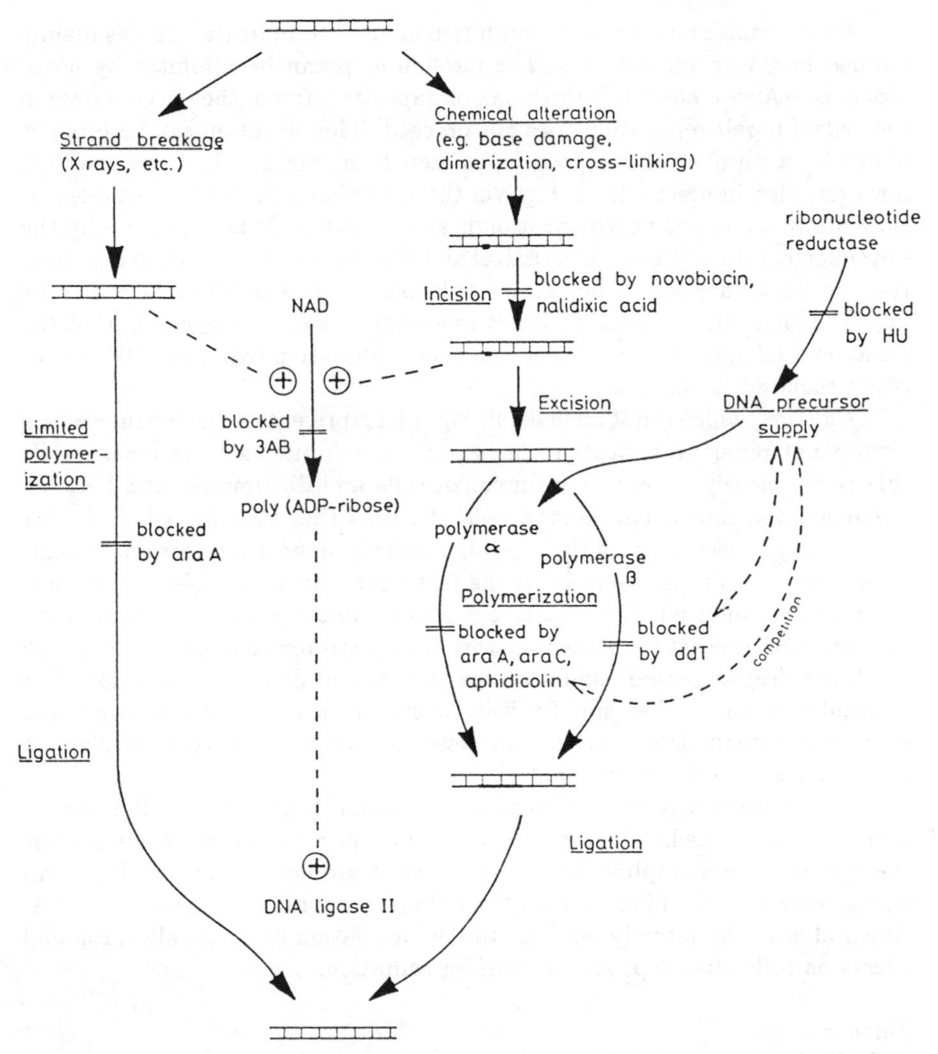

Figure 14.11. Schematic survey of the action of various repair inhibitors (after COLLINS et al. 1984)

like novobiocin or NA thus preventing the relaxation of DNA superstructure. AraA interferes with the polymerization step. The closing of gaps depends obviously on the availability of a suitable enzyme which in mammalian cells is assumed to be *ligase II*; it is activated by *poly(ADP-ribose)*. If the synthesis of this molecule is inhibited, which can effectively and selectively be achieved by 3-AB, the speed of ligation is reduced resulting in a potentiation of damage. Although this scheme is of convincing simplicity, it can by no means be considered to be unequivocally proven. Nevertheless, it is clear that 3-AB given after X irradiation increases sensitivity, both in terms of survival and mutation induction.

Base damage repair and its inhibition is more complicated. It was mainly studied in UV-irradiated cells. The incision step can be inhibited by novobiocin or NA (see above). If the break or gaps are formed, they accumulate in the cells if repair replication does not proceed. This situation can be brought about by a number of compounds, as seen from Fig. 14.11. There may be, however, also indirect effects, e.g. via the availability of DNA precursors. If they are in shortage, polymerization is slowed down. This explains why the physiological state of the cell is critical and differences exist between exponential and stationary cells. The ligation of incision breaks in UV-irradiated cells appears not to be inhibited by 3-AB to any great extent suggesting that the basal level of ligase II is sufficient and that stimulation by poly(ADP-ribose) is not required in this case.

Caffeine, which is not included in Fig. 14.11, presents a further interesting example of repair modification. In bacteria it inhibits excision repair while this is not clearly proven in mammalian cells and differences exist between UV-irradiated rodent and human cells. It seems that this agent is in higher cells mainly concerned with the gap filling process in post-replicational repair. A very interesting phenomenon is the fact that caffeine is able to overcome the radiation induced G2 arrest if it is present in the growth medium. Since survival is reduced at the same time this finding strongly suggests that the G2 block is a "repair period" and not a passive unspecific cellular reaction. The molecular action mechanism of caffeine is still unclear; it has been suggested that base damage is converted more readily into double-strand breaks but the evidence is quite circumstantial.

In this section it was attempted to summarize the effects of repair inhibitors in a somewhat "integrated view" but such an approach always has the danger of oversimplification. That matters are not so easy as they seem is, e.g. shown by the findings that "base damage repair inhibitors" like araA, Apc and araC do not only sensitize to UV irradiation but have also profound effects on cells after exposure to ionizing radiation.

Further reading
BEN-HUR 1984, CLEAVER et al. 1985, COLLINS et al. 1984 a, b, 1987, SCHRAY and TODD 1975, SHALL 1984, STREFFER 1987, TOLMACH, JONES and BUSSE 1977, WALDREN and RASKO 1978

14.5 Tonicity

This section serves as an example how treatment conditions which *prima facie* may appear irrelevant can influence the outcome of an experiment and that unforeseen results may open up also new lines of investigations. The parameter in question is salt concentration in the medium at or shortly after exposure. It is normally attempted to have *isotonicity* which can be achieved with 0.16 M NaCl for mammalian cells. Variations within certain limits do not

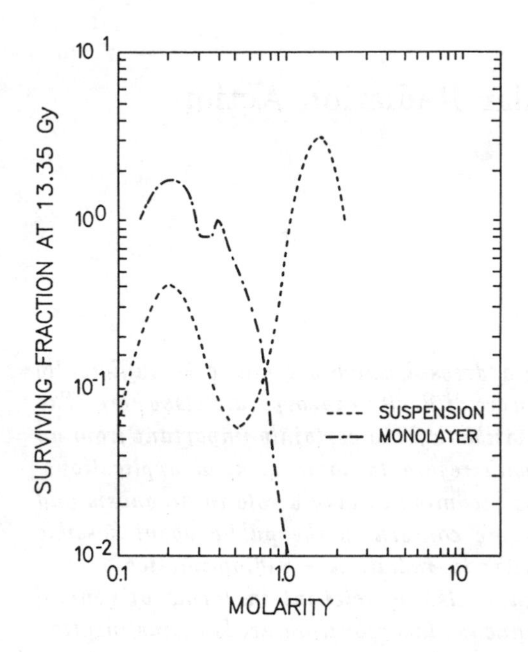

Figure 14.12. Dependence of survival after X-ray exposure on the molarity in the medium (NaCl in water) (after MOGGACH et al. 1979)

influence the colony-forming ability of unirradiated controls but it changes the radiation sensitivity, as seen from Fig. 14.12. It is interesting to note that the culture conditions are quite important, the effect is different in suspension or in cellular monolayers attached to the bottom of the dish. In both cases, one finds sensitization by *hypo-* as well as by *hypertonic* treatment, but only in attached monolayers an increase of survival after a given dose is found with high molarity. All this is not due to the specific salts since similar effects may be found if sucrose is used.

It is assumed that by anisotonic treatment the structure of the chromatin is changed in such a way that the probability of repair is reduced and the damage is rapidly fixed. This interpretation is supported by experiments demonstrating that recovery from potentially lethal damage is no longer fully possible after anisotonic treatment.

This kind of studies have so far only been performed with mammalian cells, it would be interesting to learn how other organisms – particularly those with rigid cell walls – would respond.

Further reading

ILIAKIS 1987, MOGGACH, LEPOCK and RUUV 1979

Chapter 15
Special Aspects of Cellular Radiation Action

In this chapter some questions are addressed which are felt to be sufficiently relevant to deserve mentioning but were difficult to incorporate elsewhere. The action of long-wavelength UV and visible light is certainly important from an ecological point of view but also with regard to some medical applications. Radiowaves and ultrasound are also becoming to play a role in diagnosis and sometimes therapy, and there is some concern in the public about possible hazards. Chemicals which act similar to radiation – radiomimetics – have their place in chemotherapy but may also be relevant in terms of general toxicology. The importance of radionuclide incorporation needs no justification but it is also shown that they are interesting tools to learn more about site and mode of cellular radiation action.

15.1 Near Ultraviolet and Visible Light

As shown in Chap. 8, the action spectrum for cell inactivation (as well as those for mutation and transformation) follows closely the absorption of DNA. Since this is very small in the longer wavelength region ($\lambda > 320\,\text{nm}$), one may be inclined to assume that biological damage by near UV is irrelevant. This is by no means the case. There may be effects via photosensitizers, as in principle discussed in Sect. 5.1.2, which may not only be present in the environment but also naturally in the cells. It can also not be excluded that targets other than DNA, e.g. membranes, may become more important. All this means that investigations are indicated, and not only because we all are exposed to near UV from the sun. The possible change of the solar spectrum due to ozone reduction certainly adds interest to this particular field.

The first studies were, of course, performed with bacteria. Figure 15.1 displays an action spectrum for killing, extended to longer wavelengths. It is interesting in two aspects: firstly, there is a considerably higher sensitivity beyond 320 nm than expected just from DNA absorption and, secondly, oxygen acts as a sensitizer in this range. This points to the involvement of photodynamic processes but substances responsible have not yet been unequivocally identified although there are a number of candidates.

More recently, long-wavelength UV has also been studied in mammalian cells; an action spectrum for inactivation is shown in Fig. 15.2. Although it is

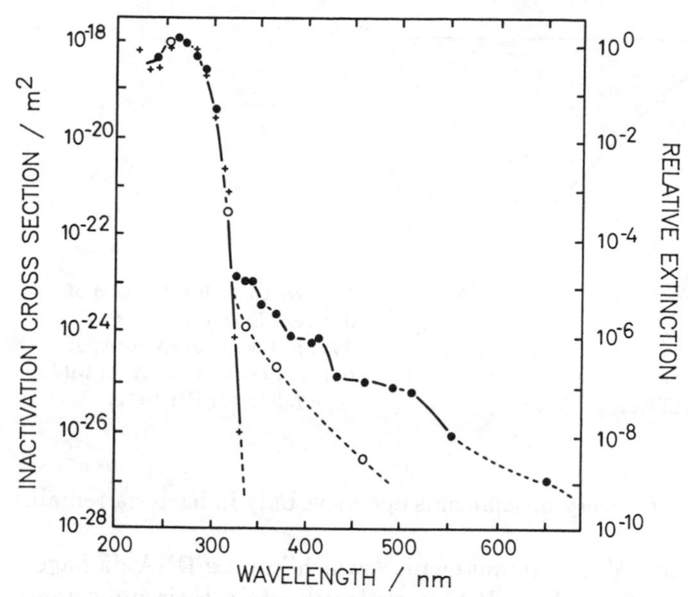

Figure 15.1. Long wavelength action spectrum for the inactivation of E. coli bacteria. Heavy line: exposure in air, broken line: exposure in nitrogen. Crosses indicate relative DNA absorption (after WEBB and BROWN 1976)

similar to that in Fig. 15.1, there are remarkable differences: mammalian cells are considerably more sensitive than bacteria in the UV-A range. Whether

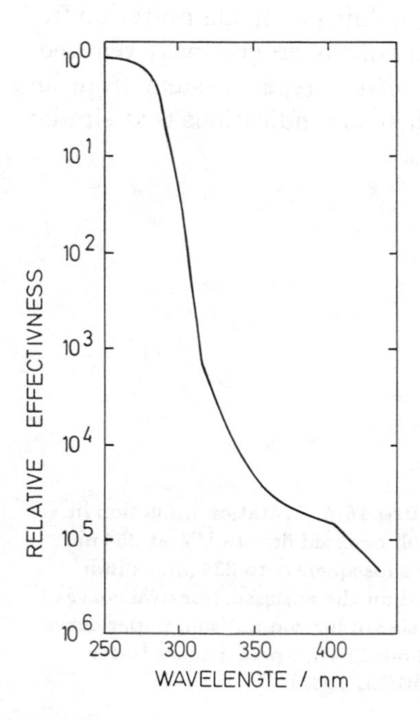

Figure 15.2. As in Fig. 15.1, but for human skin fibroblasts (after TYRELL and PIDOUX 1987)

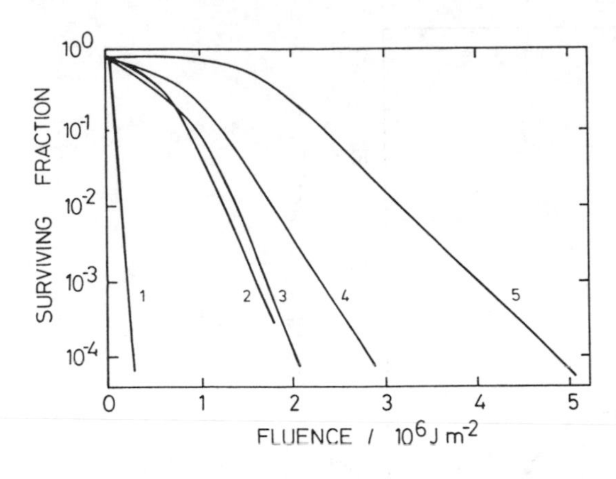

Figure 15.3. Inactivation of different E. coli mutants by 365 nm-UV. 1: uvrA recA, 2: polA, 3: recA, 4: uvrA, 5: wild type (after WEBB 1977)

this is due to specific recovery mechanisms operative only in bacteria remains to be established.

Long-wavelength UV is also mutagenic – which implies DNA damage – but considerably less than UV-C. Repair processes retain their importance with near UV as seen from Fig. 15.3, where the survival of various sensitive mutants after exposure to 365 nm is plotted. A complicating factor, however, is the necessity to apply very large fluences. They may be sufficient to damage the repair systems directly. This has, in fact, been shown, e.g. for the photoreactivating enzyme.

An interesting result was obtained with mutation induction (Fig. 15.4): If *E. coli* cells are first exposed to UV at 254 nm, yielding a certain mutation frequency and then to 334 nm, the number of mutants is progressively reduced. This suggests that near UV causes a switch of the repair system from an "error-prone" mode to an "error-free" one. There are indications that similar mechanisms are operative in mammalian cells.

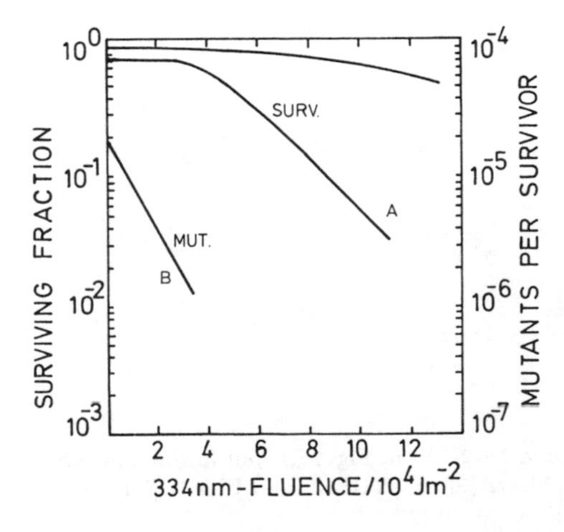

Figure 15.4. Mutation induction in E. coli exposed first to UV at 254 nm and subsequently to 334 nm as indicated on the abscissa. Survival curves are shown for comparison, upper curve without 254 nm-preexposure (after TYRELL 1980)

Table 15.1. DNA photoproducts with different wavelengths. Yields are given in $(\mathrm{J\,m^{-2}})^{-1}\,(10^8\,\mathrm{u})^{-1}$. They were calculated from the data in the references by assuming the human genome to be 3×10^{12} u.

λ/nm	dimers	6-4 products [c]	SSB [b]	crosslinks [a]
252	–	–	7.7×10^{-4}	–
254	1.73 [a]	–	–	1.1×10^{-4}
265	3 [c]	0.6	12×10^{-4}	–
280	1.83 [c]	0.27	9.3×10^{-4}	–
289	0.69 [c]	9.6×10^{-2}	4.3×10^{-4}	–
290	–	–	–	2.8×10^{-5}
297	0.11 [c]	1.8×10^{-2}	1.6×10^{-4}	–
302	0.015 [c]	2.8×10^{-3}	5.5×10^{-5}	–
313	8.7×10^{-3} [a]	2.6×10^{-5}	1.4×10^{-5}	5.3×10^{-7}
334	–	–	3.3×10^{-6}	6.2×10^{-8}
365	1.2×10^{-6} [a]	–	1.3×10^{-6}	4×10^{-9}
405	–	–	6.4×10^{-7}	9.2×10^{-9}

References:
a) PEAK et al. 1985
b) ROSENSTEIN and DUCORE 1983
c) ROSENSTEIN and MITCHELL 1987

The molecular nature of near UV damage is difficult to assess. Table 15.1 summarizes data for a few DNA photoproducts. Although the relative ratios change with wavelength, it cannot be said with certainty which of the species listed is the important one in the UV-B and UV-A region. In bacteria it is clear, however, that even at 365 nm pyrimidine dimers play a role since photoreactivability has been demonstrated. It should be reiterated that apart from DNA, other cellular targets may become more important although their involvement in lethal radiation action could not yet be shown.

Further reading
WEBB 1977, TYRELL 1984

15.2 Other Types of Radiation

15.2.1 Preliminary Remarks

Although this book is devoted to the action of ultraviolet and ionizing radiations a few words are in place about other radiation types. There are two reasons for this, firstly, the general discussion about radiation hazards does commonly not differentiate clearly between different physical agents and secondly, some of the approaches developed in radiation biology may be useful also for the investigation of other potentially hazardous influences. This is particularly so in view of the widespread medical and technical applications of ultrasound and microwaves.

The treatment has necessarily to be short and somewhat superficial, a more comprehensive review will be found in WHO (1982).

15.2.2 Ultrasound

Ultrasound is, of course, not of electromagnetic nature and not even radiation if one adheres to the definition proposed in the beginning of this book. Nevertheless it is included here since some of the questions raised about its action are very similar to those in radiation biology and the same experimental approaches may be used. Ultrasound starts – by definition – at a frequency of 2×10^4 Hz and extends to about 10^7 Hz. "Dosimetry" is not easy, and power levels inside a specimen are difficult to determine. For this reason they are commonly given for the surface; the unit is Wm^{-2}. The maximum total power in medical diagnostic application is about 40 mW but temporal-spatial peak intensities of $1000\,\mathrm{kW\,m}^{-2}$ may be reached in short pulses. The primary effects may be grouped into three classes: local heating, local pressure and "cavitation". Ultrasound propagates in matter as a pressure wave which is attenuated in the medium leading to local heating. The momentum changes at internal interfaces causing pressure differentials, and hence larger forces may be exerted. "Cavitation" is a resonance phenomenon which may be build up to considerable intensities, thus causing the rupture of molecular and supramolecular structures.

Heat damage and cellular disintegration are rather unspecific effects, they depend on frequency and power and can, in principle, easily be controlled. More important is cavitation since it acts at the molecular level. *In-vitro* studies have clearly shown that free radicals may be generated by this process which, in this respect, has some similarities with radiation action. There are many studies on mutation induction and formation of chromosome aberrations by ultrasound (survival is not a suitable parameter because of the mentioned unspecific effects). If the literature is reviewed critically, it has to be concluded that there is no solid evidence for any genetic effect of ultrasound. A possible synergistic action between ultrasound and radiation has also been suggested but also here a direct proof cannot be found, especially since the influence of hyperthermia cannot easily be excluded.

In summary, it may be said that the cellular effects of ultrasound may be essentially traced back either to mechanical damage or increased temperature. They are, therefore, distinctly different from radiobiological phenomena which does, however, not mean that ultrasound is intrinsically "safe" but there appears to be little reason to assume that late effects may be caused. Acute damage must be controlled by appropriate limitations of power levels.

15.2.3 Radiofrequency and Microwaves

Radiofrequency (RF) and microwaves (MW) belong to the spectrum of electromagnetic radiation but the frequencies involved are much smaller as with UV or even X rays.

RF radiation extends from 300 kHz to 300 MHz, microwaves cover the range from 300 MHz to 300 GHz. Their mode of action is quite different from that of optical or ionizing radiation. They induce electric or magnetic fields in the medium leading either to the movement of free ions (*conduction loss*) or molecular rotations (*dielectric loss*). Although resonance phenomena can in principle not be excluded, they appear to be rather improbable since the postulated oscillators are highly dampened in biological tissue. There are a number of reports in the literature where biological effects exhibited a sharply defined dependence on certain frequencies – suggesting resonance effects – but they are in general only poorly reproducible. A common problem with this type of studies is that the applied field is modified by the equipment used, and the probability of outside interference is very high. Temperature control is essential since all the energy is ultimately dissipated in the form of heat. Local temperature gradients may be generated in a frequency dependent manner so that the effects found may be quite different from a situation where the sample is uniformly warmed up.

Taking all this into account it is seen that there is no unequivocal proof for specific cellular actions of RF and MW. Chromosome aberrations and mutations have been reported but these investigations are very often only performed by single laboratories and were not reproduced by others. A clear picture is still missing but there appears to be a general tendency to assume that RF and MW radiations do not cause any specific, i.e. non-thermal, effect at the cellular level. Influences on the nervous system cannot be excluded but they are outside the scope of this brief section.

Further reading
BERNHARDT 1988, BROWN and CHATTOPADHYAY 1988, ROBERTS et al. 1986, ROBINSON 1985, WHO 1982

15.3 Incorporated Radionuclides

The special attractiveness of radionuclides lies in the fact that they are specifically incorporated into biomolecules and cell components. They may be retained for long times and create specific spatial exposure conditions which are quite different from irradiations from the outside. In addition, there are further effects to be taken into account as shown below.

The general dosimetric problems have already been treated in Sect. 4.3.5, the specific issues of whole body incorporation will be discussed in Chap. 21. Here only the cellular aspects are addressed.

If a radionuclide is homogeneously distributed without being bound to specific structures, e.g. like ^3H in water, its action is in principle the same as with external radiation. The situation is different if it is incorporated into a biomolecule. The dose distribution is no longer homogeneous but may be, e.g. in the case of α and β emitters, concentrated in small volumes close to the incorporation site. It is clear that this has severe consequences if the nuclide is part of the DNA.

Another important parameter is the retention time which may be characterized by a *biological half-life*. It is determined by the stability of the molecule, with other words, how fast it is degraded and excreted. This rate is high with mRNA and proteins, but low with DNA. Biological degradation and physical decay act in concert so that for the number N of the remaining nuclide atoms can be written as:

$$-\frac{dN}{dt} = (\lambda_B + \lambda_P)N \quad . \tag{15.1}$$

λ_B and λ_P are biological and physical "decay" coefficients, respectively. From Eq. (15.1) it follows that:

$$N = N_0 e^{-(\lambda_B+\lambda_P)t} \tag{15.2}$$

and for the *effective* half-life τ_{eff}:

$$\tau_{eff} = \frac{\ln 2}{\lambda_B + \lambda_P}$$

or

$$1/\tau_{eff} = 1/\tau_B + 1/\tau_P \quad . \tag{15.3}$$

Equation (15.3) demonstrates that with large differences of τ_B and τ_P, the value of τ_{eff} is essentially determined by the shorter one.

The emitted radiation is, however, not the only action by the disintegration of an incorporated radionuclide. Depending on the decay type, the atomic nucleus may gain a considerable recoil energy sufficient to cause the breakage of molecular bonds. With β emitters which are particularly relevant within this context, there are two additional effects: A new element is created by the disintegration ("transmutation") with different chemical properties so that it does no longer "fit" in the original molecular structure. Furthermore, it is ionized because of the change in nuclear charge. In the case of electron capture, there may be also AUGER electrons (Sect. 3.2.1.5) which increase the local energy deposition and may cause multiple ionizations. The energies expected with the described processes are summarized for a number of important nuclides in Table 15.2.

The decay of incorporated nuclides leads in principle to the same cellular effects as with external radiation, namely loss of colony-forming ability, chromosome aberrations, mutations and neoplastic transformations. The quantitative extent depends on the site where the nuclide is incorporated. ^3H (tritium) is often used in the form of ^3H-thymidine as part of DNA. If it is attached to the C_5 position, the mutation yield per decay is higher than at the C_6-position or if it is in the methyl group. This is presumably due to

Table 15.2. Decay data of some radionuclides, which can be incorporated into DNA. (After FEINENDEGEN, 1978)

nuclide	mean decay energy keV	electron-range µm	recoil-energy keV	energy from change in charge keV
^{3}H	5.7	0.8	0.0036	0.011
^{14}C	49.3	33	0.0071	0.045
^{32}P	695	2600	0.078	
^{125}I [a]	19.2	13	0.0001	-

[a] K-capture and emission of AUGER-electrons

the transmutation effect. ^{32}P is another important critical nuclide. Since it is an integral constituent of the DNA backbone, each disintegration causes a strand break. Furthermore, the high energy leads to the irradiation of other nearby cells so that the total effect is far larger than that of transmutation alone. ^{14}C acts also mainly via the emitted electrons.

^{125}I presents an interesting case. This element is, of course, not a constituent of DNA but it may be incorporated at the site of the thymidine CH_3 group in the form of *Iododeoxyuridine* (^{125}IUdR). Chemically it is similar to BUdR (Chap. 13) and may act as a radiosensitizer by interfering with repair. ^{125}I decays via electron capture to ^{125}Te with emission of characteristic X rays and several (6–10) AUGER electrons. This creates a very high local energy density; furthermore, the resulting ^{125}Te atom is multiply ionized. ^{125}I decay in the DNA shows all the features of high LET radiation, one finds purely exponential survival curves without shoulders and a reduction of the oxygen effect. At the molecular level, the yield of double-strand breaks is greatly increased compared to X rays, virtually each decay produces a DSB.

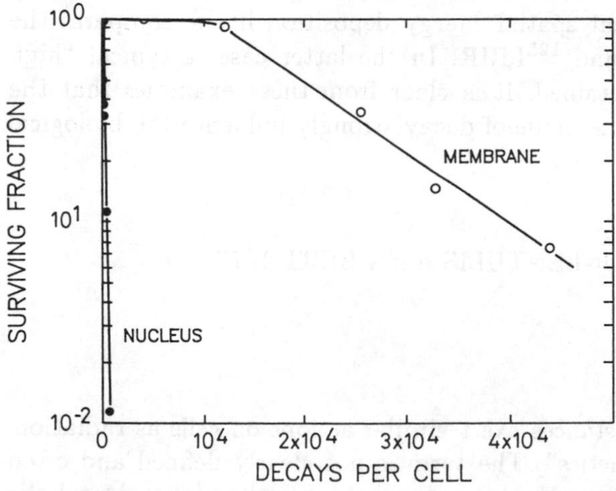

Figure 15.5. Inactivation of mammalian cells by 125-Iodine incorporated in different parts of the cell (after HOFER et al. 1977)

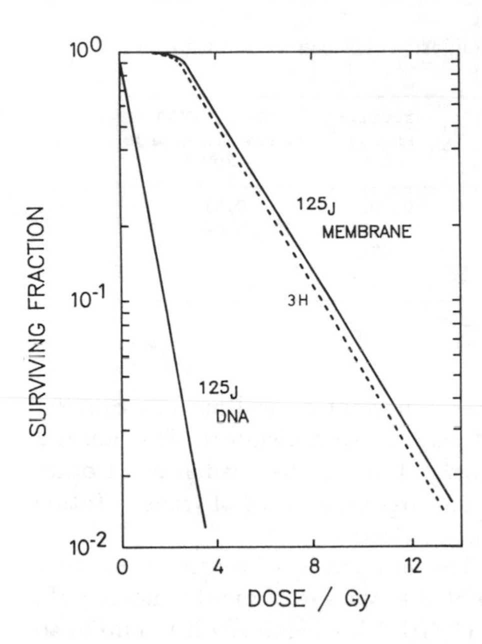

Figure 15.6. As in Fig. 15.5. but as a function of dose in the nucleus (after HOFER et al. 1977)

The range of AUGER electrons is relatively short which explains the larger local energy density. This fact may be used to identify the cellular structures whose damage is responsible for the loss of colony-forming ability by comparing the action of ^{125}I either incorporated into DNA or the cellular membrane. The result of such an experiment is shown in Fig. 15.5: It is obvious that decays in the nucleus are far more efficient than those in the membrane. If the *doses* in the nucleus are calculated according to the methods outlined in Sect. 4.2.5, the curves in Fig. 15.6 are obtained. It is seen that ^{125}IUdR in the nucleus is not only more effective because of the higher dose but obviously also because of the different spatial energy deposition if one compares the effectiveness of ^{3}H-IdUR and ^{125}IdUR. In the latter case, a typical "high LET" survival curve is obtained. It is clear from these examples that the site of incorporation and the mode of decay strongly influence the biological outcome.

Further reading
FEINENDEGEN, TISLJAR-LENTULIS and EBERT 1977

15.4 Radiomimetics

A number of chemical substances exert similar actions on cells as radiation, they are called "radiomimetics". The term is not clearly defined and often used in a uncritical manner to denote several kinds of cell poisons. Metabolic inhibitors, for instance, do certainly not belong to this category. An indication

– which is, however, not sufficient – may be taken from the shape of survival curves. If it resembles that of pharmacological agents (Sect. 8.1), different action mechanisms can be assumed. A better criterion may be gained from interaction in combination experiments where at least additive action should be found. The occurrence of chromosomal aberrations and mutations gives another clue. Systematic investigations covering *all* these aspects are rarely available, they may be only of academic interest. Practically, radiomimetics are important in tumour therapy and as potential environmental hazards.

The number of compounds whose action is thought to be radiomimetic is very high indeed and is steadily increasing. An already outdated list (FISHBEIN, FLAMM and FALK 1970) gives 110 substances; it would be considerably longer today. This means that here only a few important examples can be discussed, the selection being quite subjective.

An important class are the alkylating agents (Fig. 15.7) possessing an alkylating group which may be easily transferred to other molecules. This

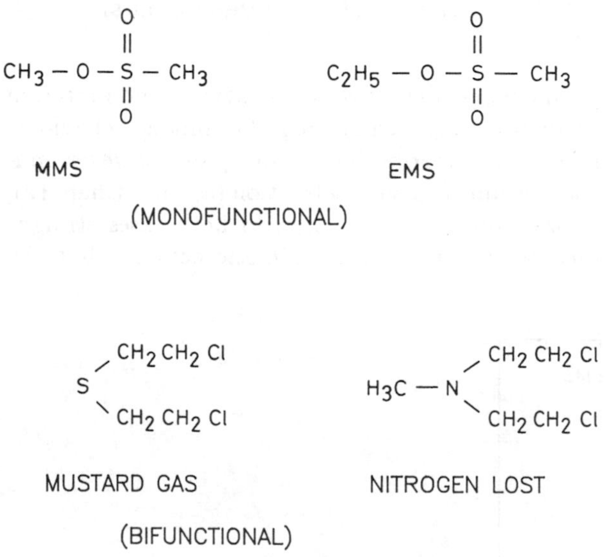

$CH_3-O-S(=O)_2-CH_3$ MMS

$C_2H_5-O-S(=O)_2-CH_3$ EMS

(MONOFUNCTIONAL)

$S(CH_2CH_2Cl)_2$ MUSTARD GAS

$H_3C-N(CH_2CH_2Cl)_2$ NITROGEN LOST

(BIFUNCTIONAL)

Figure 15.7. Structure of some radiomimetics

is, of course, particularly relevant with DNA where secondary reactions may then lead to base abstraction, breaks and crosslinks. Depending on the number of transferable alkyl groups, we distinguish between mono-, bi- or polyfunctional agents. The latter two are mainly responsible for crosslinks. Well-known prototypes of alkylating agents are nitrogen lost and mustard gas whose practical significance is, however, small.

Of greater experimental relevance are *ethyl methylsulfonate* (EMS) and *methyl methylsulfonate* (MMS). Both are monofunctional transferring only one alkyl group ($-C_2H_5$ or $-CH_3$, respectively) and affecting – in contrast

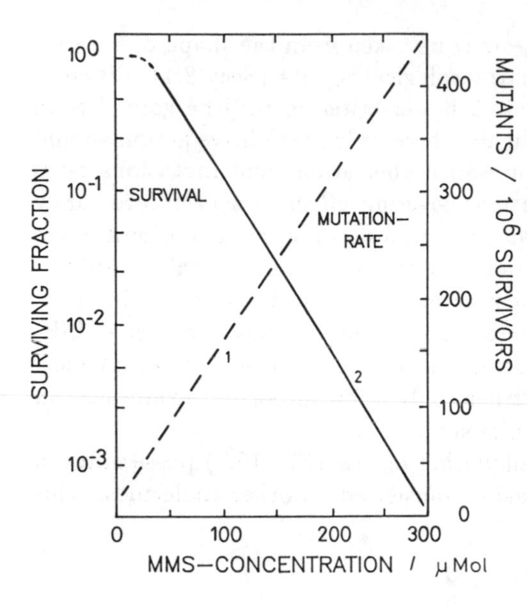

Figure 15.8. Survival and mutation induction after MMS-treatment in Chinese hamster cells (after COUCH et al. 1978)

to radiation – mainly the purine bases. Fig. 15.8 shows survival and mutation induction curves for MMS demonstrating clearly the radiomimetic behaviour. For EMS, qualitatively similar results are obtained. If one plots, however, mutation frequency versus the logarithm of surviving fraction (see also Chap. 12), an important difference becomes obvious (Fig. 15.9). In both cases straight lines are obtained – another indication for radiomimetic action – but the

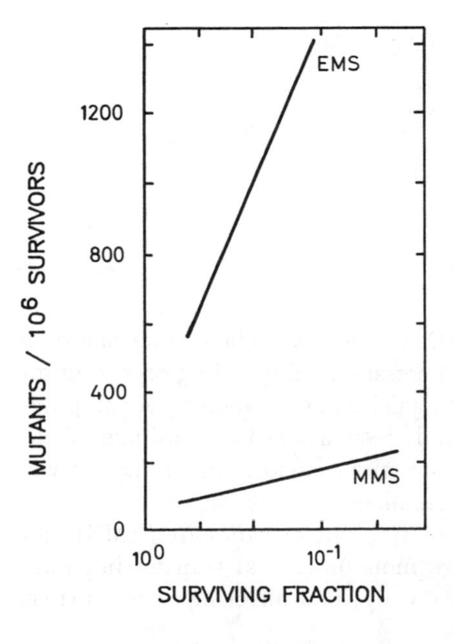

Figure 15.9. Mutation induction as function of survival with EMS and MMS in mammalian cells (after COUCH et al. 1978)

slopes are different. The "mutagenicity" of EMS is considerably greater than that of MMS. The two closely related substances resemble the differences between UV and ionizing radiation. The exact molecular mechanisms are not yet sufficiently explored.

Similar differences are found with other compounds. At first it appears that the number of alkylated bases is the relevant parameter but this is not the case since it is about equal in both cases. The exact molecular nature of the alteration seems to be more important.

Two other compounds should be mentioned which also play some role in cancer therapy. One is *bleomycin* which essentially acts through DNA strand-break induction, constituting thus a "real" radiomimetic. The other one is *neocarcinostatin* which has very similar properties.

Alkylation products and other chemically induced lesions may also be repaired, the description of these processes is beyond the scope of this book.

Further reading

HOLLAENDER 1971, LAWLEY 1966, MÜLLER and ZAHN 1977, RIEGER and MICHAELIS 1967, ROBERTS 1978, SINGER 1975, VON SONNTAG 1987, VERLY 1974

Chapter 16
Theoretical Models of Cellular Radiation Action

This chapter describes some of the theoretical approaches to describe quantitatively the action of radiation on cells. The aim of these attempts is not predominantly to bring radiation biology from a descriptive to a higher theoretical level but more to allow estimates in dose ranges where direct experimental determinations are not feasible or even possible. A well known and important example is the behaviour at very low doses and dose rates. All models may be roughly grouped into two categories: in the first, the emphasis is laid on the distribution of initial lesions and their interactions, while in the second, repair and its efficiency are the focal points. Examples of both types are given.

16.1 Target Theory

The basic concepts of target theory were already introduced in Sect. 7.1 but with the simplifying assumption that one hit suffices for inactivation. In the general case to be treated now it is supposed that a critical hit number m exists below which no radiation action is seen. If the hit number is equal to or larger than m the object will be inactivated. If this is so, then the surviving fraction y becomes:

$$y = \sum_{\nu=0}^{m-1} e^{-(vD)} \frac{(vD)^\nu}{\nu!} \tag{16.1}$$

where D is the dose and vD the mean hit number. Plotting ln y versus D in the usual way yields continuously bending curves, with the exception of m = 1 ("single-hit mechanism") where the relationship is purely exponential. Equation (16.1) gives generally poor fits to experimental results with cells so that another approach is required: It is now assumed that there are n sensitive targets in the cell which all have to be inactivated to cause cell killing. The critical hit number shall be the same for each target and they may act independently of each other. In this case, Eq. (16.1) is applicable for each individual target; thus, for the whole cell it follows that:

$$y = 1 - \left(1 - \sum_{\nu=0}^{m-1} e^{-(vD)} \frac{(vD)^\nu}{\nu!}\right)^n \tag{16.2}$$

if the sensitivity coefficient v is also taken as equal. More elaborate formulae are, of course, possible if the above-listed restrictions are relieved but this does not add to the understanding of the basic ideas.

The situation is particularly simple if m = 1, leading to:

$$y = 1 - (1 - e^{-vD})^n \quad . \tag{16.3}$$

This expression which may be termed a "single-hit multitarget curve" shows the features exhibited by many survival curves, namely a shoulder in the low dose region and an exponential terminal part. This is seen if the bracket is expanded. For high doses it is seen that $e^{-vD} \ll 1$; thus:

$$y \approx 1 - (1 - ne^{-vD}) = ne^{-vD} \quad . \tag{16.4}$$

The extrapolation number formally introduced in Sect. 8.1 has here a special meaning, namely the number of sensitive targets. In mammalian cells quite often a value close to 2 was found suggesting that two chromosomes have to be inactivated to cause cell death. Later it turned out that n is by no means constant and that it may be manipulated by changing the culture conditions. The simple picture fails also with microorganisms where sometimes very high values of n are found which cannot be accommodated in any biological model. Nevertheless, Eq. (16.4) has been proved to be of great heuristic value since many survival curves show – at least qualitatively – the described relationship.

The underlying assumptions are certainly to mechanistic to be adequate for the biological situation. A fixed critical hit number is hardly justifiable and biological systems cannot be expected to react in an "all-or-none" fashion. The other inherent features, however, namely the stochastic interaction with sensitive targets and the existence of more than one target are incorporated, more or less explicitly, also in most of the other models.

Before concluding this section it may be illustrating to consider Eq. (16.4) from a more quantitative viewpoint. Regardless of any special model, the parameter v gives the slope of the terminal part of the survival curve, or the inverse of D_0. D_0 is hence the dose which leads on an average to one hit. For mammalian cells, D_0 is about 1.3 Gy, i.e. 8×10^{18} eV/kg. With a total DNA content of 1.2×10^{-14} kg and an assumed energy per hit of 60 eV, calculation yields the very high number of 1600 hits per genome or – in the case of man – still 35 per chromosome. This simple calculation shows that either the energy required per inactivating hit is very large or that most of the hits are ineffective. It is obvious that the simple model is not suited to give a better insight into the mechanism of cell inactivation. Different approaches are evidently required.

16.2 The Two-Lesion-Model (Neary 1965)

Although the following model is about twenty years old, it is still quite important since it contains many features which can also be found in more recent formulations. The main postulate is the assumption that a biologically significant lesion is formed by the interaction of two primary hits or sublesions. The initial physical processes are – as before – considered to be statistically independent. If two hits are to interact with each other, the distance between must not be too large or, with other words, they have to occur within a specified volume, the "interaction site". There may be many of these sites in a cell containing sensitive structures where the primary alterations take place.

This model can be treated quite formally. As above, the interacting sublesions are called "hits", the final biological damage is called "lesion". The probability for a hit to occur is p, the number of hits per unit distance s and the mean pathway in the sensitive structure t. With this, the probability K for the formation of one hit by the passage of one particle may be written as:

$$K = 1 - \sum_{\nu=0}^{\infty} (1-p)^{\nu} \frac{(st)^{\nu}}{\nu!} e^{-st} = 1 - e^{-pst} \qquad (16.5)$$

As usual, it is assumed that the primary processes follow a POISSON distribution. A hit may be caused either by the passage of one particle or to the interaction of several particle passages. These are also POISSON-distributed so that the probability p_{ν} of hit formation by ν passages becomes:

$$p_{\nu} = 1 - (1-K)^{\nu} \quad . \qquad (16.6)$$

If m is the mean number of passages, then the total probability W_m that a hit is caused will be:

$$W_m = \sum_{\nu=0}^{\infty} p_{\nu} e^{-m} \frac{m^{\nu}}{\nu!}$$

or – with Eq. (16.6):

$$W_m = 1 - e^{-mK} \quad . \qquad (16.7)$$

A lesion may be caused if at least two hits are formed in the sensitive structure within the interaction site. This may occur either by single or multiple particle passages. To treat this formally, two targets X and Y are assumed along with a probability f indicating that the particle passes either through X or Y. The probability g that the two targets are intersected by the particle path is then obviously $g = 1 - f$. Since all events are statistically independent one finds for the mean number W_m of particle passages per target:

$$W_m = e^{-(1-g)} \frac{m[(1-g)m]^x}{x!} \cdot e^{-(1-g)} \frac{m[(1-g)m]^y}{y!} \cdot e^{-gm} \frac{(gm)^z}{z!} \qquad .(16.8)$$

Here x is the number of passages through X only, y the respective number for Y and z the mean number of passages through both targets. This means that there are $(x + z)$ passages through X and $(y + z)$ through Y.

The probability P_{xy} that a hit is caused both in X and Y is then, with Eq. (16.6):

$$P_{xy} = \sum_{x=0}^{\infty}\sum_{y=0}^{\infty}\sum_{z=0}^{\infty}[1-(1-K)^{x+z}]\cdot[1-(1-K)^{y+z}]\cdot W_m \tag{16.9}$$

$$= 1 - 2e^{-mK} + e^{-2mK+mgK^2} \quad . \tag{16.10}$$

The total number of lesions Y is obtained by multiplying by the site number N and the interaction probability E, so that:

$$Y = N \cdot E \cdot P_{xy} = N \cdot E(1 - 2e^{-mK} + e^{-2mK+mgK^2}) \quad . \tag{16.11}$$

Since both K and g are small, the exponential may be expanded for not too high values of m. Thus, it follows that:

$$Y \approx N \cdot E(mgK^2 + m^2K^2) \tag{16.12}$$

if higher powers than 2 in K and g are neglected.

The mean number m of passages depends on the particle fluence ϕ and the target cross section σ:

$$m = \sigma\phi$$

or with Eq. (4.20)

$$m = \frac{\sigma \cdot D \cdot \varrho}{L}$$

where D is the dose, ϱ the density and L the LET (assumed to be single-valued).

Hence:

$$Y \approx NEK^2\left(\frac{\sigma D \varrho}{L}g + \frac{\sigma^2 D^2 \varrho^2}{L^2}\right) \quad . \tag{16.13}$$

Equation (16.13) states a "linear-quadratic" relationship between the number of lesions and dose. Such a behaviour is, in fact, frequently found as shown several times before, e.g. for chromosome aberrations and mutations.

The probability g that two targets are hit when a particle traverses a site is approximately equal to the ratio of target and site cross sections. If the latter is termed S, then:

$$Y \approx NEK^2\left(\frac{\varrho D}{L}\cdot\frac{\sigma^2}{S} + \frac{\sigma^2\varrho^2 D^2}{L^2}\right) \quad . \tag{16.14}$$

The ratio between the linear and the quadratic term depends (apart from ϱ which is constant) only on S and L so that estimates of S are possible if L is known. The dose D′ where both terms are equal is:

$$D' = \frac{L}{S} \quad . \tag{16.15}$$

This may be used to get an estimate of S.

The LET dependence warrants a more detailed consideration. It is evident that the mean number of hits per unit path length s must be related to radiation quality. In simply assuming that a constant energy E_0 is deposited, this becomes:

$$s = \frac{L}{E_0} \quad . \tag{16.16}$$

Eq. (16.14) takes then the form:

$$Y = NE\left(1 - e^{-Lpt/E_0}\right)^2 \cdot \left(\frac{\sigma^2 D\varrho}{\varrho L} + \frac{\sigma^2 \varrho^2 D^2}{L^2}\right) \quad . \tag{16.17}$$

For small L, the exponential may be expanded:

$$\begin{aligned} Y &\approx NE\frac{L^2p^2t^2}{E_0^2}\left(\frac{\sigma^2\varrho D}{SL} + \frac{\sigma^2\varrho^2 D^2}{L^2}\right) \\ &= NE\frac{p^2t^2\sigma^2}{E_0^2}\left(\frac{\varrho L}{\varrho}D + \varrho^2 D^2\right) \quad . \end{aligned} \tag{16.18}$$

It is seen that – under these simplifying conditions – only the linear term depends on LET. Eq. (16.17) states that the yield of lesions as a function of LET has to pass a maximum. It is also clear – both from Eqs. (16.14) and (16.18) – that the relevance of the quadratic term decreases with higher LET values. This means that the linear part extends to higher doses – in accordance with experimental findings.

So far it has been neglected that radiations with unique LET values are hardly found in reality. Therefore, their statistical distributions must be incorporated into the model. The treatment must be limited to small LETs where the δ-electron problem does not play an important role. In this case, Eq. (16.12) may be written, taking Eq. (16.18) into account, as:

$$Y = NE\frac{L^2p^2t^2}{E_0^2}(mg + m^2)$$

or

$$Y = NE\frac{L^2p^2t^2}{E_0^2}(\sigma\phi g + \sigma^2\phi^2) \quad . \tag{16.19}$$

The fluence ϕ is now a function of LET. To obtain the total yield, we integrate over all contributions:

$$Y = A\left[\sigma g\int L^2\phi(L)\,dL + \sigma^2\left(\int L\phi(L)\,dL\right)^2\right] \tag{16.20}$$

where $A = NEp^2t^2/E_0^2$.

With Eq. (4.22), this becomes:

$$Y = A(\sigma g\varrho L_D \cdot D + \sigma^2\varrho^2D^2)$$

or with $g = \sigma/s$

$$Y = A\sigma^2\left(\frac{\varrho L_D}{s}D + \varrho^2D^2\right) . \tag{16.21}$$

The site cross section is then:

$$s = \frac{L_D}{D'} \tag{16.22}$$

where D' signifies, as before, the dose where the linear and the quadratic term give the same contribution.

The main result of this short discourse is that the dose average LD is the important parameter. This is in accordance with microdosimetric treatments which are given in the next section. They are more appropriate for the real situation since the assumption of infinitely thin particle tracks which was inherent in the present model is obviously to crude.

16.3 Theories Based on Microdosimetric Considerations

The foregoing model attempted to explain the formation of chromosome aberrations as a function of radiation quality by postulating interacting sublesions formed in microscopically small volumes. In these dimensions we are particularly confronted with the "δ-ray problem" and it is extremely difficult to assess the spatial pattern of energy deposition. It is, therefore, necessary to replace the LET concept by microdosimetric considerations. Since the fundamentals have already been outlined in Chap. 4, the treatment can be brief here.

Analogous to the preceding section, it is assumed that a biologically significant lesion is created via the interaction of <u>two</u> sublesions in a critical site. The number of the latter is proportional to the specific energy deposited in the site and the probability of their formation is independent of radiation quality. The number of *lesions* is then proportional to the square of the number of sublesions per site and hence to the square of the specific energy.

Averaging over many regions, and using the notation of the preceding section, the relationship becomes:

$$Y = K'\overline{Z^2} \tag{16.23}$$

where K' is a proportionality coefficient.

Using Eq. (4.35) then gives:

$$Y = K'(\overline{Z_{1D}}D + D^2) \tag{16.24}$$

or, using Eq. (4.37) this becomes:

$$Y = K'\left(\frac{\overline{y_D}}{S\varrho}D + D^2\right) \tag{16.25}$$

This fundamental relationship is very similar to Eq. (16.21) if L_D is replaced by y_D and $K' = A\sigma^2\varrho^2$. It predicts also a linear-quadratic dependence of lesion yield on dose. The site dimensions may be estimated similarly as before but the result will be different since z_{1D} does not only depend on radiation quality but also on site dimensions. The calculations have, therefore, to be based on experimental values of z_{1D} which are available for a great number of radiation types. Some examples are given in Chap. 4.

The assumption of a fixed site diameter is rather rigid, difficult to reconcile with the structure of biological reactions and also at variance with, e.g. experimental results with soft X rays. A better and more refined approach constitutes the "distance model" (KELLERER and ROSSI 1978) which retains the interaction-sublesions hypothesis but assumes that the probability for lesion formation depends on the distance between them. The original derivation of the mathematical formalism is rather involved and cannot be repeated here (see ZAIDER and ROSSI 1986). It leads also to a linear-quadratic dose response curve, namely:

$$Y = k'(z'D + D^2) \tag{16.27}$$

where:

$$z' = \int_0^\infty \frac{t(x)f_p(x)g(x)}{4\pi x^2 \varrho}\,dx \bigg/ \int_0^\infty f_p(x)g(x)\,dx \quad . \tag{16.28}$$

The three functions in the first integral reflect the different contributions: $t(x)$, the "proximity function" (see below) depends only on the radiation type, $f_p(x)$ is essentially the distribution of distances between sites in the sensitive volume where sublesions may be formed, i.e. it characterizes the biological object, and $g(x)$ gives the probability that two sublesions interact to create a lesion if they are a distance x apart.

$t(x)$, merely depending on physical parameters, can in principle be determined. It is defined in the following way:

$$t(x)\,dx = \sum_k \left[\frac{\varepsilon_k}{E} \cdot \sum_i \varepsilon_i \right] \tag{16.29}$$

with:

$$E = \sum_{k=1}^{N} \varepsilon_k \quad .$$

This considers single energy transfers ε_k whose total number is N and records all transfers ε_j which occur in a spherical shell of thickness dx and radius x around ε_k. All possible pairs are summed and weighted by the total energy transferred E. $f_p(x)$ and g(x) cannot separately be determined. One may, therefore, write Eq. (16.28) in a different way:

$$z' = \int_0^\infty t(x) \cdot \gamma(x)\,dx \quad . \tag{16.30}$$

The newly introduced function $\gamma(x)$ may be interpreted as the probability that two energy transfers which are separated by a distance x interact to form a lesion. $\gamma(x)$ can be extracted from experiments with different radiation types; an example for Chinese hamster cells is shown in Fig. 16.1. It

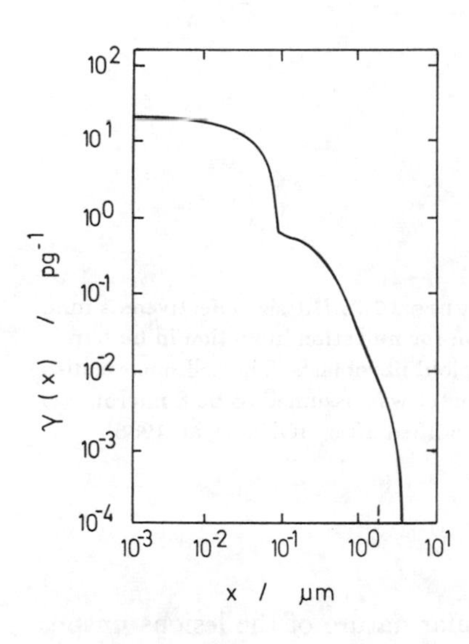

Figure 16.1. Biological interaction function according to the size model as obtained from survival curves of Chinese hamster cells exposed to different types of radiation (after ZAIDER and ROSSI 1988)

demonstrates that very small distances are involved and also the superiority of the "distance" over the "site" model.

It is assumed throughout the theory of dual radiation action that the proportionality coefficient K′ is constant and does not depend on radiation quality. This leads to discrepancies with the experiment in the case of very densely ionizing radiations.

It had also been shown in Chap. 8 that the relative biological effectiveness decreases again with very high LET values, which is not predicted by the theory. In order to remedy this, saturation of sublesions is assumed which may be approximated by a simple exponential relationship. In this case, Eq. (16.23) takes the form:

$$Y = K' z_0^2 \left(1 - e^{-\overline{z^2}/z_0^2}\right) \tag{16.31}$$

which approaches Eq. (16.23) for $\overline{Z_{1D}}/z_0 \ll 1$.

Microdosimetry may be applied in a different, more heuristic way. By postulating that the biological effectiveness depends only on the energy deposited per single traversal in the sensitive volume, the response is given by the product of the "hit size effectiveness" and the probability of the "hit size". "Hit size effectiveness functions" can be extracted from experimental results if assumptions are made of the dimensions of the gross sensitive volume. A few examples are shown in Fig. 16.2.

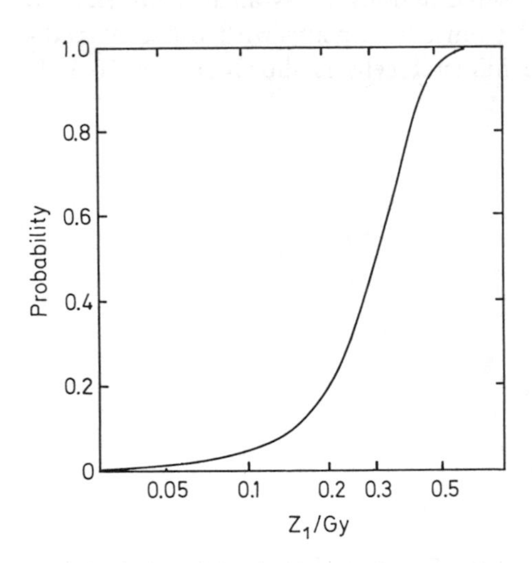

Figure 16.2. Hit size effectiveness function for mutation induction in human diploid fibroblasts. The cell nuclear diameter was assumed to be 8 micrometer (modified after BOND et al. 1988)

16.4 The "Molecular Theory"

The models described so far left the molecular nature of the lesions unspecified. They may be tentatively identified with obviously serious DNA alterations, namely double-strand breaks (DSB). Within the framework of the "molecular theory" it is assumed that they are either formed by a single event or via the interaction of two closely spaced single-strand breaks (SSB). If Y_0 is the number of initial DSB, it can be shown that:

$$Y_0 = \alpha_0 D + \beta_0 D^2 \quad . \tag{16.32}$$

The derivation, which shall not be reported here follows essentially the lines indicated in Sect. 16.2. Both SSB and DSB may be repaired by the cell with probabilities $(1 - f_1)$ and $(1 - f_0)$, respectively. f_1 is thus the fraction of SSB available for interaction. The final yield Y of DSB is then:

$$Y = f_0(\alpha_0 D + f_1 \beta_0 D^2) \tag{16.33}$$

which may be abbreviated to:

$$Y = \alpha D + \beta D^2 \quad . \tag{16.34}$$

The two parameters α and β both depend on radiation quality and also on environmental factors in a complex manner. Although the final result (Eq. (16.34)) is formally equal to Eq. (16.21) and Eq. (16.24), it has to be stressed that the underlying assumptions are quite different. This is particularly so with regard to the interaction distances. This issue will be taken up again in the last section of this chapter.

16.5 Track Structure and Action Cross Sections

The energy deposition by particle radiation is quite inhomogeneous. As stated several times before, the "dose" quantity is inadequate here and it cannot be expected that dose-response curves are identical for different types of particles. The concept of "inactivation cross section" where "particle fluence" is the independent variable, introduced in Sect. 8.3.2, is obviously more appropriate.

In a first approximation – to be modified later – it may be assumed that the survival probability of a cell depends only on the amount of energy deposited in the mass of its sensitive site. There are good reasons for this supposition since even with heavy particles most of the local dose is due to electrons, as it is the case also with γ or X rays. The main difference is the variation in electron and energy density.

If z is the specific energy in the cell's sensitive site and y(z) the survival probability as a function of z, the surviving fraction of an irradiated population y may be written as:

$$y = \int y(z) f(z) \, dz \tag{16.35}$$

where f(z), the density distribution of z, depends, of course, both on particle type and fluence. Equation (16.35) may be evaluated further if special assumptions are made for y(z). The simplest case is exponential survival.

Furthermore, it has to be taken into account that the number of hits follows a POISSON distribution; thus:

$$y = \sum_{\nu} \int (\sigma\phi)^{\nu}/\nu! e^{-\sigma\phi} e^{-kz} f_{\nu}(z,\phi)\, dz \tag{16.36}$$

in which:

σ: "hit" cross section (see below)
ϕ: particle fluence
K: sensitivity parameter ($K = 1/D_0$ for X rays)
$f_{\nu}(z,\phi)\, dz$: distribution of specific energy with v hits

It had already been shown in Sect. 4.2.3 that $f_{\nu}(z,\phi)$ is the ν-fold convolution of the single traversal distribution $f_1(z)$, so that:

$$y = \sum_{\nu} \left[(\sigma\phi)^{\nu}/\nu! \cdot e^{-\sigma\phi} \cdot \int f_1^{*\nu}(z) e^{-Kz}\, dz \right] \tag{16.37}$$

The integral fortunately happens to be a LAPLACE transform (see Appendix I.5) with the important property that the transform of a ν-fold convolution equals the ν-fold product of the transform of the function. Thus, it follows that:

$$\begin{aligned} y &= \sum_{\nu} (\sigma\phi)^{\nu}/\nu!\, e^{-\sigma\phi} \left(\int f_1(z) e^{-Kz}\, dz \right)^{\nu} \\ &= \exp\left[-\sigma\phi \left(1 - \int f_1(z) e^{-Kz}\, dz \right) \right] \quad . \end{aligned} \tag{16.38}$$

The survival curve of the total population is again exponential and may be characterized by an inactivation cross section σ_i:

$$\sigma_i = \sigma \left(1 - \int f_1(z) e^{-Kz}\, dz \right) \quad . \tag{16.39}$$

The main task lies now in the evaluation of the integral. This may be achieved using track structure models as described in Sect. 4.2.4. Figures 16.3 and 16.4 give a few examples of calculated energy depositions by some heavy ions in spherical sites of cellular dimensions as a function of the distance between the target center and the ion path. It is clearly seen that – depending on ion type and energy – considerable local doses may be caused even by traversals outside the target. This is due to the action of "penumbra" electrons and has to be included in the calculations. It means that σ is <u>not</u> just the target cross section but that of a circle whose radius is the sum of target and penumbra radius, r_T and r_P, respectively. The latter depends – as explained in Sect. 4.3 – on ion energy:

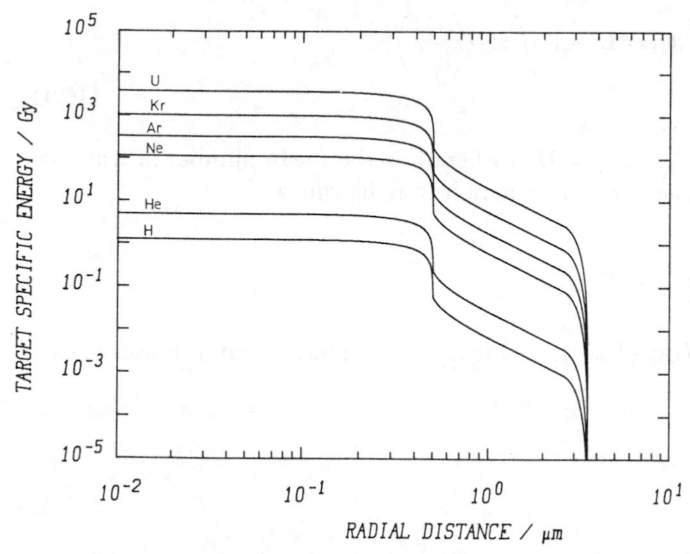

Figure 16.3. Specific energies deposited by various ions (energy 10 MeV/u) in spherical sites of 1 micrometer diameter (calculated according to KIEFER and STRAATEN 1986)

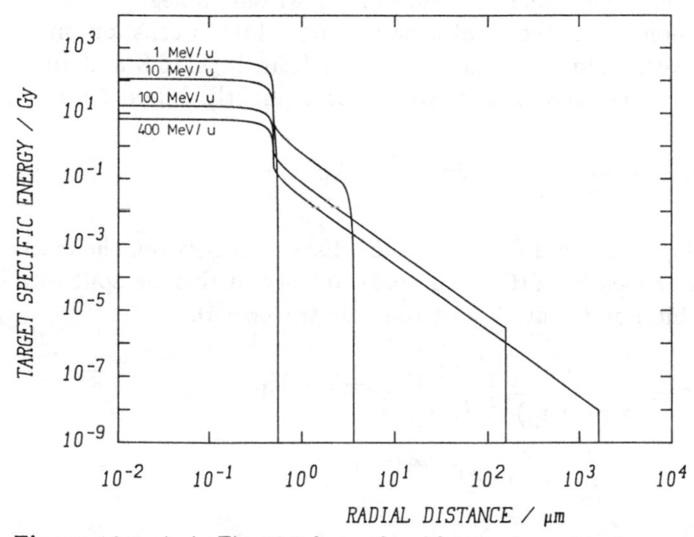

Figure 16.4. As in Fig. 16.3. but only with neon ions of various energies as indicated

$$\sigma = \pi(r_T + r_p)^2 \tag{16.40}$$

In the general case, the integral in Eq. (16.39) has to be evaluated numerically. For some special situations approximations are possible:

1. Dose approximation: If the energy deposition by a single particle is small, the exponential term may be expanded:

$$\sigma_i \approx \sigma\left(1 - \int f_1(z)\,dz + K\int z f_1(z)\,dz\right)$$
$$= K\sigma\overline{z_1} \quad . \tag{16.41}$$

According to Sect. 4.2.3, $z_1 = D/n$ where n is the mean number of traversals which equals $\sigma\phi$. The surviving fraction thus becomes:

$$y = e^{-\sigma\phi \cdot K\overline{z_1}}$$
$$e^{-\sigma\phi \cdot KD/n} = e^{-KD} \quad . \tag{16.42}$$

Since $D = L_T \cdot \phi/\varrho$ (Eq. (4.20)), the inactivation cross section increases linearly with LET:

$$y = e^{-\frac{K\overline{L_T}}{\varrho}\cdot\phi}$$
$$\sigma_i = \frac{K\overline{L_T}}{\varrho} \quad . \tag{16.43}$$

2. LET approximation: For lighter ions and not too sensitive systems, the inactivation probability by outside traversals may play only a negligible role. Also it may be assumed that the local dose by direct hits is constant since the variations are rather small as may be judged from Figs. 16.3 and 16.4. The integral in Eq. (16.41) may then be written in a slightly different form:

$$\int f_1(z)e^{-Kz}\,dz = \frac{1}{\pi(r_T + r_p)^2}\int 2\pi x e^{-Kz_1}\,dx \tag{16.44}$$

where $z_1(x)$ is the local dose as a function of the distance x between the track and the target center (see Figs. 16.3 and 16.4). It may further be split into an "inner" (direct hits) and "outer" part (outside traversals):

$$\int f_1(z)e^{-Kz}\,dz = \frac{1}{\pi(r_T + r_p)^2}\left(\int_0^{r_T} 2\pi x e^{-Kz_1(x)}\,dx + \int_{r_T}^{r_T+r_p} 2\pi x e^{-Kz_1(x)}\,dx\right) \quad .$$

If for direct hits $e^{-kz(x)}$ is assumed to be constant e^{-Kz} and unity for outside traversals, then:

$$\int f_1(z)\,dz = \frac{1}{\pi(r_T + r_p)}\left(e^{-Kz_c}(\pi r_T^2) + \pi(r_T + r_p)^2 - \pi r_T^2\right)$$

and for the inactivation cross section:

$$\sigma_i = \pi r_T^2(1 - e^{-Kz_c}) \tag{16.45}$$

zc may be approximated by:

$$z_c \approx \frac{\overline{L_T}}{\varrho \pi r_T^2}$$

thus yielding:

$$\sigma_i = \pi r_T^2 \left(1 - e^{-K \frac{\overline{L_T}}{\varrho \pi r_T^2}}\right) \quad . \tag{16.46}$$

For small values of the exponent, Eq. (16.46) approaches Eq. (16.43). The treatment just described is not limited to exponential survival but may easily be generalized to all cases where y(z) can be written as a series of exponential functions. As an example, the multi-target single-hit curve is used:

$$\begin{aligned} y(z) &= 1 - (1 - e^{-Kz})^n \quad \text{(see also Eq. (16.3))} \\ &= \sum_{i=1}^{n} \binom{n}{l} (-1)^{l+1} e^{-lKz} \quad . \end{aligned} \tag{16.47}$$

Inserting this into Eq. (16.35) and using the same formalism, yields for the surviving fraction:

$$y = \sum_{i=1}^{n} \binom{n}{l} (-1)^{l+1} \exp\left[-\sigma\phi \left(1 - \int f_1(z) e^{-lKz}\, dz\right)\right] \quad . \tag{16.48}$$

Within the limitations of the LET approximation, it approaches exponential survival for high LET values, namely:

$$y \approx e^{-\pi r_T^2 \cdot \phi} \quad . \tag{16.49}$$

This is the case if $Kz_c \gg 1$ so that the exponential term in Eq. (16.46) may be neglected and penumbra effects do not contribute significantly.

Although the theoretical treatment described appears to be consistent, there are still a few remaining problems. Nothing has been said so far about the sensitivity parameter K. As a first approximation, it may be taken from X ray survival curves. This is, however, at variance with experimental results as can be seen from Fig. 16.5 where theoretical calculations and experimental determinations of cross sections for haploid yeast are compared. It is obvious that in a certain LET range particle radiation is more effective than predicted. This discrepancy demonstrates the RBE problem, the higher effectiveness of densely ionizing radiation. Numerous approaches are available in the literature to resolve this on the basis of theoretical models, some are described in this chapter. It appears that biologically significant lesions are formed more effectively if the ionization density is increased. These may be,

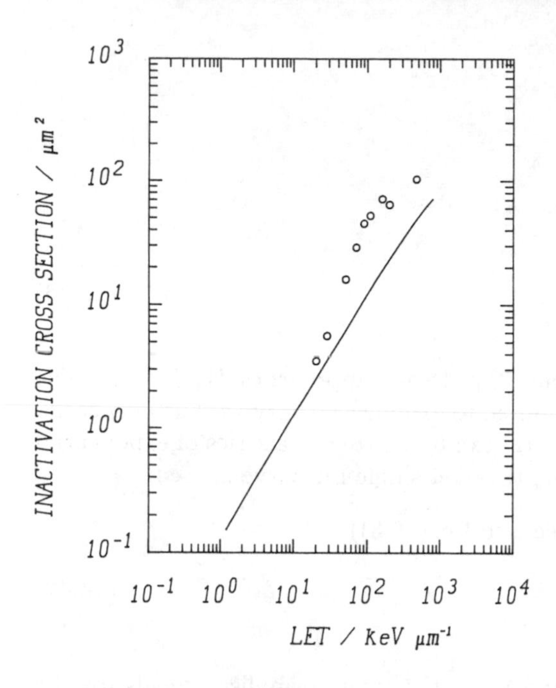

Figure 16.5. Comparison between theoretical (line) and experimental cross sections for the survival of human diploid fibroblasts (data from COX and MASSON 1979)

e.g. double strand-breaks. GUENTHER and SCHULZ (1983), following the lines described in this chapter, were able to obtain good agreement between theory and experiment for a number of systems by incorporating the dependence of DSB formation on the spatial distribution of energy deposition. Their theory is quite elaborate and cannot be repeated here, the interested reader is referred to their book.

A further hidden assumption is that the survival-curve shape does not change with radiation quality. This must not be necessarily so. If, for example, repair processes are less effective with more densely ionizing radiations, the function y(z) may change. Loss of the shoulder is in itself, however, not a sufficient proof for this to occur as can be seen from Eq. (16.49) which demonstrates that for $Kz_c \gg 1$ always exponential survival is obtained, even in the case of a multi-target single-hit model.

Equation (16.42) states that for high LET values (and neglecting penumbra effects) a limiting inactivation cross section is approached which should be equal to that of the sensitive site, e.g. the cell nucleus. The experimental values, however, are quite often considerably smaller (see Fig. 8.13). This is still an open question.

Figure 8.13 shows also that the inactivation cross sections increase further with very heavy ions. This is in qualitative agreement with the theory of this section but the quantitative comparison has still to be worked out.

16.6 Repair Models

The models described so far dealt exclusively with the action of ionizing radiation centered about the spatial pattern of energy deposition. Repair processes were not really included as determining parameters. They play a dominant role in other approaches to be discussed in this section. One advantage of this type of model lies in the fact that they are also applicable to UV, an aspect largely neglected so far although the shapes of dose response curves are quite similar to that of X rays. Apart from this, it is self-evident (see Chap. 13) that repair plays an essential role for the degree of damage expression.

The treatment in this section must be restricted and may seem somewhat elusive. A few examples can only be given which are thought to be representative. There are many more in the literature, a complete coverage would require a book of its own. Nevertheless, a model which incorporates most of the repair phenomena described in Chap. 13 in a mathematically stringent form is still lacking.

If – and this is the underlying theme in this section – the "shoulder" in the survival curves is due to repair processes, it is clear that their effectivity must decline or saturate with higher doses or fluences. It does not matter in this respect whether this is due to an exhaustion of the system or a radiation-induced destruction. The simplest form to describe this behaviour is:

$$\mathrm{Y = aD - bR(D)} \tag{16.50}$$

where Y is (as before) the number of lesions, D the dose, a and b parameters and R(D) a dose-dependent "repair-function". To account for saturation, one may approximate it by an exponential expression yielding:

$$\mathrm{Y = aD - b(1 - e^{-\tau D})} \tag{16.51}$$

with τ as an additional parameter.

For small doses, a linear-quadratic relationship is obtained as may be seen by the expansion of the exponential:

$$\begin{aligned} \mathrm{Y} &\approx \mathrm{aD - b}\left[1 - \left(1 - \tau \mathrm{D} + \frac{\tau^2 \mathrm{D}^2}{2} \cdots\right)\right] \\ &= (\mathrm{a - b}\tau)\mathrm{D} + \frac{\mathrm{b}\tau^2}{2}\mathrm{D}^2 \quad . \end{aligned} \tag{16.52}$$

Although the qualitative behaviour is correctly predicted, the approach appears rather simplistic.

A more refined hypothesis postulates – as several times before – that lesions are formed by the binary interaction of sublesions. Additionally, it is assumed that primary hits may be repaired in a time-dependent manner. To account for the obviously reduced repair with higher doses, it is postulated that repair is linearly related to the number of hits, "misrepair" yielding biological lesions – quadratically. If $U(t)$ is the number of sublesions – or "uncommitted" lesions as they are called –, the following differential equation for their time dependence becomes:

$$\frac{dU(t)}{dt} = -K_1 U(t) - K_2[U(t)]^2 \tag{16.53}$$

(K_1 and K_2 are parameters).

The first term describes repair, the second misrepair. This approach is, therefore, known as the "repair-misrepair model" ("RMR model") (TOBIAS et al. 1980).

The solution is:

$$U(t) = \frac{K_0 e^{-K_1 t}}{1 + \frac{K_2}{K_1} U_0 (1 - e^{-K_1 t})} \tag{16.54}$$

with $U(0) = U_0$.

The number of lesions $R_L(t)$ is given by:

$$R_L(t) = \int_0^t K_2 (U(x))^2 \, dx \tag{16.55}$$

and hence, with (16.54) it follows that:

$$R_L(t) = \frac{U_0 \left[1 + \frac{K_2}{K_1} U_0 (1 - e^{-K_1 t})\right]}{1 + \frac{K_2}{K_1} U_0 (1 - e^{-K_1 t})} - \frac{K_1}{K_2} \ln \left[1 + \frac{K_2}{K_1} U_0 (1 - e^{-K_1 t})\right] \quad . \tag{16.56}$$

After long periods of time all initial hits have vanished, they were either repaired or transformed into lesions. Their final number Y is then:

$$Y = U_0 - \varepsilon \ln \left(1 + \frac{U_0}{\varepsilon}\right) \tag{16.57}$$

with $\varepsilon = K_1/K_2$.

If the reasonable assumption is made that U_0 is proportional to dose, i.e. $U_0 = aD$ (better would be specific energy), then:

$$Y = aD - \varepsilon \ln \left(1 + \frac{aD}{\varepsilon}\right) \quad . \tag{16.58}$$

A shortcoming of this model is that for low doses, a purely quadratic relationship is obtained:

$$
\begin{aligned}
Y &\approx aD - \varepsilon\left(\frac{aD}{\varepsilon} - \frac{a^2D^2}{2\varepsilon^2}\cdots\right) \\
&= \frac{a^2}{2\varepsilon}D^2 \quad .
\end{aligned}
\tag{16.59}
$$

This is obviously at variance with experimental results. The situation may be remedied by replacing D^2 by z^2 which is (Eq. (4.35)) related to the dose in a linear-quadratic fashion.

Alternatively, one may postulate that repair has only a limited fidelity f_R (≤ 1). If this is incorporated into the formalism, instead of Eq. (16.58) one finds:

$$
Y = aD - f_R D \ln\left(1 + \frac{aD}{n}\right) \tag{16.60}
$$

and for small doses:

$$
Y = a(1 - f_R)D + \frac{a^2 f_R D^2}{2n} \quad . \tag{16.61}
$$

The model described does not postulate "irreparable hits", lesions are formed only by "misrepair".

With the assumption that a certain fraction of the initial alterations cannot be repaired at all, constituting "lethal" lesions right from the beginning due to their specific molecular structure, and by considering repair (and misrepair) for the remainder, the problem can essentially be treated in a similar fashion (CURTIS 1986). The final result for the surviving fraction is then:

$$
y = e^{-(a_L + a_{PL})D}\left(1 + \frac{a_{PL}D}{\varepsilon}\right)^{\varepsilon} \tag{16.62}
$$

where a_L and a_{PL} are the induction coefficients for lethal and potentially lethal damage, respectively.

This may be compared to Eq. (16.60) by postulating that survival and lesion number are related by an exponential relationship:

$$
y = e^{-Y} \quad .
$$

The main difference between the two approaches is that in the first case the initial linear part is due to infidelity of repair, while in the second irreparable damage is assumed.

16.7 Comparisons

16.7.1 Behaviour at Low Doses

Virtually all the models include a linear term which means that the slope of the dose-effect curve is non-zero near the origin. Unfortunately, this is of little predictive value since the actual figures depend on assumptions which are not directly testable ("fidelity fraction", "fraction of irreparable damage", etc.).

The situation is slightly better for the comparison of different radiation types and their RBEs.

The linear-quadratic relationships lead to conclusions which are amenable to experimental verification. RBE is defined (Chap. 8) as the ratio of doses which yield the same effect with the reference (^{60}Co γ rays or 250 kV X rays) and the test radiation. This means that:

$$a_x D_x + b_x D_x^2 = a_n D_n + b_n D_n^2 \tag{16.63}$$

where the subscripts x and n indicate the reference and the test radiation, respectively.

Since RBE $= D_x/D_n$, using Eq. (16.63) leads to:

$$\text{RBE} = \frac{2(a_n + b_n D_n)}{a_x + [a_x^2 + 4b_x D_n(a_n + b_n D_n)]^{1/2}} \quad . \tag{16.64}$$

With D ?? 0 it follows that RBE $= a_n/a_x$ as expected; with very high doses RBE approaches unity. The behaviour is depicted in a normalized manner in Fig. 16.6 for different parameters a_n/a_x.

The region of medium doses is particularly interesting. Here the quadratic component is negligible for the test radiation while it is dominating for the reference radiation. One may, therefore, approximate:

$$b_x D_x^2 \approx a_n D_n$$

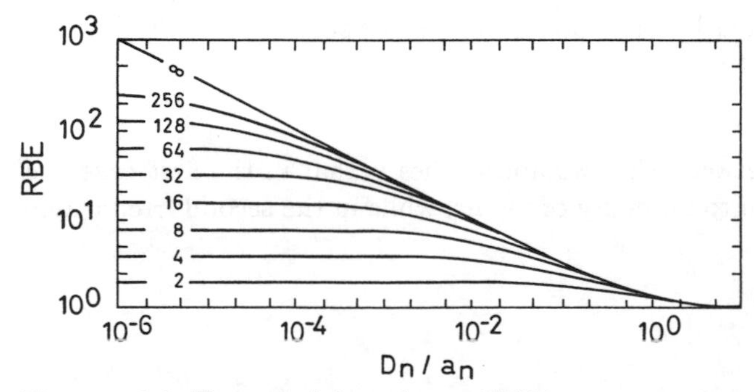

Figure 16.6. Theoretical dependence of RBE on neutron dose with various ratios a_M/a_X (after KELLERER and ROSSI 1972)

and hence

$$\mathrm{RBE} \approx \frac{a_n}{b_x D_n} \quad . \tag{16.65}$$

This means that the RBE should be proportional to $D_n^{-1/2}$ over a certain range. This prediction has been tested with neutrons for a number of systems; the results are summarized in Fig. 16.7.

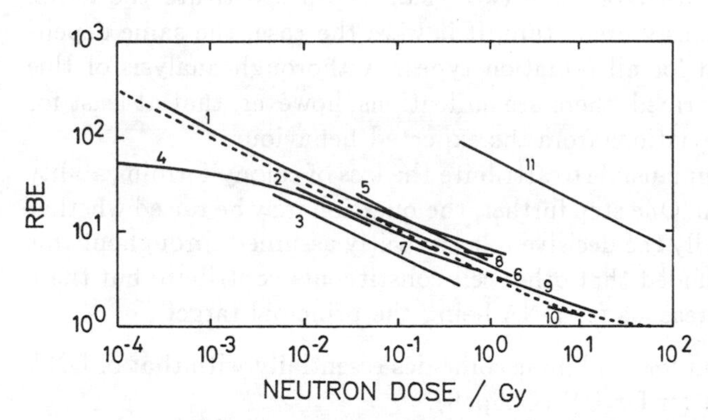

Figure 16.7. RBE as a function of neutron dose with various test parameters. 1-3: eye cataracts in mice, 4: mutations in Tradescantia, 5: mammary carcinoma in mice, 6: chromosome aberrations in human lymphocytes, 7,8: growth reduction in bean roots, 9: skin damage in man, rats, mice and pigs, 10: crypt cell inactivation in mice, 11: various effects in maize seeds (after KELLERER and ROSSI 1972, original references therein)

But there are further conclusions: In the framework of the microdosimetric analysis it is found that $b_x = 1$ and $a_n = z_{1Dn}$. Equation (16.64) may then be written as:

$$\mathrm{RBE} \approx \frac{\sqrt{z_{1Dn}}}{D_n} \quad . \tag{16.66}$$

It is thus possible to derive z_{1Dn} experimentally and to estimate the size of the sensitive site by comparison with microdosimetric measurements (see Sect. 16.8.4).

In the "distance" model, z_{1D} has to be replaced by z', as defined by Eq. (16.30). Equation (16.66) may then be used to estimate $\gamma(x)$ if $t(x)$ is known.

16.7.2 "Critical Lesion" and Survival

The discussion so far concentrated mainly on lesion formation, e.g. chromosome aberrations or other mutations, leaving aside survival. Assuming that a single lesion suffices to kill a cell and, further, if a POISSON distribution is used, an exponential relationship between surviving fraction y and the mean number of lesions Y is obtained:

$$y = e^{-Y} \quad . \tag{16.67}$$

If this were true, there should be a linear relationship between the logarithm of surviving fraction and mean lesion number, irrespective of the particular mechanism. This has in fact been shown for chromosome aberrations (Fig. 11.8) and mutations (Fig. 12.10) after X ray exposure. This may be considered as proof that either of these two lesion types constitute the lethal events but this is obviously premature. If it were the case, the same dependence should be found for all radiation types. A thorough analysis of this kind has not been performed, there are indications, however, that at least for mutations there are deviations from the expected behaviour.

It appears to be implausible to attribute the loss of colony-forming ability to a single type of lesion. One step further, the question may be raised whether DNA damage plays really the decisive role as tacitly assumed throughout this book. It cannot be excluded that other cell constituents contribute but there are a number of good reasons for DNA being the principal target.

1. The action spectrum for cell killing coincides essentially with that of DNA absorption, at least for far UV (Chap. 8).
2. Halogenated base analogues like BUdR sensitize only if they are incorporated into DNA (Chap. 9).
3. ^{125}I is dramatically more efficient if it is incorporated into DNA (Chap. 15).
4. The sensitivity to X rays depends on the DNA content per chromosome (Chap. 8) over several orders of magnitude.
5. The just discussed relationship between survival and chromosome aberrations or mutations – even if not identical for different radiation types – argue for a common target.
6. The dependence of cell survival on DNA repair (Chap. 13) demonstrates the principal importance of this molecule.
7. The overlap between the action of UV and ionizing radiation (Chap. 8) points to a common target which for UV has been shown DNA.

All the evidence listed above does, of course, not rule out that other cellular components – e.g. nuclear proteins or the nuclear membrane – participate in the reactions but it is very improbable that they play the essential role.

This section is to be closed with a critical remark about the argumentative strength of curve fitting on the basis of a given model.

Figure 16.8 shows theoretical survival curves calculated by assuming experimental points at about 90% and 1% survival. Over about three decades – which reflects the experimental reality in many cases – the deviations are not very large and within the limits of experimental errors. This situation makes a decision about a given model only on the basis of curve-fitting virtually impossible. It demonstrates that conclusions drawn from such exercises are not really well-founded.

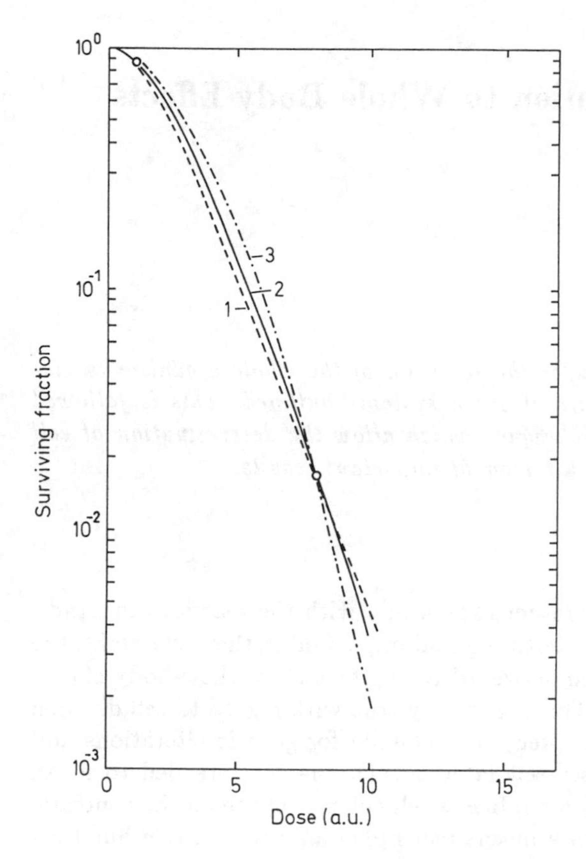

Figure 16.8. Theoretical survival curves obtained by adjusting different exporessions to two predetermined points as indicated. 1: multi-target model, 2: repair-misrepair model, 3: linear-quadratic dependence

Each model has to be checked by various experimental approaches and compared critically with its specific predictions. The discussion in this chapter has shown that none of the given examples is without shortcomings – the "all explaining" model still waits for its discovery.

In spite of this sobering statement, the development of quantitative theoretical models should not be regarded merely as a playground for physicists in biological research. There is a great practical interest in a well-founded theoretical description which may be safely extrapolated. Furthermore, the various attempts here led to new experimental problems and methods and opened avenues to a deeper insight into basic mechanisms, not only in radiation biology but also in the "parent disciplines" physics and biology.

Further reading

ALPER 1975, 1979, CHADWICK and LEENHOUTS 1981, CHAPMAN 1980, CURTIS 1986, DERTINGER and JUNG 1969, GOODHEAD 1980, 1985, 1987, GUENTHER and SCHULZ 1983, HAYNES and ECKARDT 1979, HUG and KELLERER 1966, KAPPOS and POHLIT 1972, KATZ et al. 1971, KELLERER and ROSSI 1972, 1978, KIEFER 1982, 1988, TOBIAS et al. 1980, ZAIDER and ROSSI 1983, 1986

Chapter 17
Cell Survival in Relation to Whole Body Effects

The relevance of cell division for the function of the whole organism is discussed, and the basic structure of some systems outlined. This is followed by the description of some techniques which allow the determination of cell survival in vivo together with a review of important results.

17.1 General Aspects

The preceding chapters dealt rather extensively with the reactions in irradiated cells. Although they are interesting and important in their own right, the approach owes its justification above all to the fact that whole-body effects start from the cellular level. This is not only true with regard to cell division – as detailed below in this chapter – but equally for genetic alterations and carcinogenesis, which are discussed later on. This is not intended to mean that a human or animal body resembles a cell culture. Quite on the contrary, higher principles and regulatory mechanisms play an essential role but they act again on the cellular level. The interplay, however, is much more complex than can be modelled in a culture dish. Nevertheless, it is still true that disturbances of cellular functions by radiation determine to a great extent the fate of the whole organism. Cell division plays a key role in this respect.

17.2 Renewal Systems

Most of the cells in the human or animal body are specialized for certain functions which is reflected in their appearance in the microscope. These properties are gained through differentiation processes during which most – but not all – of the cells lose the ability to divide. The life-span of fully differentiated cells is limited so that there would be a continuous loss if it were not counteracted by cell division. Most organs possess a pool of undifferentiated "stem" cells from where differentiation starts if the need arises and which is kept constant by division. The system in its intricate interplay of different – mostly humoural – regulations is finely tuned to safeguard its proper function. This is especially the case with regard to division activity. If the tissue fails to respond to the regulatory signals, uncontrolled growth – cancer – may result.

The general scheme may be depicted as a "block diagram" of cell compartments containing cells of the same property (Fig. 17.1). This abstract representation must not be mistaken with an anatomical delineation, quite on the contrary, a morphological differentiation is in many cases not possible.

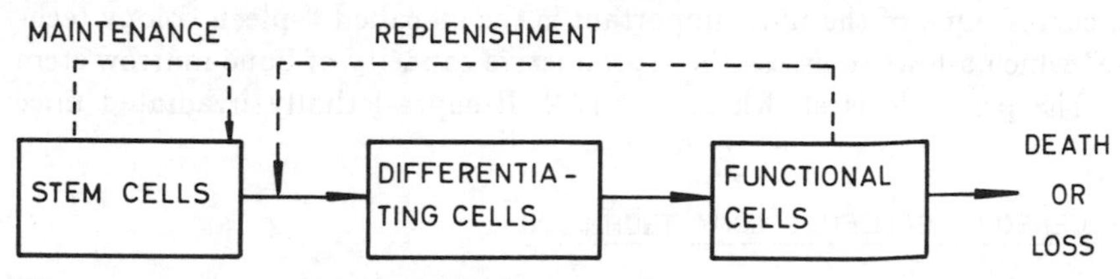

Figure 17.1. Basic scheme of a renewal system

The first compartment consists of undifferentiated stem cells with unlimited division potential. Their number is – on an average – kept constant by internal regulation. If the number of mature cells is reduced, stem cells are called upon to enter the differentiation compartment. There may be also a limited number of divisions during this process which are, however, restricted to early stages. The mature cells usually no longer possess the ability to divide. Since their life-span is limited, there must be a constant flow through the system: the loss at one end is compensated by division at the other. Organs of this kind are called "renewal tissues". Not all body cells are regenerated in this manner, important exceptions are nerve and some muscle cells. Replenishment is particularly essential in those organs which undergo heavy wear and tear. These are internal and external surfaces like skin and the alimentary tract but also the blood forming system.

Table 17.1. Renewal systems in the human body. (After FLIEDNER and NOTHDURFT, 1979)

system	production rates cells/d * 10^9	cells/70y * 10^{14}	kg/70y
skin	0.7	0.18	86
stomach-intestine	56	14.31	6850
lymphocytes	20	5.11	275
erythrocytes	200	51.10	460
granulocytes	120	30.66	5400
thrombocytes	150	38.33	40

The impressive capacity of some representative renewal tissues is illustrated by the data in Table 17.1. They demonstrate that cell division is indispensable for whole body function. Some special examples are described in Chaps. 18 and 19 where also some of the kinetic parameters are given.

17.3 Cell Survival *in vivo*

Cell survival of cells cultured *in vitro* has been extensively discussed in Chap. 8. Since it is not *a priori* clear whether the results are applicable to the whole body situation, several attempts were made to determine "*in vivo*" cell survival curves. One of the most important is the so-called "spleen colony technique" which allows to assess the reproductive capacity of bone marrow stem cells. The principle is shown in Fig. 17.2: If supra-lethally irradiated mice

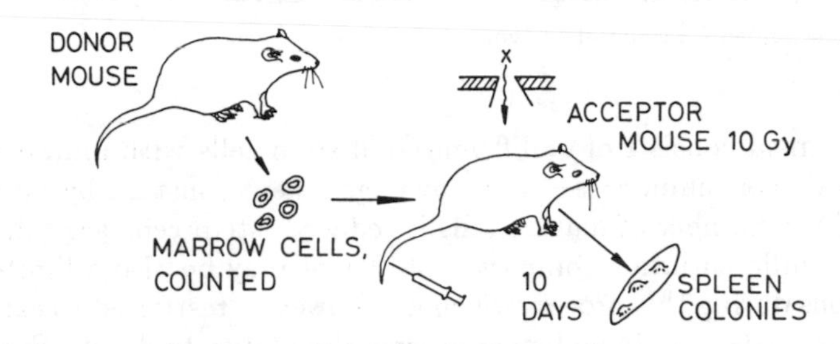

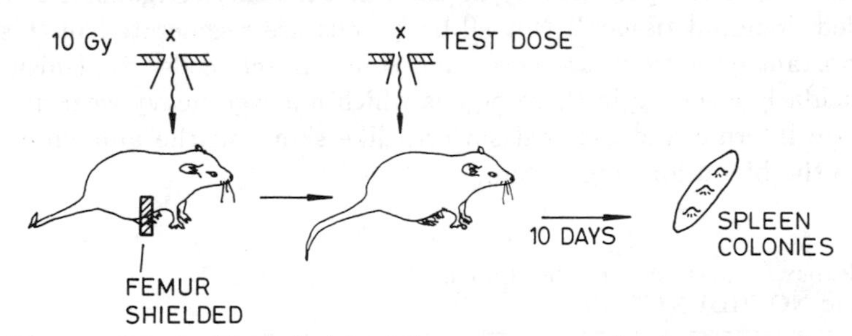

Figure 17.2a,b. The principle of the exogeneous and endogeneous spleen colony assay

(about 10 Gy) are injected with bone marrow cells from an isogenic donor, distinct nodules on the spleen of the exposed animal are to be found after about ten days. It could be shown that they are derived from injected bone marrow cells, the number of these "colonies" is proportional to the number of vital cells injected. Since it is not known which cell type is responsible, one prefers to speak of "spleen colony forming units" (S-cfu). Survival curves obtained in this way are qualitatively similar to those determined *in vitro*.

The method described suffers from the disadvantage that the marrow cells have to be removed from the donor mouse and injected into the acceptor. The handling procedure may traumatize them and modify the result in an unknown manner. This complication is avoided by the "endogenous spleen

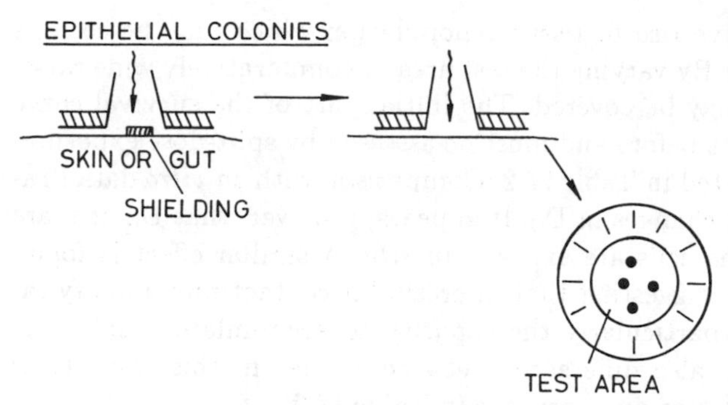

Figure 17.3. An assay for colony forming ability in epithelial tissue

colony technique" (Fig. 17.2, lower panel). In this case, the same mouse serves both as donor and acceptor: During the first lethal dose a part of the bone marrow is shielded – e.g. the femur – and receives only the second test dose. Repopulation starts from the remaining surviving cells and leads also to the formation of spleen nodules. Since the original number of marrow cells cannot be manipulated, reliable results can only be obtained with comparatively high doses where the number of colonies is small enough so that they can be distinguished. The initial part of the survival curve can thus not be determined. An estimate of the shoulder width may be obtained by fractionated exposure (see also Chap. 13). If the total dose is split into two fractions given with an interval allowing full recovery, a survival curve is found which is displaced to the right by the "quasi-threshold" dose D_q which can thus be estimated.

The method can be adapted to other tissues, e.g. skin or gut (Fig. 17.3): A circular area is heavily irradiated (about 30 Gy) with the inner part shielded. This procedure causes a complete sterilization of the outer ring. The centered area then receives the test dose which must be high enough to leave only few

Table 17.2. Cell survival curve parameters for tissue cells after X or γ irradiation. (From HENDRY, 1985, original references therein)

Organ	D_0 /Gy	n	animal	assay
bone marrow	0.7 – 1	2.4 – 5.2	mouse, rat	transplantation
mammary epithelium	1.3	5	rat	"
thyroid	2	4	rat	"
liver	2.7	0.9	rat	"
spleen CFU [a)]	0.6 – 0.9	1	mouse	in situ
cartilage	1.7	6	rat	"
gastric mucosa	1.4	52 – 83	mouse	"
epidermis	1.4	20 – 70	"	"
spermatogonia	1.4 – 1.8	92 – 5	"	"
jejunum	0.9	1300	"	"

a) CFU = colony forming units

surviving cells. They give rise to tissue repopulation which may be seen as distinguishable nodules. By varying the test area, a comparatively wide range of surviving fractions may be covered. The initial part of the survival curve remains undetermined as before and must be assessed by split-dose exposure.

Some results are listed in Table 17.2. Comparison with *in vitro* data (Table 8.1) reveals no great changes in D_0. It appears, however, that D_q or n are significantly larger if the cells are exposed *in situ*. A similar effect is found in spheroids (Chap. 24) suggesting that intercellular contact may modify radiation sensitivity and particularly the capacity to accumulate "sublethal" damage. There may be also differences between tissues in this respect, as seen from the data on testes and cartilage in Table 17.2.

Further reading

BOND, FLIEDNER and ARCHAMBEAU 1965, HENDRY 1985, WITHERS 1975a

Chapter 18
Acute Radiation Damage

Acute radiation reactions in special organs or systems (skin, eye, immune system) are treated first, both with UV and ionizing radiation. The main part of the chapter is devoted to the acute radiation syndrome as it develops after whole-body exposure to ionizing radiation and which is mainly caused by the impairment of cell division in the renewal systems of the blood-forming organs and the alimentary tract. In the last section, the acute "radiation sickness" in its clinical form is briefly described.

18.1 General Aspects

Acute radiation effects in the organism develop over the course of minutes to several weeks and are essentially reversible if they do not lead to death. They are not only important in the case of radiation accidents and nuclear warfare but also in radiation therapy to assess reactions of healthy tissue whose exposure can generally not be avoided. Acute damage is mainly seen in actively dividing tissues of renewal systems where cell replacement is diminished. This has long been known and led to the formulation of the "law" of BERGONIÉ and TRIBONDEAU (1906), stating that the radiation sensitivity of a tissue depends directly on the rate of cell division and is inversely related to the duration of the mitotic phase and the degree of differentiation. Although this generalization cannot be accepted today, the essence of the rule is still true.

Acute effects are characterized by threshold doses, i.e. below a certain dose level no reaction can be recorded. The actual values vary widely and may be quite low in certain instances.

18.2 Skin

18.2.1 Erythema and Related Responses

There are two reasons to begin with the skin. Firstly, it undergoes considerable wear and tear and depends, therefore, critically on functioning cell renewal. Secondly, it is the entrance organ in all cases of external radiation exposure if not properly shielded. It is also – apart from the eye – the only tissue where ultraviolet radiation has to be taken into account.

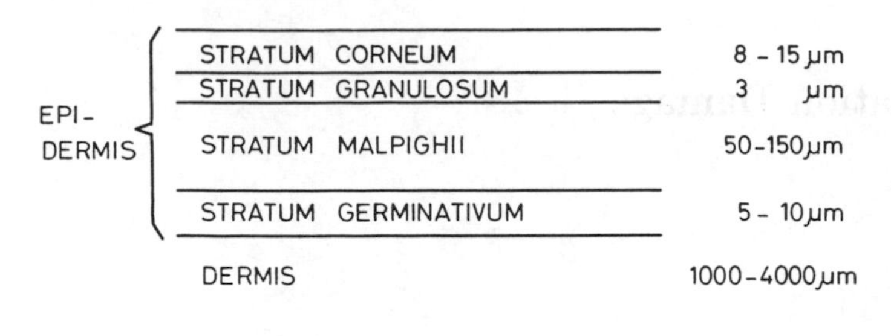

Figure 18.1. Schematic cross sectiom of human skin

Figure 18.1 gives a schematic cross section of the human skin: The uppermost layer – *stratum corneum* – consists of dead cells which are replenished from deeper layers (*keratinocytes*). Differentiation takes place in the *Malpighii* zone while the dividing stem cells are found in the *stratum germinativum*. The total epidermis has a thickness of 0.1 to 0.2 mm so that even weakly penetrating radiations are able to reach the germinative layer.

The sensitivity to UV radiation is determined by the optical properties of the outer layers, mainly absorption, but also scattering and reflection. They may be strongly influenced by skin pigments, the most important being *melanin*. It is formed by specific cells (*melanocytes*) and incorporated as small particles (*melanosomes*) in the skin. This process is stimulated by UV exposure and leads to tanning.

The absorption of some skin components is shown in Fig. 18.2. It is generally high below 300 nm which is due to the protein moiety. The absorption of melanin is generally stronger and extends far into the near UV region.

UV damage to the skin may be divided into several dose-dependent stages:

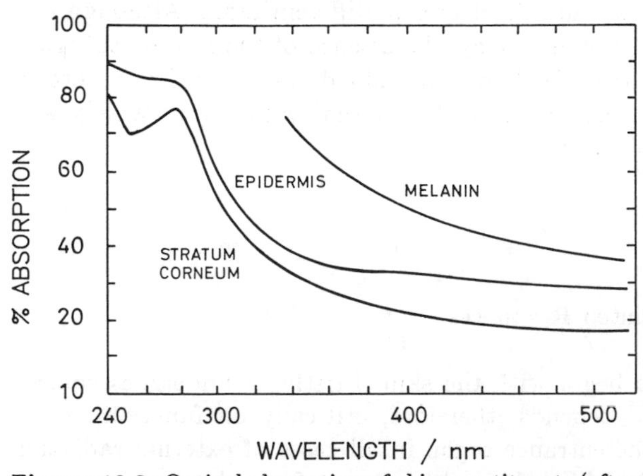

Figure 18.2. Optical absorption of skin constituents (after PARRISH et al. 1978)

- reddening (erythema)
- inflammation
- ulceration
- necrosis

The erythema may already be seen minutes after exposure but generally it takes hours to develop fully and may last for days. It is easy to detect and, therefore, a popular parameter for the quantification of radiation effects in skin. Its exact evaluation, however, is not a simple task and may be influenced by subjective judgement, both on the side of the investigator and the object.

This is reflected in the fact that there is still not yet a general agreement about the exact shape of the action spectrum. The wavelength-dependence is obviously a function of the degree of damage which means that a true action spectrum cannot be given (see Sect. 5.1.3). To illustrate the situation, action spectra for different degrees of erythemae are given in Fig. 18.3. E_0 refers to a reddening which is just detectable. The fluence to produce it is commonly called the "minimal erythema dose" (MED). It varies, of course, excessively among different individuals depending, e.g., on the degree of pigmentation. In white people, a value of about $100\,J/m^2$ has been suggested for untanned parts of the body with UV between 240–290 nm.

Figure 18.3 demonstrates also that erythema formation is not restricted to UV of short wavelength but may also be caused by UV-A. An action spectrum which is proposed as a mean effectivity curve is given in Fig. 18.4.

The complete mechanism of erythema formation is still unclear. Damage to the dividing cells in the germinative layer plays certainly a decisive role but the contribution of other tissue factors, e.g. of hormonal nature, remains to be clarified.

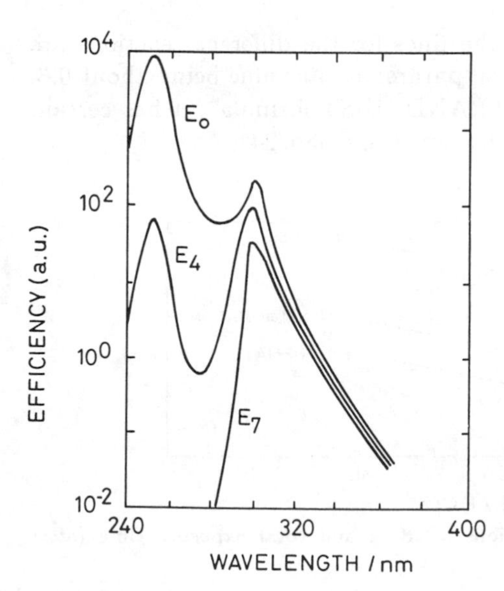

Figure 18.3. Erythema action spectra for different degrees of severity (in ascending numbers) (after TRONNIER 1977)

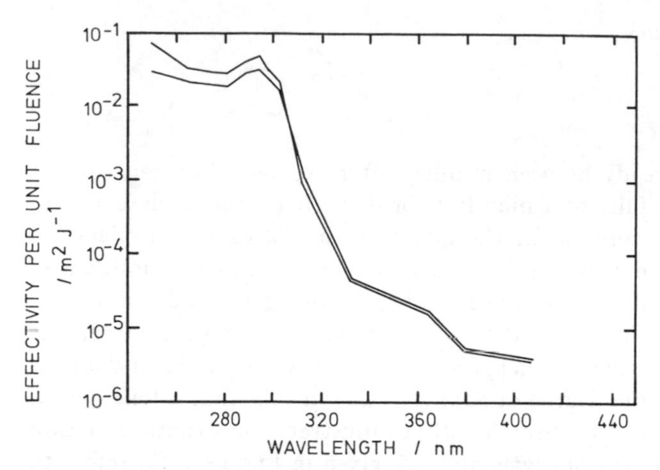

Figure 18.4. Erythema action spectra of human skin based on minimal (threshold) ethythema fluence, recorded either after 8 (upper curve) or 24 hours (lower curve) (after PARRISH et al. 1982)

Erythemae may also be caused by ionizing radiation with a threshold dose of several Gray. Since the skin is often the limiting organ in radiation therapy, its reactions have been investigated in some more detail, particularly with regard to the time pattern of exposure. Plotting the dose necessary to cause a given reaction versus total irradiation time (Fig. 18.5) gives straight lines if logarithmic scales are used for both variables. This means that effective dose D′ and time t are related by a power function:

$$D = K \cdot t^n \tag{18.1}$$

with K and n as parameters. Since the lines for the different reactions are parallel, it appears that n is a universal parameter, its value being about 0.3. Equation (18.1) is known as the "STRANDQUIST formula"; it has certain applications in radiation therapy as discussed in Chap. 24.

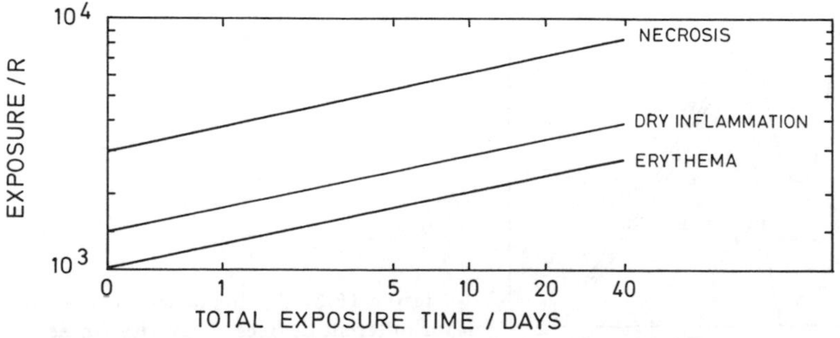

Figure 18.5. Dependence of skin reactions on dose and total exposure time (after STRANDQUIST 1944)

18.2.2 Other Photoinduced Alterations

UV light may cause – apart from erythema – a number of other clinical symptoms, commonly summarized under the name "photodermatoses". They differ from erythema – discussed in the previous section – that they are not always found and have often a complicated etiology. The following list gives the most important examples without claiming comprehensiveness:

- *solar urticaria*
- *photoallergy*
- *lupus erythematosus*
- *pemphigus*
- *polymorphic lighteruption*
- *actinic reticuloid-photosensitive eczema*

Only a short description will be given here, more details may be found in the literature (see end of Sect. 18.2). It is common to all these diseases that immunological reactions have been implicated in their etiology but rarely unequivocally demonstrated.

Solar urticaria is characterized by hives and skin eruptions which develop shortly after exposure to UV at the site of irradiation. It may be accompanied by headache and general ill-feeling. The spectral dependence is not clear and may vary widely between different patients.

Photoallergy is a disease where an allergic reaction to a certain chemical compound is enhanced or even prompted by exposure to light. The list of such substances is very long. In many cases they act as photosensitizers, and the action spectrum resembles closely their absorption behaviour. Others may interfere with the immune system. Only these may be classified as "photoallergens" *sensu strictu* although a clear distinction is generally not easy.

Lupus erythematosus (LE) is an autoimmune disease and may affect many organs, not only the skin. Whether UV exposure plays a role in its etiology is unclear although most patients exhibit a pronounced light sensitivity. It has been suggested that antibodies to UV-irradiated DNA are involved but there is no proof for this. The idea is tempting because LE is very often – but not always – characterized by the appearance of anti-DNA antibodies in the serum.

Pemphigus which shows up as intraepidermal blisters belongs also to the class of autoimmune diseases. As with LE, the role of light as causative agent is not clear although the reaction can be enhanced locally by exposure to UV.

Polymorphic lighteruption (PLE) which is a delayed skin response to light exposure appearing from hours to days after irradiation is even less understood. It may involve photoallergic reactions with unidentified compounds but little is really known.

The same is true for *actinic reticuloid-photosensitive eczema*, a chronic dermatosis developing on light-exposed skin areas. The involvement of unknown environmental agents is suggested by the fact that the disease exhibits clear geographic variations.

18.2.3 Photoimmunology

The skin is the organ which acts as the interface between the body and its environment. Its role in mechanical and chemical protection from possibly hazardous external influences is obvious but it has only become clear quite recently that it constitutes also an essential immune barrier, both locally and systemically. The mechanisms are not yet quite clear but there seems to be no doubt that the LANGERHANS cells play an important role. They comprise about 3% of the epidermal cell population and are thought to act in antigen presentation to T lymphocytes, similarly as macrophages in the normal immune response. This view has been substantiated both by *in vivo* animal and *in vitro* studies. UV irradiation causes morphological changes in LANGERHANS cells and reduces their immunological competence. Whether this is the sole cause for the immune defects reported below or whether other cell types – particularly T lymphocytes – are also involved cannot unequivocally be decided.

Immune suppression in skin by UV exposure is demonstrated by essentially two effects, the reduction of *contact hypersensitivity* (CHS) and the diminished rejection of skin tumour transplants.

CHS which is clinically also referred to as "allergic contact dermatitis" is a skin reaction which develops after exposure to certain compounds, the prototype of which in humans is poison ivy (*Rhus dermatitis*) but many other materials – both natural and artificial – are known. UV is able to suppress the skin alterations although not completely. It can also prevent their development if given before the contact but high fluences are required. The wavelength dependence of these phenomena is not yet clearly established but the studies suggest that the region below 320 nm is most effective.

The experiments with UV-induced skin cancers appear to be even more important. They started with the finding that these tumours are very difficult to transplant in experimental animals, mostly mice. The rejection is reduced by UV irradiation of the host. That immune responses play a role here is also demonstrated by the fact that other immune-suppressive agents like X rays lead to similar effects. These findings may have important implications also for the occurrence of skin cancers if UV does not only act as inducing agent but may also diminish the body's immune response.

Further reading
MORISON et al. 1985, PARRISH et al. 1978, 1983, POTTEN 1985, TRONNIER 1977

18.3 Eye

The eye is one of the organs which is mostly endangered by radiation, both by UV and ionizing radiation. This is particularly true for acute effects but also for late damage, namely cataract formation discussed in Sect. 20.3.

UV of short wavelength is largely absorbed by the *cornea*. It may, however, still easily reach the connective tissue and lead to *conjunctivitis* which develops after a few hours and is fully reversible. Although exact figures are not available, current laboratory experience shows that fairly low exposures to 254 nm suffice.

Photokeratitis is better characterized. Threshold doses have been measured for a number of animals and also humans (Fig. 18.6). There are large

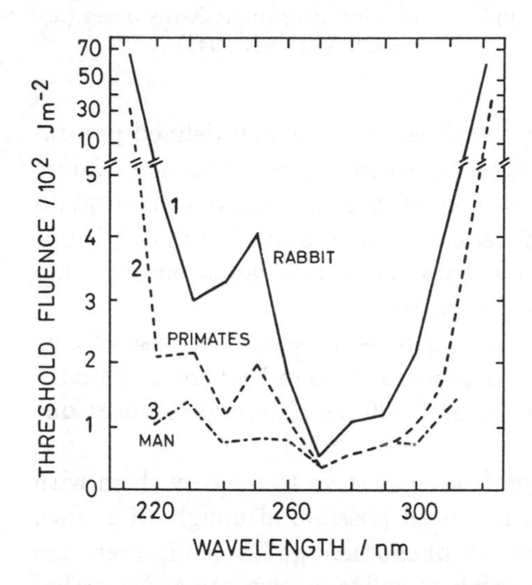

Figure 18.6. Threshold fluence for keratitis in different animals and man (after PITTS 1974)

differences, with humans obviously reacting most sensitively. The curve is essentially flat from 200–300 nm and then rising steeply. Photokeratitis disappears after a few days, remaining damage has not been reported.

Ionizing radiation – in contrast to UV – can reach all parts of the eye because of its greater penetration power. The most sensitive part appears to be the lense where opacity may be caused. This cataract formation is discussed in Sect. 20.3.

18.4 Radiation Syndromes and Lethality

18.4.1 Survival

Since the lifetime of an organism is limited, any assessment of radiation effects on survival has to take into account the time factor. This is different from

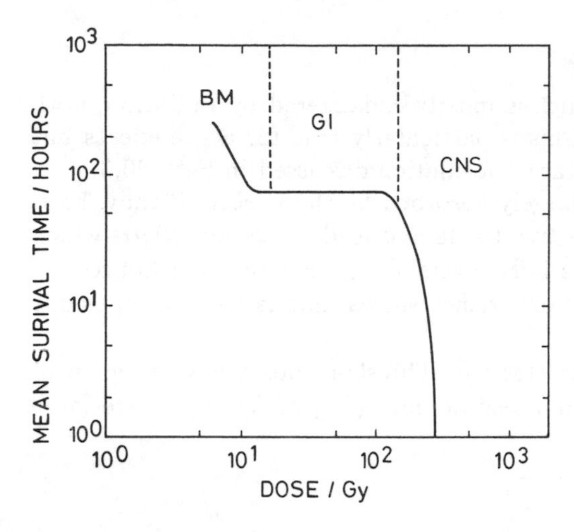

Figure 18.7. Mean survival time of mice after single X-ray doses (after QUASTLER 1945)

cellular studies where colony-forming ability serves as a well-defined parameter. Figure 18.7 shows the mean survival time of mice as a function of dose in a double-logarithmic plot. Similar studies with other animals yield qualitatively comparable results. Human data are not available (fortunately!) but the experience with the atomic bomb victims and a few radiation accidents indicate that a similar behaviour is to be expected.

Figure 18.7 has a remarkable shape: There is a steady decrease of survival time with lower doses but then a plateau is found where it remains constant at about 70 hours between 10 and 100 Gy. For higher doses one finds a precipitous fall.

There are clearly three regions and it is suggestive to identify them with different underlying mechanism. This is indeed possible although, of course, radiation induced death is a very complex phenomenon. Clinically there are a number of symptoms, a situation which is called a *syndrome*. Nevertheless, it can be said that with a given dose level, damage to a certain organ constitutes the limiting and decisive event. Up to about 10 Gy, i.e. before the plateau is reached, this is the blood forming system, mainly the bone marrow. The intermediate region can be correlated with the destruction of the digestive tract, mainly the small intestine (*jejunum*) while the very short survival times after excessively high doses are due to damage to the central nervous system. It has thus become customary to speak of the "bone-marrow syndrome", the "gastro-intestinal syndrome" and the "central-nervous-system syndrome". This classification is, of course, rather crude and indicates only roughly the general behaviour but it constitutes a useful guideline. It must not be taken to mean that other organs are not affected but their malfunction does normally not limit the survival probability.

The curve in Fig. 18.7 ends at about 3 Gy, there are essentially no acute deaths below this value. The critical time appears to be about 30 days so that it has become customary to record survival after this time span. A survival

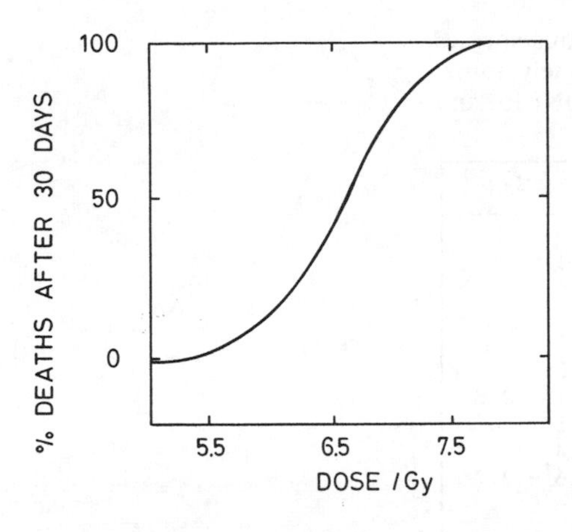

Figure 18.8. Survival of mice 30 days after X-ray exposure as a function of dose

curve with this parameter is shown in Fig. 18.8 for mice. It has again a typical shape which is quite different from cell survival curves but similar to those obtained after pharmacon action. A closer analysis reveals that it may be described by a GAUSSIAN distribution function as:

$$y = 1 - \frac{1}{\sigma\pi/2}\int_0^D \exp\frac{(LD_{50,30} - x)^2}{2\sigma^2}\,dx \qquad (18.2)$$

where y stands for surviving fraction, D for dose, σ is a coefficient and $LD_{50,30}$ the "mean lethal dose". If $1 - y$ is plotted on probability paper or subjected to a "probit" transformation (Appendix I.5), a straight line is obtained from which $LD_{50,30}$ is easily read (Fig. 18.9); It is the dose where the average sur-

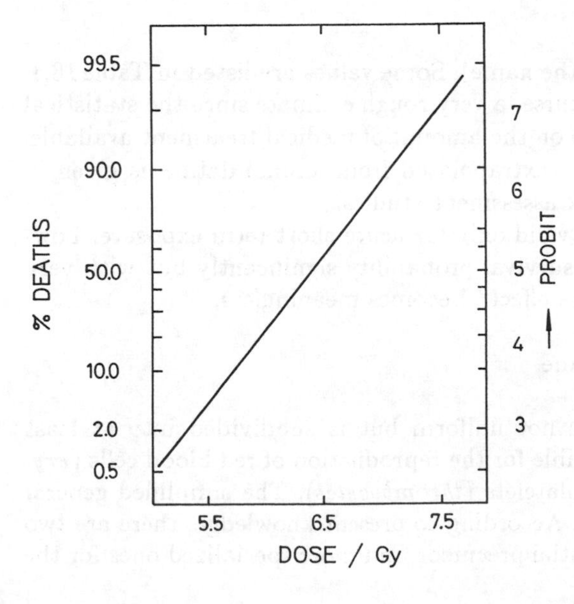

Figure 18.9. As in Fig. 18.8 but in probit representation

Table 18.1. 'Mean lethal doses' of various species (observation period 30 days) with sparsely ionizing radiation. (After BOND, FLIEDNER and ARCHAMBEAU, 1965)

species	$LD_{50,30}$/Gy
mouse	6.4
rat	7.1
dog	2.5
monkey (Macaca mulatta)	6
rabbit	7.5
guinea-pig	4.5
hamster	6.1 - 8.6
pig	2.5
goat	2.4
donkey	2.6 - 3.7
man	3 - 5

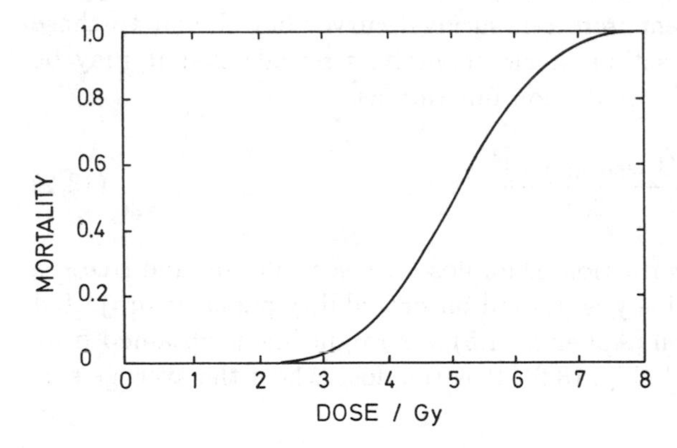

Figure 18.10. Assumed survival curve for human (after DEUTSCHE RISIKOSTUDIE 1979)

vival equals just 50% (hence the name). Some values are listed in Table 18.1. The value for humans is, of course, a very rough estimate since the statistical basis is small. It depends also on the amount of medical treatment available. An estimated survival curve – extrapolated from animal data – is given in Fig. 18.10 as it is used for risk assessment studies.

All data given so far are valid only for acute short term exposure. Lowering the dose rate increases survival probability significantly but with very low dose rates the term "acute effect" becomes meaningless.

18.4.2 Bone-Marrow Syndrome

The blood-forming system is not uniform but is subdivided into at least three connected parts responsible for the reproduction of red blood cells (*erythrocytes*), *granulocytes* and platelets (*thrombocytes*). The simplified general scheme is shown in Fig. 18.11. According to present knowledge, there are two kinds of stem cells, pluripotential precursor and more specialized ones for the

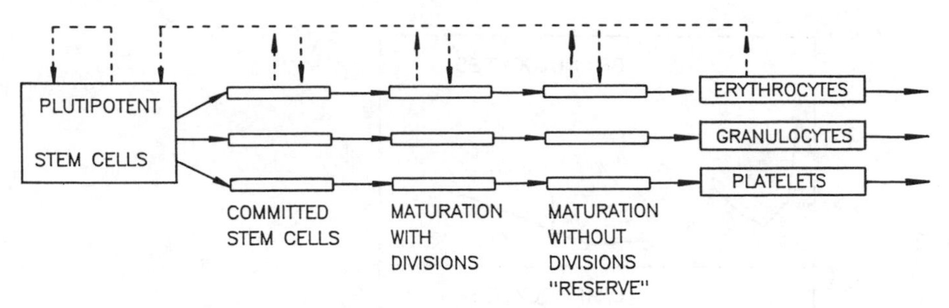

Figure 18.11. Schematic structure of the blood forming system (after FLIEDNER and NOTHDURFT 1979)

three branches of hemopoesis. All stem cell compartments are intrinsically regulated in such a way that the average cell number remains constant. Differentiation involves also a limited number of divisions so that an amplification effect results. The loss of fully differentiated cells is compensated from the pool of immature precursors. Table 18.2 lists the kinetic parameters. While

Table 18.2. Kinetic parameters of hemopoiesis in man. (After BOND, FLIEDNER and ARCHAMBEAU, 1965)

cell type	life-span days	maturation time days
erythrocytes	109 - 127	4 - 7
granulocytes	6 - 7 [a]	9 - 10
thrombocytes	8 - 9	4 - 10

[a] half-life, since excretion occurs at random

erythrocytes and platelets have a well-defined life span, the granulocytes die in a stochastic manner which can only be characterized by a half-life value, similar to radioactive decay. The difference in residence times is seen to be quite remarkable, the maturation times, on the other hand, are comparable.

A whole body radiation interferes with cell division at all stages. This results in a rather fast reduction of the short-lived components in the peripheral blood (granulocytes and thrombocytes) while there is initially little effect on the number of red blood cells. Even if the supply were completely stopped, their daily loss amounted to less than 1%.

Another important component are the *lymphocytes*, white blood cells, formed similarly – although less well understood – by the lymphatic system. They play a very important role in the immune response.

Figure 18.12 depicts the time course of blood cell number after about 2 Gy X ray dose.

There is a fast drop in lymphocyte and platelet number, which persists in the first case while it is transient in the second. With the granulocytes, an initial increase is seen which is presumably due to mobilization from storage

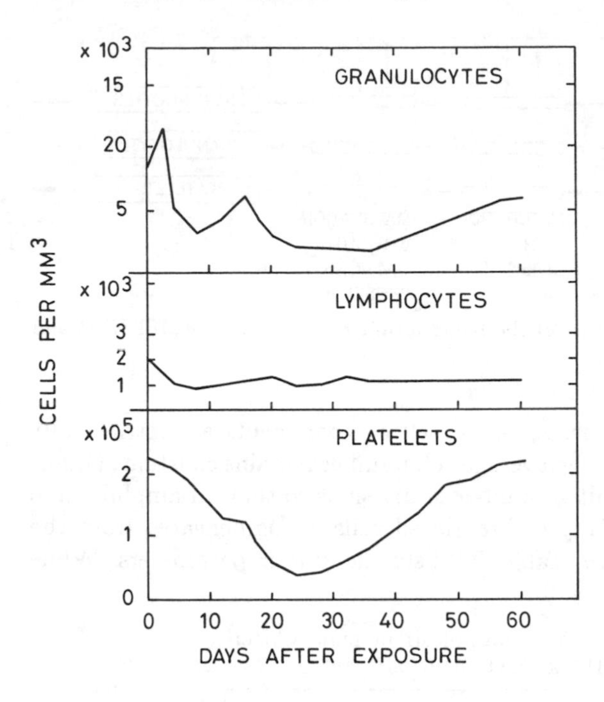

Figure 18.12. Time course of blood cells count after about 2 Gy X-rays (after CRONKITE and FLIEDNER 1972)

compartments. The long-lasting reduction of lymphocytes has severe consequences for the immune competence of the body. It can be suppressed by even smaller doses as this cell type is obviously very sensitive to radiation. This has practical consequences in a number of respects: Radiation may be used to overcome the host-versus-graft reaction with organ transplantations. It makes also bone-marrow transfusions as a therapeutic means after a radiation accident feasible. Course and outcome of the radiation syndrome depends critically on the immune system. Its misfunction is to a great extent responsible for radiation-induced death, the life-limiting factor is very often an internal infection.

18.4.3 The Gastro-intestinal Syndrome

The gastro-intestinal tract consists of many parts which form an "inner surface" and are, therefore, subject to considerable stress. The degree is, however, variable as documented by the different cell turnover times listed in Table 18.3. Generally they are rather short, especially so in the small intestine. It plays evidently a decisive role which is in accordance with the radiobiological experience that damage to it determines essentially course and outcome of the GI syndrome. The discussion is, therefore, restricted to this organ to illustrate the main points. The small intestine consists of several parts which are, however, quite similar in structure and function: *duodenum*, *jejunum* and *ileum*. A schematic cross section is depicted in Fig. 18.13: the intestinal lumen

Table 18.3. Mean renewal times in the gastrointestinal tract. (After BERTALANFFY and LAU, 1962)

organ	turnover time (days)
lip	14.7
oral cavity	4.3
stomach	9.1
small intestine	1.3 - 1.6
colon	6.2 - 10

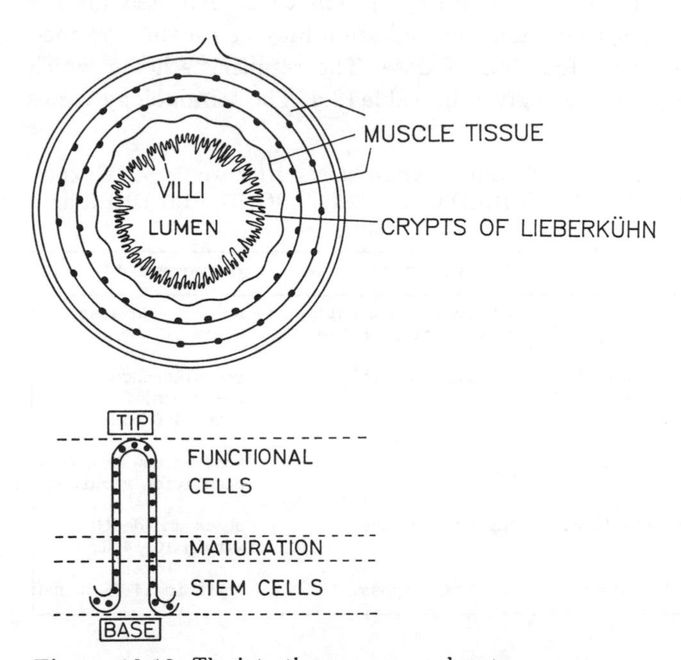

Figure 18.13. The intestine as a renewal system

is surrounded by an epithelium with a villous structure. The cells of these villi are the functional units. Since their life-time is limited, they are replaced from the base where the *crypts* reside. They represent the stem-cell compartment. For maturation, the cells travel up the villus until they reach the top from where they are shed. Each villus is hence an autonomous renewal system with a structure as shown in Fig. 18.13. There are about 100 stem cells per crypt with a mean generation time of 24 hours in humans. The mature cells reach the top within 80 hours, the total traversal time is 180 hours. If cell division is completely stopped by a sufficiently high radiation dose, there will be no functional epithelial cells left after about $3\frac{1}{2}$ days which coincides with the plateau region of the survival time curve in Fig. 18.7, giving some affirmation to the described course of events.

18.5 Course and Therapy of Radiation Sickness

It is not intended to describe comprehensively the clinical symptomatology of the radiation sickness but rather focus on a few points in an exemplary manner. Right from the beginning it should be clear that the classification into well-discernible symptoms does not reflect the real clinical situation but serves only as a didactic tool, in practical cases there are large regions of overlap. The present knowledge about radiation sickness in humans stems mainly from a number of well-documented radiation accidents and – although to a lesser degree – from the bomb victims in Hiroshima and Nagasaki.

Survival probabilities and the main symptoms to be expected after a whole-body exposure to sparsely ionizing radiation may be roughly grouped into several categories as a function of dose. The resulting scheme which is only intended as a guideline is given in Table 18.4. The terminology varies

Table 18.4. Classification of the radiation sickness stages after whole body exposure (schematically). (After [a]) MAXFIELD u.a., 1973; [b]) BOND, FLIEDNER and ARCHAMBEAU, 1965)

dose/Gy	category [a]	category [b]	prognosis [a]
2	I: subclinical	survival practically certain or probable	no acute damages
2 - 4	II: moderate hemopoietic damages	survival possible	convalescence after 5-6 weeks, finished after 4-6 months
4 - 6	III: serious hemopoietic damages		bone-marrow transplantation required
6 - 10	IV: gastro-intestinal damages	survival unlikely	shock and death within 10-14 days
10	V: cerebral damages	survival impossible	death in 14-36 hours

with different authors, two examples are listed. A flow-diagramme which may be helpful in diagnosis is depicted in Fig. 18.14.

Death is certain in the highest category (more than 10 Gy). The damage to the central nervous system is then so severe that the survival time is less than two days. No treatment, except for pain relief, is possible.

In the next class (6–10 Gy), which is dominated by the GI syndrome, one has to expect a combination of all possible symptoms. The blood count drops rapidly with a virtually complete disappearance of lymphocytes. The decrease in thrombocytes causes internal bleeding, the damage to the intestinal epithelium leads to a dramatic loss of fluids, nausea and diarrhoea. There is heavy general pain and a strong reduction of coordination ability. Depending on dose, death is expected to occur within 14 days. Any treatment has to concentrate – apart from pain relief – on the replacement of fluids. Possible infections – which may also be caused internally – may be counteracted by

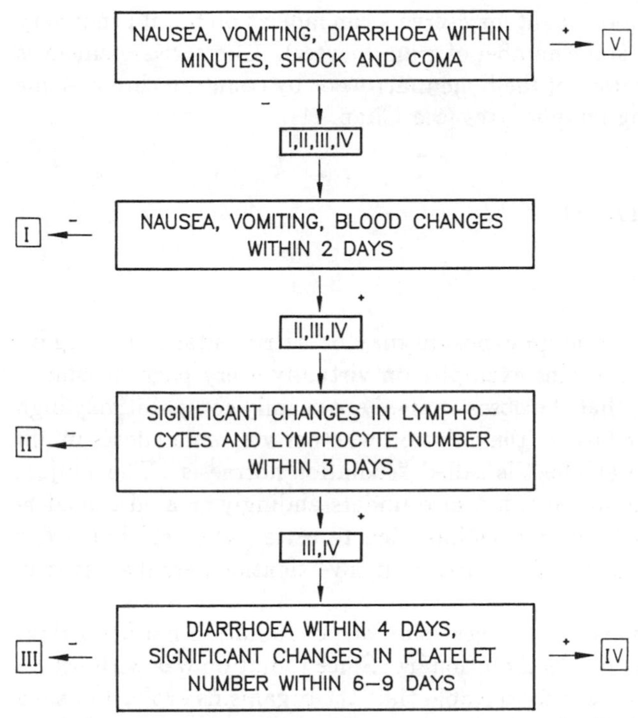

Figure 18.14. Flow diagramme for the classification of radiation sickness categories (after MAXFIELD et al. 1973)

high doses of antibiotics. In less severe cases, bone marrow transfusions may be considered to increase the platelet number and to stimulate the regenerative potential of the blood-forming system. The general prognosis, however, remains poor.

In the medium category, possible infections pose the main problem. They may be fought by massive treatment with antibiotics but this must be done in a careful way. They should be applied only if a strong rise in body temperature indicates that an infection has occurred. A premature medication could otherwise lead to the selection of resistant bacteria which are more difficult to overcome. The blood count is also reduced here, but less drastically. Blood and bone-marrow transfusions may help to reconstitute the function of the blood-forming system but particularly the latter are not without problems. The reader is referred for this to the special literature (CRONKITE and FLIEDNER 1972; MICKLAM, LOUTIT and FORD 1966).

In the low-dose category (less than 2 Gy) there are only mild changes in the blood picture. A special therapy – continuous supervision implied – does not appear necessary. Possible infections must be carefully monitored, psychological problems have to be taken into account.

The flow diagramme of Fig. 18.14 may help to estimate the severity of the damage. In accident situations correct – or even approximate – dosimetry is seldom available. The blood count may serve as an indication but it is not very reliable (the best suited is the number of granulocytes). A better assessment is possible by the investigation of the bone marrow or by counting chromosome aberrations in remaining lymphocytes (see Chap. 11).

18.6 Radiation Hormesis

18.6.1 General Aspects

There is no doubt that radiation exposure may be detrimental to the organism exposed, this book contains examples on virtually every page. It cannot be overlooked, however, that the doses applied are usually comparatively high and it is certainly allowed to ask the question whether very small doses might be even beneficial. Such an effect is called "radiation hormesis". The subject has recently raised some interest, but also understandingly created considerable controversies. It is by no means intended to review the field but a few comments are in place. Some of the pertinent investigations are discussed in the next section.

There are two reasons not to reject the idea of radiation hormesis right from the beginning. The first is evolutionary: Since radiation was with life on Earth since its inception, it is conceivable that the organisms evolved in such a way that they not only are able to counteract the harmful influences, e.g. by repair, but that they could turn bad into good by making use of the energy supplied. Photosynthesis is, of course, an eminent example, the formation of vitamin D another, but it is very difficult to imagine a mechanism for ionizing radiation. The lack of imagination, however, is not a sufficient proof for the non-existence of such a principle.

The second point in favour for radiation hormesis is the well-known fact that many substances indispensable for proper body function are toxic if overdosed. Is it possible that also radiation belongs to this category?

With regard to the first argument – the evolutionary one – it should be stated that there is already at least one benefit of radiation, namely mutation and recombination – both being radiation-inducible. These processes have been quite important for evolution and it appears that radiation played its part in it. This, however, can be hardly classified as "hormetic" since it is not concerned with unknown mechanisms.

The comparison with chemical agents overlooks the very fact that radiation interaction is quantal in nature and quite different from that of any pharmacon (with the possible exception of radiomimetics, see Chap. 15). The basic response is hence discontinuous. As pointed out in Chap. 4, lowering the dose does not mean – in the low dose region – that all cells receive less radiation but only that fewer cells are hit. This is a fundamental difference which cannot be circumvented.

Keeping this in mind one has to admit that stimulatory effects of radiation can be imagined as part of a general stress response. The mobilization of granulocytes (Sect. 18.4) is just an example.

18.6.2 Hormetic Responses

Table 18.5 lists a few examples of "hormetic" effects. It is seen that they were reported for microorganisms, plants, animals and even human beings. Nearly

Table 18.5. Some reported examples of radiation hormesis

Organisms	response	dose-level	ref.
protozoa, bacteria	growth stimulation	< 50 mGy/year	1
mice	immune reaction	0.025 - 0.1 Gy	2
human	cellular immunity	< 0.5 Gy	3
plant seeds	growth stimulation	variable	4,5
human	cancer incidence	0.03 μSv/year	6

References:
1. PLANEL et al. 1987
2. LIN et al. 1987
3. BLOOM et al. 1987
4. SHEPPARD and REGITNIG 1987
5. MILLER and MILLER 1987
6. NAMBI and SOMAN 1987

all investigators state that special conditions are mandatory, also there is a general tendency of poor reproducibility. Nevertheless, in certain instances stimulatory effects of low-level radiation is well documented. This is particularly so for early plant growth. Although the mechanism is not clear, this must not necessarily represent "hormesis" but may just be a change of regulation as a consequence of tissue damage. The same applies to findings about improved immune reactions in animals or humans. Reports of lower cancer incidence in areas with higher natural radiation background have to be treated with some caution because of the considerable statistical and epidemiological difficulties involved. The group of people which is undoubtedly best supervised are the survivors of Hiroshima and Nagasaki. A beneficial effect of low radiation doses could here not be detected.

In conclusion, it may be stated that radiation-induced damage may well lead to changes in control systems in plants or animals which appear to be beneficial. It is, however, – on the basis of the present evidence – at least premature to postulate a general hormetic effect of radiation.

Further reading

BOND, FLIEDNER and ARCHAMBEAU 1965, BROERSE and MACVITTIE 1984, CRONKITE and FLIEDNER 1972, CRONKITE and WALKER 1987, DALRYMPLE et al. 1973, DUNCAN and NIAS 1977, HALL 1978, LETT and ALTMAN 1987, LUCKEY 1980, PIZZARELLO 1982, SAGAN 1987

Chapter 19
Radiation Effects and Progeny

This chapter summarizes all the effects which influence – either directly or indirectly – the children of the exposed individual. These are disturbances of fertility, intrauterine death, teratogenesis and genetic changes. The experimental methods for hazard assessment are introduced and the basis for risk estimates discussed.

19.0 Preliminary Remarks

Possible effects on the progeny – either in the first or later generations – cause undoubtedly great concern. They may be acute or express themselves only after very long times. They are treated here together under a common heading because in this way the interrelationships will become more obvious although there are clearly systematic differences.

19.1 Fertility

A discussion of radiation effects on fertility has to start with a description of germ cell development. It is different in males and females which has consequences for fertility impairment and its recovery after radiation exposure.

In males germ cell production represents a typical renewal system (Fig. 19.1): The stem cells are *type A spermatogonia* which differentiate into spermatozoa via a number of intermediary steps all involving further divisions (*type B spermatogonia*, *primary* and *secondary spermatocytes*, *spermatids*). The most important step is the formation of secondary spermatocytes since it occurs via *meiosis* (see Appendix II.2) so that the originally diploid chromosome set is halved, all the following stages are *haploid*. This event is particularly critical for genetic alterations. The total maturation time is about 72–74 days in man but only 35 days in mice. Sperm cell production in humans begins early in life around an age of ten years and continues normally until death.

Germ cell development in females proceeds in an entirely different fashion (Fig. 19.2): It starts from *oogonia* as precursors which give rise to *primary oocytes*. This process is already completed at birth when 400 000 primary

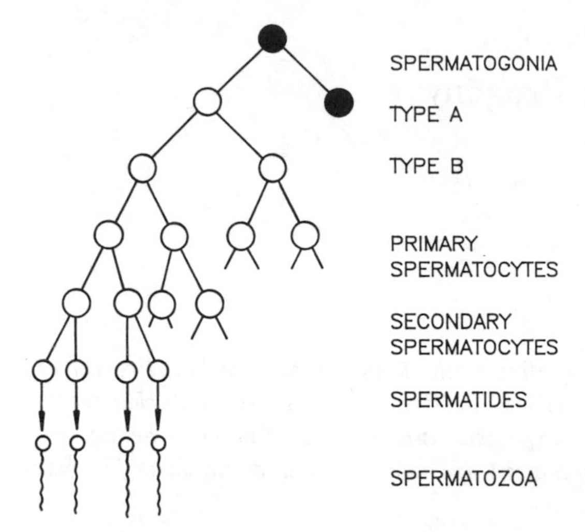

Figure 19.1. Scheme of male germ cell production

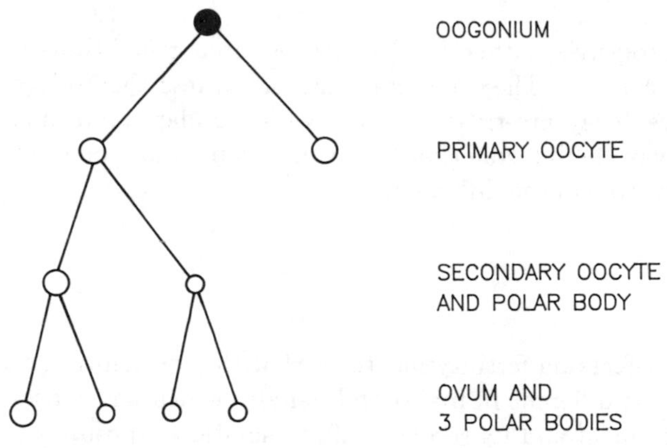

Figure 19.2. Scheme of female germ cell production

oocytes are present. They develop into about 380 mature *follicles* during the course of the life. The first step is an asymmetric reductional division generating one *secondary oocyte* which contains most of the protoplasm and the much smaller polar body. The final step of ovulation is again an asymmetrical division yielding finally one mature egg cell and three polar bodies. It is clear from this description that the female germ cell production does not represent a renewal system, the different stages are already preformed and only one egg cell is produced per cycle. This has consequences for the effects of radiation.

An exposure of the male gonads to moderate doses causes at first a reduction of sperm cell number. This process is very sensitive, an effect may already be seen after 0.1 Gy. Higher doses (more than 0.5 Gy) lead to sterility which is only slowly regained and may – depending on dose – persist for years. Gonad doses over 5 Gy result in permanent sterility. The critical cell population in this respect is that of the type A spermatogonia. As seen from Table 17.3, their radiation sensitivity lies in the normal range. There are indications, however, that the population is not homogeneous. The test assay measures essentially the sensitivity of the resistant part since only few cells are sufficient to secure the regenerative ability of the system. Sperm cell reduction, on the other hand, reflects presumably the sensitive fraction. These questions are not yet completely answered, there are reasons to believe that quiescent and actively dividing spermatogonia differ widely in radiation sensitivity which could at least partly account for the mentioned discrepancies.

Female fertility is an even more sensitive parameter. The main reason lies in the fact that the inflicted damage to the system cannot be overcome by division of surviving cells, there is no regenerative potential. The loss of division ability in primary oocytes is, therefore, permanent. A dose of 4 Gy leads to irreversible sterility in humans. It is – differently from males – also coupled with hormonal alterations which may be characterized as "premature climacterium". Considerably lower doses ($\approx$ 0.5 Gy) are sufficient to cause short-term sterility. Because of the special structure of the system, this is not really completely recoverable (another difference to males), it may recur in a random manner since oogonia once damaged are not replaced.

19.2 Prenatal Radiation Effects on the Embryo

The developing embryo consists in the view of the radiobiologist of a number of rapidly dividing cell populations. One would, therefore, expect *prima vista* a high radiation sensitivity. This is, in fact, found although with considerable variations depending on the actual parameter. The main effects are: miscarriages, still-births, teratogenic alterations and changes in postnatal development. Genetic damage will be discussed in the following section.

All radiation effects on the embryo display a distinct dependency on exposure time at different stages of pregnancy. Since human data are rarely available, one has to recur to animal experiments. Because of the obviously existing physiological differences, it is necessary to give a short comparative synopsis of embryonal development in humans and some laboratory animals: Three main phases of embryogenesis are distinguished: the *preimplantation phase* from fertilization to the egg's implantation in the uterus, *organogenesis* where most of the organs are formed and the *foetal phase* where the prenatal growth processes are completed. The durations differ, of course, considerably in different species; a summary is given in Table 19.1. If comparisons are made between laboratory animals and humans, they have to be taken into account. Intrauterine lethality as a result of radiation is found predominantly early in the development. Figure 19.3 shows determinations in hamsters and dogs from where it may be seen that implantation is obviously the most critical event. In humans it is around the 7th day after conception.

Table 19.1. Phases of embryogenesis in various species (days after conception). (After UNSCEAR, 1977)

type	preimplantation	organogenesis	fetal period
hamster	0 - 5	6 - 12	13 - 16.5
mouse	0 - 5	6 - 13	14 - 19.5
rat	0 - 7	8 - 15	16 - 21.5
rabbit	0 - 5	6 - 15	16 - 31.5
guinea-pig	0 - 8	9 - 25	26 - 63
dog	0 - 17	18 - 30	31 - 63
man	0 - 8	9 - 60	60 - 270

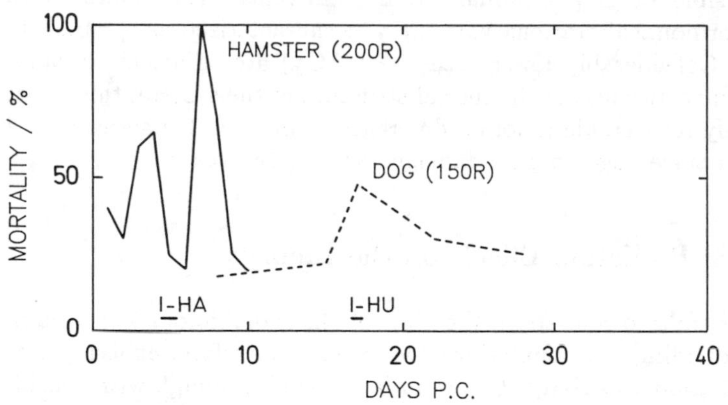

Figure 19.3. Intrauterine deaths after radiation exposure as a function of time between conception and irradiation in hamsters and dogs (after UNSCEAR 1977)

Teratogenic effects depend to an even higher degree on the time of exposure. Human data are again very scarce but some quite tentative conclusions appear possible. The most sensitive organs are those which are formed at early times of the development when the embryo consists of a comparatively small number of cells. They include the brain, the central nervous system, the eyes and the skeleton. Brain damage may lead also to mental incapacities. Apart from these specific effects, a reduction in body-size and postnatal growth retardation have been recorded. The critical period (in humans) appears to be between the 6. and 15. week of gestation.

The human data base is quite small and the doses involved were comparatively high (> 0.5 Gy in all cases) so that any risk estimate is rather uncertain. The present figure is 10^{-3}/Gy for all teratogenic effects. Animal studies indicate, as the examples in Table 19.2 demonstrate, that malformations may occur after very low doses so that it is prudent to protect pregnant women from all avoidable radiation.

Table 19.2. Lowest doses, after which malformations were observed (selection). (After UNSCEAR, 1977)

affected organ	mouse time [a]	mouse dose [b] Gy	rat time [a]	rat dose [b] Gy
brain (exencephaly)	0.5–1.5	0.15–0.20	-	-
brain (hydrocephalus)	8	0.25	-	-
brain on the whole	-	-	8–9	0.36–0.40
skeleton	7.5	0.05	8	0.125
eye	8.5	0.50	8–9	0.36–0.40
spine	8	0.25	9	0.50

[a] days after conception
[b] approximate doses original data are given as exposure in "Roentgen"

The problem of the existence of threshold doses in teratogenesis is still controversial although most experts agree that it may be assumed. This conclusion is based on many animal investigations with large populations. But even if there is a threshold, it depends on the type of damage and may be very low.

19.3 Genetic Effects

Genetic alterations form undoubtedly one of the major possible hazards of radiation. While in the beginning the fruit fly *Drosophila melanogaster* constituted the main research object, the emphasis has recently shifted to rodents, especially mice. Only the latter investigations will be reported here, a more comprehensive account including the earlier work may be found in AUERBACHs book (1976).

The discussion is centered here around the possible radiation hazard which means that one is particularly interested in small doses and low doserates. In this case, the expected effect level is small so that large animal populations have to be studied. The necessary experimental effort is quite considerable but it has been possible by these large-scale investigations to arrive at approximate risk estimates. This is the more valuable since human data are virtually missing. This is not surprising in view of the fact that most mutations are *recessive*, i.e. they show up normally only in later generations.

Quite specific approaches allow, however, to determine their rate of occurrence in experimental animals. It is instructive in this respect to give not only the results but to describe also the methods involved. The most important is the "specific locus method" which is schematically depicted in Fig. 19.4. It makes use of specially inbred mice strains where the males are ho-

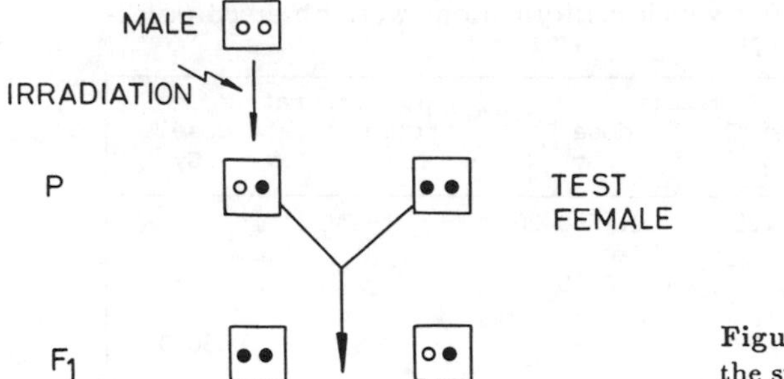

Figure 19.4. The basic scheme of the specific locus method

mozygous for the dominant, the females homozygous for the recessive marker. On irradiation of the males, a certain proportion of the germ cells will mutate. After fertilization, a few egg cells will be homozygous for the recessive allele so that it will be phenotypically expressed in the offspring. The mutation rate can thus be easily measured.

Another important genetic trait is "autosomal recessive lethality" which characterizes a mutation on autosomes (i.e. chromosomes other than the sex chromosomes) leading to intrauterine death in homozygous embryos. The experimental test which is sketched in Fig. 19.5 requires the investigation of

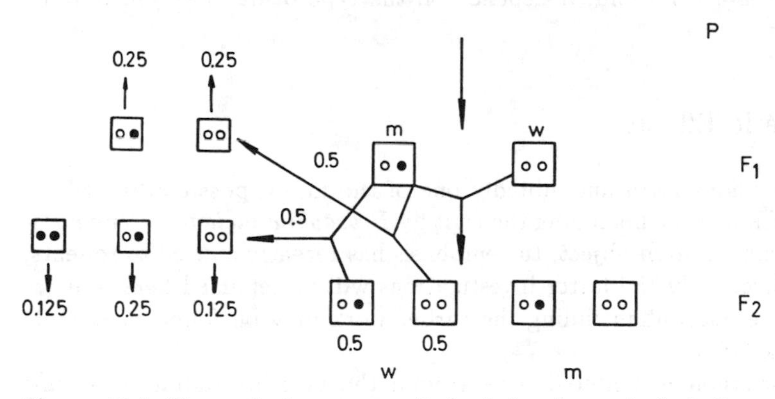

Figure 19.5. The method to assess radiation induced recessive lethality

several offspring generations. If the parents' germ cells, which are initially free of these lethal factors, are irradiated, a certain fraction of the F1 offspring will harbour the lethal marker but it will not be expressed since it is recessive. If the females are now backcrossed to their irradiated fathers, 12.5% of the following generation will be homozygous for recessive lethality if the parent father carried such a mutation, as may be read off the figure. The original mutation rate may thus be determined. This method is not without complications: The assumption that there are no "spontaneous" (i.e. without radiation) mutations is obviously too simple, one has, therefore, to correct for this background (around 8%). A numerically exact determination of intrauterine deaths is very difficult, and there are possibly other causes for this to occur than autosomal lethal factors. Since the actual numbers are quite small, a careful statistical analysis is required taking into account the statistical distributions, not just the expectation values. The details are not to be given here, they may be found in the literature (see, e.g. LUNING 1975).

Dominant mutations are more easily scored because they are seen already in the first filial generation.

Two endpoints are used in the mouse, namely eye cataract formation and that of skeletal abnormalities. The rates found were 0.5×10^{-4}/(Sv[1] and gamete) and 1×10^{-3}/(Sv and gamete), respectively. These experimental figures may be used to estimate frequencies in children born to mothers exposed in pregnancy in Hiroshima and Nagasaki. The conclusion is that less than 1 cataract and about 11 skeletal mutations were expected which is far smaller than the statistical variation. This means that the Japanese data are insufficient to check the validity of the estimate but they are also not in contradiction to it.

Structural chromosome aberrations – if they do not cause lethality – have also genetic consequences. An important class are *reciprocal translocations* where segments of genetic material are exchanged between different

Table 19.3. Estimated mutation induction rates in mice. (After SANKARANARAYAN, 1974; UNSCEAR, 1972; [a]) SEARLE, 1974, 1977)

type	dose rate	rates (frequency/Sv) male	rates (frequency/Sv) female	spontaneous [a]) rates per germ-cell
specif. locus	high	$1.7 \cdot 10^{-5}$	$5.4 \cdot 10^{-5}$	$1.4 \cdot 10^{-6}$ (oogonia)
	low	$0.5 \cdot 10^{-5}$	$0.2 \cdot 10^{-5}$	$8.3 \cdot 10^{-6}$ (spermatogonia)
recessive lethal mutations	high	$0.9 \cdot 10^{-2}$	–	$2.9 \cdot 10^{-4}$
	low	$0.3 \cdot 10^{-2}$	–	
dominant-visible	high	$5.0 \cdot 10^{-5}$	–	0.67
	low	$1.7 \cdot 10^{-5}$	–	
skeleton-alterations	high	$1.1 \cdot 10^{-3}$	–	$6 \cdot 10^{-2}$
	low	$0.4 \cdot 10^{-3}$	–	
translocations	high	$0.3 \cdot 10^{-2}$	$0.3 \cdot 10^{-2}$	$10.4 \cdot 10^{-4}$
	low	$3.3 \cdot 10^{-4}$	–	
loss of X-chromosome	high	–	$15 \cdot 10^{-4}$	$5.1 \cdot 10^{-4}$
	low		$6.5 \cdot 10^{-4}$	
dominant lethal mutations	high	$8.6 \cdot 10^{-3}$	$9 \cdot 10^{-2}$	

[1] The unit "Sievert" is explained in Chap. 23

Table 19.4. Estimates of genetic risk in man based on the direct method (low dose-rate, low-LET radiation). (After UNSCEAR, 1986)

damage	expected frequency per 0.01 Gy and 10^6 live births	
	males	females
mutation with dominant effects	10 - 20	0 - 9
recessive mutations [a)]	0	0
unbalanced products of reciprocal translocations	1 - 15	0 - 5

a) The figures apply to the first generation, 1 case per 10^6 is expected in the following ten generations and about 10 per 10^6 further on.

chromosomes. There is usually no effect in diploids but at meiosis an unequal distribution may occur. If this happens the cell is no longer viable. Since the probability for this *unbalanced distribution* is 50%, the fertility is reduced to one half: Reciprocal translocations cause *semisterility*.

The critical cells for genetic risk are the spermatogonia in males since they are the precursors of the mature sperm cells, and the oocytes in females. Table 19.3 summarizes a number of results obtained with different endpoints in mice after exposure to sparsely ionizing radiation. The spontaneous rates are given for comparison. The data demonstrate clearly the influence of the dose rate: if it is high, the mutant yield is increased by about a factor of three.

There are essentially two ways to deduce hazard estimates for humans from animal data. The first one relies only on dominant mutations, and the data are directly extrapolated. Since only one particular mutation type is considered, it is necessary to apply multiplication factors to include all possible dominant mutations. They are estimated to be 41 and 5 for cataracts and skeletal abnormalities, respectively. The total expected rates are then $2\text{–}5 \times 10^{-3}$/Sv based on the data given above. Taking into account all the

Table 19.5. Natural prevalence of diseases with genetic traits. (After UNSCEAR, 1986)

monogenic	1.25 %
chromosomal	0.4 %
congenital anomalies	5.85 %
multifactorial [a)]	60 %

a) The heritability if these diseases lies between 0.3 (diabetes) and 0.9 (systemic lupus erythematosus)

Table 19.6. Estimated values for doubling doses with mice (X- or γ-radiation). (After SEARLE, 1977)

sex	mutation type	dose rate 10^{-5} Gy min^{-1}	doubling dose Gy
m	dominant, visible	1 - 8	0.7
m	specif. locus	1 - 9	1.2
w	specif. locus	3 - 9	1
m,w	recessive lethal mutations	1	1.1
m,w	recessive lethal	0.3	0.7
m	translocations	7	2.3
m	translocation	3	1.9
m	translocations	4	2.6
m	translocations	20	1.8
m	translocations	90	0.4
w	loss of the X-chromosome	6	1

available evidence, UNSCEAR (1986) recommends the figures given in Table 19.4.

Another approach to deal with the problem of human hazard estimation is the "doubling dose method". The philosophy behind this is that there are too many diseases of genetic origin to be separately assessed and it would be a better way to give a figure which indicates the estimated increase of *all*

Table 19.7. Doubling doses for male mice with densely ionizing radiation. (After SEARLE, 1977)

mutation type	radiation type	dose rate 10^{-5} Gy min^{-1}	doubling dose Gy
dominant lethal	fission neutrons	1 - 2	0.025
specif. locus	fission neutrons	1 - 2	0.07
translocations	fission neutrons	1	0.08
translocations	4.1 MeV neutrons	1	0.13
translocations	14.5 MeV neutrons	10	0.5
translocations	^{239}Pu-α	0.1	0.12

Table 19.8. Estimated genetic effect of 0.01 Gy (low LET, low dose-rate) on genetic damage occurrence per 10^6 live births, based on a doubling dose of 1 Gy. (After UNSCEAR, 1977)

type	spontaneous number	first generation	equilibrium
autosomal dominant + X-chromosome associated	10 000	20	100
recessive	1 100	few	very slow increase
chromosomal	4 000	38	40
other anomalies	90 000	5[a]	45[a]
SUM	105 000	63 (0.06%)	185 (0.17%)

[a] assuming that 5% are due to mutations

of them. There are, however, two hidden assumptions: The first is that the experimentally studied mutation types are representative for all of them and secondly that the spectrum of spontaneous and radiation-induced mutations are essentially equal. Both suppositions are difficult to verify. Also, to use the doubling dose in a quantitative way for human risk assessment the spontaneous rates have to be known. There are a few recent studies which show that they are comparatively high. A distinction must be made between *congenital anomalies* which are inherited according to classical MENDELIAN rules and multifactorial diseases with hereditary components.

The presently available estimates are listed in Table 19.5.

Doubling doses for low LET radiations and neutrons as obtained in the mouse are given in Tables 19.6 and 19.7. A mean value of 1 Sv may be deduced from the data which is the currently recommended value also for humans. The margin of uncertainty is, however, rather wide.

The method described must not be applied in a naive fashion. For example, it cannot be assumed that the already very high rate of multifactorial diseases is just doubled but there is certainly reason for some concern. On the basis of very careful considerations the estimates summarized in Table 19.8 were arrived at. It must be pointed out that the new figures on multifactorial diseases were not yet incorporated for reasons just discussed.

Further reading

AUERBACH 1976, CZEIZEL and SANKARANARAYANAN 1984, DALRYMPLE et al. 1973, JENSH and BRENT 1987, KRIEGEL et al. 1986, NEEL and MOHRENWEISER 1984, SEARLE 1974, 1987, UNSCEAR 1972, 1977, 1986, WOLSKI 1982

Chapter 20
Late Somatic Effects

This chapter deals with those biological consequences in the mammalian organism which show up long times after a radiation insult. The eye reacts rather sensitively in this respect by cataract formation. This is an important example of late effects other than cancer. Radiation-induced tumours form, of course, the most important class. They are also related to radiation-induced life shortening, as discussed in some detail. The ways and means of tumour risk assessment are described.

20.1 Eye Cataracts

Ionizing radiation may cause an opacification of the eye's lens which is termed "cataract"; it is comparatively easy to detect already in early stages with appropriate instrumentation and represents one of the most sensitive indicators of radiation damage. Eye cataracts are presumably initiated by a disturbance of cell division in the lense epithelia. The doses required may be quite small as seen from the data in Table 20.1. It is also quite obvious that remark-

Table 20.1. Minimal doses for eye cataract formation (most sensitive methods). (After VOGEL, 1973)

animal	dose with sparsely ionizing radiation / Gy	dose with fast neutrons / Gy
mouse	0.33	0.013
rat	2.4	0.12 - 0.37
guinea-pig	1.2	0.9
rabbit	0.75	0.02 - 0.07
dog	3	-
monkey	5	0.75
goat	≈ 4	4.66
man	2	-

able interspecies differences exist, both in terms of sensitivity and RBE for fast neutrons. There is, however, a general tendency for a markedly higher effectiveness of densely ionizing radiations so that the recommended quality factors (Chap. 23) appear to be too low for this kind of damage. This has been incorporated in the regulations as discussed later in Chap. 23. There appears to be a threshold-dose although its exact value is still under debate.

Eye cataracts have even been diagnosed in occupationally radiation-exposed workers. There is an important lesson to learn, namely that the protection of the eye should be given the necessary attention. This is not always the case, particularly in the field of medical radiation application. The most recent radiation protection recommendations incorporate, therefore, special regulations for the eye as a particularly sensitive organ.

The latency times for cataract formation – i.e. the period between irradiation and the first clinically detectable symptoms – vary considerably, depending on dose and species. They extend from a few months up to twelve years.

20.2 Radiation-induced Life-shortening

This is a difficult problem which is only to a certain extent accessible to scientific exploration since too many factors are involved. It is clear that an agent like radiation which causes lethal diseases reduces the mean life span. Tumours play the most prominent role in this respect, the next section is entirely devoted to this question. It remains to be discussed whether radiation may lead to unspecific life shortening in the sense of premature aging. This is difficult to assess by animal experiments since their average lifetime is usually much shorter than that of humans and it is by no means clear whether a simple "scaling" is allowed. If, on the other hand, animals with lifetimes comparable to that of humans were studied, the results would not become available for decades. There are two approaches to resolve this dilemma: Comparative analysis of animal data might show whether there is a relationship between mean life expectancy and radiation-induced shortening and whether extrapolation to humans is possible. This, however, could never give a "proof" but only some kind of contingency check. The other way is to study carefully the fate of people exposed to radiation, either professionally or by accident.

The most comprehensive studies have been performed with large mouse colonies which were exposed to daily γ doses throughout their life-time. The range was 0–0.5 Gy/day (see GRAHN et al. 1978) with 0.003 Gy/day being the lowest level. This corresponds to about the 20-fold of 0.05 Gy/year which is the recommended limit to radiation workers.

The experiments show (Fig. 20.1) that with low dose rates the reduction in life expectancy is solely a function of daily dose according to the relationship:

$$\mathrm{MAS_D} = \mathrm{MAS_0}e^{-\beta D_a} \tag{20.1}$$

where $\mathrm{MAS_D}$ is the mean survival time after the start of exposure to daily

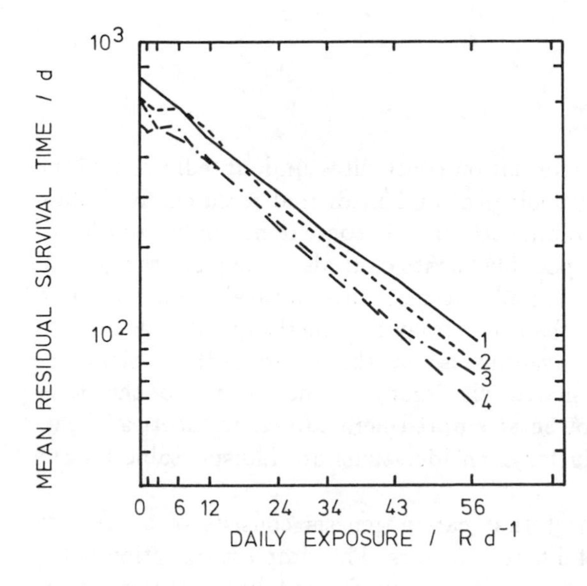

Figure 20.1. Mean residual survival time of mice exposed to daily doses as given on the abscissa. The numbered curves refer to different mouse strains (after GRAHN 1970)

doses D_a. MAS_0 is the average lifetime of unirradiated controls, β a species-specific coefficient. For the mouse it was shown that $\beta = 4$ days/Gy. By comparison with other species (dog, guinea pig) one is led to the conclusion that β is proportional to the mean life span. The respective ratio is about 30 between mice and humans suggesting that:

$$\beta_{man} \approx 120\,\text{days/Gy} \tag{20.2}$$

An example may serve as an illustration: A radiation worker is assumed to start his occupation at an age of 20 – his statistical life expectancy is then another 55 years – and may be exposed to the limit of 0.05 Gy/year for the following 45 years. The daily dose under these condition is 1.4×10^{-4} Gy. Calculations with Eqs. (20.1) and (20.2) give an expected lifetime of 54.1 years, i.e. the statistical reduction is about 330 days which will be in fact lower since the exposure stops at the retirement age of 65. The assumptions made are rather unrealistic and only intended to demonstrate the orders of magnitude. It should be added that a continuous exposure to the upper limit of 0.05 Gy/year is explicitly not permitted by protection regulations.

The life-span reduction in mice is due to a number of effects. The analysis shows that about 80% of premature deaths are caused by tumours which constitute obviously the most important hazard. They are discussed extensively in the next section.

Whether an unspecific radiation-induced senescence exists is still under debate. Neither animal nor retrospective human studies are sufficient to give a clear-cut answer. There is, however, no doubt that unspecific effects – if they exist – play only a minor role compared to cancer, so that a further discussion appears to be rather academic.

20.3 Cancerogenesis

20.3.1 General Aspects

The understanding of tumour formation constitutes undoubtedly one of the central themes of present-day biological and medicinal research. Radiobiological investigations figure prominently in this context not only in order to delineate extents and limits of possible hazards but also to explore the general underlying mechanisms. The great advantage of this physical agent compared to others lies in the fact that the action may be exactly quantified, both in time and space. The great disadvantage is that the nature of the initial damage is not known but this is shared with many – if not most – of the other known carcinogens. It will not be attempted here to review theories of carcinogenesis, but a few introductory considerations are indispensable for the discussions to follow.

There is general agreement that carcinogenesis consists of a complex interplay of a large number of different factors. The simplest reduction is the "two stage theory": The primary alterations are formed during the *induction* phase. They may remain latent or even removed via repair processes. During the *promotion phase* the initial lesions develop to a neoplastic transformation which will ultimately lead to the clinically observable cancer. Both stages may be influenced by external factors. A number of chemicals are known which act selectively either as inducers or promoters. The situation is less clear with radiation but it has to be assumed that it acts on both levels.

There appears to be little doubt that the primary event occurs in the cellular DNA. This must not be taken to mean that cancer is initiated by somatic mutations – this may be the case in certain instances – chromosomal rearrangements like translocations may be even more important. Radiation may also activate latent "oncogenes" and stimulate their expression similarly to prophage induction in bacteria (Sect. 7.2.2). It is certainly prudent to identify the induction process not with one clearly defined event but to treat it as a spectrum of a great number of different alterations. The understanding of the promotion stage is even less developed. Hormonal, immunological and tissue-specific factors have additionally to be taken into account. That radiation interferes here also is clear from the finding that it leads to a shortening of the latency period, i.e. the time between induction and tumour expression ("acceleration").

A further complicating aspect is that induced cells may be inactivated by radiation so that a complex dose-relationship results. This may be illustrated by a simple model. The probability P(D) for tumour formation as a function of dose may be written as:

$$P(D) = f_i(D) \cdot f_s(D) \cdot f_p(D) \cdot f_{org}(D) \qquad (20.3)$$

with:

$f_i(D)$: induction probability per cell
$f_s(D)$: surviving probability per cell
$f_p(D)$: "promotion probability", probability that a tumour is formed from one induced cell
$f_{org}(D)$: all organic influences

To gain a first approximate of the expected dose dependence the last two factors are set to unity. For $f_i(D)$ a linear-quadratic relationship – as found in the in-vitro system – is assumed and an analogous expression for survival.

This yields:

$$P(D) = (a_1 D + a_2 D^2) e^{-\alpha D - \beta D^2} \quad . \tag{20.4}$$

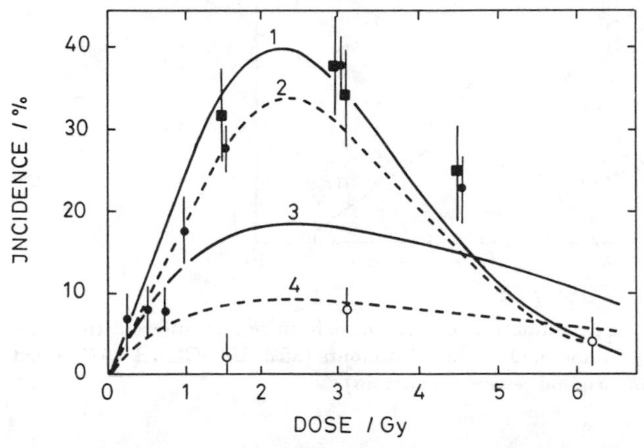

Figure 20.2. Schematic expected dose relationships for cancer induction and comparison with experimental data (leukemia in mice). Upper curves: 250 kV X-rays, high dose rate; lower cures: gamma-rays, low dose rate. Curves 1 and 3 are based on linear-quadratic, curves 2 and 4 on linear induction kinetics (after BARENDSEN 1978)

The resultant curve is shown in Fig. 20.2 together with some experimental animal data which demonstrate that the predicted behaviour is qualitatively correct. The actual shape depends critically on the relations between the coefficients which vary considerably with tumour type. It may even occur that radiation leads only to a reduction of an already high tumour incidence as illustrated in Fig. 20.3. This is a rare case but it should not be overlooked.

There are large differences between tumour types even in one animal species.

So far "tumour incidence" has been used as the fraction of irradiated animals which will develop at least one tumour during their lifetime. This is an integral quantity where no attention is paid to the time dependence. Another way to analyze the experiments is to give the number R(t) of tumours per animal as a function of time. In this way age specific variations may be detected. If a POISSON distribution of induced tumours is assumed, then the time dependent incidence I(t) becomes:

$$I(t) = 1 - e^{-R(t)} \quad . \tag{20.5}$$

For small R(t) it is found that $I(t) \approx R(t)$.

The dose-effect curves in Fig. 20.3, which should be taken as representative examples, demonstrate a typical difficulty in the interpretation of ex-

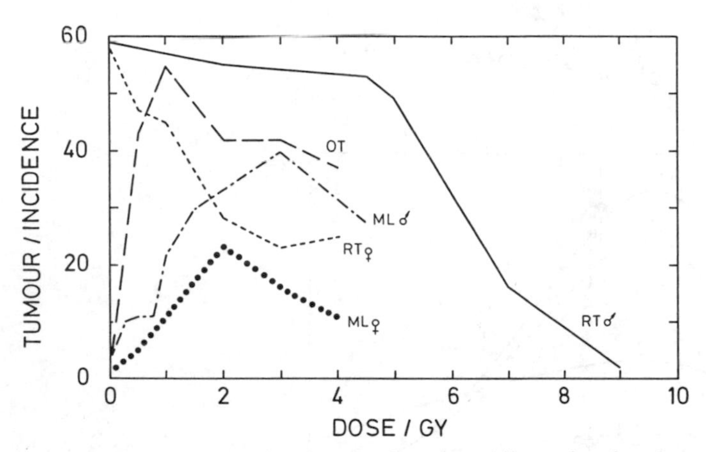

Figure 20.3. Cancer induction as a function of X-ray dose in mice and different tumours. ML: leukemia, RT: reticulary tumours, OT: ovarial tumours (after UNSCEAR 1977 based on studies by various authors, original references therein)

perimental data: The relationship between dose and tumour incidence is not unambiguous. If only isolated points – usually in the higher dose region – and not the complete curve is known, an extrapolation is necessarily bound to be erroneous. It is here that reliable – and proven – theoretical descriptions would help.

The experiments described so far were restricted to X and γ rays. There is a great influence of radiation quality, also on the parameters in Eq. (20.4). These questions are discussed in Sect. 20.3.3.

The extrapolation from animals to humans encounters another difficulty. As already said, there is – depending on tumour type – a fairly long time span between induction and clinical manifestation. Examples of these *latency periods* in humans are listed in Table 20.2. The times involved are all much longer than the mean life span of common laboratory animals. Since they do exhibit radiation-induced tumours during their life, one has to conclude that the mechanisms are different. The promotion phases may be shorter and the spectrum of tumour types different. Reliable data for human risk estimates cannot be obtained in this way. The use of longer-living primates could only to a certain extent remedy this situation – apart from ethical

Table 20.2. Average latency period of tumours in man after radiation exposure. (after: UNSCEAR 1977)

organ or tumour type	latency period years
leukemia	10 - 15
thyroid	20.3
bladder	20.7
breast	22.6
head and neck	22.8 - 24.1
larynx and pharynx	23.4 - 27.3
skin	24.5 - 41.5

and cost considerations: The data are required now and not after decades of experimentations. This all means that most of the emphasis has to be laid on human epidemiology.

Even if the quantitative assessment is unreliable, animal experiments are indispensable to study the basic mechanisms. The difficulty is, however, to select the appropriate objects. For example, some strains possess a high incidence of spontaneous tumours. A well-known example is the female *Sprague-Dawley rat* which will develop mammary tumours with nearly 100% probability during its lifetime. Radiation reduces the age at which they are found but not the overall incidence. This shortening of latency times appears to be a general phenomenon. It is clear in this special case that there is no additional induction but only radiation-induced acceleration. The statistical interpretation is greatly impeded by this further complication.

There are also, on the other hand, animal strains where the rate of spontaneous tumours is negligible. This may be due to the fact that the latency period is significantly larger than the mean lifetime.

If tumours are found after irradiation, this could also be caused by acceleration but also by induction. It is very difficult to distinguish between these alternatives. If the time between tumour manifestation and irradiation does not depend on dose, then real induction may take place.

This introductory section was meant to illustrate some of the difficulties of interpretation and to create a feeling of understanding why the present knowledge is still rather rudimentary in spite of the remarkable experimental and theoretical efforts.

20.3.2 Cancerogenesis by Optical Radiation

The only organ at risk is the skin. Skin tumours are relatively easy to cure (if not detected too late) with up to 95% chance for complete healing. The only exception is the *melanoma* which is particularly aggressive and spreads metastases at a very early stage. The problem of skin carcinogenesis plays an eminent role in the discussion of the ozone problem (Sect. 22.2). It has to be assumed that a shift of the solar spectrum on Earth leads to a substantial increase in skin cancer rates.

An action spectrum is not yet available which is understandable in the light of the great experimental effort involved. A certain estimate may be gained by comparing the carcinogenic effectivity of different lamp types. In this way an approximate action spectrum may be obtained which appears to be quite similar to *edema* formation and not very different from the erythema action spectrum. It is shown in Fig. 20.4.

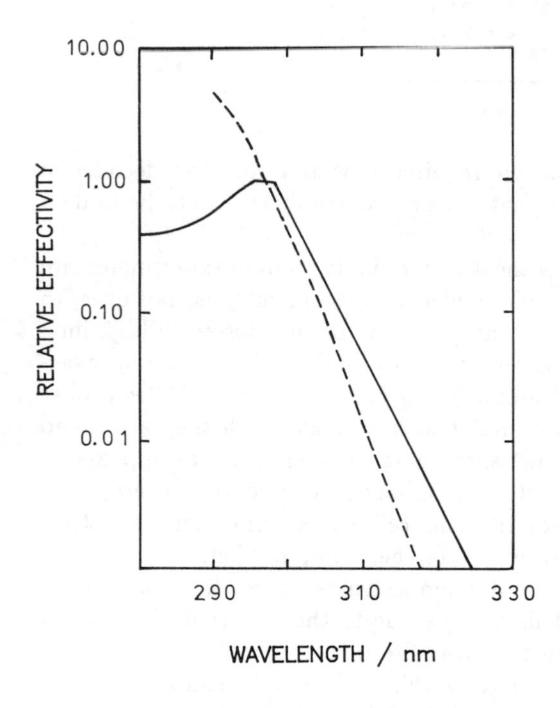

Figure 20.4. Estimated action spectrum for skin tumour formation. The broken line indicates DNA absorption (after COLE et al. 1983)

The time course of skin tumour formation has been studied in hairless albino mice. Figure 20.5 shows results obtained with different daily doses from a UV-B lamp. If the time of occurrence tm is plotted versus daily dose D in a double-logarithmic plot, straight lines are obtained. That means that the relationship may be described by a power function:

$$t_m \sim \dot{D}^{-0.6} \quad . \tag{20.6}$$

The exponent was read off the graph.

Eq. (20.6) has interesting implications. It may be rearranged as:

$$Dt_m \sim \dot{D}_{cum}\dot{D}^{0.4} \tag{20.6a}$$

where D_{cum} is the total accumulated fluence. Because of the positive exponent, D_{cum} increases with D, or, in other words, small daily doses are more

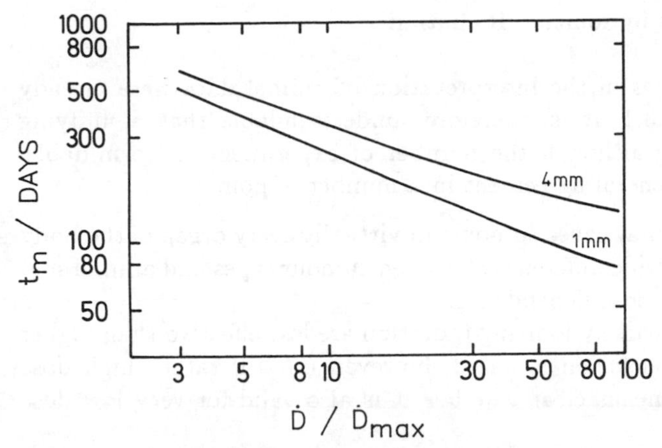

Figure 20.5. Time of skin tumour appearance as a function of relative fluence rate for two diameters of tumours (after DEGRUIJL et al. 1983)

effective in terms of total fluence if given over a sufficiently long time period. This is clearly a deviation from the reciprocity rule. There is, of course, a natural threshold, namely if the time to detection is longer than the life expectancy.

The same data may be used to estimate tumour *incidence* I. It is also given by a power function:

$$I \sim D^{a}_{cum} \tag{20.7}$$

with the exponent being close to 2.

The above, however, only applies to non-melanoma skin tumours. For melanomas there are neither data about the spectral dependence nor about the incidence-fluence or incidence-fluence-rate relationship. Human data for skin cancer are scarce and difficult to obtain, partly because of the high curing rates. Present estimates range to around 1.7×10^{-3} cases per year (non-melanoma; USA), the respective figure for melanomas is 4×10^{-5} cases per year. There are definite indications from epidemiological surveys that skin tumour incidence is positively correlated to duration and intensity of solar exposure.

It is not clear at present whether immunological factors (Sect. 18.2) play a role in skin tumourigenesis although the experiments with transplanted tumours point strongly to a possible involvement.

Further reading
URBACH 1987, van der LEUN 1984, 1987

20.3.3 Cancerogenesis by Ionizing Radiation

The principal difficulties in the interpretation of animal data were already alluded to in Sect. 20.3.1. It is, therefore, understandable that a unifying picture is still missing although the number of experiments is formidable. There appears to be general agreement in a number of points:

1. External radiation may cause tumours in virtually every organ of the body although there are wide differences between tumour types and animal species in terms of the dose dependence.
2. Low dose rates of sparsely ionizing radiation are less effective than higher dose rates. This conclusion is based, however, on still rather high dose rates. It is by no means clear whether it is also valid for very low dose rates.
3. Irradiation may shorten the latency period ("acceleration").
4. The probability of tumour formation depends on age at exposure.
5. Densely ionizing radiation is more effective than sparsely ionizing radiation.

The shapes of the dose-effect curves vary (see Fig. 20.3), and there is no agreement about a generally applicable mathematical formalism. In many cases an expression like Eq. (20.3) is at least compatible with the data. With high LET the linear component dominates which is in line with the cellular models (Chap. 16). As a typical example, Fig. 20.6 shows the tumour formation in rat skin after exposure to electrons (sparsely ionizing) or α particles (densely ionizing).

The initial part can be approximated by a linear-quadratic dose depen-

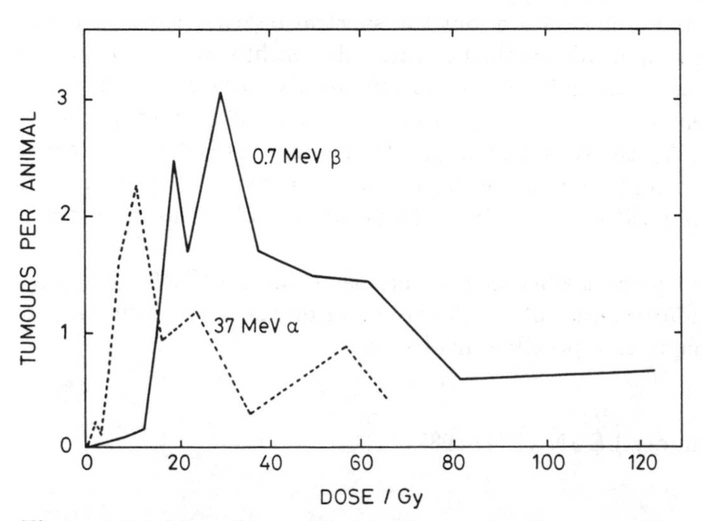

Figure 20.6. Mean skin tumour rates in rats after electron and alpha exposure as function of dose (after BURNS et al. 1968)

dence for β-radiation and by a purely linear one for α particles. This is in line with the usual assumptions. Interpreting the decreases with higher doses is less simple: If it is due to cell killing as normally postulated, the α curve should be much steeper – according to the higher RBE. In fact, both curves are nearly parallel.

The most recent UNSCEAR report (1986) reviewed existing mathematical models for cancer incidence as a function of dose. The conclusion is that a universal formula cannot be given although the linear-quadratic relationship seems to be applicable in a number of cases. Purely quadratic expressions appear to be inappropriate. Extensive comparisons may also be found in the BEIR report (1980).

RBE plays an important role in radiation protection. The animal experiments do not allow definite conclusions. There are examples in the literature of RBE values near 100 for neutrons. The dependence of RBE on dose creates here a special problem. It is not at all clear whether the cellular models discussed in Chap. 16 are applicable to tumour formation. The question of RBE is also especially pertinent to assess the action of radionuclide incorporation, which is treated in Chap. 21.

A realistic hazard evaluation has to rely on human data. There are a number of population groups from which they may be drawn.

1. Radiation workers
2. Radiologists
3. Uranium miners
4. Patients who were treated by radiotherapy or underwent diagnostic examinations employing radionuclides
5. Survivors of nuclear warfare or test explosions.

The common problem with all these retrospective investigations is that the dose received is generally not known and must be estimated. This can be done reliably in the case of radionuclide incorporation since the physical half-life and the physiological behaviour are known. This method fails, of course, with short-lived isotopes. Good estimates are available from (mostly female) workers who painted watch dials with a mixture of Radium and ZnS in the twenties of this century. Another, rather sad and bad example are "Thorotrast" patients who received Thorium-containing contrast media for diagnostic X ray examinations. This – in the light of today's experience – irresponsible practice did not cease before the forties in some countries. By far the most data, however, were obtained from the survivors of Hiroshima and Nagasaki. A more detailed discussion is, therefore, in place.

Before doing this, another general remark has to be made: Since there is always a certain rate of spontaneous tumours, the statistical analysis has to concentrate on the calculation of the *excess* rate. The choice of the correct control population is, therefore, extremely critical. Spontaneous tumour rates depend not only on the general environment but also on the ethnic background, living habits and even the profession not to speak of age and sex. A comparison with the general population in the country may lead to quite misleading results. Local variations have to be appropriately taken into account. This is by no means a simple task! A great deal of disagreement about risk factors stems from different choices of control populations.

Already the definition of "risk" is not unambiguous and implies certain model assumptions. The two most commonly used approaches are the "relative" and the "absolute" risk model which are sketched in Fig. 20.7. In the first

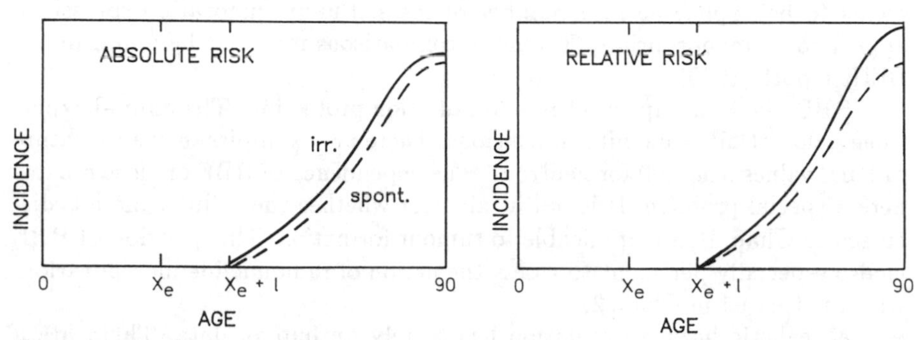

Figure 20.7. Absolute risk and relative risk in comparison (l: latency time) (after UNSCEAR 1986)

case, it is assumed that the spontaneous tumour rate is increased by a certain dose-dependent factor, i.e. radiation-related incidence is proportional to the background frequency. The second model postulates additivity, i.e. radiation leads to a certain additional *number* of tumour cases which is independent of the spontaneous incidence.

A more formal way of description is:

$$\mathrm{TR} = \lambda + \lambda \cdot g_r \cdot \mathrm{D} \tag{20.8}$$

("relative risk model")

$$\mathrm{TR} = \lambda + g_a \cdot \mathrm{D} \quad . \tag{20.9}$$

("absolute risk model")

TR is the total tumour rate, λ the spontaneous incidence, D the dose, g_r and g_a risk coefficients which depend on a number of parameters. These include type of tumour, sex and age of exposure. A general decision which approach may be best cannot be made. Figure 20.8 displays relative risk factors for leukemia and all the other tumours as estimated from data obtained in Hiroshima and Nagasaki. It is clear from the different curve shapes that the "relative risk model" is inappropriate for leukemia but may be applicable to the other types.

Unfortunately, there is no generally accepted way to present risk estimates, and different figures are found in the literature. This creates some

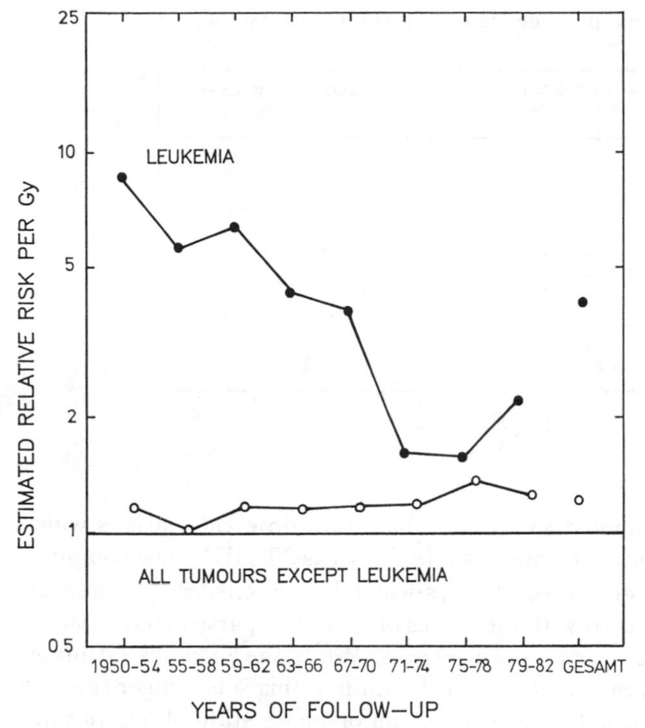

Figure 20.8. Relative risk factors from survivors of Hiroshima and Nagasaki (after PRESTON et al. 1987)

confusion which is particularly unfortunate in a field that is already in the center of public discussion.

The *lifetime* risk for radiation-induced cancer mortality was estimated by the "International Commission on Radiological Protection" (ICRP) in 1977 as listed in Table 20.3. These figures are still the basis for regulatory recommendations although they are likely to be changed (see below). They are related to "dose equivalent" which is defined in Chap. 23. According to Table 20.3, the total cancer risk adds up to 1.25×10^{-2}/Sv (equivalent to Gy for low LET radiation) which is commonly rounded to "1 percent per Sievert". This is a weighted average over both sexes and all age groups.

Another way of data presentation is to give the risk per year and unit dose. These figures based on follow-up studies of the Japanese bomb survivors are compiled in Table 20.4. Comparing the relative weights of the various organs with regard to radiation induced tumours, shows obvious discrepancies between the two tables which require clarifications in the future. Nevertheless it is clear that leukemia poses the greatest risk, followed by cancers of the lung and the female breast. The high number of stomach cancers may be particular for Japanese conditions since the spontaneous rate is also greater than in other countries.

Table 20.3. Estimated tumour incidence and mortality risk factors for different organs

organs	incidence a) 10^{-4} Sv^{-1}	mortality risk b) 10^{-4} Sv^{-1}
breast	100	25
red bone marrow c)	20 - 50	20
lung	25 - 50	20
thyroid	100	5
bone	2 - 5	5
jejunum, pancreas, rectum	2 - 5	-
stomach, liver, colon	10 - 15	-
remainder	-	-
total	not given	125

a) after UNSCEAR 1977
b) from ICRP 26 (1977)
c) essentially leukemia

All the estimates quoted so far were based on dose calculations which were made in the sixties and commonly referred to as "T65D". New computations have been carried out in recent years which led to considerable changes. They are not only concerned with the values of doses but particularly with the relative contributions of γ rays and neutrons. While the previous estimates indicated a larger neutron component in Hiroshima, this is no longer the case with the new calculations. This is a rather unfortunate state of affairs since the previously found difference formed the basis to estimate RBE factors for humans which were used for the definition of "quality factors" (see Chap. 23). They are now less well founded and may eventually be changed.

Table 20.4. Mortality risk and deduced estimated incidence of the most important radiation-induced cancer types (After PRESTON, KATO et al., 1987, mortality risks after BEIR, 1980), averaged over all age groups. Values given for 10,000 persons per Gy and year

tumour type, organ	risk 10^{-4} Gy^{-1} y^{-1}		mortality %		incidence (estimated) 10^{-4} Gy^{-1} y^{-1}	
	male	female	male	female	male	female
leukemia	1.95	1.20	100	100	1.95	1.20
oesophagus	0.14	0.22	100	100	0.14	0.22
stomach	0.9	1.07	75	78	1.2	1.37
colon	0.19	0.37	52	55	0.37	0.67
liver	0.11	0.06	100	100	0.11	0.06
lung, bronchi	0.78	0.92	83	75	0.94	1.23
bladder, kidney and other	0.27	0.23	37	46	0.73	0.5
myeloma	0.07	0.06	?	?	-	-
breast	-	0.65	-	39	-	1.67
total	5.24	5.62				

The dose determinations which have to be performed for each bomb survivor separately involve several steps. The first one which has to take into account the exact location relative to the explosion epicenter, shielding by the environment and housing yields the KERMA free in air at the body surface. From that organ, doses are calculated which depend not only on the KERMA values but quite considerably also on the composition of the radiation.

Comparisons between the two dose computations in terms of risk estimates have been performed for leukemia and all other cancers. Figs. 20.9 and 20.10 show the results for the first case. It is seen that in terms of KERMA there are rather large differences suggesting a considerable increase in risk. It is smaller if organ doses are compared, but still substantial.

It is at present not at all clear what this may mean for general risk esti-

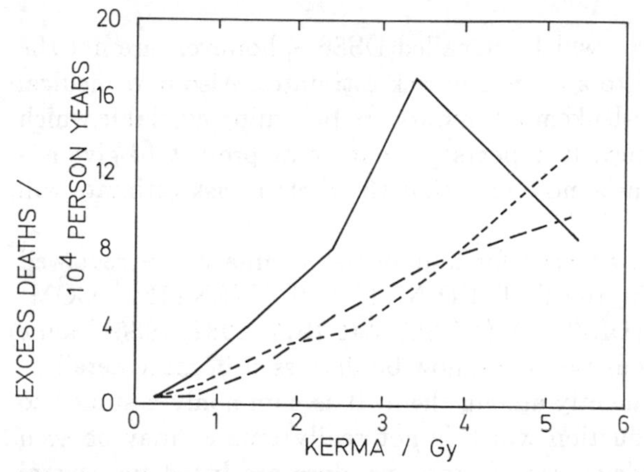

Figure 20.9. Death rates by leukemia as a function of estimated KERMA with different calculations. Heavy line: DS 86, heavy broken line: T65D, light broken line: T65D but only for the subgroup for which DS 86 estimates are available (after PRESTON and PIERCE 1988)

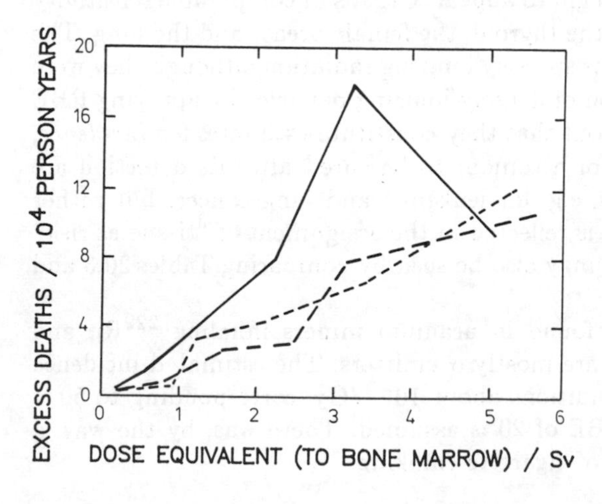

Figure 20.10. As in Fig. 20.9 but for organ dose (after PRESTON and PIERCE 1988)

Table 20.5. Increase of risk factors on the basis of new dose calculations with different assumed RBE-values for neutrons. The comparison is based on DS-86 values below 4 Gy (After data from PRESTON and PIERCE, 1988) and organ doses

RBE	non-leukemia	leukemia
1	0.97	1.18
5	1.13	1.46
10	1.32	1.8
20	1.72	2.36
30	2.11	2.82

mates. Since there are now great uncertainties about RBE factors, different assumptions influence the calculations greatly. This is shown by the figures in Table 20.5.

The new dose estimates – which are called DS86 –, however, are not the only factors which may lead to a change in risk estimates. Also new medical data, particularly about non-leukemia tumours are becoming available which also show an upward tendency. It appears premature at present to give numerical figures but there can be no doubt that the current risk estimate will have to be corrected.

All the data which may be used for human risk estimates are reviewed and summarized regularly by the UNITED NATIONS SCIENTIFIC COMMITTEE ON ATOMIC RADIATION (UNSCEAR 1977, 1982, 1986). Some of the more important tumour types will now be discussed in more detail.

Leukemia figures prominently among them. It is commonly assumed to have the highest rate of induction which is not really true as may be seen from Table 20.6, where incidence risk factors per dose are listed for several tumours. It gains its exceptional position mainly from the short latency period (Table 20.2). Because of this, it is seen considerably earlier than other cancers. This is also the reason why the most reliable data are available for leukemia, for other tumours they just begin to appear. Organs of comparable sensitivity are – in terms of incidence – the thyroid, the female breast and the lung. The data in Table 20.6 are given for sparsely ionizing radiation although they were partly derived from the action of densely ionizing particles by applying RBE factors. It should be pointed out that they constitute estimates for *incidence*, not mortality. The chances for a tumour to be cured after its detection are quite variable. They are low, e.g. for leukemia and lung cancer, but rather high for thyroid cancer. This is reflected in the assignment of "tissue at risk" as discussed in Chap. 23 and may also be seen by comparing Tables 20.3 and 20.6.

Lung cancer was often found in uranium miners inhaling ^{222}Rn and its daughter products which are mostly α emitters. The estimated incidence risk rate was under these conditions about 10^{-2}/Gy, corresponding to 50×10^{-4}/Gy for X rays if an RBE of 20 is assumed. There was, by the way, a definite synergistic influence of cigarette smoking.

Liver tumours were found in "Thorotrast" patients with a rate of about 10^{-2}/Gy, mainly caused by α particles.

The high sensitivity of the female breast warrants special attention, particularly in view of current diagnostic X ray practice. Experience from this field but also from Hiroshima and Nagasaki points to a significant age dependence of susceptibility. The period of maximum sensitivity appears to be between 10 and 35 years.

Children which were exposed *in utero* have a significantly higher incidence of malignant diseases than adults. This is true for all tumour types, but leukemia is again especially noticeable because of the short latency period. The total risk is about 2–2.5 times that of adults underlining again the particular radiation sensitivity of the growing embryo.

Further reading

BOICE and FRAUMENI 1984, BROERSE et al. 1985, FREY et al. 1970, HOEL 1987, IAEA 1978, MOSSMAN 1986, PRESTON et al. 1987, PRESTON and PIERCE 1988, SHIGEMATSU and KAGAN 1986, ULLRICH 1982, UNSCEAR 1977, 1982, 1986

Chapter 21
Effects of Internal Exposure

The incorporation of radionuclides plays an important role in the assessment of radiation risks. The basic considerations, the ways of incorporation and the principles of computational methods are introduced. Incorporation may take place either by breathing ("inhalation") or eating or drinking ("ingestion"). For the estimation of the radiation load it is necessary to distinguish between the deposition organs which may – apart from being exposed themselves – act as radiation sources from which other parts of the body ("target organs") may be irradiated. The calculations lead to the recommendation of 'annual limits of intake' for which examples are given.

21.1 Uptake and Distribution of Radionuclides

The uptake of radionuclides may either take place by breathing ("inhalation") or together with the food ("ingestion"). The main ways in the body are depicted in Fig. 21.1: The entry organs are either the lung or the digestive tract. Only in the case of tritium, the skin may also play a – although minor

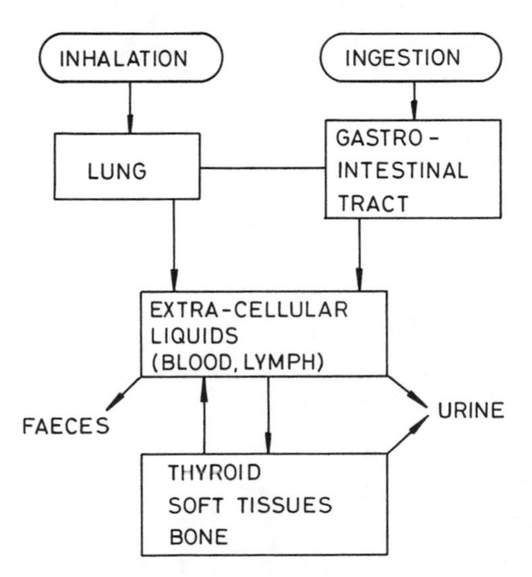

Figure 21.1. The pathways of radioactivity in the human body (after ICRP 30, 1979)

– role. The radioactivity is then distributed via the blood or the lymph fluid to the deposition organs where it remains for some time – depending on its chemical nature – before it is excreted. As a first approximation, the whole system may be described by a set of linear differential equations which may be used to calculate the time course and the total amount of irradiation in the different parts of the body. In order to do this, the transfer coefficients must be known, they have to be determined experimentally.

The *biological half-life* (see Chap. 15) plays a central role in this context. It is related to the excretion rate and depends only on the chemical – and not on the radiobiological – nature of the incorporated compounds. The total effective half-life τ_{eff} depends both on the physical decay – characterized by τ_D – and the biological half-life τ_B according to (see also Eq. (15.3)):

$$\frac{1}{\tau_{eff}} = \frac{1}{\tau_B} + \frac{1}{\tau_D} . \tag{21.1}$$

It is clear from this relationship that τ_{eff} is essentially determined by the smallest of the two components.

A simplified example may serve to illustrate the ways of calculation (Fig. 21.2): A system of one entry organ and one deposition organ is considered. There is an activity uptake A_0 at $t = 0$ and it is assumed that excretion may occur from all organs which are indicated by subscripts according to the figure. The transfer coefficients are λ_{12} and λ_{23}, the excretion coefficients λ_1, λ_2, λ_3 and the physical decay constant λ_D. The activities in the organs are $q_1(t)$, $q_2(t)$ and $q_3(t)$, respectively. On the basis of linear time kinetics, the following equations are obtained:

$$\begin{aligned} \dot{q}_1(t) &= -(\lambda_1 + \lambda_2 + \lambda_D) q_1(t) \\ \dot{q}_2(t) &= \lambda_{12} q_1(t) - (\lambda_{23} + \lambda_2 + \lambda_D) q_2(t) \\ \dot{q}_3(t) &= \lambda_{23} q_2(t) - (\lambda_3 + \lambda_D) q_3(t) \quad . \end{aligned} \tag{21.2}$$

With the initial condition $q_1(0) = A_0$, the solutions are:

$$\begin{aligned} q_1(t) &= A_0 e^{-\lambda_D t} \cdot e^{-\lambda_{12}\lambda_1} \\ q_2(t) &= \frac{\lambda_{12} A_0 e^{-\lambda_D t}}{\lambda_{12} + \lambda_1 - \lambda_{23} - \lambda_2} \left(e^{-(\lambda_{23}+\lambda_2)t} - e^{-(\lambda_{12}+\lambda_1)t} \right) \\ q_3(t) &= \lambda_{12} \cdot \lambda_{23} A_D e^{-\lambda_D t} \left(\frac{e^{-(\lambda_{12}+\lambda_1)t}}{(\lambda_{12} + \lambda_1 - \lambda_3)(\lambda_{12} + \lambda_1 - \lambda_{23} - \lambda_2)} \right. \\ &\quad - \frac{e^{-(\lambda_{23}+\lambda_2)t}}{(\lambda_{12} + \lambda_1 - \lambda_{23} - \lambda_2)(\lambda_{23} + \lambda_2 - \lambda_3)} \\ &\quad \left. + \frac{e^{-\lambda_3 t}}{(\lambda_{12} + \lambda_1 - \lambda_3)(\lambda_{23} + \lambda_2 - \lambda_3)} \right) \end{aligned} \tag{21.3}$$

Depending on the actual values of the coefficients, the activity remains in the body for a given time which is important for the calculation of the total

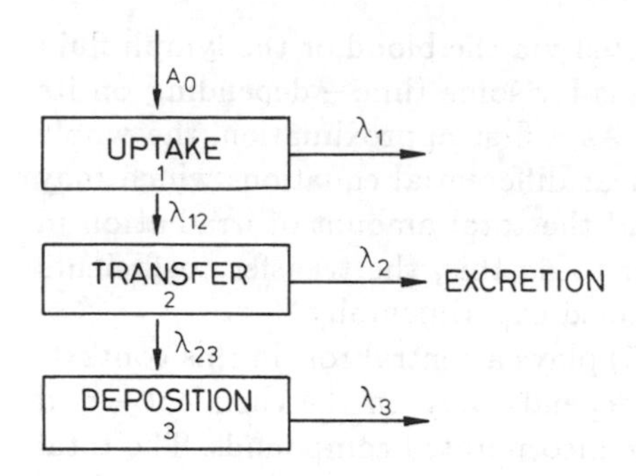

Figure 21.2. Simplified diagramme for activity distribution in the body

hazard. The total number of decays is obtained by integration of Eqs. (21.3). For radiation protection purposes, usually a time span of 50 years is chosen.

If the radionuclides are part of a decay series, the contribution of the daughter products have to be taken into account. This is in principle not difficult although the calculations may become rather involved. The same is true with branched deposition pathways.

The example given should only demonstrate the fundamentals of the approach. In reality most of the organs have also to be subdivided into several compartments. Also the chemical nature of the incorporated compounds has to be appropriately considered.

The two principal ways of incorporation shall now be discussed in more detail. Inhalation of radioactive gases plays only a minor role since they are mostly immediately exhaled. By far more important are those gaseous radionuclides which decay to solid daughter products which adsorb to water droplets in the air or to dust particles ("aerosols") because they may be deposited in the pulmonary tract and remain there for a certain time. This results firstly, of course, in a radiation exposure of the entry organ but it is also possible that they may be transferred to other parts of the body. Radon-222 has to be specially mentioned in this context. Inhalation is, however, not restricted to gases but may also occur if solid radionuclides are finely dispersed in the air.

For a more specific treatment, the pulmonary tract has to be subdivided into *nasal passage* (N-P), *trachea* (T), the *broncheal tree* (B) and the inner surface of the lung, the *pulmonary parenchyma* (P). The deposition probability in these regions depends on aerosol size which is characterized by the "aerodynamic diameter" which is not necessarily identical to the real dimensions. Figure 21.3 gives the deposition probabilities for the different parts of the pulmonary tract. It is seen that particularly the small aerosols are deposited in the lung while the larger ones are already captured in the nasal passage. The next important parameter is the residence time after deposition which depends to a great extent on solubility. It is customary to define several classes of aerosols according to their half-life which is determined by the transport out of the lung (P). The radioactivity is then transferred either to

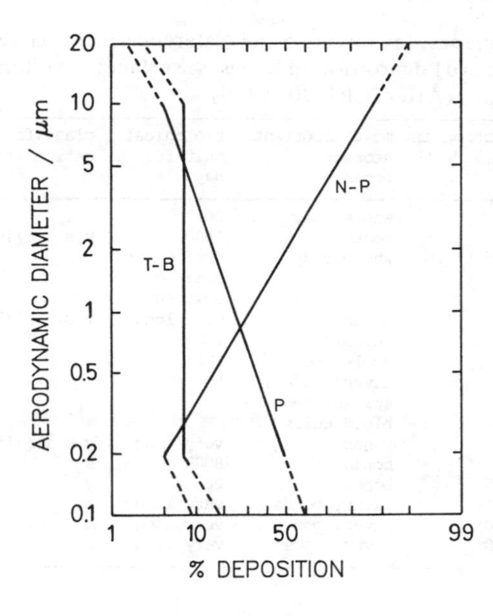

Figure 21.3. Deposition probability of aerosols in the different parts of the broncheal tract as a function of aerodynamic diameter. NP: nasal passage, TB: trachea and bronchial tree, P: pulmonary parenchyma (after ICRP, 1979)

the body fluids – if the aerosol is absorbed – or to the gastrointestinal tract if it is expelled by the cilia of the broncheal system. It is obvious that damage to the entry organ is more and more relevant if the residence times become longer. Three types are to be distinguished:

D (<10 days), W (10–100 days) and Y (>100 days). This classification refers to pulmonary parenchyma, in other parts the times may be quite different.

With *ingestion* two processes are particularly important, damage to the gastrointestinal tract and the distribution by the body fluids to other organs. The mean residence time of food in the body is 42 hours with 1 hour in the stomach, 4 hours in the small intestine, 13 hours in the upper and 24 hours in the lower large intestine. It is possible to give the fraction f_1 of an element which is transferred to the body fluids if it is taken up with the food. Examples are listed in Table 21.1.

The last station on the way is represented by those organs where the nuclides are finally deposited and where they remain according to their biological half-life. Exact calculations are only available for bones which are, of course, important because of their proximity to the blood-forming system. It is also relevant to note that the compounds may be either incorporated on the surface or homogeneously in the total bone volume.

Some metabolic data estimated for an average human of 70 kg ("reference human", Table 21.2) are summarized in Table 21.1.

With the particularly important earth alkali metals and phosphorus it is assumed that nuclides with half-lives longer than 15 days are homogeneously distributed in the body; for the other elements no such general rule can be given. A special note has to be added for decay series. In this case also all daughter products have to be taken into account. This is again interesting

Table 21.1. Metabolic data of some elements; (a) refers to the 'reference man', largest values; (b) see text; (c) no single halflife; (d) deposition in bones; s: surface; v: volume; (e) inorganic Co; (f) organic-complexed Co. (After ICRP 30, 1979)

element	mass in body [(a)]	daily uptake kg d^{-1}	fraction in blood f_1 [(b)]	most important storage organ	biological halflife days	classification [(d)]
hydrogen	7	0.35	1	whole body	20	
phosphorus	0.78	$1.4 \cdot 10^{-3}$	0.8	bones	1500	v,s (τ_D>15)
cobalt	$1.5 \cdot 10^{-6}$	$3 \cdot 10^{-7}$	0.05 [(e)] 0.3 [(f)]	whole body	60%: 6 [(c)] 20%: 60 20%: 800	
strontium	$3.2 \cdot 10^{-4}$	$1.9 \cdot 10^{-6}$	0.3	bones	very long	v,s (τ_D>15)
iodine	$1,1 \cdot 10^{-5}$	$2 \cdot 10^{-7}$	1	thyroid	120	
caesium	$1.5 \cdot 10^{-6}$	$1 \cdot 10^{-8}$	1	whole body	110	
polonium	-	-	0.1	liver, kidney spleen, red blood cells	50	
radium	$3.1 \cdot 10^{-14}$	$2.3 \cdot 10^{-15}$	0.2	bones	very long	v,s (τ_D>15)
thoron	-	$3 \cdot 10^{-9}$	$2 \cdot 10^{-4}$	bones	8000	s
uranium	$9 \cdot 10^{-9}$	$1.9 \cdot 10^{-9}$	0.05	bones	20	v
plutonium	-	-	10^{-4}	liver, bones	very long	s
americium	-	-	$5 \cdot 10^{-4}$	liver, bones	very long	s
californium	-	-	$5 \cdot 10^{-4}$	liver, bones	very long	s

Table 21.2. Organ- and tissue-masses of the reference man. (After ICRP 23)

organ	mass/kg
whole body	70
ovaries	0.011
testes	0.035
muscles	28
red bone-marrow	1.5
lungs	1
thyroid	0.02
liver	1.8
skin	2.6
corticular bones	4
trabecular bones	1
bone lining	0.12

with ^{222}Rn which is – being a noble gas – not expected to participate in the body's metabolism. Experimental findings indicate, however, that it may be stored in fatty tissue (retention fraction 0.3) so that the daughter products may enter the metabolism.

The data in Table 21.1 may also be used to gain a lower limit of the biological half-life: If it is assumed that the amount of a compound which is taken up is entirely added to the body's inventory – which is generally not true –, the following relation holds:

$$\frac{dm}{dt} = \dot{m} - \lambda_B m \tag{21.4}$$

where m is the total mass (column 2), m the daily intake (column 3) and λ_B the biological "decay" coefficient. In the steady state of an adult organism, the time derivative vanishes so that:

$$\lambda_B \approx \frac{\dot{m}}{m}$$

and:

$$\tau_B \approx \frac{\dot{m} \ln 2}{m} \quad . \tag{21.5}$$

The values estimated in this way are usually too small because of the simplifying assumptions but they may provide a guideline.

21.2 Dose Estimates

The linear models described in the preceding section may be used to determine the activity distributions after incorporation of radionuclides in the different organs of the body. They represent internal radiation sources ("source organs" SO). In many cases they are identical with those where most of the biological damage is recorded. These are called "target organs" (TO). The doses

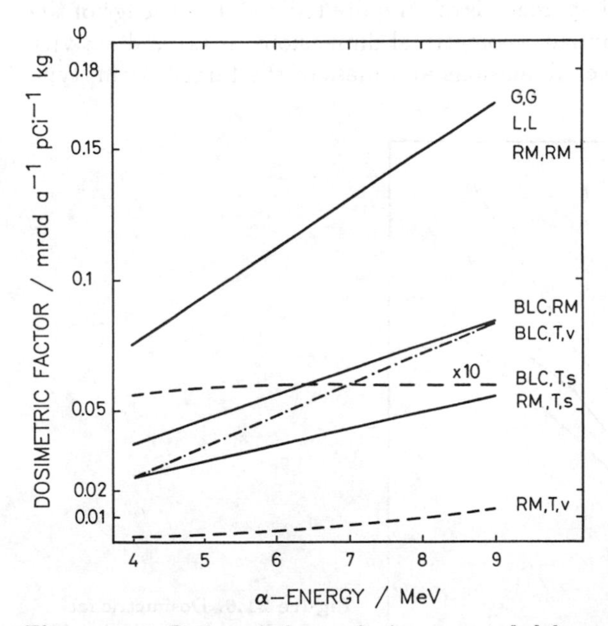

Figure 21.4. Dosimetric factors for incorporated alpha emitters. The first letter indicates the source-, the second the target organ. G: gonads, L: lung, RM: red bone marrow, BLC: bone lining cells, T: trabecular bones. In the last case, factors are given for deposition on the surface (s) and in the volume (v) (after UNSCEAR 1977)

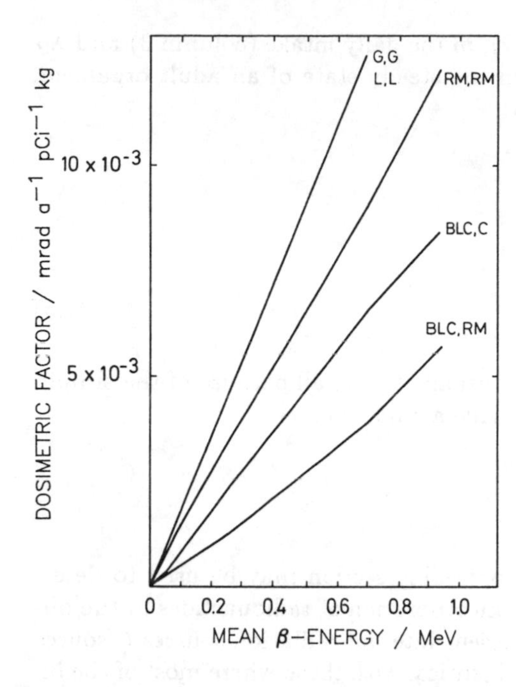

Figure 21.5. Dosimetric factors for beta emitters (see legend to Fig. 21.4). C: compact bone (after UNSCEAR 1977)

may be calculated in principle according to the rules described in Chap. 4 although this straightforward approach is often quite tedious. Knowledge of the following parameters is required: geometrical dimensions, mass and activity of the source organ, distance, dimensions and mass of the target organ, type

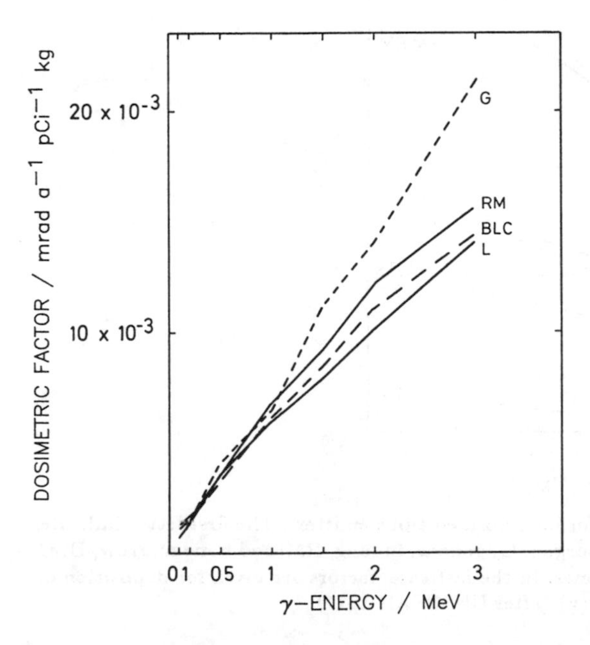

Figure 21.6. Dosimetric factors for homogeneously districuted gamma emitters with different target organs (after UNSCEAR 1977)

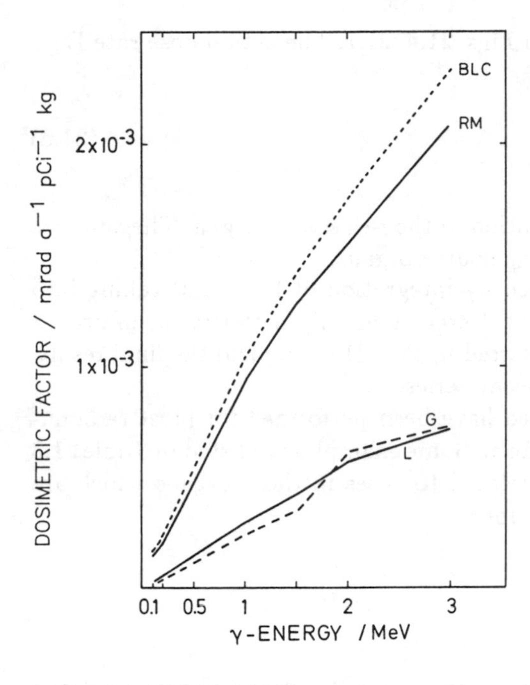

Figure 21.7. Dosimetric factors for gamma emitters homogeneously distributed in the skeleton with different target organs (after UNSCEAR 1977)

and energy of radiation. Table 21.2 lists the masses of some key organs of the "reference human".

To facilitate the practical computations, conversion functions ψ were derived which relate source organ activity and target organ dose for a few

Table 21.3. Dose factors for conversion of radioactivity taken up by ingestion in effective whole body or organ doses. Calculations for adults (a) (NOSSKE, GERICH and LANGER, 1985) and for children (c) of 1 year (HENRICHS et al., 1985)

nuclide	whole body/Sv Bq^{-1} a	c	organ/Sv Bq^{-1} a	c	affected organ
^{3}H	1.6×10^{-11}	5.2×10^{-11}	-	-	-
^{32}P	2.4×10^{-9}	2×10^{-8}	8.1×10^{-9}	7.9×10^{-8}	bone marrow
^{60}Co	7.3×10^{-9}	8.4×10^{-8}	1.3×10^{-8}	1.1×10^{-7} a)	colon
^{90}Sr	3.5×10^{-8}	1.1×10^{-7}	1.7×10^{-7}	1×10^{-6} b)	bone marrow
^{125}I	9.4×10^{-9}	4.4×10^{-8}	3.1×10^{-7}	1.3×10^{-6}	thyroid
^{129}I	6.7×10^{-8}	2.8×10^{-7}	2.2×10^{-6}	8.5×10^{-6}	thyroid
^{131}I	1.3×10^{-8}	1.1×10^{-7}	4.3×10^{-7}	3.5×10^{-6}	thyroid
^{137}Cs	1.4×10^{-8}	9.3×10^{-0}	-	-	-
^{210}Po	5.1×10^{-7}	3.3×10^{-6}	2.5×10^{-6}	1.2×10^{-5}	kidney
^{226}Ra	3.6×10^{-7}	2.6×10^{-6}	6.8×10^{-6}	4.6×10^{-5}	bone lining
^{232}Th	7.4×10^{-7}	1.5×10^{-6}	1.6×10^{-5}	3.5×10^{-5}	bone lining
^{235}U	7.2×10^{-8}	3×10^{-7}	1×10^{-6}	3.5×10^{-6}	bone lining
^{239}Pu	1.2×10^{-7}	3.6×10^{-7}	2.1×10^{-6}	4.8×10^{-6}	bone lining
^{241}Am	5.9×10^{-7}	1.7×10^{-6}	1×10^{-5}	2.5×10^{-5}	bone lining
^{252}Cf	1.4×10^{-7}	9.3×10^{-7}	2.1×10^{-6}	1.2×10^{-5}	bone lining

a) red bone marrow
b) bone lining

important cases. They are given in Figs. 21.4–21.7. The actual dose rate D_{TO} in a target organ is then:

$$\dot{D}_{TO} = \sum_i \psi_{i(so\ \ to)} \cdot C_{iso} \qquad (21.6)$$

where C_{iso} is the activity concentration in the i-th source organ. The sum has to be extended over all contributing source organs.

The long-term load is obtained by integration of Eq. (21.6) taking into account that C_{so} is not constant but decreases usually with time as indicated by Eqs. (21.3). They have to be inserted in Eq. (21.6) to yield the final result. Attention has also to be paid to decay series.

The calculation just mentioned have been performed for most radionuclides and are available in tabular form. Some examples are listed in Table 21.3 where incorporated activities are related to doses in those organs which are most relevant for the particular element.

21.3 Special Actions

In principle one would, of course, not expect qualitatively different reactions of incorporated radionuclides compared to external irradiation. The special spatial situation, particularly the often short distances, creates, however, a unique exposure geometry and the specificity of incorporations makes some sites in the body more damage prone than others.

Acute radiation effects are practically negligible because of the usually low doses so that only genetic alterations and carcinogenesis have to be discussed. The intimate contact between source and target increases the significance of α- and β-emitting radionuclides which play a minor role with external irradiation because of their small penetration ability. In addition, the high RBE of α particles increases their risk potential even further. This is at least the case for carcinogenesis while for genetic actions β and γ radiation are more important since α-emitting nuclides are not selectively incorporated in the gonads.

With regard to carcinogenesis, there is a certain organ specificity of some radionuclides which is related to their uptake and final distribution in the body. Table 21.4 lists a few examples.

The lung carries, of course, the highest risk with inhalation. α-emitters play here an eminent role. Calculating the absorbed dose using the models described and comparing the action with that of inhaled β or γ emitters, leads to a RBE of about 10 which is compatible with the results of cellular studies.

Lung cancer as a consequence of the inhalation of radioactive particles constitutes a prominent hazard in uranium mining. Historically, this was one of the earliest examples to demonstrate the life-threatening danger of incor-

Table 21.4. Tumours found in experimental animals after nuclide uptake from the blood. (After UNSCEAR, 1977)

nuclide	organ or tumour	active radiation
^{90}Sr	bones	β
^{45}Ca	bones	β
^{226}Ra	bones	α
^{232}Th	liver, spleen, lung	α
^{239}Pu	bones	α
^{131}I	thyroid	β, γ

porated radioactivity ("Schneeberger disease", first detected in the mines of Joachimsthal, today in Czechoslowakia). The long-term investigation over decades provided quite reliable data for risk estimates. The most important single component is ^{222}Rn, whose solid daughter products attach to aerosols. Because of historical reasons the exposure level is given as "working level month" (WLM) with one month taken as 170 working hours (see also Chap. 1).

$$\begin{aligned} 1\,\mathrm{WLM}\,(^{222}\mathrm{Rn}) &= 100\,\mathrm{pCi\,dm^{-3}} \cdot 170\,\mathrm{h} \\ &= 1.7 \times 10^4\,\mathrm{pCi\,dm^{-3}} \cdot \mathrm{h} \\ &= 6.3 \times 10^5\,\mathrm{Bq} \cdot \mathrm{m^{-3}} \cdot \mathrm{h} \quad . \end{aligned}$$

It is thus estimated that 1 WLM is equal to about 0.01 Gy of an α dose in the broncheal epithelium. The studies revealed a lung cancer incidence of 200–450 $\times 10^{-6}$ per WLM over 40 years. With a RBE of 10 this figure coincides with the data given in Table 20.3. Incidentally, the risk was higher by about a factor of 8 in heavy smokers (more than 20 cigarettes per day).

21.4 Conclusions

The incorporation of radionuclides, if they enter the food chain, constitutes without doubt a major potential hazard, both for those occupationally exposed and for the general public. The models described and the calculations based on them as well as empirical data from animals and humans – e.g. by investigating the distribution of natural radioactivity – allow a comparatively reliable estimate of retention probabilities in the body and the doses expected. Uncertainties seem still to exist for transuranium elements although more data are now coming in, particularly for plutonium.

All this is, of course, of utmost importance in radiation protection. It is thus possible to estimate the limit of incorporation beyond which the recommended dose limits (see Chap. 23) are surpassed. Since the activity remains in the body for some time, one has to integrate over a reasonable time interval which is usually taken as 50 years. This leads to "annual limits of intake"

Table 21.5. Assumed average consumption habits in the Federal Republic of Germany. (After BONKA, 1982)

assumed consumption per year	
vegetable products without leaf vegetables	210 kg
of that root-crop	(87 kg)
cereals	(96 kg)
fruit and fruit juice	(27 kg)
leaf vegetables	21 kg
milk	110 l
(milk for small children	300 l)
meat etc.	75 kg
fish (fresh-water)	1.3 kg
drinking-water	440 l
breathing rate	0.83 m^3/h
breathing rate of small children	0.22 m^3/h

(ALI) (Chap. 23). To relate them to activity concentrations in food, water or air, assumptions have to be made about consumption habits. A certain guideline is given in Table 21.5 which is thought to be more or less representative for the Federal Republic of Germany. There are, of course, wide variations which depend on personal life-style and regional habits. Because of these uncertainties, limiting values for concentrations in drinking water are generally no longer recommended (they may still exist on a national basis). This reservation is not valid for air, "derived air concentrations" may be calculated on the basis of the average breathing volume. The whole issue will be taken up again in Chap. 23.

Further reading

BEIR 1974, BONKA 1982, ICRP 1979, 1980, PRICHARD and GESELL 1984, UNSCEAR 1977, 1982

Chapter 22
Radioecology

This chapter deals with radiation in the environment. The treatment requires a systematic approach whose fundamental principles are outlined. Because of the manifold interrelations it becomes necessary to use simplifying models and calculational methods. Specifically, possible changes in the solar spectrum due to atmospheric reactions are described as well as natural and anthropogenic sources of ionizing radiation. The distribution of radioactivity with particular emphasis on the food chain is discussed laying thus the basis for realistic risk estimates.

22.1 General Aspects

The interaction between radiation and life has been part of organic evolution on Earth from the very start; including that of human beings. Radiation protection can, therefore, not be seen as an isolated discipline but only in the context of all the physical and chemical reactions proceeding constantly in the environment. The development of our planet and biological evolution have created a finely tuned steady state whose change – even at a seemingly irrelevant component – may lead to consequences which would not have been predictable by an isolated approach. Radiation ecology is, therefore, system theory at a global scale; consequently, it also makes ample use of the mathematical methods developed in this – originally technical – discipline.

To illustrate the situation, Fig. 22.1 gives a very simplified diagramme of the interacting network between radiation and life. Mankind is put into the center because the hazard to humans is the main concern of this book. Only the most important influences are sketched. The sun is the most important source of radiation, not only because it enables life through heat production and via photosynthesis but also because it causes ozone formation via photochemical reactions in the atmosphere, thus providing a protection shield against the damaging short-wave UV fraction of the spectrum. Humans are – and also have been – exposed to ionizing radiation in many ways, from outer space and the sun, from soil and building materials, from radioactive substances in food and air and from radiation sources in technology and medicine.

Risk estimates require, in principle, the complete knowledge of all inter-

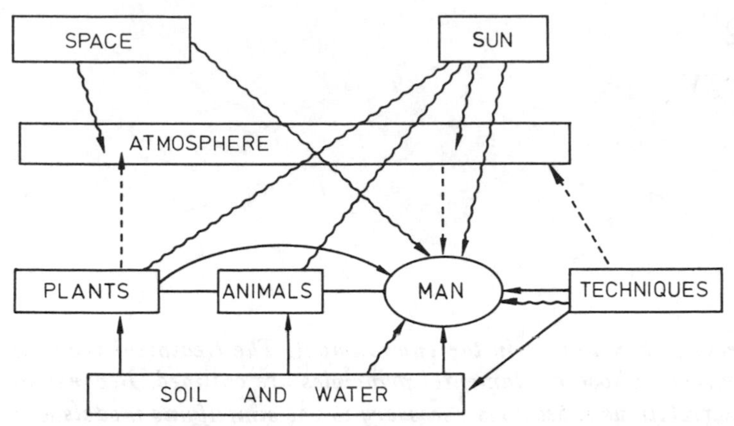

Figure 22.1. Radiation in the ecosphere. Wavy lines: direct radiation, heavy lines: radionuclide transport, broken lines: transport of chemicals with radioecological relevance

acting transfer pathways, both qualitatively and quantitatively. Since this is obviously not achievable, the only approach is by model calculations. These must be continuously checked against the real developments. If they prove to be reliable to some extent, they may serve to estimate exposure levels. The biological risk may then be assessed on the basis of radiobiological data – if available, although this final step involves extrapolation over orders of magnitude and is not possible without model assumptions. Examples are numerously found throughout this book.

Radiation hazards are quite often – particularly in the public domain – only discussed with regard to nuclear energy. This falls quite short of reality and does not give justice to the real problems. It is fair to say that many decades of radiobiological research laid the basis for a very sophisticated system of hazard evaluation which is unsurpassed by any other environmental impact. The regulations are constantly revised but the sensitivity of measuring devices and the solid body of radiobiological data already allow a high degree of reliability. This statement, however, should not obscure the fact that radioecology is an enormously complicated branch of science. An impressive example is the transfer in air or water which is treated in more detail in Sect. 22.3.3.

22.2 Optical Radiation

Radioecological problems involving optical radiation are restricted to photochemical processes in the atmosphere. Ozone (O_3) plays here a central role. The most important reactions were already discussed in Sect. 5.1.4. O_3 is photochemically formed by light absorption in the wavelength range around 240 nm. The maximum of its absorption spectrum (Fig. 22.2) coincides roughly with that of nucleic acids; compared to O_2 it is shifted towards

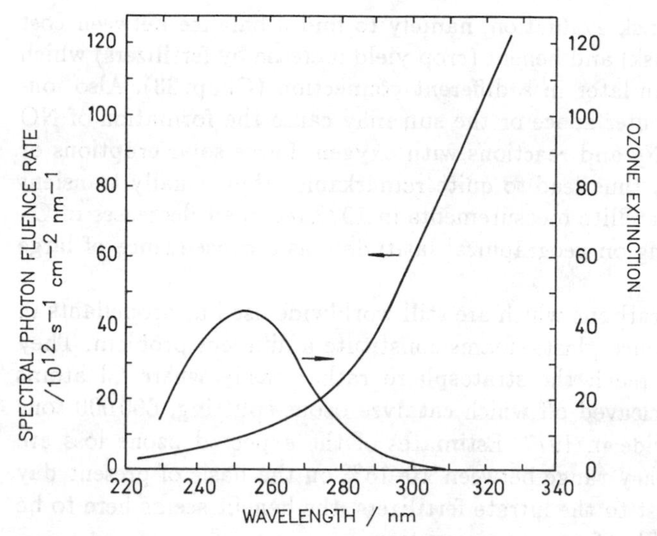

Figure 22.2. Absorption spectrum of ozone and the expected solar spectrum on earth in the complete absence of ozone (after SELIGER 1977)

longer wavelengths. This means that ozone formation is favoured in the upper layers of the atmosphere while its destruction prevails in lower regions since both reactions are part of a complete steady state. The Earth's ozone layer equals about a pressure of 3 mm mercury. On a global scale, it reduces solar radiation at 265 nm by a factor of 10^{40}! A reduction of the ozone belt would lead to a shift of the solar spectrum on the Earth's surface towards shorter wavelengths. It is estimated that a decrease by 10% would double the fraction of solar radiation at 290 nm. Figure 22.2 shows the expected solar radiation fluence rate on Earth if the ozone were completely destroyed. A comparison with Fig. 1.3 illustrates the impressive changes.

Ozone destruction is caused by a number of substances, with nitrous oxides and chlorofluorocarbons being the most important. The former are produced by combustion processes, the decay of nitrates by plants and also by photochemical reactions in the air. It was also feared that high-flying supersonic aircraft may inject large masses of NO into the stratosphere thus causing further ozone decay. This problem appears to be no longer relevant since supersonic transport is no longer developed at a larger scale.

The artificial fertilizers pose a more real and quite relevant problem. The increased application of nitrate fertilizers leads to a higher and higher release of N_2O on a global scale. It moves upward to the stratosphere where it reacts with excited oxygen to NO which in turn catalyzes O_3 decomposition. Changes in the ozone layer take very long times but once initiated they cannot be stopped. Careful estimates forecast O_3 reduction between and 1% and 10% for the years 2025 to 2050. Taking all this into account, it is obvious that the expected changes are by no means negligible. This illustrates a central problem of risk evaluation, namely to find a balance between cost (increased radiation risk) and benefit (crop yield increase by fertilizers) which will be taken up again later in a different connection (Chap. 23). Also ionizing radiation from outer space or the sun may cause the formation of NO via the excitation of N_2 and reactions with oxygen. Large solar eruptions or supernova events may thus lead to quite remarkable – but usually transient – ozone reductions. Satellite measurements in 1972 recorded decreases of O_3 up to 20% – depending on geographical latitude – as a consequence of large solar flare of protons.

The chlorofluorocarbons which are still worldwide used as propellants in spraycans and to produce plastic foams constitute a different problem. They are quite stable and reach the stratosphere rather easily where Cl atoms are photochemically cleaved off which catalyze ozone-splitting. 600 000 tons were released worldwide in 1977. Estimates of the expected ozone loss are again very difficult, they range between 10–15% on the basis of present day production. In contrast to the nitrate fertilizers, the benefit seems here to be quite clearly on the side of ozone preservation.

Estimating the biological risk is not easy. There are a number of studies on the effect of increased short wavelength UV in plant communities and marine ecosystems. The results are so far inconclusive, it appears that there is a high buffering capacity for varying amounts of UV light. The expected increase of skin tumours in humans appears to be the most important risk. Quantitative figures are difficult to obtain since the incidence-fluence relationship is not well known for the wavelengths involved. A certain estimate may be obtained if the incidence of skin tumours is compared in countries with naturally different sunlight conditions. An example of such a study is shown in Fig. 22.3 where tumour incidence is plotted versus yearly fluence of UV-B. There is an obvious relationship but it must be pointed out that it is only valid for non-melanoma tumours.

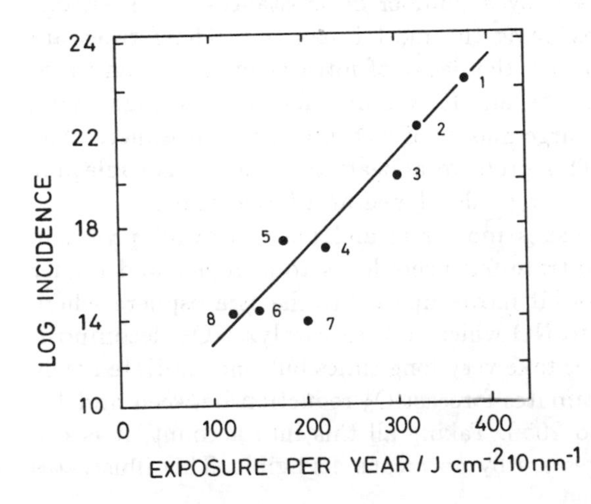

Figure 22.3. Skin cancer incidence as a function of UV-B exposure in different countries. 1: Australia, 2: Texas, 3: South Africa, 4: Nevada, 5: Canada, 6: England, 7: Germany, 8: Scotland (after GORDON 1976)

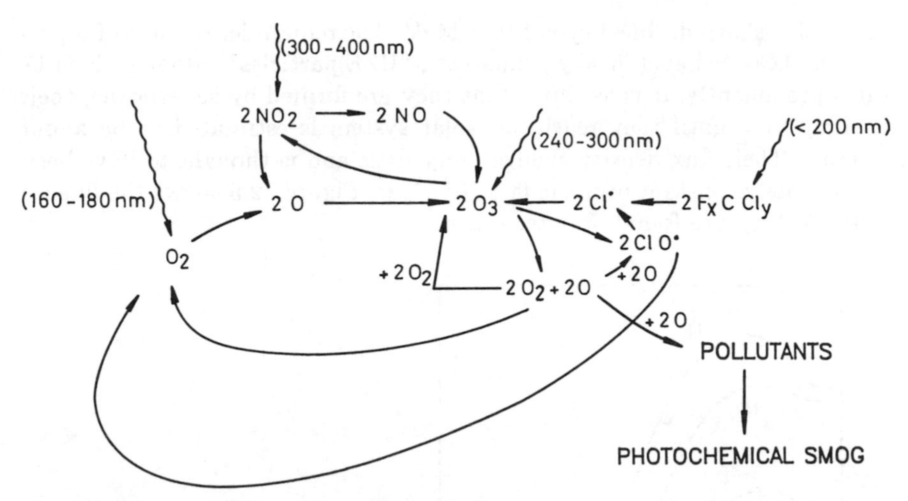

Figure 22.4. A simplified scheme of photochemical reactions in the atmosphere

Another consequence of atmospheric photochemical reactions is the formation of "photosmog". It occurs in the lower layers of the atmosphere and starts with the photolysis of NO_2. The atomic oxygen which is thus formed reacts readily with pollutants like NO_2, CO, SO_2, etc. starting chain reactions whose final products (H_2O_2, HNO_3, aldehydes, ketones and others) are noted as "smog".

Figure 22.4 makes an attempt to summarize the photochemical pathways just described in a simple way.

Further reading
BJOERN and BORMAN 1983, FABIAN 1980, WORREST and CALDWELL 1986, SELIGER 1977

22.3 Ionizing Radiations

22.3.1 Natural Radiation Burden

22.3.1.1 Cosmic Rays

Cosmic rays are an important component of environmental radiation. They originate either in the universe or the sun. Upon entry into the atmosphere nuclear reactions occur which give rise to a cascade of secondary particles. Apart from this, a number of radionuclides are formed which contribute to the terrestrial radiation.

The galactic component consists to about 90% of protons covering an energy range between 1 and 10^{14} MeV with a broad maximum around 300 MeV and a rather sharp decline beyond 1000 MeV. The remainder is made of α particles and heavier ions ("heavy primaries", "HZE particles") among which Fe figures prominently. It is assumed that they are formed by supernovae, their travelling time until they reach our solar system is estimated to be about 10^7 years. Their flux density changes very little and is thought to have been approximately constant over the last 10^9 years. Figure 22.5 shows the fluence spectra as they are found in outer space.

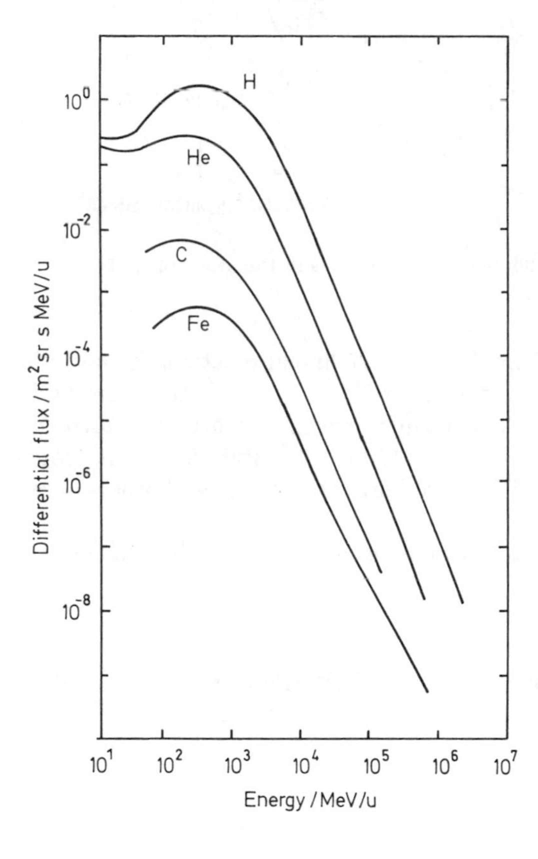

Figure 22.5. Differential fluence spectra for ions in outer space (after SIMPSON 1983)

The solar component consists also mainly of protons but with a considerably smaller maximum energy of about 40 MeV; additionally α particles are found. The intensity is not constant but varies with the solar cycle. Because of the lower particle energy, most of the radiation does not reach the Earth's surface so that the contribution to the environmental radiation is small. Its significance increases with height. Solar ionizing radiation has, therefore, to be taken into account with higher altitude flights and particularly with manned space missions. The solar cycle also influences the galactic component. It is modulated by the strong electric and magnetic fields in such a way that its intensity is smaller at a solar maximum. Apart from this regular and largely predictable behaviour, there are also sudden eruptions ("solar flares") during which the flux of solar particles is dramatically increased. These may pose a serious hazard to astronauts.

The magnetic field of the Earth exerts significant influence on the strength and the composition of cosmic radiation. Particles of low energy are reflected before they enter the atmosphere. Others are captured in the inhomogeneous field leading to two radiation belts ("van-Allen belts") at altitudes of 1.2 and 8 Earth radii (7600 and 51 000 km, respectively). Since the field strength varies with geographical latitude, this is also the case with the particle density within the belts. More important for the terrestrial exposure is the fact that the non-captured more energetic particles are deviated towards the poles resulting in a strong dependence on geographical latitude, particularly at higher altitudes. The fact that the magnetic field is not symmetrical around the Earth leads to perturbations also in the particle distribution so that there is an additional variation with geographical position. A famous example is the increased particle density above the South Atlantic ("South Atlantic anomaly").

The primary cosmic radiation plays only a negligible role at the Earth's surface, the products of the nuclear reactions in the atmosphere are considerably more important. They may be subdivided into a directly ionizing (electrons, μ mesons) and an indirectly ionizing component (mostly neutrons). The average value of the first amounts to about 2.1 ionizations per cm^3 and

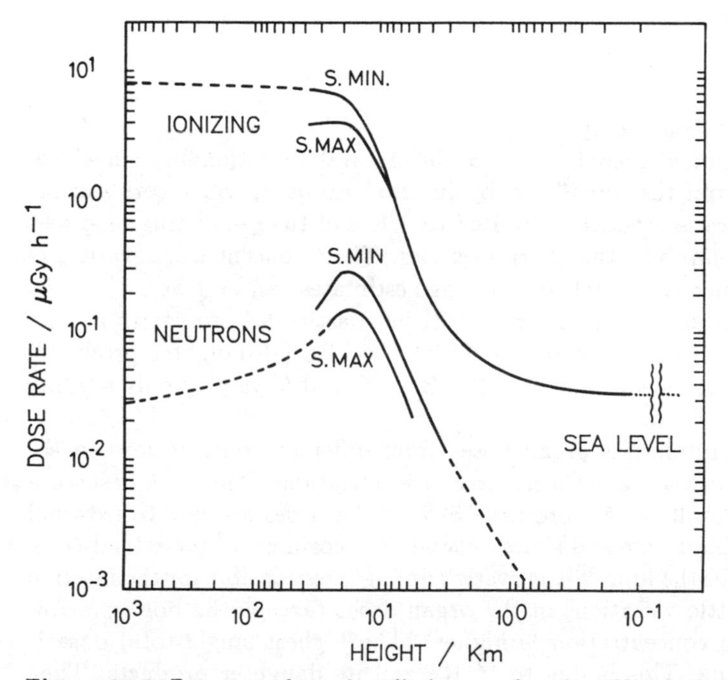

Figure 22.6. Dose rate of cosmic radiation as a function of height for periods of high and low solar activity for a latitude of 50 degrees. Measurements from 1969 (max) and 1945 (min) (after UNSCEAR 1977)

second at sea level. Since 33.7 eV are spent per ionization in air, the mean dose rate is calculated as $3.2 \times 10^{-8}\,\mathrm{Gy\,s^{-1}} = 2.8 \times 10^{-4}\,\mathrm{Gy\,y^{-1}}$. Neutrons do not play a significant role at sea level but only at higher altitudes. Figure 22.6 summarizes the relations, it shows also the variations with the solar cycle.

22.3.1.2 Terrestrial Radiation

Terrestrial radiation consists of two components. One stems from those radionuclides which were formed at the origin of our planet and persisted because of their long half-lives plus their decay products (*primordial radionuclides*), the other contribution is due to those nuclides which are formed by nuclear reactions between atmospheric atoms and cosmic particles (*cosmogenic nuclides*). The first group consists of the natural decay series (see Chap. 1) – except the neptunium series which is extinct because of the short half-life of the starting product – and a few others, essentially only ^{40}K and ^{87}Rb ("radiofossiles").

The most important cosmogenic products are tritium (^{3}H) and radiocarbon (^{14}C). They are formed via the following reactions:

$$^{14}_{7}\mathrm{N} + \mathrm{n} \quad \rightarrow \quad ^{12}_{6}\mathrm{C} + ^{3}_{1}\mathrm{H}$$

and

$$^{14}_{7}\mathrm{N} + \mathrm{n} \quad \rightarrow \quad ^{14}_{6}\mathrm{C} + ^{1}_{1}\mathrm{H}$$

depending on neutron energy.

Terrestrial nuclides contribute to the natural radiation burden either by irradiation from the outside or by internal exposure via incorporation. External exposure is practically limited to ^{40}K and the γ-emitting members of the two uranium and the thorium series. Their concentration varies, of course, geographically so that only average estimates can be given.

Internal exposure is due again to ^{40}K but particularly to ^{222}Rn and its daughter products, less important are ^{226}Ra, ^{220}Rn (+ daughter products) and ^{87}Rb. The cosmogenic nuclides ^{3}H, ^{7}Be, ^{14}C and ^{22}Na play only a minor role.

Table 22.1 summarizes organ doses from different contributions in so-called "normal" areas, i.e. with no excessive deviations. The data deserve a few comments: In all cases more than 50% of the doses are due to external radiation with about an equal share between the cosmic and terrestrial component. Except for the lung ^{40}K is overwhelmingly responsible for the internal exposure with little variations in the organ doses (except the bone marrow where potassium concentration is higher). The highest single total dose is found for the lung. This is due to ^{222}Rn and its daughter products. They do not only add to the dose but they change also significantly the radiation quality since they are mostly α emitters (see last line of the table).

Table 22.1. Organ doses from natural environmental radiation background in 'normal areas' (10^{-5} Gy a^{-1}). (After UNSCEAR, 1977)

	gonads	lung	bone liming	red bone marrow
external exposure				
cosmic radiation:				
ionizing component	28	28	28	28
neutrons	0.35	0.35	0.35	0.35
terrestrial γ-radiation	32	32	32	32
internal exposure				
cosmogenic radionuclides:				
$^{3}H(\beta)$	0.001	0.001	0.001	0.001
$^{7}Be(\gamma)$	–	0.002	–	–
$^{14}C(\beta)$	0.5	0.6	2.0	2.2
$^{22}Na(\beta+\gamma)$	0.02	0.02	0.02	0.02
primordial radionuclides:				
$^{40}K(\beta+\gamma)$	15	17	15	27
$^{87}Rb(\beta)$	0.8	0.4	0.9	0.4
^{238}U–$^{234}U(\alpha)$	0.04	0.04	0.3	0.07
$^{230}Th(\alpha)$	0.004	0.004	0.8	0.05
^{226}Ra–$^{214}Po(\alpha)$	0.03	0.03	0.7	0.1
^{210}Pb–$^{210}Po(\alpha+\beta)$	0.6	0.3	3.4	0.9
^{222}Rn–$^{214}Po(\alpha)$ inhalation	0.2	30	0.3	0.3
$^{232}Th(\alpha)$	0.004	0.04	0.7	0.004
^{228}Ra–$^{208}Tl(\alpha)$	0.06	0.06	1.1	0.2
^{220}Rn–$^{208}Tl(\alpha)$ inhalation	0.008	4	0.1	0.1
sum (rounded)	78	110	86	92
% densely ionizing contribution	1.2	31	8.5	2.1

As stated above there are quite considerable geographical variations of the natural radiation environment. One reason is the dependence of cosmic radiation on latitude but more important is the varying content of radionuclides of soils and minerals. There are areas, e.g. in India, Brazil and China, where the environmental radiation background is extremely high. It would be of great interest to study health effects in the indigenous populations. Unfortunately, there are only very few usable epidemiological investigations of this kind. The limited data available do not indicate higher morbidity or mortality rates compared to control populations.

22.3.2 Anthropogenic Radiation Sources

22.3.2.1 Nuclear Energy

The scientific basis and the technical realizations of nuclear energy production will not be described here. All reactors make use of nuclear fission as energy source, the actual constructions vary. Nearly all of the commercial plants – at least in the Western Hemisphere – have water as moderator and primary coolant, the most important types are the "boiling-water reactor" (BWR) and the "pressurized-water reactor" (PWR). A few new versions are presently being developed but they are still in the prototype stage and will not be considered any further. A general discussion of nuclear safety is outside the scope of this book, only radiobiological aspects are addressed.

Although the fission process is potentially the largest source of radioactivity, the contributions of fuel production and reprocessing must not be neglected. The various stages of the nuclear fuel cycle are depicted in Fig. 22.7. There is a number of nuclides involved, the more important ones are listed in Table 22.2. To give an account of the radioactivities, the average emissions are normalized to 1 MW (megawatt) electrical energy. Respective values as calculated from worldwide determinations are given in Table 22.3, limited to the more important nuclides.

Although the released radioactivity is certainly important the half-lives have also to be taken into account. ^{129}I has to be particularly mentioned

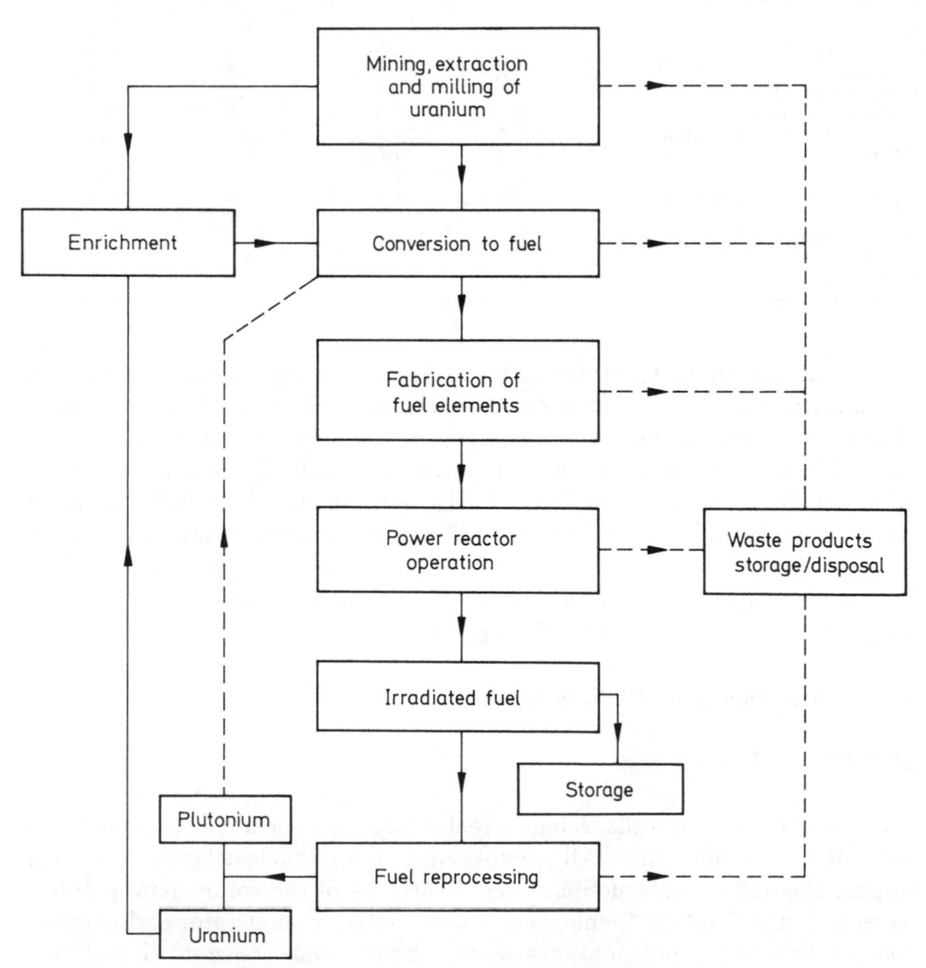

Figure 22.7. The nuclear power fuel cycle (after WHO 1978)

Table 22.2. Radionuclides playing a role in the nuclear fuel cycle. (After UNSCEAR, 1977)

nuclide	half-life	origin
^{3}H	12.5 y	nuclear reaction in the reactor reprocessing
^{85}Kr	10.8 y	nuclear fission, reprocessing
^{90}Sr	27.7 y	nuclear fission
^{95}Zr	65.6 d	nuclear fission
^{129}I	1.7 x 10^{7} y	nuclear fission, reprocessing
^{131}I	8.1 d	nuclear fission
^{133}Xe	5.3 d	nuclear fission
^{137}Cs	30 y	nuclear fission
^{222}Rn	3.8 d	fuel production
^{226}Ra	1600 y	fuel production
^{230}Th	8 x 10^{4} y	fuel production
^{237}Np	2.1 x 10^{6} y	nuclear reactions
^{239}Pu	2.4 x 10^{4} y	nuclear reactions, reprocessing
^{241}Am	458 y	nuclear reactions, reprocessing

here. It is liberated in small quantities during reprocessing (no significant quantities could be found during reactor operation), namely about 10^7 Bq y^{-1} per 1 MW electrical energy, but it is dispersed globally and has a half-life of 1.7×10^7 years.

Table 22.3. Released activity per year per 1 MW electrical power in the fuel cycle (Bq y^{-1}). (After UNSCEAR, 1977)

	air		water	
fuel production	6.3 x 10^{9}	^{222}Rn	3.1 x 10^{6}	U
			3.7 x 10^{5}	^{234}Th
			3.7 x 10^{5}	^{234}Pa
BWR-reactor a)	3.5 x 10^{13}	noble gases (^{135}Xe, ^{88}Kr, ^{131}Xe, ^{138}Xe)	3.7 x 10^{9}	^{3}H
			9.3 x 10^{8}	^{137}Cs
			6.4 x 10^{8}	^{134}Cs
	1.9 x 10^{9}	^{3}H	2.3 x 10^{8}	^{60}Co
	6.7 x 10^{8}	^{135}I		
	5.6 x 10^{8}	^{133}I		
PWR-reactor b)	5.6 x 10^{11}	noble gases (^{133}Xe)	4.4 x 10^{10}	^{3}H
			8.5 x 10^{7}	^{131}I
	7 x 10^{9}	^{3}H	6.3 x 10^{7}	^{137}Cs
	1.8 x 10^{7}	^{131}I	4.8 x 10^{7}	^{133}I
			3.7 x 10^{7}	^{134}Cs
reprocessing	1.5 x 10^{13}	^{85}Kr	5.1 x 10^{11}	^{137}Cs
	3.1 x 10^{10}	^{3}H	5 x 10^{11}	^{3}H
			4.5 x 10^{11}	^{106}Ru
			3.8 x 10^{11}	^{95}Zr
			3.5 x 10^{11}	^{95}Nb
			2.3 x 10^{11}	^{90}Sr
			2 x 10^{11}	^{144}Ce
			8.5 x 10^{10}	^{134}Cs

a) BWR: boiling water reactor
b) PWR: pressurized water reactor

The estimation of expected doses to the general population is complicated. Complex transfer and distribution models are necessary which have especially to take into account that the nuclides are incorporated in the food chain. Because of the different half-lives one has to differentiate between *local* (i.e. in the immediate vicinity of the plant), *regional* and *global* effects. A useful quantity in this context is the *collective dose* (more fully discussed in the next chapter). It is calculated by multiplying the dose value of a specified area by the number of people affected and summing over all areas considered.

Table 22.4. Normalized collective dose commitments with nuclear reactor operation (man Gy per 1 MW electrical power per year). (After UNSCEAR, 1977)

	gonads	lung	thyroid	bone marrow	bone lining cells	% a)
regionally						
ore processing:						0.02
U, Ra, Th	10^{-7}	4×10^{-7}	10^{-7}	2×10^{-7}	2×10^{-6}	
Rn	8×10^{-6}	10^{-3}	8×10^{-6}	8×10^{-6}	8×10^{-6}	
fuel production (U)	-	10^{-7}	-	4×10^{-8}	2×10^{-7}	10^{-4}
operation: air:						5.6
Kr, Xe, ^{41}Ar	2×10^{-3}	2×10^{-3}	2×10^{-3}	2×10^{-3}	2×10^{-3}	
^{3}H	4×10^{-5}	4×10^{-5}	4×10^{-5}	4×10^{-5}	4×10^{-5}	
^{14}C	6×10^{-6}	7×10^{-6}	6×10^{-6}	2×10^{-5}	2×10^{-5}	
^{131}I	-	-	1×10^{-3}	-	-	
Cs, Sr, Co, Rn	6×10^{-5}	6×10^{-5}	6×10^{-5}	6×10^{-5}	6×10^{-5}	
operation: water:						1.1
^{3}H	3×10^{-4}	3×10^{-4}	3×10^{-4}	3×10^{-4}	3×10^{-4}	
Cs, Co, Mn, I	10^{-4}	10^{-4}	2×10^{-4}	10^{-4}	10^{-4}	
reprocessing: air:						0.3
^{85}Kr	7×10^{-6}	2×10^{-5}	7×10^{-6}	10^{-5}	10^{-5}	
^{3}H	2×10^{-6}	2×10^{-6}	2×10^{-6}	2×10^{-6}	2×10^{-6}	
^{14}C	10^{-5}	2×10^{-5}	10^{-5}	6×10^{-5}	5×10^{-5}	
^{131}I, ^{129}I	-	-	2×10^{-3}	-	-	
Cs, Ru, Sr	2×10^{-6}	4×10^{-6}	2×10^{-6}	6×10^{-5}	8×10^{-5}	
processing: water:						6.3
^{3}H	4×10^{-4}	4×10^{-4}	4×10^{-4}	4×10^{-4}	4×10^{-4}	
^{129}I	-	-	3×10^{-3}	-	-	
Cs, Ru, Sr	9×10^{-4}	9×10^{-4}	9×10^{-4}	2×10^{-3}	2×10^{-3}	
transport:						0.08
globally						
operation and processing:						85.5
^{3}H	10^{-3}	10^{-3}	10^{-3}	10^{-3}	10^{-3}	(2.6)
^{85}Kr	9×10^{-4}	2.5×10^{-3}	9×10^{-4}	1.5×10^{-3}	1.5×10^{-3}	(3.9)
^{14}C	9×10^{-3}	9×10^{-3}	9×10^{-3}	3×10^{-2}	3×10^{-2}	(79)
^{129}I			5×10^{-3}			
sum:	1.5×10^{-2}	1.7×10^{-2}	2×10^{-2}	3.8×10^{-2}	3.8×10^{-2}	

a) % contribution to bone marrow dose

The unit is "man Gy". Respective values estimated worldwide for various nuclides and the different stages of the nuclear fuel cycle are given in Table 22.4. They show that the collective doses are small regionally but this is somewhat misleading since only small populations are involved. Nevertheless, more than 70% of the calculated burden is due to the long-lived ^{3}H, ^{85}Kr, ^{14}C and ^{129}I. The last one is – as pointed out before – an interesting example: Although it is emitted in very small amounts, it contributes to 79% of the collective bone-marrow dose.

The maximum collective dose per 1 MW electrical energy is 0.038 man Gy (bone marrow and bone lining). There was a total installed power of 80 000 MW worldwide in 1977. On an assumed basis of 100% operation (70% is probably an upper limit), a total collective bone-marrow dose of 3×10^3 man Gy is obtained by calculation. It is interesting to compare this with the natural radiation: The average value is 9.2×10^{-4} Gy *per person*. With an assumed world population of 4×10^9 this results in 3.9×10^6 man Gy. The ratio is 8×10^{-4}, in other words, the yearly radiation burden of *all* power reactors corresponds to about 7 hours in the natural environment.

The input values for the estimation are measured radioactivity emissions, the subsequent conclusions are, of course, hampered by the uncertainties of model assumptions but it is not expected that they are completely wrong. It must be reemphasized, however, that they are only valid for "normal" operation.

22.3.2.2 Other Civilisatory Radiation Sources

The most important contribution comes from the medical application of radiation, particularly X-ray diagnosis. It is difficult to give here quantitative figures which are generally applicable because of great variations in techniques and habits. The following considerations which are based on data for the Federal Republic of Germany (1977) should, therefore, be taken as an illustrating example although it can be assumed that the conditions are typical for most modern countries: On an average each person underwent per year 1.7 diagnostic X-ray applications (including dental examinations) with an average gonad dose of 4.8×10^{-4} Gy. The dose variations are considerable with maximum values around 0.01 Gy (urological examinations). Since carcinogenesis is the more important risk, the organ doses are also to be considered as they may be as high as 0.04 Gy. With examinations of the thorax, which comprise 43% of all cases, values vary between 10^{-4} and 2×10^{-3} Gy. Taking all this into account, calculations reveal an average dose for the whole population of 5×10^{-4} Gy per year and person, i.e. nearly 50% of that caused by natural radiation.

Radiation therapy and nuclear medicine contribute only little because the number of people involved is small, the individual doses, however, may be quite high. This is a different situation since the therapeutic application of radiation is given to people which are already severely ill and thus, the usual considerations of radiation protection do not make much sense under such circumstances.

The diagnostic use, however, particularly if it is performed on a large scale, e.g. as a screening programme, is a real case for cost-benefit considerations. This means that the expected gain by early detection of diseases has to be weighted against the possible risk. With a total population of 60 million the yearly collective dose for the Federal Republic of Germany is 3×10^4 man Gy. With a risk factor of 0.01/Gy (Chap. 20), one would expect additional 300 deaths by cancer which has to be compared to the "natural" rate of 144 000 per year, i.e. it amounts to about 0.2%. This admittedly crude illustration does not take into account that certain diagnostic techniques yield considerably higher doses but it may suffice to show the orders of magnitude. That pregnant women have to be spared from radiation – if ever medically possible – has already been stated in Chap. 19 but it should be reemphasized here. In summary, medical radiation usage contributes significantly to the exposure of the general public and definitely more than nuclear energy. The development of new diagnostic techniques appear to be very welcome also in this respect.

Radiation exposure due to fallout from nuclear explosions become fortunately less important although the long half-lives of the nuclides involved are still responsible for a yearly dose of about 10^{-5} Gy. If no new military developments interfere, this value is bound to decrease in future although slowly.

Interestingly, the contribution of natural radiation may be increased by civilisatoric influences. The best example is flying at high altitudes. Since the cosmic component is here quite considerable, the passengers are subjected to higher doses than on Earth. The actual values are, however, small: At 11 000 m, the dose rate is about 4×10^{-6} Gy h^{-1}, in other words, the mean annual dose from natural sources is doubled by a flight time of 275 hours. Taking into account, however, the large number of passengers, the total burden is not so small. With 10^9 passenger hours per year (1975), a collective dose of 4×10^3 man Gy is obtained which happens to be roughly the same as expected from the worldwide use of nuclear energy (3×10^3 man Gy, see above).

Another, by no means negligible, factor is the radioactivity content of building materials. ^{226}Ra is the most important contribution. It decays to ^{222}Rn which emanates from the walls, and is inhaled by the inhabitants. It contributes quite significantly to the natural radiation burden and leads to the somewhat paradoxical situation that the indoor dose is higher by a factor of about 1.3 than outside.

22.3.3 Radioactivity Transfer in the Environment

The pathways of radionuclides from their source to human beings are rather complex. This section is aimed to illustrate the principles of analysis by simple examples. Figure 22.8 gives a rough scheme how radioactivity in air may

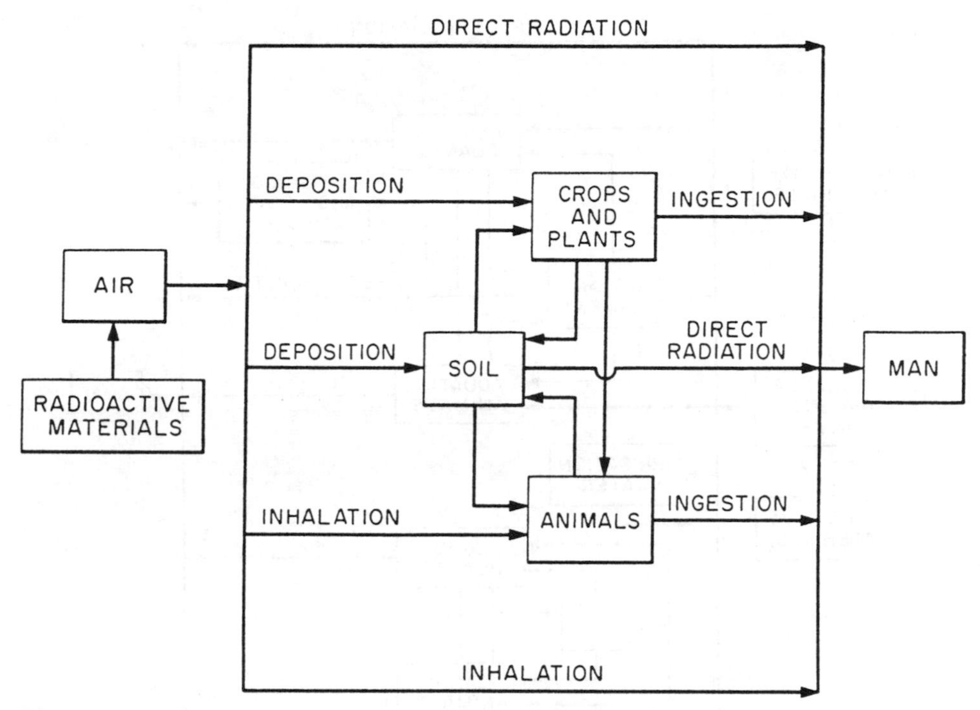

Figure 22.8. The effects of radioactivity in the air (after ICRP 29, 1979)

affect people; Fig. 22.9 illustrates in the same way the situation with water. The diagrammes demonstrate, however, that not only direct effects have to be taken into account but that also – even more importantly – indirect hazards via the food chain have to be considered. This is achieved by establishing the relations as shown in the figures in mathematical terms yielding a large set of coupled differential equations. To illustrate the procedure – which is very similar to that discussed in Chap. 21 for the human body – a simple example is chosen (Fig. 21.10), namely the way from soil to humans. As usual, linear kinetics are assumed, so that the following set of differential equations is obtained:

$$\dot{q}_2 = \tau_{12}q_1 - (\lambda_D + \lambda_2)q_2$$
$$\dot{q}_3 = \tau_{13}q_1 - (\lambda_D + \lambda_3)q_3$$
$$\dot{q}_4 = \tau_{24}q_2 - (\lambda_D + \lambda_4)q_4$$
$$\dot{q}_5 = \tau_{35}q_3 + \tau_{45}q_4 - (\lambda_D + \lambda_5)q_5$$

τ_{ij} are the "transfer coefficients", λ_D is the physical decay coefficient of the nuclide in question, and λ_i the "loss" factors to describe biological decomposition and excretion. q_i is the radioactivity in the i-th component as a function of time.

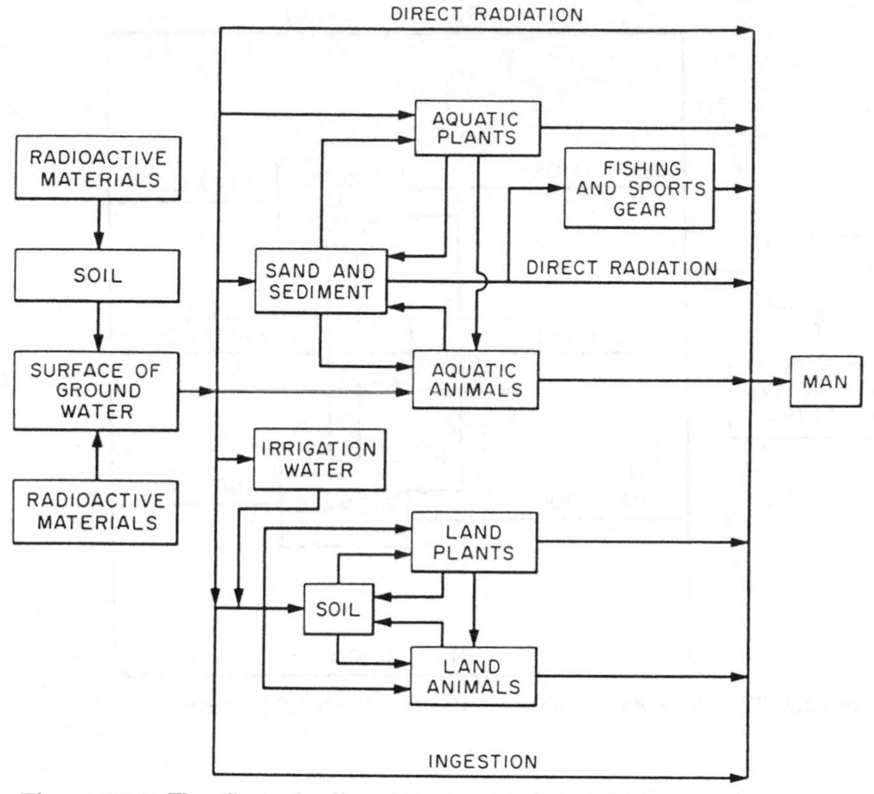

Figure 22.9. The effects of radioactivity in water (after ICRP 29, 1979)

The system can be solved, yielding sum expressions of exponential functions for all q_i.

The transfer coefficients are the most important parameters. They depend only on the chemical nature of the compound. Some representative examples are listed in Table 22.5 for parts of the food chain. Together with the data of Table 21.5 ("consumption habits") they may be used to assess possible hazards from environmental contamination.

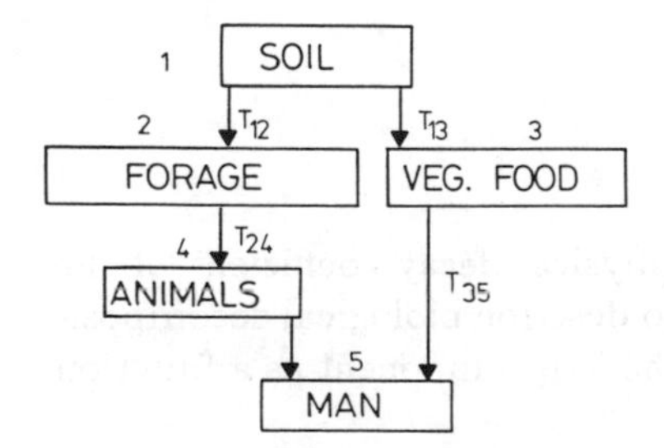

Figure 22.10. A simplified scheme for the effects of groundwater radioactivity on human beings

Table 22.5. Transfer coefficients for different elements. For the passage soil to plant they refer to equal masses of soil and non-dried plant material. For the other cases the reference quantity is "daily uptake" (= 55 kg wet plants). (After BONKA, 1982)

element	soil-plant kg/kg	plant-meat kg/day	plant-milk kg/day
Na	0.05	0.03	0.04
Fe	0.00004	0.04	0.0012
Sr	0.2	0.00006	0.003
Zr	0.00002	0.034	0.0000005
Ru	0.011	0.4	0.0000001
I	0.02	0.003	0.01
CS	0.05	0.04	0.012
Ba	0.005	0.0032	0.00004
U	0.003	0.000034	0.00005
Pu	0.000025	0.0000015	0.0000005

Further reading

BONKA 1982, COUNCIL ON SCIENTIFIC AFFAIRS 1987, EISENBUD 1987, UNSCEAR 1977, 1982, 1986, WHO 1978

Chapter 23
Principles of Radiation Protection Regulations

It is the aim of this chapter to give a certain insight into the considerations which lead to recommendations for radiation protection and which form the body of most national and supra-national regulations. The relevant new quantities and units are introduced and discussed.

23.1 Ionizing Radiation

23.1.1 General Comment

Radiation protection rules have the purpose to protect individuals and the population as a whole from the harmful effects of radiation. They apply only to man-made radiation sources, the natural component is excluded.

An international body, the "International Commission on Radiological Protection" (ICRP) regularly publishes recommendations which form the basis of national legislation in most countries. They are based on radiobiological concepts and experiments as described in this book, but most of the results are not directly applicable, especially so because the doses used are in most cases far beyond the levels which are acceptable in terms of radiation protection. Extrapolations are, therefore, unavoidable.

23.1.2 Concepts, Quantities and Units

To establish the framework of quantitative radiation protection, a few special concepts and units have to be defined:

1. Detriment: This expression summarizes all kinds of damage incurred by radiation action and is not restricted to health effects, although they are the most important. The biological effects are subdivided – for the purpose of radiation protection – into two categories: *stochastic* and *non-stochastic* damage. The first class is governed by probabilistic influences: only the probability of occurrence is a function of dose and there are no degrees of severity. Either the effect is found or not (a "yes-or-no" event). Examples are mutations and carcinogenesis. It is generally assumed that no threshold exists, although this is not inherent in the concept.

Non-stochastic effects are those where the extent of damage is a function of dose, e.g. erythema and radiation sickness. Threshold doses are generally found, i.e. below certain dose levels the degree of damage is too small to be detectable. But as above, this is not necessarily part of the general concept but depends on the practical limits of resolution.

2. Dose equivalent: This quantity has been introduced to account for the different biological effectiveness of different radiation types. It is *not* a dose, not even a physical quantity, and cannot be directly measured.

The dose equivalent H is defined by:

$$H = N \cdot Q \cdot D \tag{23.1}$$

where D is the physical dose (energy absorbed per mass), Q a "quality factor" typical for a given radiation, and N an additional factor to account for special exposure patterns or specific properties of a particular organ. It is set to unity in most cases, a noticeable exception is with non-stochastic effects in the eye where it is set to 3.

If D is measured in Gy the unit of dose equivalent is "Sievert" (Sv); the old unit, related to "rad" was "rem" ("rad equivalent man"), with 1 Sv = 100 rem.

The quality factor Q is a *recommended* value, with other words, a convention whose value is based on radiobiological RBE determinations. It is only to be used in the context of radiation protection. Q is so far related to the linear energy transfer (LET) of the radiation, although other ways of definition are being discussed. Figure 23.1 shows the recommended relationship given in 1977. Because of the uncertainties in neutron RBE, which were brought about by the reevaluation of the dose levels in Hiroshima and Nagasaki, it was felt in 1985 that Q for fast neutrons should be increased to 20 which is somewhat at variance with Fig. 23.1. Approximate values for different radiation types are given in Table 23.1, which may serve as a guideline.

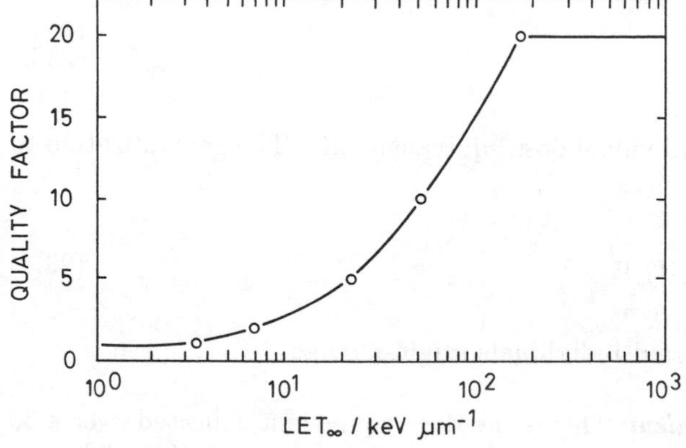

Figure 23.1. The recommended relationship between quality factor and LET (after ICRP 26, 1977)

Table 23.1. Quality factors for different radiation types

radiation type	quality factor
X-rays, γ-rays, electrons	1
protons, singly charged particles (m > 1 u)	10
neutrons, α-particles and other multiple charged particles	20

Values in excess of 20 are so far not recommended. The change of Q to 20 for neutrons, as adapted in 1985, appears to be not quite consistent since that for protons was at the same time left at the original 10 although energy deposition by neutrons is largely via recoil protons.

3. Collective dose equivalent: To estimate the collective risk to a population it is useful to calculate the average dose equivalent per person and multiply it by the number of people involved. The quantity thus obtained is the *collective dose equivalent* S. It is given by:

$$S = \sum H_i P_i \quad . \tag{23.2}$$

Here H_i is the dose-equivalent to the i-th subgroup and P_i the number of persons in that group. The unit to be used is "man-Sv" (formerly "man-rem").

4. Dose equivalent commitment: Radiation exposure extends commonly over long time periods and the rate may be changing. This is particularly the case when radionuclides, either incorporated or in the environment, are involved. To account for this, the total accumulated dose equivalent has to be computed to give the *dose equivalent commitment* H_c. It is defined as:

$$H_c = \int_0^\infty \overline{\dot{H}(t)}\, dt \tag{23.3}$$

where $\dot{H}(t)$ is the mean individual dose equivalent rate. The generalization to populations is obvious:

$$S_c = \sum \int_0^\infty \dot{H}_i(t) P_i(t)\, dt \tag{23.4}$$

where the number of affected individuals might change.

5. Committed dose equivalent: This is the dose equivalent collected over a 50 year period which is taken to be representative for a whole working life:

$$H_{50} = \int_{t_0}^{t_0+50\,a} \dot{H}(t)\,dt \quad . \tag{23.5}$$

6. Tissues at risk: As has been shown before, the radiation risk, particularly in terms of carcinogenesis, is not the same for all body organs. This is taken into account by assigning weighting factors w_T which are listed in Table 23.2.

Table 23.2. Weighting factors W_T for tissues at risk. (After ICRP 26, 1977)

tissue	W_T
gonads	0.25
breast	0.15
red bone marrow	0.12
lung	0.12
thyroid	0.03
bone lining	0.03
others	0.30

The line "Others" represents the five tissues not included in the list (e.g. skin – receiving the highest doses) each with a weighting factor of 0.06.

7. Effective whole body dose equivalent: If parts of the body are exposed to different doses an effective whole-body dose equivalent can be calculated by adding the *weighted* contributions:

$$H_{Wb} = \sum w_{Ti} H_{Ti} \tag{23.6}$$

where w_{Ti} are the organ weighting factors and H_{Ti} the dose equivalents in these tissues.

8. Reference man: To estimate the distribution of radionuclides and the expected doses, physiological data like tissue mass, distribution coefficients, etc. are needed. In order to have a common base line, the "reference man" was introduced which is assumed to be a good average of the general adult population. It has already been used in Chap. 21 where also some typical data are found in Table 21.2.

9. Dose equivalent limit: This is presumably the practically most important quantity. It signifies the level of radiation exposure which is considered to be "acceptable" based on radiobiological experience but also taking other nonscientific factors into account (see below). The recommended limits are by no means considered as firm values, they are arrived at by weighting of many different aspects and they may change as a consequence of new scientific findings or even practical developments. The recommended figures are upper limits which must not be exceeded. This is quite different from older concepts of "tolerance doses" which suggested that certain amounts of radiation may be "tolerated". The present view is that every dose may be harmful and that any risk incurred has to be weighted against other undesirable factors. Dose equivalent limits are recommended only for the purpose of radiation protection and originally also only for those people being professionally exposed. They are, however, in some countries also used to derive limits for the general public. They are not applicable for the medical use of radiation in patients but, of course, must be kept for the personnel, both technical and medical. The natural environmental background is as a rule not included in the calculations.

Limits are separately given for *stochastic* and *non-stochastic* effects. The former refers always to whole body exposure or its effective equivalence (see above); the latter is applicable to any part of the body, tissue or organ. The presently recommended threshold limits per year (a) ("for radiation workers") are:

stochastic:	0.05 Sv/a	(whole body)
non-stochastic:	0.5 Sv/a	(all tissues except the eye)
	0.3 Sv/a	(eye only)

The weighting procedure described above implies that the organs have higher limits for stochastic damage if irradiated separately, namely 0.2 for the gonads, 0.33 for the mammary gland, 0.42 for bone marrow and the lung, and as high as 1.66 for the thyroid and the bone lining (all in Sv/a). These values show that under certain conditions the limit for stochastic damage may significantly exceed that for non-stochastic impairment. The recommendations state clearly that neither of the two should be exceeded. This means that both of the following conditions have to be fulfilled:

$$\sum \mathrm{w_{Ti} H_{Ti}} \leq \mathrm{H_{WB,L}} \quad \underline{\text{and}} \quad \mathrm{H_T} < \mathrm{H_{ns,L}} \tag{23.7}$$

where $\mathrm{H_{Wb,L}}$ and $\mathrm{H_{ns,L}}$ are the limits for stochastic and non-stochastic damage, respectively.

10. Annual limits of intake: The foregoing considerations are valid so far only for external radiation excluding the possible additional incorporation of radionuclides. To account for the latter, annual limits of intake (ali) are recommended for all possible nuclides, as already mentioned in Chap. 21. Their values are based on the limits for the organs involved and the expected relationship for the committed dose equivalent. The ali values are defined for the incorporation of only one particular nuclide disregarding any dose contribution from the outside. Taking all influences together, requires Eq. (23.7) to be modified as:

$$\sum \frac{w_{Ti} H_{Ti}}{H_{Wb,l}} + \sum \frac{I_j}{ali_j} \leq 1 \tag{23.8}$$

where I_j is the annual intake of the j-th radionuclide, and ali_j the respective limit value. Eq. (23.8) shows that with multiple radiation loads, it is not sufficient to control just the annual limits individually. It should also be remembered that under the given principles, the natural background is excluded.

Annual limits of intake are often used to arrive at certain limiting radionuclide concentrations in food. These calculations are necessarily based on a number of assumptions, not only physiological data but also life-style, i.e. eating and drinking habits. Because of these uncertainties, the ICRP does no longer recommend such values although they are still widely used in national regulations.

Some representative ali values are given in Table 23.3.

Table 23.3. Annual limits of intake (ALI) for some radionuclides (Bq y^{-1}) and occupationally exposed persons (ICRP recommendations). (After ICRP 30, 1978)

nuclide	oral uptake	inhalation	DAC [a] Bq m^{-3}	aerosol class
^{3}H (water)	3×10^{9}	3×10^{9}	8×10^{5}	-
^{32}P	2×10^{7}	1×10^{7}	6×10^{3}	W
^{60}Co (f=0.3)	7×10^{6}	1×10^{6}	1×10^{6}	Y
^{90}Sr	1×10^{6}	1×10^{5}	6×10^{1}	Y
^{125}I	1×10^{6}	2×10^{6}	1×10^{3}	D
^{129}I	2×10^{5}	3×10^{5}	1×10^{2}	D
^{131}I	1×10^{6}	2×10^{6}	7×10^{2}	D
^{137}Cs	4×10^{6}	6×10^{6}	2×10^{3}	D
^{210}Po	1×10^{5}	2×10^{4}	1×10^{1}	D,W
^{226}Ra	7×10^{4}	2×10^{4}	1×10^{1}	W
^{232}Th	3×10^{4}	4×10^{1}	2×10^{2}	W
^{235}U	5×10^{5}	2×10^{3}	6×10^{-1}	Y
^{239}Pu	2×10^{5}	2×10^{2}	8×10^{-2}	W
^{241}Am	5×10^{4}	2×10^{2}	8×10^{-2}	W
^{252}Cf	2×10^{5}	1×10^{3}	4×10^{-1}	W,Y

a) "derived air concentration"

11. Derived air concentrations: There is one exception to the above-said in the case of air. This is a slightly simpler case since the volume of breathing varies only over a comparatively small range. For the purpose of calculations it is assumed that the typical radiation worker inhales 0.02 m^3 and is exposed 40 hours per week and 50 weeks per year. Derived air concentrations (DAC) are also given in Table 23.3. In some cases the particular physical state (e.g. aerosol size) or chemical composition has to be especially considered.

23.1.3 General Regulations and Principles

The quantities and considerations introduced in the foregoing paragraphs constitute the basic system for radioprotection. Its essence is straightforward as summarized by the ICRP in these three lines (ICRP 26, 1977):

(a) no practice shall be adopted unless its introduction produces a positive net benefit,
(b) all exposures should be kept as low as reasonably achievable, economic and social factors taken into account, and
(c) the dose equivalent to individuals shall not exceed the limits recommended for the appropriate circumstances by the Commission.

A few comments are in place: It is explicitly stated that only *useful* applications of radiation are allowed so that positive benefit is produced (a). This may be medical or technological. If the same goal can be reached without radiation and at the same cost, this method has to be preferred. The second line (b) is commonly known as the ALARA ("as low as reasonably achievable") principle. Its implementation may be rather difficult in practice since it involves cost-benefit calculations whose significance is always debatable if hazards to human beings are involved. This is clearly not a purely scientific matter and will not be discussed any further. In the light of the ALARA principle, it is immediately clear that the ICRP recommends *upper limits* which must not be exceeded while all possible attempts should be made to reduce any exposure. The limits given as well as the entire scope of these regulations applies only to so-called "radiation workers", i.e. those people who have the benefit to earn their living but have at the same time bear a certain risk as it is always found – although different in nature – in any profession (see below).

The ways and means of risk estimates are, of course, of some concern. The threshold doses for non-stochastic effects can be rather precisely determined (with the possible exception of teratological alterations in the developing brain; see Chap. 19) so that the discussion is more concentrated on stochastic damage where no threshold doses are assumed to exist. These are genetic hazards and carcinogenesis. If a doubling dose of 1 Gy (Chap. 19) for hereditary defects is accepted, the calculated limit of 0.2 Gy to the gonads (as computed from the whole body limit), if exposed alone, corresponds to 1/5, or in other words, the "natural" rate of about 10% would increase to 12% if a person would receive this dose or dose equivalent. Although this additional risk is not low, it is well within the natural variation of hereditary defects and considerably less than that due to the age of the mother when a child is born.

Carcinogenesis appears to be of greater public concern. One reason is that this risk is not restricted to people in their reproductive period but extends over the whole life. The overall mortality rate is estimated to be 1% per Sv for adults (Chap. 20). In this context, it makes no difference whether this figure has to be increased by a factor of 2 or more as recently proposed, the question is how it can be used for risk estimates with long term exposures at low dose rates and low doses. This brings up the problem of extrapolation since the original data were obtained under different circumstances. Since experiments are ethically forbidden and – if one would think of primates – technically not feasible, one has to rely on model calculations. Here, the considerations presented in Chap. 16 find their practical justification. It is commonly – but not universally – assumed that a linear extrapolation to low doses and neglecting any dose-rate effect yields "conservative" risk estimates. Whether this is really true depends on the "real" shape of the dose-response curve, as illustrated in Fig. 23.2. If it were to exhibit an upward bent cur-

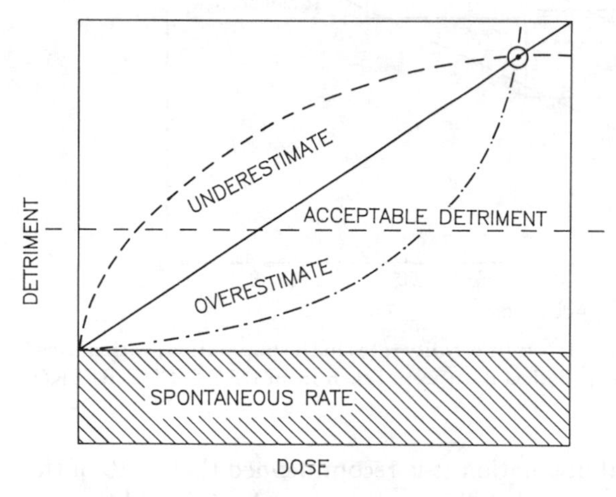

Figure 23.2. The influence of different assumed dose-effect relationships on the estimation of detriment. Acceptable detriment is set here as the twofold spontaneous incidence

vature, the actual risk would be underestimated. There are a few examples where such a shape has been found (see particularly Sect. 14.2).

The definition of "acceptable risk" is certainly debatable. It is intimately tied to the problem of risk-benefit calculations. There should be no discussion that any risk which is not connected to a positive benefit is not acceptable but an agreement may be very difficult to reach in practice because personal judgement is involved. Benefit may be technological, medical or scientific in nature and people will widely disagree whether there is a real advantage in a certain procedure. Complicating is the fact that cost and benefit are usually not shared by the same people. Since "radiation workers" have at least the benefit of their income, it is logical that the radiation protection recommendations apply in the first place to this group of the population. "Acceptable risk" has to mean then that the risk should be about the same as in other professions where there is an exposure to hazards. The limits recommended were calculated under these premises.

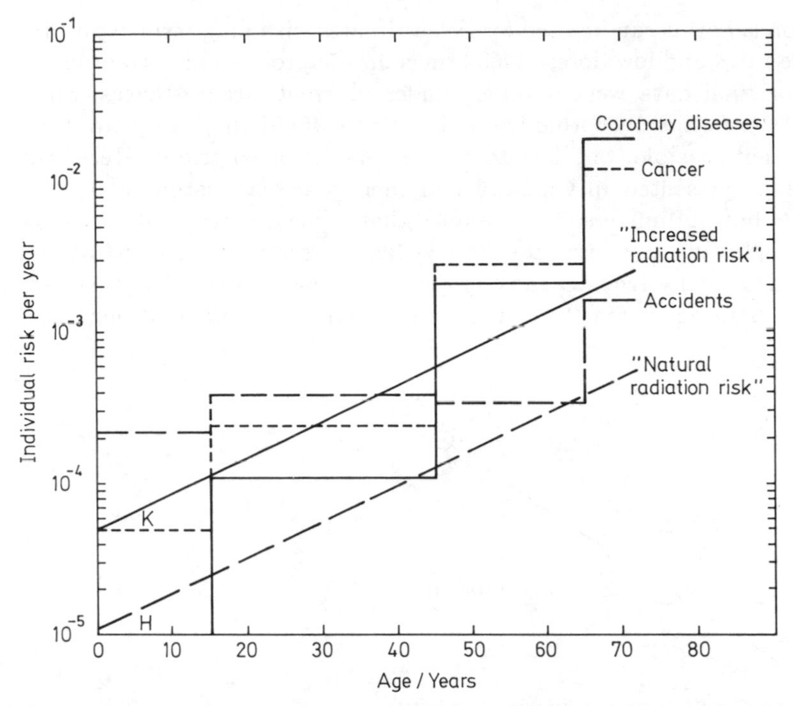

Figure 23.3. A comparison of natural risks (mortality rates in the population) with natural and increased radiation risk. The statistical data were taken from DEUTSCHE RISIKO-STUDIE (1979)

In the case of the general population it is recommended that 1/10 of the "workers limit" should not be exceeded. The hazards involved should then be compared with the "natural" every day risk. Figure 23.3 attempts to illustrate this situation where mortality rates due to different causes are given as function of age. The yearly individual risk is shown as well as the probabilities of fatal cancer either caused by the natural radiation background or by a dose equivalent of 0.005 Sv/a (i.e. 1/10 of the recommended limit for workers). It is quite clearly seen that even the "enhanced" radiation is smaller than the total risk due to other causes, i.e. accidents and coronary diseases. Although the uncertainties are still high, it is at least obvious that current radiation protection rules appear to be based on reasonable assumptions supported by a considerable body of radiobiological experience. Such an approach is rather more the exception than the rule in environmental protection.

23.2 Optical Radiation

Although there has been less concern about the hazardous action of ultraviolet light, there can be no doubt that protection is necessary for workers who may be exposed. An organization like the ICRP does not exist for this field but the issue has been taken up by the World Health Organization WHO in a fairly recent report (WHO 1982). The guidelines differ from country to country, they cannot be discussed here. The main concern is obviously carcinogenesis but acute effects to the skin and particularly to the eye are also of utmost importance. As an example, limiting fluences as a function of wavelength are plotted in Fig. 23.4 as they were suggested by the American Conference of Governmental Industrial Hygienists (ACGIH). For wavelengths between 320 and 400 nm, it is recommended that the total irradiance must not exceed 10 W/m^2 if the exposure time is longer than 1000 s (about 17 minutes).

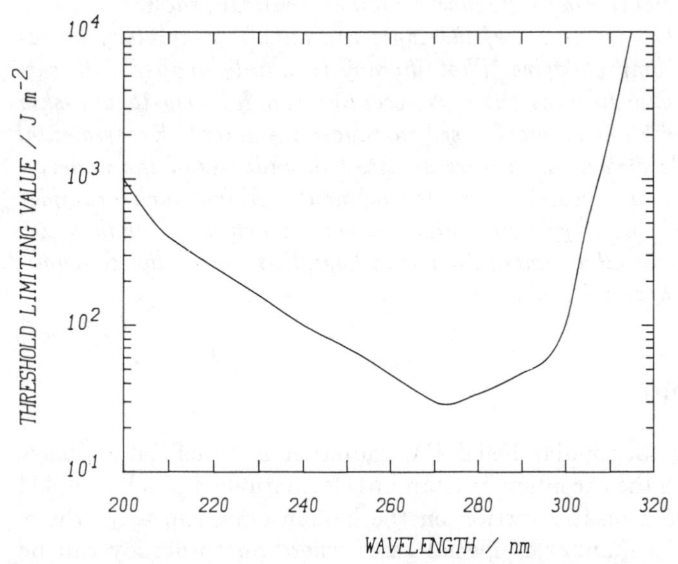

Figure 23.4. Threshold limiting fluence values for a 8-hour day (after WHO 1982)

The relationship depicted in Fig. 23.4 applies to monochromatic sources, but it may be used also to calculate threshold fluences for polychromatic lamps. For the latter, independent action of the different wavelengths is assumed.

Further reading
All publications of the ICRP, particularly ICRP 26, 1977, All publications of the NCRP, particularly NCRP 91, 1987, also: SINCLAIR 1987

Chapter 24
Radiobiology in Radiation Therapy

Radiobiological experience may influence radiotherapeutical practice and lead to new modalities. The treatment of the topic can only be exemplary in nature and cannot be comprehensive. Phototherapy is mainly applied with certain nonmalignant skin diseases but it is recently extended also to non-skin cancers. Ionizing radiation is widely used in cancer treatment. Experimental model systems are described which were designed to understand the underlying principles and to test possible new developments. A few such examples are discussed: namely new types of radiation, particularly those with better depth dose distributions, the combination with hyperthermia or the concomitant application of chemical agents.

24.1 Phototherapy

Although according to popular belief UV radiation is beneficial to one's health, there is – with the exception of vitamin-D formation – no well-founded evidence that proves a positive action on the human organism as a whole. There are, however, a number of special cases where phototherapy can be successfully applied.

Hyperbiliburinemia is a condition sometimes occurring in newborns characterized by an increased blood biliburin level, a decay product of hemoglobin. It may be treated by exposure to blue light which is thought to stimulate the further decomposition of this metabolic product.

The UV treatment of skin tuberculosis is today only of historical interest. It was introduced by the Dane Niels FINSEN who was awarded the Nobel prize in 1903. More important is the treatment of two particular skin diseases: *vitiligo* and *psoriasis.* The first one expresses itself in the form of nonpigmented circumscribed areas while the other is typified by sharply defined red covered spots which are caused by a misregulation of epithelial cell divisions. It is estimated to be found in 1–5% of the population. The classical way to treat these diseases was to deliver series of UV-B exposures; nowadays *photochemotherapy* is preferred applied. A photosensitizer is given – either by local injection or orally –, and the skin is exposed to UV having a wavelength coinciding with the maximum of the sensitizers action spectrum. Tar preparations were formerly used, but today psoralens are exclusively applied. The best results appear to be obtained with 8-methoxypsoralen (8-MOP). The mostly used wavelengths are around 360 nm, i.e. in the UV-A region, hence the name "PUVA" (= psoralen + UV-A) which was given to the method. The chemical and cellular principles have been discussed in Sects. 6.1.2 and 9.1. Since PUVA has been shown to cause mutations in in-vitro systems, a possible carcinogenic action must be taken into account. However, there has been no evidence of skin tumours from observations on a large number of patients. Nevertheless, further careful supervision is warranted particularly in view of the very long latency periods involved.

Hematoporphyrin derivative (Hpd, see Sect. 8.1) is selectively concentrated in many tumours (but not in all; a noticeable exception being melanoma of the eye). Upon exposure to suitable optical radiation, it kills the cells via photodynamic action. Although the absorption maximum is around 400 nm, the weaker band around 600 nm is commonly used. One reason for this is the higher transparency of tissues in this range, the other the availability of laser sources. Using a combination of a laser and fiber optics, it is possible to treat even deep-seated tumours, e.g. in the bladder, with apparently good success. Reports on more than 1500 patients have up to now been given in the literature. Hpd is also available commercially under the name "Photofrin".

Further reading

ANDREONI and CUBEDOU 1984, GOMER 1987, PARRISH et al. 1978, VAN DER LEUN and WEELDEN 1984

24.2 Tumour Therapy by Ionizing Radiation

24.2.1 General Aspects

The therapeutic potential of X rays was realized very early – already less than one year after their discovery a tumour was "treated" by the American physicist E.H. Grubbe (1896 in Chicago). Radiation therapy developed for a long time on the basis of clinical experience although in cooperation with physicists and engineers who constructed and improved the equipment. The history is certainly not free of errors and – as seen from today – wrong approaches but it documents also an overwhelming wealth of experience. Even nowadays, radiation therapy constitutes – apart from surgery and chemotherapy – still one of the main and successful ways to fight and in many cases cure cancer. Radiation biology as a discipline developed later and to some extent separately. It was not before the 1940s that both disciplines approached each other. An honest account cannot escape the conclusion that it is still not really understood in detail how and why a tumour is cured by the common treatment schedules. This is not intended to mean that nothing is known about the radiation biology of tumours – quite on the contrary – but a comprehensive and convincing theoretical foundation is still lacking. Perhaps this is in fact asking too much. Initial assumptions, e.g. that tumour cells are more sensitive to radiation than their normal counterparts proved wrong upon closer inspection. It is likely that no responsible physician would have dared to apply radiation therapy having known about the uncertainties of tumour radiation biology. This is of course exaggerated but it demonstrates the dilemma that there is still a long way to go until the clinical situation is really understood.

Radiation biology, on the other hand, has certainly contributed to the understanding of certain aspects. This has led to new approaches which have been introduced to clinical practice and are already yielding some impressive results. Such attempts can only be successful with a continuous cooperation and exchange of ideas. Both – the practically working physician and the theoretically minded scientist – can learn from each other. A good way to achieve this is the establishment of radiation biology laboratories as part of cancer hospitals, as it is done in many countries.

It has always to be kept in mind that a tumour is not an isolated system but part of the body. The real problem of therapy is not to kill the tumour cells – this can always be achieved by sufficiently high doses – but to spare the surrounding normal tissue to secure its proper function. The key problem is, therefore, *selectivity*.

The following sections thus serve multiple purposes: One is to give the radiation biologist a certain "feeling" for the practical aspects and an introduction into "applied" radiation biology. This is also meant to demonstrate, in a very modest way, the complexity of the problem. Sometimes simplified experimental or theoretical models may help to clarify the situation. A few examples will be described. The clinician will be inclined to view these as oversimplifications but he should realize that this is the only way to arrive at new approaches which might ultimately lead to an improvement in treatment schedules (as always in this book only the principles are outlined). The experimental research scientist in the laboratory must bear in mind that his "clean" systems give at best shadow images of the reality.

24.2.2 Tumours as Experimental Objects in Radiation Biology

Cancer is characterized by cell proliferation which is no longer under the control of systemic regulation. The starting point is neoplastic transformation of progenitor cells but this is not sufficient to produce a tumour, as discussed in Chap. 20. A particular property of malignant neoplasia is the spreading of daughter tumours so that the disease does not remain localized but new foci may be formed elsewhere in the body. Contrary to popular belief, there are many quite slowly growing tumours. To demonstrate this, some kinetic data of human tumours are listed in Table 24.1. Lack of differentiation cannot be considered as a universal criterion, there are well-differentiated tumour types, they are, however, not or only poorly integrated in the surrounding tissue.

Schematically, a tumour may be similarly described as a *renewal system* (see Sect. 17.2): The nucleus consists of stem-cells with indefinite division

Table 24.1. Kinetic data of some human tumours. (After MALAISE, CHAVANDRA and TUBIANA, 1973)

type	doubling time days	growth fraction	loss fraction
embryonal	27	0.9	0.94
reticulo	29	0.9	0.94
sarcoma	41	0.11	0.68
carcinoma	58	0.25	0.9
adenocarcinoma	83	0.06	0.71

potential which develop into more-or-less differentiated cells which die after some time either because of physiological aging or of nutritional restrictions. In contrast to the normal renewal systems, the proliferation rate is not under the regulating control of the body. This does not mean, however, that all stem-cells are constantly in the divisional cycle. The actively proliferating subpopulation is called the "growth fraction" g. Also there may be extensive cell loss, characterized by the "loss fraction" l. Growth can thus be described by the following differential equation:

$$\frac{dN}{dt} = g \cdot KN - lN \quad . \tag{24.1}$$

N is the cell number and K the growth coefficient which is related to the cycle time τ_c as:

$$K = \frac{\ln 2}{\tau_c} \quad . \tag{24.2}$$

Equation (24.1) is, of course, only applicable as long as limiting influences, e.g. lack of nutrients, can be neglected. Within this range, exponential growth is found:

$$N = N_0 e^{(gK-l)t} \quad . \tag{24.4}$$

The rate is obviously determined by the relation between growth and loss fraction. The "tumour doubling time" t_D, an easily obtainable parameter, is:

$$t_D = \frac{\ln 2}{g \cdot K - l} \quad . \tag{24.5}$$

Both, g and l, are not invariable quantities, they may change with time and tumour size, particularly at later stages.

The unregulated proliferation leads to tissue deorganization whose immediate consequence is usually an insufficient blood supply because the distances between cells and capillaries are becoming too large. This situation is schematically illustrated in the "tumour nodule model" in Fig. 24.1: The

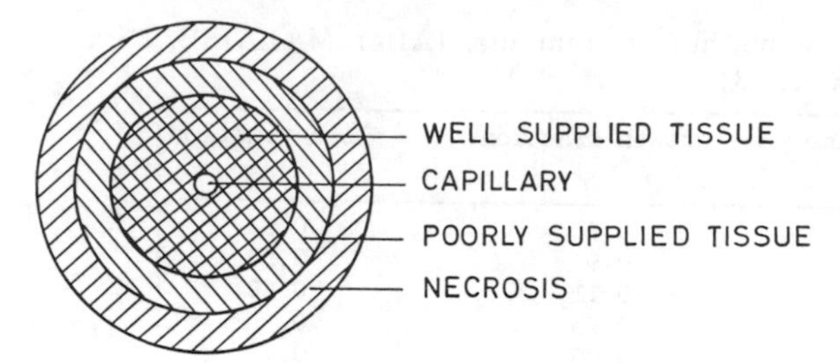

Figure 24.1. The model of a tumour nodule

cells in the immediate vicinity of the capillary are well supplied. The next region consists of cells which have only limited access to nutrients and oxygen but they are still viable although not actively proliferating. The outer rim is a zone of necrotic cells. Such a simple model immediately provokes radiobiological considerations, which are discussed below.

As stated before, the real problem of radiation therapy is not the killing of tumour cells but the sparing of the surrounding tissue, the "tumour bed". In most cases this is not just a geometrical question of spatial dose distribution since tumour and tumour bed can only rarely clearly be separated. Great improvements have been made in recent years with the aid of modern imaging techniques like computer tomography and nuclear magnetic resonance tomography. The data obtained is stored in a computer which then, in turn, controls the mode of exposure. But even these sophisticated techniques do not eliminate the damage to the normal tissue. All considerations for improving radiation therapy center around the question of selectivity, either by physical, chemical or biological means.

The necessary input parameter is always the tolerance of normal tissues which may vary widely between different organs. Table 24.2 lists a few "tolerance doses" $TD_{5/5}$. They are defined as those doses which lead – based on clinical experience – within 5 years to complications in 5% of all cases. The actual values depend on the type of radiation and the application scheme. The examples are given for photon irradiation with maximum energies between 1 and 6 MeV and fractionated exposure (10 Gy per week in five fractions, no irradiation during the weekend). An additional factor is the field size which is also given. The table shows that there are large differences as expected. The growing organism as well as the gonads and the eye are particularly sensitive while muscle tissue appears to be quite resistant. This is certainly related also, but not exclusively, to the proliferation activity in the organ concerned.

It is standard practice in radiation therapy to give the total dose in a number of fractions to spare the healthy tissue. The application schemes vary considerably between different treatment centers, and a general agreement about the underlying principles seems not to be in sight. There are a number of attempts to find a rationale for the procedure. One of the first approaches is based on skin tolerance. Since it is the entry organ in most treatment

Table 24.2. Tolerance doses of organs ($TD_{5/5}$, explanation see text): $TD_{5/5}$: Occurrence of the given symptoms to 1 – 5 % within five years. (After RUBIN and CASARETT, 1973)

organ	damage after 5 years	$TD_{5/5}$ Gy	radiation-field or -volume
skin	ulceration	55	100 cm³
intestine	ulceration	45	100 cm³
liver	loss of function	35	total organ
kidney	scelerosis	23	total organ
testes	sterilisation	5 - 15	total organ
ovary	sterilisation	2 - 3	total organ
breast (child)	no development	10	5 cm³
breast (adult)	necrosis	> 50	total organ
lung	inflammation, fibrosis	40	one side
heart	inflammation	40	total organ
bones (child)	growth arrest	20	10 cm³
bones (adult)	necrosis	60	10 cm³
cartilage (child)	growth arrest	10	total tissue
cartilage (adult)	necrosis	60	total tissue
bone marrow	hypoplasia	20	locally
eye lens	cataract	5	total organ
fetus	death	2	totally

schemes, its reaction is certainly relevant although not always limiting. The starting point is the "STRANDQUIST curve" (Sect. 18.1, Fig. 18.5) which demonstrates that the dose causing a certain effect and the time over which it is spread are related by a power function (Eq. (18.1)). A similar relationship exists for the cure of skin tumours although with different parameters. Figure 24.2 summarizes the situation for erythema formation, skin tolerance and tumour cure. It is seen that for the latter the curve is less steep than for the "physiological" responses so that both intersect. This means that with a certain time schedule it is possible to achieve tumour cure without exceeding

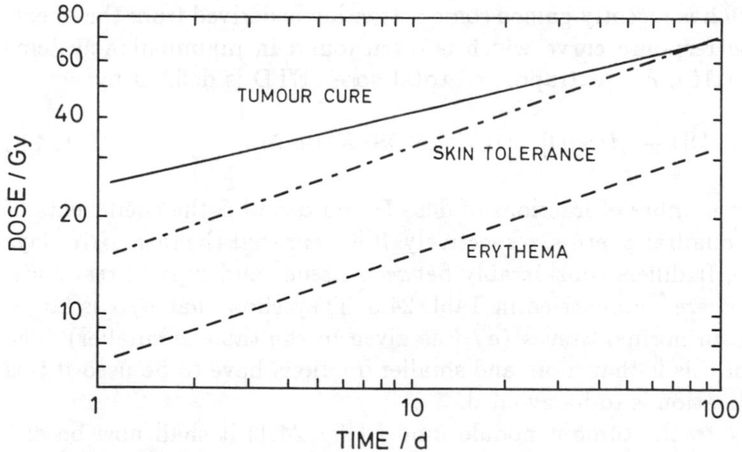

Figure 24.2. STRANDQUIST curves for erythema, skin tolerance and tumour cure (after ELLIS 1968)

the skin tolerance dose. It is assumed that regeneration is more efficient in the healthy tissue as compared to the tumour. The primary damage per fraction, on the other hand, is the same in both. In the tumour, the total damage depends only on the number of fractions while there is an additional time factor in normal skin. With these postulates the following can be written:

$$\begin{aligned} \text{tumour}: \quad & D_T \sim N^a \\ \text{skin}: \quad & D_S \sim N^a \cdot t^\beta \quad . \end{aligned} \tag{24.6}$$

N is the number of fractions; t the total exposure time; D_T and D_N are doses required to cause the effect.

For five fractions per week, Fig. 24.2 gives $a = 0.22$, i.e. corrected for daily exposures $a = 0.24$. The skin tolerance curve also gives $a + \beta = 0.33$, hence $\beta = 0.11$ (since total time is considered here, no correction is required). This leads to the relationship for the tolerance dose:

$$D_S = \text{NSD} \cdot N^{0.24} \cdot t^{0.11} \tag{24.7}$$

where NSD is a proportionality factor which is numerically equivalent to a "nominal single dose". Equation (24.7) allows to estimate the normal tissue reaction, it is known as the "ELLIS formula" or "NSD concept".

The derivation was given in some detail to explain the underlying concept, not to "prove" a general applicability. A number of assumptions are inherent in the approach which are difficult to justify. It has to be borne in mind that only acute skin damage is considered. A generalization to late affects and to other organs is – at least – questionable. Nevertheless, the formula has been proven useful for the situation for which is was devised. There is an extensive number of examples in the literature where it did not work and many formulations are based on similar concepts.

One which has recently gained some attraction is derived from the linear-quadratic dose response curve which is often found in mammalian systems (see, e.g. Chap. 16). An "extrapolated total dose" ETD is defined by:

$$\text{ETD} = N \cdot D[1 + (\beta/\alpha)D] \quad \text{(BARENDSEN 1982)} \tag{24.8}$$

where N is the number of fractions of dose D and α and β the coefficients of the linear and quadratic terms, respectively. It is clear that the ratio β/α plays a decisive role. It differs considerably between tissues and type of reactions. Some examples are summarized in Table 24.3. They show that β/α is larger for late effects in normal tissues (α/β as given in the table is smaller). The consequence of this is that more and smaller fractions have to be used if this type of complication is to be avoided.

Returning to the tumour nodule model (Fig. 24.1) it shall now be discussed what happens in such a system when exposed to fractionated irradiation: The well-supplied cells close to the capillary are to a great extent

Table 24.3. α/β-ratios (in Gy) for different tissues and endpoints from animal experiments. (After FOWLER, 1984, original data and reference therein)

clamped tumours: [a]	
local control	9 - 18
growth delay	7 - 19
acute effects in normal tissues	7 - 26
late effects	1 - 6.3

a) The tumours were clamped to make them anoxic in order to avoid irregularities due to different hypoxic fractions.

divisionally active while the others in more distant regions are likely not to proliferate because of nutrient and oxygen limitations. The population is quite heterogeneous, also in terms of radiation sensitivity. A first dose will preferably kill the cells around the capillary, mainly because of the oxygen effect. If they are eliminated, the more distant parts will have better access to nutrients and oxygen so that the survivors may start to proliferate giving rise to new tumour growth. Their selective resistance is not only due to the oxygen effect but also to recovery from potentially lethal damage (Chap. 13) since they remain for some time under growth limiting conditions, a situation which is typical for this process. The second and all following dose fractions hit, therefore, a population of different structure. The overall sensitivity is then determined – apart from split-dose recovery (Chap. 13) – by a better oxygen supply and an increased fraction of proliferating cells. Cells which were initially not in the division cycle will enter it after the first exposure. These factors have been summarized – somewhat ironically, at least for the English – as the "four R's of radiation therapy", namely *recovery*, *reoxygenation*, *redistribution* (of cell cycle phases) and *regeneration*. Which of these parameters is finally decisive for the outcome of treatment cannot be said.

The presence of hypoxic regions constitutes a serious problem in radiation therapy – whether it actually limits radiocurability, however, is not clear. A number of approaches have been suggested to overcome it, e.g. the use of radiation "hypoxic" sensitizers or different radiation types (see Sect. 24.3). It is also clear, however, that fractionation alone reduces its weight, as just discussed, since initially hypoxic cells are in a state of improved oxygenation at the time of the second dose fraction.

Another problem is obviously tumour regeneration which can only be fought by repeated exposures. Regeneration in the tumour bed which is desired and necessary competes, however. A cure is only possible if it can be selectively suppressed in the tumour without se ere damage to the healthy tissue. The interplay of these factors ultimately determines the radiocurability.

24.2.3 Experimental Techniques and Model Systems

There is no doubt that clinical experience is of utmost importance but the data do normally not allow to draw conclusions which may be generalized from one system to the other. They have to be supplemented by animal experiments, particularly if new modalities are to be tested. Experimental tumours are a very useful model system. They can be established in mice, rats or other rodents by injection of tumour cells. There are, however, a number of reservations which have to be borne in mind before results obtained are extrapolated to the human situation. The kinetic parameters are of major concern. Table 24.4 lists data of some typical experimental tumours.

Table 24.4. Kinetic data of some experimental tumours. (After DUNCAN and NIAS, 1977)

animal	code	type	doubling time /hours	growth fraction	loss fraction
mouse	L1210	leukemia	10	0.95	0.05
rat	R1B5	fibro-sarcoma	24	0.45	0
rat	R1	rhabdomyo-sarcoma	66	0.29	0.62
mouse	C_3H	carcinoma of the breast	110	0.3	0.7

Comparison with data in Table 24.1 immediately reveals quite considerable discrepancies. This is the old problem in cancer research: the only good model system for humans is the human being! It cannot be circumvented, even not by experiments with primates, quite apart from considerations of animal protection.

Another point is the role of immune reactions. They are activated against injected tumour cells and may change the results. The errors may be minimized by using isogenic cell lines, i.e. those which were derived from the same inbred animal strain. They change their properties during subcultivation, however. Another possibility is the use of tumours spontaneously arising in the animals, but here one has the difficulty that only those strains are of practical value where the incidence rate is comparatively high, i.e. which are "cancer prone". Their reactions again may not be typical.

The most obvious and easiest measurable parameter is tumour growth, i.e. tumour volume as a function of time. A typical experiment is shown in Fig. 24.3 where growth of an experimental neck tumour after a single dose of 20 Gy X rays is plotted versus time: An initial lag is followed by a nearly normal increase of size due to the proliferation of regenerating surviving cells.

It is, of course, of great interest whether new foci may arise after irradiation. This depends on the survival behaviour of the population irradiated *in*

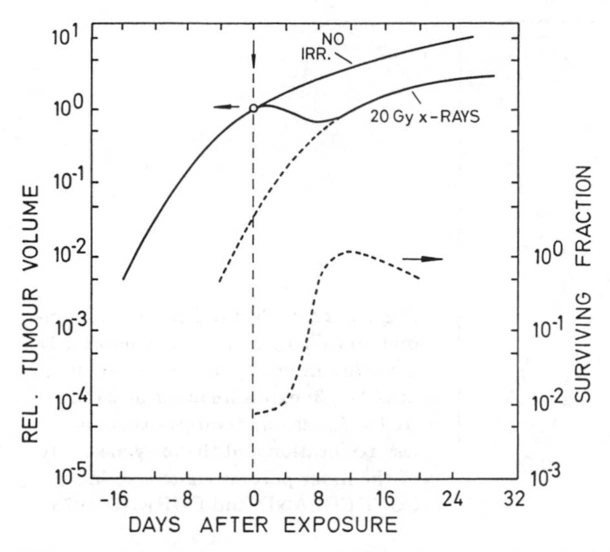

Figure 24.3. Growth of a transplanted tumour (rat rhabdomyosarcoma) with and without X-ray exposure (20 Gy). The surviving fraction of tumour cells is also given (after HERMENS and BARENDSEN 1969)

situ. In some experimental tumours it is possible to perform *in vivo* and *in vitro* experiments in parallel so that the surviving fraction can be determined separately. The results of such an experiment are also given in Fig. 24.3: The surviving fraction was measured concomitantly with tumour growth. One sees that the surviving fraction increases with time up to about 100%, which demonstrates clearly that tumour regrowth is caused by regeneration initiating from surviving cells.

Although already simplified, the situation in an animal tumour is still rather complex. Simpler experimental model systems would be a great help to understand certain aspects of basic mechanisms. An interesting approach is the use of spheroids: Under special culture conditions, mammalian cells in tissue culture form multicellular aggregates which are stable and their size can to some extent be manipulated. They resemble "tumour nodules" but with the supply from the outside. The effect of oxygen and nutrient deprivations can easily be studied in such a model.

Figure 24.4 shows the survival of cells exposed in spheroids and separated before plating: The influence of oxygen deprivation was suppressed by incubation at 24°C where respiration stops. Spheroid cells irradiated in air at 37°C display a survival behaviour typical of a heterogeneous population, indicating a substantial proportion of hypoxic cells as can be seen from the comparison with anoxic spheroids (curve 1). This break is lost at 24°C but the curve is still slightly complex, its shape must be attributed to other additional factors. One such factor could be interaction between cells in the tissue-like aggregates. If the survival curves of single-cell suspensions and of cells ex-

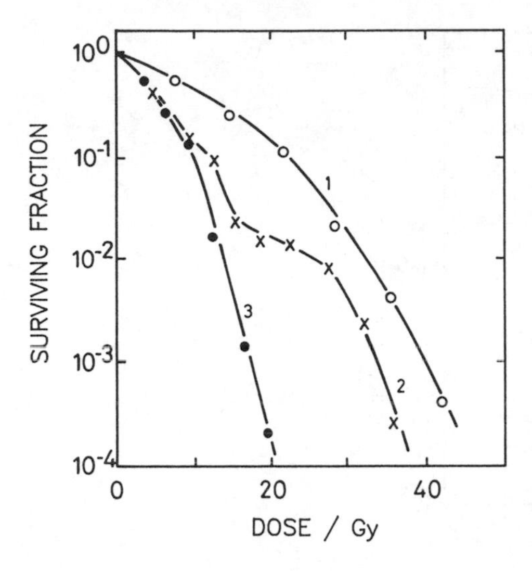

Figure 24.4. Survival curves of mammalian cells irradiated in spheroids. 1: exposure in nitrogen, 2: exposure in air at 37°C, 3: exposure in air at 24°C. At the lower temperatures there is less respiration and the oxygen supply of the inner parts is enhanced. (after SUTHERLAND and DURAND 1973)

posed in spheroids are compared (Fig. 24.5), the terminal slopes are found to be identical but exposure "in situ" leads to much larger shoulders, i.e. higher n or D_q values. A similar phenomenon was also found in animal tissues, as described in Sect. 17.3. It appears as if intercellular contact favours recovery but this is still a speculation, especially it is not at all clear which factors might be involved. Recovery from potentially lethal damage may play a role but even this has to be shown. Nevertheless, the experiments demonstrate that spheroids mimic the *in vivo* situation at least in certain aspects.

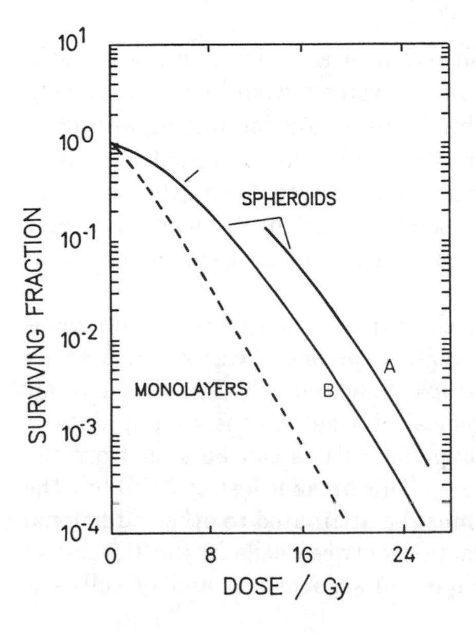

Figure 24.5. Comparative survival of cells exposed as monolayers or in spheroids. Curve A gives the survival of spheroids, curve B the calculated survival of single cells but exposed in spheroids (after SUTHERLAND and DURAND 1973)

Further reading
BARENDSEN 1982, DALRYMPLE et al. 1973, DUNCAN and NIAS 1977, ELLIS 1968, FOWLER 1984, ORTON 1986a, ORTON and COHEN 1988, SUTHERLAND and DURAND 1976, SUTHERLAND 1988, STEEL 1977, WITHERS 1975, WITHERS et al. 1982

24.3 Modifications and New Modalities in Radiation Therapy

24.3.1 Radiation Quality

As repeatedly stated, the ultimate aim of every new modality is to improve the differential action between tumour and normal tissue cells, in other words, to increase selectivity. With regard to radiation quality, there are three approaches which are not mutually exclusive:

1. Increase of the *dose differential* between the tumour and normal tissue.
2. Increase of *relative biological effectiveness* (RBE) selectively in the tumour.
3. Decrease of the *oxygen enhancement ratio* (OER) within the tumour.

It would be – in principle – desirable to apply radiation to the tumour only and to spare the healthy tissue completely. This goal can be approached by suitable depth dose distributions but it is never completely achievable because of physical and biological limitations. Even with very sharp depth dose maxima there is still a substantial dose deposited to the surroundings. Furthermore, tumours are only very rarely sharply defined, there is always some normal tissue in between. This means that the ideal situation does not exist; nevertheless, improvements are possible. Depth dose curves have been discussed in Sect. 4.2.2 where it was shown that high energy photon and electron beams deliver the maximum dose at a certain depth, depending on energy, but the peak is not very sharp and the dose differential rather small. The best radiation type in this respect are protons, but also other accelerated ions and π^- mesons. Low-energy X rays and neutrons, having an approximately exponential dependence of dose on depth, are least usable except for surface exposure where very low energy X rays or electrons may be applied.

Spatial distribution constitutes, however, only one aspect, although a very important one. The biological quality is an additional factor which is related to LET. It is larger with neutrons so that a higher RBE and a lower OER is to be expected. This, however, is not restricted to a certain part of the irradiated volume but found everywhere. A gain is, therefore, only expected if the oxygen effect plays an important role in the particular tumour. Whether this is really the case, especially in view of fractionation schemes favouring reoxygenation, is still debated. Clinical results with neutrons do not give a coherent picture. π^- mesons and accelerated ions appear to be the method of choice. They have a favourable depth dose distribution (although not as good as protons) and a high LET in the dose maximum while it is lower in the "plateau" part. This means that all three postulates listed above can be met at the same time. The high technical effort which is required to produce these particles and hence the costs involved limit their widespread application. It remains still to be demonstrated by ongoing clinical trials whether the results in terms of improved cure rates justify the effort.

The radiation types just mentioned cannot be used clinically in their pure monoenergetic form since the dose peaks are so sharp that only a minor part of the tumour would be covered. This is remedied either by using special attenuation filters which can be "tailored" to the actual situation or by modulating the beam energies. This results in a certain loss of beam quality but the advantages are still considerable. The increase of ionization density in the tumour causes also a change of survival curve shape – from shouldered to exponential – with the consequence that split-dose recovery does not occur and cell cycle dependent sensitivity variations are diminished. This leads to an additional increase in selectivity.

The actual gain to be expected depends on the particular radiation type and the contribution of the oxygen effect. Some data are listed in Table 24.5

Table 24.5. RBE- (10 % survival) and OER-values outside and inside the tumour with various radiation types. (After [a]) RAJU and RICHMAN, 1972; [b]) BLAKELY, TOBIAS et al., 1980)

type	energy	outside		inside		ratios		dose-ratio
	MeV	RBE	OER	RBE	OER	RBE_i/RBE_a	OER_a/OER_i	D_i/D_s [c)]
^{60}Co-γ [a)]	1.3	1	3	1	3	1	1	0.5
fast neutrons [a)]	14	3	1.8	3	1.8	1	1	0.4
π⁻-mesons [a)]	65 *)	1	3	2	1.8	2	1.7	2.5
protons [a)]	150 *)	1	3	1	3	1	1	2
He-ions [b)]	600 *)	1	3	1.3	2	1.3	1.5	2
C-ions [b)]	4800	1	2.7	2	1.6	2	1.7	3.5
Ne-ions [b)]	8500	1.3	3	2	1.2	1.5	2.5	4.4
Ar-ions [b)]	22800	3.8	2.3	1.2	1.2	0.3	1.9	3.5

assumed tumour depth: 15 cm
*) not given, estimated
c) D_i : Dose at tumour depth, D_s : Surface dose

for an assumed tumour at 15 cm depth. They illustrate quantitatively the aspects discussed. The radiation energy was always chosen so that the dose maximum was in the center of the tumour. A closer inspection of the figures is interesting: There is no differential effect with γ rays and neutrons and the dose ratio is unfavourable. Only the OER is generally reduced with neutrons. π^- Mesons and carbon ions appear to be the best suited, particularly the latter, since RBE- and dose ratios are high and the OER reduced. Heavier ions like argon show a worse performance with a RBE ratio below unity. The situation for heavy ions can be graphically summarized in a two-dimensional display (Fig. 24.6) where the expected gains are plotted with various ions tak-

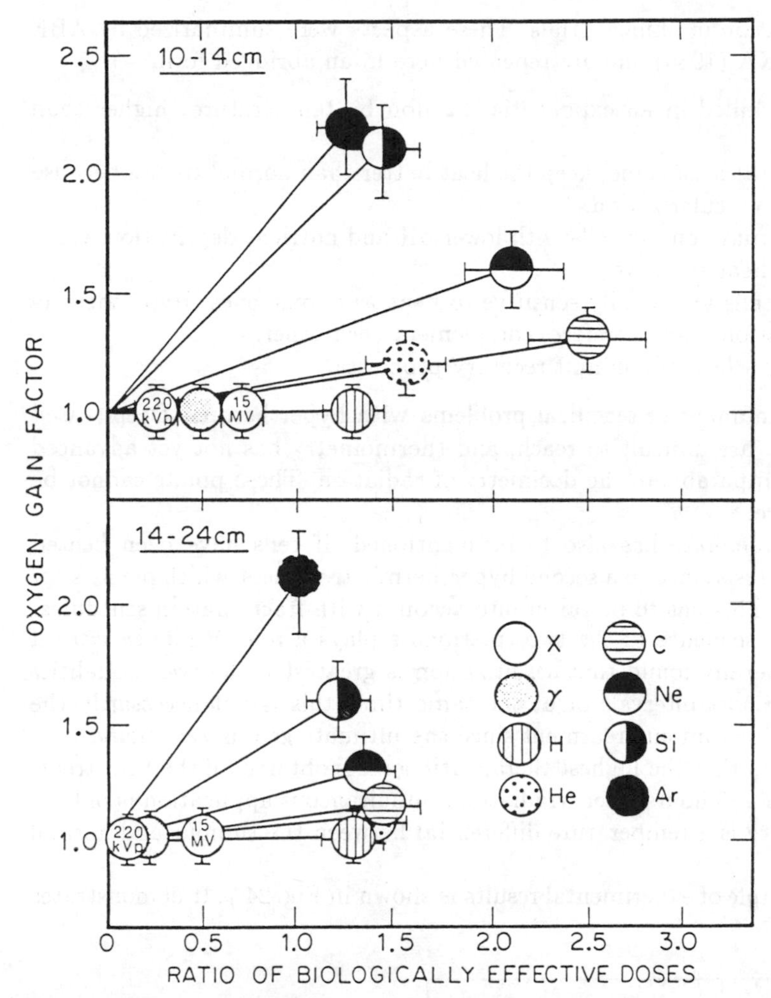

Figure 24.6. Benefit diagramme for the application of heavy ions in radiotherapy. Upper panel: small radiation field, lower panel: large radiation field. Ordinate indicates the gain due to the reduced oxygen effect, the abscissa increased RBE in the tumour (from BLAKELY et al. 1984, with permission)

ing into account both RBE and OER. It follows that ions of medium nuclear mass have the greatest potential. This is welcome since their production and acceleration does not require excessively high costs.

24.3.2 Hyperthermia and Radiation

In vitro studies have clearly shown that cells are sensitized to radiation action by hyperthermia (Sect. 14.3). This fact taken alone would not suffice to also apply heat in radiation therapy but there are a number of factors which justify this combination in clinical trials. These aspects were summarized by ABE and HIRAOKA (1985) and are repeated here in an abridged form:

1. Cells are killed in an exponential fashion by temperatures higher than 42°C.
2. Tumours (at least some) keep the heat better than normal tissues because of poorer vascularization.
3. Tumours may contain cells with lower pH and nutrient deprivation which are more heat-sensitive.
4. Hypoxic cells are equally sensitive to heat as anoxic ones. Hyperthermia and radiation may, therefore, complement each other.
5. Heat may selectively inhibit recovery processes.

There are a number of technical problems with hyperthermia. Deep-seated tumours, e.g. are difficult to reach, and thermometry has not yet advanced to a stage comparable to the dosimetry of radiation. These points cannot be discussed here.

Thermotolerance has also to be mentioned. If cells have been heated they develop resistance to a second hyperthermic treatment which persists for many hours. This has to be taken into account with fractionation schedules.

Also the sequence of the two treatments plays a role. While *in vitro* it has been generally found that sensitization is greatest if the two modalities are given in short intervals or at the same time, this is not necessarily the best scheme for tumour treatment since the ultimate goal is *selectivity*.

It appears that the highest therapeutic gain is obtained if the heat treatment is given 2–6 hours after irradiation. Simultaneous application is only of benefit if there is a temperature differential between the tumour and normal tissue.

An example of experimental results is shown in Fig. 24.7. It demonstrates

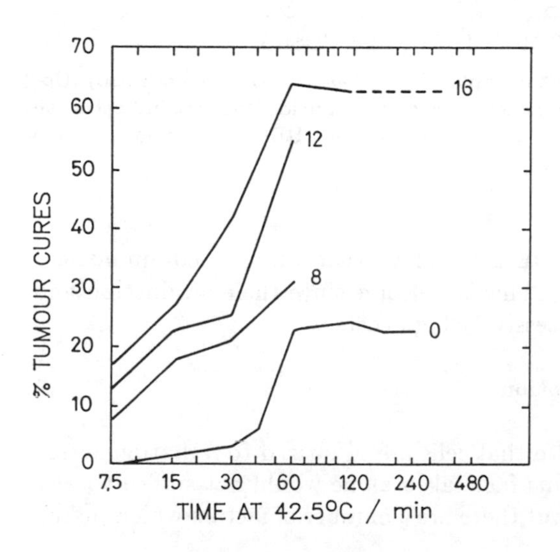

Figure 24.7. Cure rates of an experimental tumour by a combination of hyerthermia and X-rays (doses approximately in Gy). No cures were found with radiation alone (after OVERGAARD 1978)

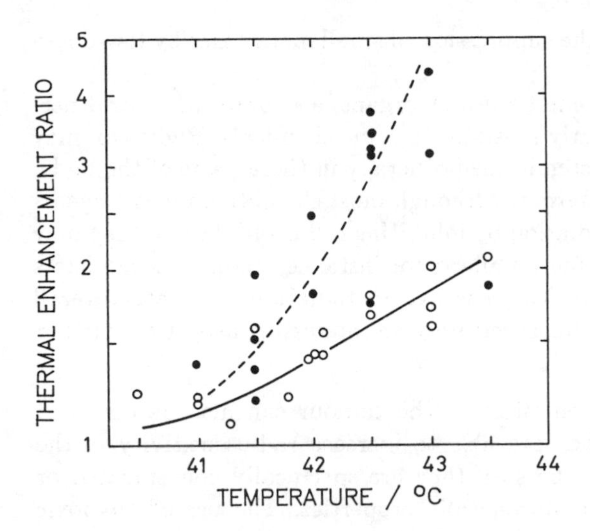

Figure 24.8. Thermal enhancement ratios for tumours (•) and normal tissue (○) (after OVERGAARD 1978)

the superiority of the combined treatment at least in this system. Figure 24.8 illustrates the points listed in the beginning of this section: there appears to be a higher radiation sensitivity of tumours compared to normal tissues if the temperature is higher than 42°C.

Clinical trials seem to prove the usefulness of the combination heat plus radiation. This is certainly true for superficial tumours where heating does not pose problems. The situation is less clear with deep-seated tumours, mainly because of the technical difficulties to achieve sufficient localized hyperthermia. It is expected, however, that this situation will change in the future. Since the equipment used is comparatively inexpensive, hyperthermia may be used more than special radiation types which require costly installations.

24.3.3 Combination with Chemotherapy

This is a wide yet promising field but difficult to summarize in a systematic way; again only principles will be outlined. It is to be expected that the envisaged approaches will give new impetus to radiation therapy as also documented already by some clinical trials.

In combination therapy a distinction has to be made between *cytotoxic* and *non-toxic* chemical agents. There may be – and generally is – also toxicity with the second group but this is not part of the therapy and an unwanted side effect.

The rationale to use cytotoxic compounds in conjunction with radiation may be summarized in three points:

1. *Local radiotherapy plus systemic chemotherapy*: By the additional application of chemotherapeutic agents, tumour cells may be killed if these cannot be treated by radiation. An example is the radiation therapy of

the primary tumour and the suppression of small metastases by low-dosed radiomimetics.

2. *Adjuvant radiotherapy*: Some tissues or organs, e.g. brain or central nervous system, are particularly sensitive to some chemicals. Radiation may be used to "booster" a systemic chemotherapy in these parts of the body.
3. *Interactive combination therapy*: Although most chemotherapeutic agents are similar in action to radiation by inhibiting cell proliferation, they may differ in detail. If radiomimetics affect, for instance, mainly those phases of the cell cycle which are less sensitive to radiation, a greater overall effect could be achieved. This is certainly an interesting aspect for further systematic studies.

A potentiation of the radiation effect in the tumour can also be obtained by non-toxic substances if they are able to increase radiosensitivity in the tumour *selectively*. This could be so if they are specifically concentrated or if their action depends on tumour-specific properties. The use of "hypoxic sensitizers" is a typical example. If recovery from potentially lethal damage plays a role in tumour therapy – which still remains to be unequivocally demonstrated –, inhibitors of this process may be helpful.

Another approach being to some extent a mirror image of the one just outlined, is the selective protection of normal tissues by radioprotectants which are found at lower concentrations in the tumour because of the poorer vascularization. In this case, higher doses could be applied without side effects.

Figure 24.9 attempts to give a schematic classification of the various approaches in combination therapy; some actual examples are listed in Table 24.6 which, of course, cannot be comprehensive. Some of the substances

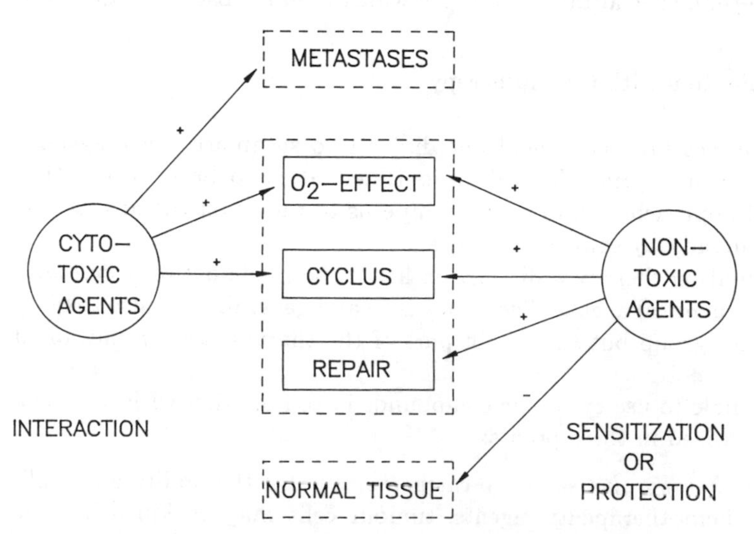

Figure 24.9. An attempt to schematize combination therapy with radiation and chemicals

Table 24.6. Examples of combined radiationchemotherapy

radiomimetics	vincristin, cyclophosphamid, bleomycin, adriamycin methotrexat, vinblastin
'hypoxic sensitizers'	misonidazol, metronidazol
recovery-inhibitors	actinomycin D, hydroxyurea, 5-fluoro-uracil

are being tested in clinical trials, the results obtained do not yet permit final conclusions. The "hypoxic sensitizers" which have been very much discussed recently belong to this group.

With all these considerations which may be theoretically quite attractive, the problem of unforeseen side effects must not be forgotten. Again the "hypoxic sensitizers" may serve as an example: If applied clinically, they lead to neurological disturbances, even at concentrations with low cytotoxicity. The reasons are not really understood but the very fact of their existence demonstrates again that there is always a long way from the laboratory to clinical practice.

Further reading

ABE and HIROAKA 1985, ADAMS, FOWLER and WARDMAN 1978, BELLAMY and HILL 1984, BLAKELY et al. 1984, CASTRO et al. 1983, 1987, CURTIS 1979, DIETZEL 1978, FIELD 1976, GOODMAN 1987, HAHN 1978, 1982, HAR-KEDAR and BLEEHEN 1976, MUNZENRIDER et al. 1987, RAJU 1980, STEEL 1977, STREFFER et al. 1978, SAUNDERS et al. 1986, TSUNEMOTO et al. 1987, TUBIANA 1987, VAETH 1969

Appendix I
Mathematical-Physical Relations

I.1 Polar Coordinates

The location of a point is mostly given in *cartesian* coordinates, i.e. in a plane by the abscissa value x and the ordinate value y. Any space vector *s* may then be characterized by its origin and the two components s_x and s_y (see Fig. I.1.1). For a number of problems – e. g. that treated in the next

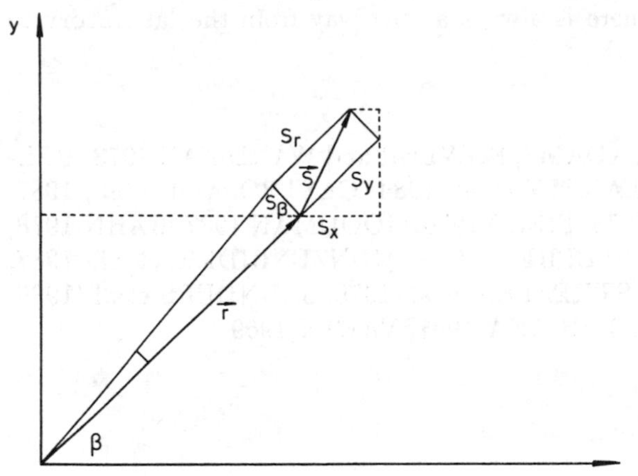

Figure I.1.1. Representation of vectors in polar coordinates

section – it is more convenient to use the distance from the origin r and the angle β which lies between the vector and the positive x-axis.

A space vector *s* may then be described by a component in the direction of the radius vector r and a tangential component perpendicular to it. From Fig. I.1.1 the following relations may be read off:

$$x = r\cos\beta \qquad y = r\sin\beta \tag{I.1.1}$$

and:

$$\begin{aligned} s_x &= s_r\cos\beta - s_\beta\sin\beta \\ s_y &= s_r\sin\beta - s_\beta\cos\beta \quad . \end{aligned} \tag{I.1.2}$$

The complementary relations are:

$$r = x^2 + y^2 \qquad \mathrm{tg}\,\beta = \frac{y}{x} \tag{I.1.3}$$

and:

$$\begin{aligned} s_r &= s_x \cos\beta + s_y \sin\beta \\ s_\beta &= s_x \sin\beta + s_y \cos\beta \quad . \end{aligned} \tag{I.1.4}$$

Very often time-dependent variations are considered so that velocity ***v*** and acceleration ***a*** are also required. The relevant expressions are obtained in a similar way by differentiation with respect to time:

$$\begin{aligned} \dot{x} = v_x &= \dot{r}\cos\beta - r\dot{\beta}\sin\beta \\ \dot{y} = v_y &= \dot{r}\sin\beta + r\dot{\beta}\cos\beta \end{aligned} \tag{I.1.5}$$

and analogously by replacing s by v in (I.1.2):

$$v_r = \dot{r} \qquad v_\beta = r\dot{\beta} \quad . \tag{I.1.6}$$

Further differentiation leads to:

$$a = \ddot{r} - r\dot{\beta}^2 \quad \text{and} \quad a_\beta = r\ddot{\beta} + 2\dot{r}\dot{\beta} \quad . \tag{I.1.7}$$

It should be noted that $a_r \neq dv_r/dt$ and $a_\beta \neq dv_\beta/dt$!

Sometimes also the area element dA is required. One finds (see Fig. I.1.1):

$$dA = r\,dr\,d\beta \quad . \tag{I.1.8}$$

I.2 Mean Pathlength in a Sphere

It is assumed that a sphere of radius r is exposed homogeneously to an isotropic particle fluence ϕ. Because of the isotropy it is sufficient to consider only one direction. The total number n of intersecting particles is:

$$n = \phi \cdot \pi r^2 \quad . \tag{I.1.6}$$

The number of particles traversing at a distance a from the center is proportional to the area of the ring $2\pi a\,da$ The pathlength l(a) is then (Fig. I.2.1):

$$l(a) = 2\sqrt{r^2 - a^2} \tag{I.2.1}$$

and hence:

$$a^2 = r^2 - \{l(a)\}^2/4 \quad . \tag{I.2.2}$$

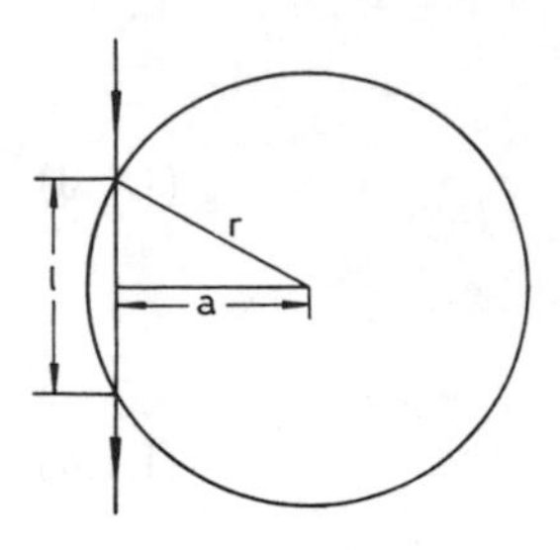

Figure I.2.1. On the tracks length distribution in a sphere

Differentiation with respect to a yields:

$$2a\,da = -\frac{l\,dl}{2} \tag{I.2.3}$$

The fraction of particles intersecting the sphere at a distance a is $2\pi a\,da/\pi r^2$ so that one obtains for the distribution density function $f(l)\,dl$ of the pathlength l:

$$f(l)\,dl = \frac{l\,dl}{2r^2} \qquad 0 \le l \le 2r \quad . \tag{I.2.4}$$

The mean pathlength is then:

$$\bar{l} = \int_0^{2r} \frac{l^2\,dl}{2r^2} = \frac{4}{3}r \; . \tag{I.2.5}$$

This relationship holds only if the range of the particle is not smaller than the sphere diameter.

For limited ranges the situation is different. It is assumed for the following that secondary particle equilibrium exists. The mean particle range is called $\overline{x}$, and the mean traversal length in the sphere for a single particle is $\bar{s}$. Equilibrium is characterized (see Chap. 3) by a situation where every stopping particle is replaced by the formation of an identical one. The total pathlength of <u>all</u> particles is $\phi\pi r^2 \cdot \bar{l}$. The number of all individual particles is the sum of those which enter the sphere and those which are newly formed in it. If n/V is the number of newly formed particles per unit volume one finds with Eq. (3.30) for the special case of a sphere:

entering particles: $\phi\pi r^2 = n/V \cdot \pi r^2 \cdot \overline{x}$
newly formed particles: $n/V \cdot 4/3\pi r^3$

The mean pathlength s of individual particles is obtained by dividing the total pathlength by the total number of particles

$$\bar{s} = \frac{n/v\pi r^2 \cdot \bar{x} \cdot \bar{l}}{n/v\pi r^2 x + n/\bar{v} \cdot 4/3\pi r^3}$$
$$= \frac{\bar{x} \cdot \bar{l}}{\bar{x} + 4/3r}$$
$$= \frac{\bar{x} \cdot \bar{l}}{\bar{x} + \bar{l}}$$

and:

$$\frac{1}{\bar{s}} = \frac{1}{\bar{x}} + \frac{1}{\bar{l}} \quad . \tag{I.2.6}$$

For large ranges l and s are identical. Equation (I.2.6) which was derived here under special assumptions is generally applicable to all concave bodies.

I.3 The "KEPLER Problem"

The problem here is the interaction between two bodies which are attracted (or repelled) by a force depending on the distance between them. The treatment is most convenient if polar coordinates are used as introduced in Sect. I.1. It is assumed that the second particle with m_2 is resting at the origin of the system ("relative system", see Chap. 2). The force may depend inversely on the square of the distance. Under these conditions one has for the radial and tangential components of acceleration:

$$a_\beta = r\ddot{\beta} + 2\dot{r}\dot{\beta} = 0 \tag{I.3.1a}$$

$$a_r = \ddot{r} - r\dot{\beta}^2 = -\frac{k}{m_1 r^2} \tag{I.3.1b}$$

k is here a proportionality coefficient. The negative sign indicates that the attracting force is directed towards the center. Equation (I.3.1a) may be written as:

$$\frac{1}{r}\frac{d}{dt}(r^2 \cdot \dot{\beta}) = 0$$

which means:

$$r^2 \cdot \dot{\beta} = \text{const.} = f \tag{I.3.2}$$

("constancy of areal velocity").

The second equation is rewritten by elimination of the time dependency and taking r as a function of β:

$$\dot{r} = \frac{dr}{d\beta} \cdot \dot{\beta} = r'\dot{\beta}$$

$$\ddot{r} = \frac{d^2r}{d\beta^2}\dot{\beta}^2 + \frac{dr}{d\beta}\ddot{\beta}$$

$$= r''\dot{\beta}^2 + r'\ddot{\beta} \quad . \tag{I.3.3}$$

With this one has for Eq. (I.3.1):

$$2r'\dot{\beta} + r\ddot{\beta} = 0 \tag{I.3.4a}$$

and

$$r''\dot{\beta}^2 + r'\ddot{\beta} - r\dot{\beta}^2 = -\frac{k}{m_1 r^2} \quad . \tag{I.3.4b}$$

Elimination of $\ddot{\beta}$ leads to:

$$r''\dot{\beta}^2 - \frac{2r'^2\dot{\beta}^2}{r} - r\dot{\beta}^2 = -\frac{k}{m_1 r^2} \tag{I.3.5}$$

and hence with Eq. (I.3.2):

$$r'' - \frac{2r'^2}{r} - r = -\frac{kr^2}{m_1 f^2} \quad . \tag{I.3.6}$$

Integration of this equation is achieved by introducing the new variable $u = 1/r$:

$$u'' + u = \frac{k}{m_1 f^2} \quad . \tag{I.3.7}$$

Equation (I.3.7) has the general solution:

$$u = A_0\{\cos(\beta + \beta_0) + \sin(\beta + \beta_0)\} + \frac{k}{m_1 f^2} \quad . \tag{I.3.8}$$

A_0 and β_0 are constants which are to be especially determined. β_0 can be chosen freely and is defined in such a way that Eq. (I.3.8) may be written as:

$$u = A_1 \cos\beta + \frac{k}{m_1 f^2} \tag{I.3.9}$$

With this, one obtains:

$$r = \frac{\frac{m_1 f^2}{k}}{1 + \frac{A_1 m_1 f^2}{k}\cos\beta} \quad . \tag{I.3.10}$$

For very large distances (r ?? ∞) there are limiting angles β_1 and β_2, given as:

$$\cos\beta_1, \beta_2 = -\frac{k}{m_1 A_1 f^2} \quad . \tag{I.3.11}$$

With Eq. (I.3.10) one obtains:

$$\begin{aligned} r'\dot{\beta} = \dot{r} &= A_1 r^2 \dot{\beta} \sin\beta \\ &= A_1 \cdot f \sin\beta \quad . \end{aligned} \tag{I.3.12}$$

At large distances the velocity has a radial direction and is equal to v_1. This means that:

$$v_1^2 = A_1^2 f^2 \sin^2\beta_1 \tag{I.3.13}$$

and with Eq. (I.3.10):

$$\cos^2\beta_1, \beta_2 = \frac{k^2}{k^2 + m_1^2 f^2 v_1^2} \quad . \tag{I.3.14}$$

Now the areal velocity f has to be determined. Since it is constant the value may be found by considering very large distances. With reference to Fig. I.3.1 one has:

$$\frac{r\,d\beta}{p} = \frac{r\,dr}{r + dr} \approx \frac{dr}{r}$$

$$dr \approx v_1 dt$$

and hence:

$$r^2 \cdot \frac{d\beta}{dt} = b v_1 = f \tag{I.3.15}$$

b is the "impact parameter", already introduced in Chap. 2. The final expression for the limiting angles is then:

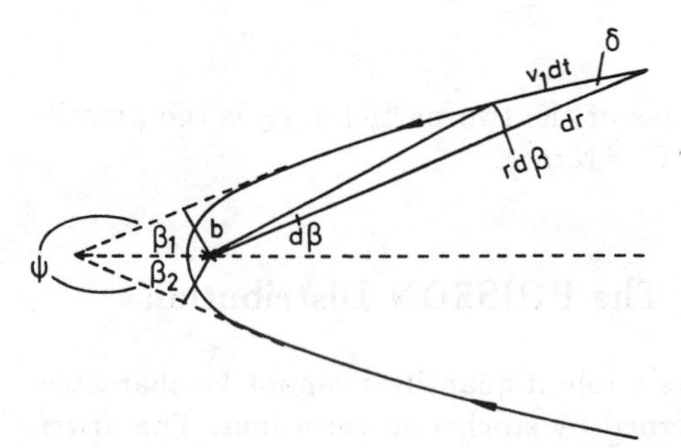

Figure I.3.1. The path of a particle under the influence of a central force

$$\cos^2\beta_1, \beta_2 = \frac{k^2}{k^2 + m_1^2 v_1^4 b^2} \quad . \tag{I.3.16}$$

The energy T_2' transferred to the second particle is according to Eq. (2.30):

$$T_2' = \frac{4AT_0}{(1+A)^2} \sin \varphi/2 \quad . \tag{I.3.17}$$

From Fig. I.3.1 one sees that:

$$\varphi = \pi - 2\beta_1$$

and hence

$$\sin\varphi/2 = \cos\beta_1 \quad .$$

Equation (I.3.16) gives for the differential action cross section:

$$(d\sigma = 2\pi b\, db = \pi\, d(b^2))$$

$$d\sigma = -\frac{4\pi k^2 AT_0}{\mu^2 v_1^4 (1+A)^2 \varepsilon^2}\, d\varepsilon$$

$$d\sigma = \frac{2\pi k^2}{m^2 v_1^2 \varepsilon^2} \quad . \tag{I.3.18}$$

Since the second particle was considered to be at rest one has:

$$T_0 = \frac{m_1 v_1^2}{2}$$

which was used to obtain the expression above.

The coefficient k is in the case of electrostatic interaction (COULOMB's law):

$$k = \frac{Z_1^* \cdot Z_2^* \cdot e^2}{4\pi\varepsilon_0}$$

where Z_1^*, Z_2^* are the effective charges of the two particles. ε_0 is the permittivity of vacuum ($4\pi\varepsilon_0 = 8.99 \cdot 10^9\, C^{-2}\, N\, m^2$).

I.4 Statistical Distributions. The POISSON Distribution

Statistical distributions play always a role if quantities cannot be characterized by a single value but are governed by stochastic variations. The *distri-*

bution function F(x) gives the probability to find values smaller x (x is here assumed to be non-negative). Differentiation yields the practically important *distribution density* f(x) dx describing the probability to find a value in the interval x ... x + dx. This means:

$$F(x) = \int_0^x f(n)\,dn \quad . \tag{I.4.1}$$

Important parameters of any distribution are the *expectation value* x and the *variance* σ^2 with:

$$\overline{x} = \int_0^\infty x \cdot f(x)\,dx \tag{I.4.2}$$

and:

$$\sigma^2 = \int_0^\infty (\overline{x} - x)^2 f(x)\,dx \quad . \tag{I.4.3}$$

Also one has:

$$\begin{aligned} \sigma^2 &= \int_0^\infty x^2 f(x)\,dx - \overline{x}^2 \\ &= \overline{x^2} - \overline{x}^2 \quad . \end{aligned} \tag{I.4.4}$$

The POISSON-distribution is a special case applicable to "rare" events, i. e. with very low probability of occurrence. It shall be exemplified here by a time-dependent process. The probability for an event to occur in an interval dt is $\lambda \cdot dt$. If the probability for events is called $p_n(t)$ one can formulate the following set of differential equations:

$$dp_n(t) = \lambda\,dt(p_{n-1}(t) - p_n(t)) \quad . \tag{I.4.5}$$

Since there are negative event numbers one has especially: ?? which has, of course, the solution:

$$p_0 = e^{-\lambda t} \quad . \tag{I.4.6}$$

Inserting this into Eq. (I.4.5) the system can be solved in a step-wise fashion:

$$p_1(t) = \lambda t \cdot e^{-\lambda t}$$

and generally:

$$p_n(t) = \frac{(\lambda t)^n}{n!} e^{-\lambda t} \quad . \tag{I.4.7}$$

The expectation value n is then:

$$\begin{aligned}\overline{n} &= \sum_{n=0}^{\infty} n p_n(t) \\ &= \sum_{n=0}^{\infty} n \cdot \frac{(\lambda t)^n}{n!} e^{-\lambda t} \\ &= \lambda t \quad .\end{aligned} \tag{I.4.8}$$

With this one may write:

$$p_n = \frac{\overline{n}^n}{n!} e^{-\overline{n}} \tag{I.4.9}$$

which is the general form of a POISSON distribution.

The variance is obtained in a similar way:

$$\begin{aligned}\sigma^2 &= \sum_{n=0}^{\infty} (n - \overline{n})^2 \frac{\overline{n}^{-n}}{n!} e^{-\overline{n}} \\ &= \sum_{n=0}^{\infty} n(n-1) + n \frac{\overline{n}^n}{n!} e^{-\overline{n}} - \overline{n^2} \\ &= \overline{n^2} + \overline{n} - \overline{n^2} = \overline{n} \quad .\end{aligned} \tag{I.4.10}$$

One sees that for the POISSON distribution expectation value and variance are identical.

I.5 LAPLACE Transforms

The mathematical formalism to be briefly introduced here is very useful in many fields of mathematical physics but particularly where convolution integrals are involved.

If an original function f(x) is given its LAPLACE transform g(s) is defined as:

$$g(s) = \int_0^{\infty} f(x) e^{-xs} \, ds \tag{I.5.1}$$

where s is another variable, the counterpart of x in the "LAPLACE space". Both functions may be considered as equivalent representations in different frames of reference:

$$g(s) \longleftrightarrow f(x) \quad .$$

If f(x) is sufficiently "well-behaved", the following relations hold also true:

$$\begin{aligned} sg(s) &\longleftrightarrow -\frac{df(x)}{dx} \\ s^2g(s) &\longleftrightarrow \frac{d^2f(x)}{dx^2} \end{aligned} \tag{I.5.2}$$

They are quite remarkable and practically important since a differentiation in the original space is equivalent to a simple multiplication in the LAPLACE space. This means that differential equations are replaced by algebraic expressions which are, of course, much easier to solve.

If f(x) is a probability distribution density the characteristic parameters can be directly obtained if only the LAPLACE transform g(s) is known:

$$\begin{aligned} \overline{x} &= \int_0^\infty xf(x)\,dx \\ &= \lim_{s\to 0}\int_0^\infty xf(x)e^{-xs}\,ds \\ &= -\lim_{s\to 0}\frac{dg}{ds} \end{aligned} \tag{I.5.3}$$

and:

$$\sigma^2 = \lim\left(\frac{d^2g}{ds^2} - \left(\frac{dg}{ds}\right)^2\right) \quad . \tag{I.5.4}$$

As pointed out above the LAPLACE transform is indispensable for the treatment of convolution integrals:

$$f(x) \star f(x) = \int_0^x f(u)f(x-u)\,du \quad .$$

After transformation one has:

$$\int_{x=0}^\infty \int_{u=0}^x f(u)\cdot f(x-u)e^{-xs}\,du\,dx$$

which now depends on the two variables x and u. This means that the first integration has to be performed for u from 0 to x and then for x from 0 to ∞. Figure I.5.1 demonstrates that the same area in u-x-space is covered if x is extended from u to ∞ and u from 0 to ∞. One may thus write also:

$$f(x) * f(x) = \int_{x=u}^\infty \int_{u=0}^\infty f(u)\cdot f(x-u)e^{-xs}d\cdot du$$

and substituting $z = x - u$

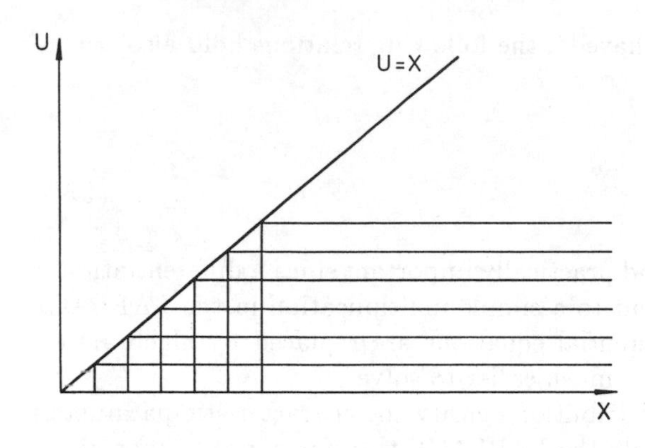

Figure I.4.1. On the derivation of the LAPLACE-transform of a convolution integral

$$f(x) * f(x) = \int_{z=0}^{\infty} \int_{u=0}^{\infty} f(u) \cdot f(z) e^{-zs} e^{-us} \, dz \, du \qquad (I.5.5)$$
$$= g(s) \cdot g(s) \quad .$$

This very important result states that a convolution in the original space corresponds to a mere multiplication of functions in the LAPLACE representation.

This is, of course, also true for an n-fold convolution:

$$f^{*n}(x) \longleftrightarrow g^n(s) \quad ,$$

where ^{*n} signifies a n-fold convolution.

If a density distribution function $f_n(x)$ is given by the n-fold convolution of a parent function f(x) one finds for expectation value x_n and variance σ_n^2:

$$\begin{aligned} \overline{x}_n &= - \lim_{s \to 0} \frac{d}{ds} g^n(s) \\ &= -n \lim_{s \to 0} g^{n-1} \frac{d}{ds} \\ &= n\overline{x} \end{aligned} \qquad (I.5.6)$$

and:

$$\begin{aligned} \sigma_n^2 &= \lim_{s \to 0} \left(\frac{d^2 g^n(s)}{ds^2} - \left(\frac{dg^n(s)}{ds} \right)^2 \right) \\ &= \lim_{s \to 0} \left(n(n-1)^{n-2} \left(\frac{dg}{ds} \right)^2 + ng^{n-1} \frac{d^2 g}{ds^2} - \left(ng^{n-1} \frac{dg}{ds} \right)^2 \right) \\ &= n\sigma^2 \quad . \end{aligned} \qquad (I.5.7)$$

Expectation value and variance are the n-fold of the respective value of the parent function.

Further reading
DOETSCH 1961

I.6 Probit Transformation

If a quantity investigated is assumed to follow a *normal* or GAUSS-distribution this may be checked by the procedure of "Probit analysis" which provides the same time estimates of expectation value and variance.

It is in this case:

$$F(x) = \frac{1}{\sigma\sqrt{2\pi}} \int_{-\infty}^{x} \exp\frac{(\overline{x}-n)^2}{2\sigma^2}\, dn = y \quad . \tag{I.6.1}$$

F(x) is a function of a rather general nature. It may be replaced by a "standard" distribution which is characterized by the expectation value 0 and variance 1. The substitution changes, of course, the limits of integration:

$$\frac{1}{\sigma\sqrt{2\pi}} \int_{-\infty}^{x} \exp\frac{(\overline{x}-n)^2}{2\sigma^2}\, dn = \frac{1}{\sqrt{2\pi}} \int_{-\infty}^{k} \exp\frac{z^2}{2}\, dz$$

with k – the new upper limit – now being a function of F(x). It is easily seen that $k = (x - x)/\sigma$ if one sets $Z = (n - x)/\sigma$.

This means that there is a linear relationship between k and x if the assumption of a normal distribution function is correct. One may even deduce x and σ from that graph. k can be extracted from tables of the standard normal distribution.

For practical purposes not K is used but K + S. This is to avoid non-negative values. The quantity thus introduced is called "Probit" Y.

$$Y = \frac{x - \overline{x}}{\sigma} + 5 \quad . \tag{I.6.2}$$

For Y = 5 the abscissa indicates the expectation value, while the slope determines the variance.

The probit analysis is very useful since it transforms the bell shaped GAUSS curve into a straight line which is, of course, much easier to check.

Further reading
FINNEY 1962

I.7 Reaction Kinetics

Since reaction kinetics is referred to several times, particularly in the first chapters of this book, a few remarks about this topic are in place.

A simple case shall be considered where a product C is formed by the reaction of two species A and B with concentrations c_A, c_B, c_C, respectively. As a first approximation one may assume that the probability of the reaction to take place is proportional to the concentrations of both partners:

$$\frac{dc_C}{dt} = k \cdot c_A \cdot c_B \quad . \tag{I.7.1}$$

The proportionality coefficient K which determines the velocity is termed the "bimolecular reaction rate constant" and has the dimension $mol^{-1}\,dm^3\,s^{-1}$. Its actual value depends on the particular type of reaction.

Those reactions which occur on "first encounter" with unit probability are of special interest. They are called "diffusion-controlled", for obvious reasons. In this case a certain estimate of K may be obtained.

One considers a spherical volume with only one molecule A in its center and containing no other reactants. Molecules B diffuse into it to react with A if a certain critical distance a is reached. Under the given premises (reaction on first encounter) the reaction rate will be proportional to the number of molecules B reaching the critical distance per unit time and also to the concentration of A.

For diffusion into a sphere the following relation holds:

$$\frac{dn_B}{dt} = D_B \cdot 4\pi r^2 \cdot \frac{dc_B}{dr} \tag{I.7.2}$$

n_B is the number of molecules B passing the sphere surface, r the radial distance and D_A, D_B the respective diffusion constants. If equilibrium is assumed dn_B/dt is constant ($=j$) and Eq. (I.7.2) can be integrated giving:

$$c_B(b) - c_B(a) = -j\left(\frac{1}{4\pi D_B b} - \frac{1}{4\pi D_B a}\right) \tag{I.7.3}$$

with b being the sphere radius and a the critical distance as before.

The total number of spheres is proportional to c_A. If this is not too high, b is much larger than a. Furthermore, $c_B(a) = 0$ because there is an instant reaction. Taking all this together, one obtains:

$$j = 4\pi D_B a \cdot c_B \quad . \tag{I.7.4}$$

An exactly analogous result is obtained for molecules B. The total reaction rate is then calculated by multiplying with the number of spheres in both cases and adding the separate terms for A and B. This gives:

$$\frac{dc_C}{dt} = 4\pi a c_A c_B (D_A + D_B) \quad . \tag{I.7.5}$$

By comparison with Eq. (I.7.1) one sees that:

$$k = 4\pi a (D_A + D_B) \tag{I.7.6}$$

a is unknown but one may estimate the order of magnitude by inserting typical values of molecule radii. For small uncharged reactants they are about $a \approx 10^{-8}$ cm and $D_A + D_B \approx 10^{-4}\,cm^2\,s^{-1}$ so that one has:

$$K \approx 10^{10}\,mol^{-1}\,dm^3\,s^{-1} \quad .$$

Appendix II
Biological Background

II.1 Structure and Properties of the Genetic Material

In cells of higher organisms the genetic information is carried by chromosomes which can readily be viewed during mitosis. All body cells (somatic cells) possess a double set of homologous but normally not identical chromosomes. The number per set is fixed and typical for a given species (23 for humans). The information for a genetically determined property – this is called the cell's "phenotype" – can accordingly be found at two corresponding sites, which are called alleles. If these are different, two ways of expression are possible: either one allele dominates the other (which is then called recessive) or less often the phenotype is a mixture of both. Obviously recessive alleles can only be expressed phenotypically if they are present on both homologous chromosomes.

There is one exception to the otherwise strict homology, the sex chromosomes in cells of males. They have an X- and a Y-chromosome, whereas cells of females have two X-chromosomes. For genetic studies sex-linked characters, determined by genes carried on the sex chromosomes, present a specially interesting case. They are called "allosomal" or "heterosomal". Characters determined by genes on other chromosomes are called "autosomal". The number of chromosome sets in a cell is referred to as "ploidy". Somatic cells are diploid.

Germ cells have only one set of chromosomes: they are haploid. After fertilization the new organism is again diploid, carrying information from both the egg and the sperm, which leads to the combination of maternal and paternal genetic properties.

Chromosomes consist mainly of proteins and *deoxyribonucleic acid* (DNA).

DNA is a large polymeric molecule, i.e. it is built from small units, the monomers, which are called nucleotides. There are four different nucleotides in DNA; however, these are constructed in the same way (Fig. II.1.1). A heterocyclic organic base is linked via a *N-glycosidic* bond to a sugar, *deoxyribose*. Phosphate residues attach to either the C_5- or C_3-atoms in the deoxyribose so that a continuous chain may be formed. The only difference between nucleotides is found in the bases which are derived either from a pyrimidine or a purine. Thymine and cytosine belong to the first group, adenine and gua-

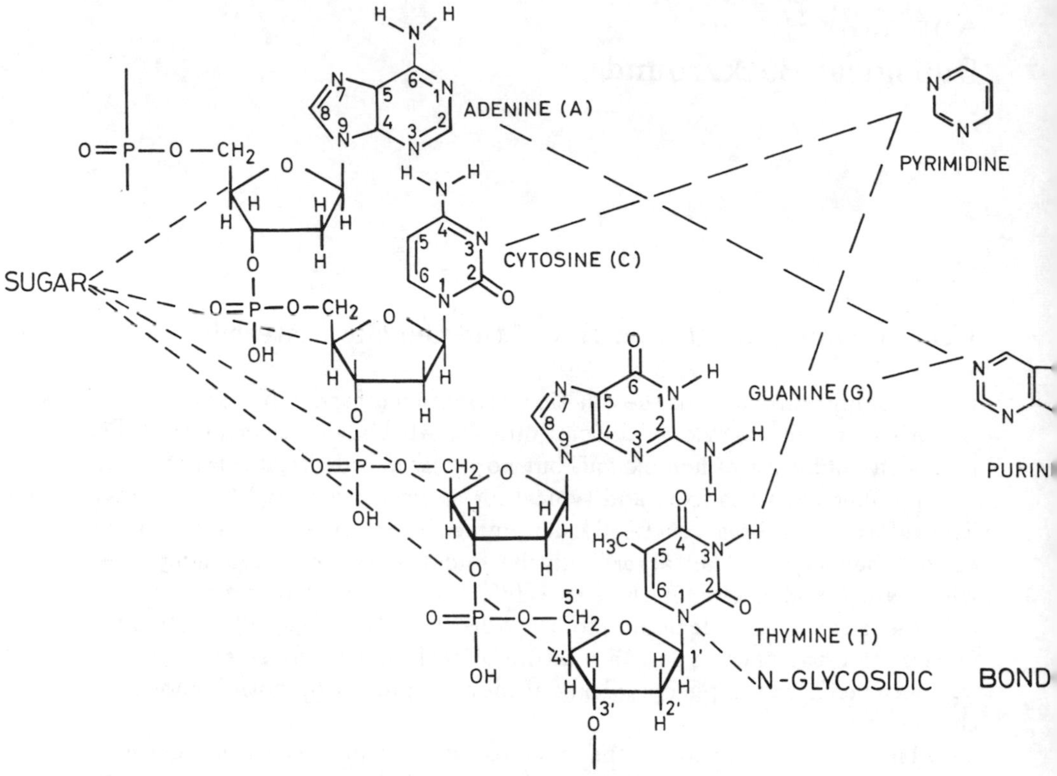

Figure II.1.1. The molecular structure of a DNA strand

nine to the second. The combination of base and sugar is called a nucleoside. Table II.1 lists the special names and their common abbreviations.

According to the position where the phosphate is attached nucleotides may be grouped into the categories 5- and 3-*monophosphates*. In the DNA chain the sequence of 5- and 3-bonds constitutes a clear polarity of the whole molecule. DNA exists normally as a double helix: adenine and thymine can form hydrogen bonds between each other as well as cytosine and guanine (two in the first case, three in the latter). so that a double-stranded struc-

Table II.1.1. Bases, nucleosides and nucleotides. (Common abbreviations are given in brackets)

base	nucleoside	nucleotide
adenine (A)	adenosine (AdR)	adenosine monophosphate (AMP)
guanine (G)	guanosine (GdR)	guanosine monophosphate (GMP)
cytosine (C)	cytidine (CdR)	cytidine monophosphate (CMP)
thymine (T)	thymidine (TdR)	thymidine monophosphate (TMP)
uracil (U)	uridine (UR)	uridine monophosphate (UMP)

[1)] in ribonucleic acid only

ture is possible where always an A is found opposite to a T, a G opposite a C and vice versa. The two strands contain the same information, like a positive and a negative picture. The whole molecule is twisted in a regular manner (Fig. II.1.2). In chromosomes higher structures can be discerned, in which proteins play a constitutive part. The most important of these are the histones, which are characterized by a relatively high content of basic amino acids. A sequence of about 140 nucleotide pairs is wound around a protein core of these histones forming a "nucleosome". These small bodies are linked via short DNA stretches ("linker DNA") where another histone (H_1) is associated in a way not yet fully understood (Fig. II.1.3). At this level

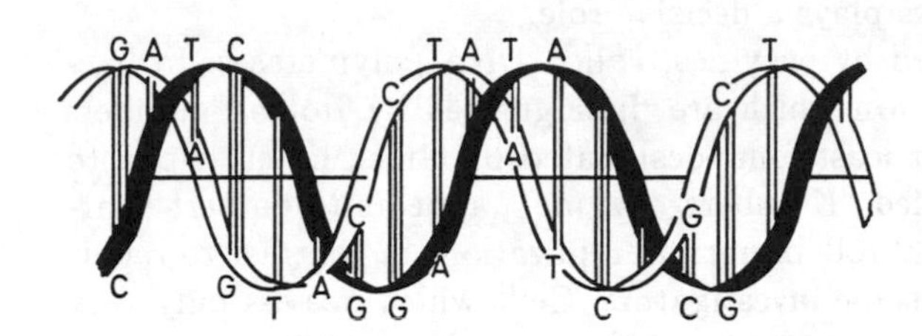

Figure II.1.2. The twisting of the DNA double helix (after LEHNINGER 1970)

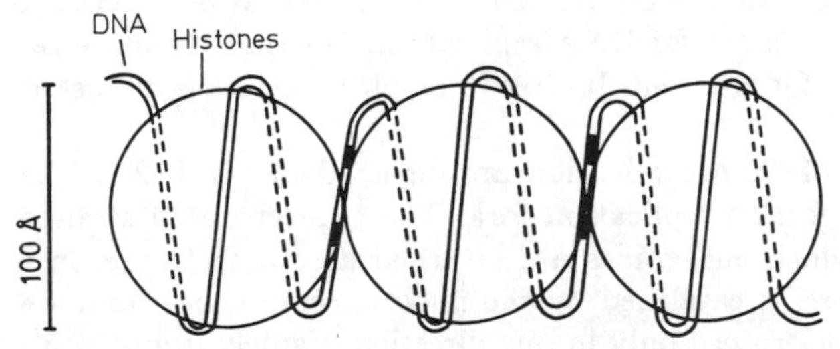

Figure II.1.3. The winding of the DNA double helix around histone bodies to form nucleosomes

the genetic material looks like a string of beads. Higher structures are formed by "super-helices" of the nucleosome chains.

It is seen from this description that DNA and proteins are in very close contact, which is certainly relevant for the understanding of radiobiological reactions, but has not yet received much attention in experimental investigations.

The optical properties of DNA are essentially determined by the bases; the other constituents contribute only indirectly. The absorption spectrum (see Fig. 6.1) is characterized by a strong, rather broad band around 260 nm. The interaction between the π^- electron systems of the bases in the regular array of the double helix reduces the extinction coefficient ("hypochromicity"). If the two strands are separated by high temperatures or extreme pH the absorption is increased.

The DNA content of a given cell type is essentially constant, but it varies widely within the plant and animal kingdoms and depends, of course on the ploidy state.

II.2 Replication of DNA

To secure the stability of genetic information over many cell generations it is obviously essential to achieve a very high fidelity in the replication process. Its mechanism is also important for the understanding of repair processes (Chap. 13). The situation is best understood in the bacterium *E. coli*: in higher organisms it is not known to the same degree. In all cases, however, it is clear that the doublestrandedness plays a decisive role.

Replication is mediated by enzymes, called DNA-polymerases. In bacteria three of them are known which are distinguished by Roman numbers (in eucaryotes, there are at least four, designated by Greek letters alpha to δ. Polymerase I, the so-called "Kornberg-enzyme", appears to be best characterized although its actual role in normal replication (in contrast to repair replication) is doubted by some investigators. Cells which possess only very small amounts can nevertheless replicate at the normal speed. They are, however, very sensitive to the action of UV light. It is believed that polymerase I where it is present is responsible for DNA replication, but that it can be replaced by other enzymes if it is absent. Its role in repair processes is discussed elsewhere (Chap. 13).

The basic features of DNA replication are depicted in Fig. II.2.1. The double strand opens to form a replication fork. The two separated strands serve as templates on which nucleotides are attached according to the base pairing rules. This process is catalyzed by the polymerases, whose common property is that they can proceed only in one direction, namely from the 5′- to the 3′-position. Since the two strands have opposite polarity a continuous synthesis is only possible on one strand. On the other one only short fragments ("Okazaki fragments") are formed which are linked together at a later stage

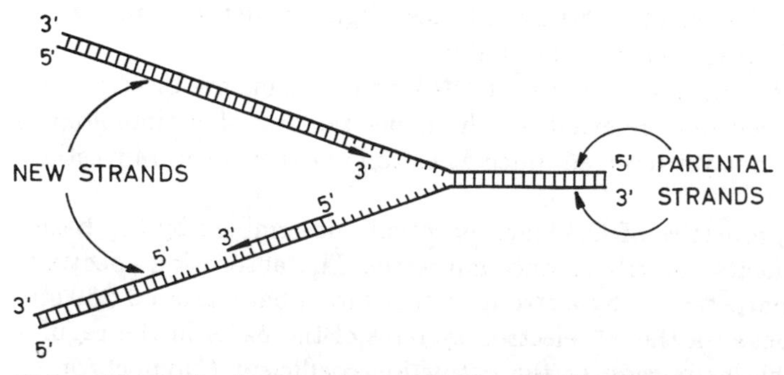

Figure II.2.1. Semi-conservative replication of DNA. The new strands are polymerized in the 5–3-direction. At least in one strand only short stretches are synthesized (OKAZAKI fragments) which are later joined by ligases

by another enzyme, called *ligase*. It cannot be ruled out that the synthesis is initially discontinuous on both strands although this is not really necessary. After completion of the whole round, two double-stranded molecules evolve which both consist of an old parent and a new daughter strand. This mode of replication is called "semi-conservative".

Polymerase I is a multifunctional enzyme. In the context of the present discussion the so-called "proof-reading" activity has to be mentioned. Strictly speaking, this is an exonucleolytic activity: The enzyme is able to chop off one nucleotide in the 3′–5′-direction, i.e. opposite to the process of replication. This occurs if the wrong nucleotide has been inserted so that the base pairing would not be correct. By this action the degree of fidelity is markedly increased, which could not be understood on merely thermodynamical grounds. Thus the polymerase fulfills at least two purposes, to catalyse replication and to safeguard its fidelity.

II.3 Information Processing

The machinery of cellular metabolism is operated by catalytic proteins, the enzymes. Their properties are determined by the sequence of amino acids which is encoded in the base sequence of the DNA. Since there are 20 different amino acids in the cell but only four bases, a "word" of three bases ("triplet") codes for one amino acid. The actual processing of genetic information proceeds in several steps (Fig. II.3.1): First the base sequence on the DNA is copied onto another nucleic acid, *ribonucleic acid* (RNA) where again

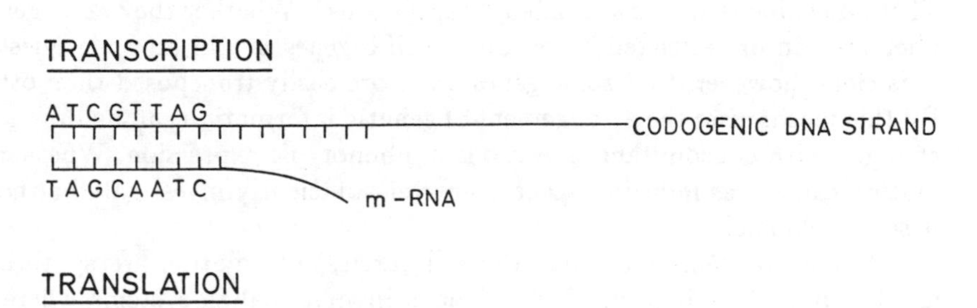

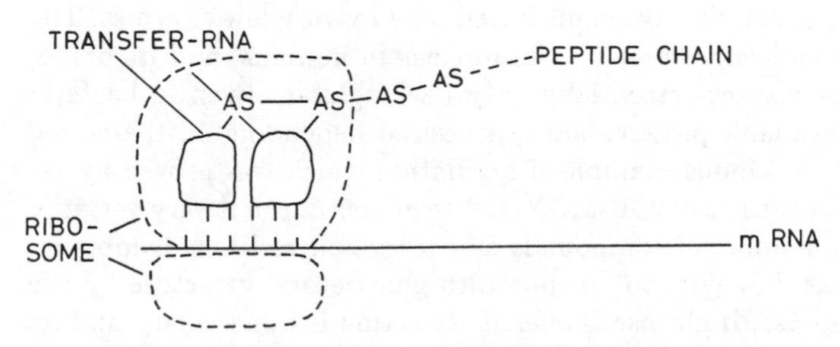

Figure II.3.1. Basic scheme of protein biosynthesis (see text)

the specific base-pairing is used (*transcription*). RNA is single-stranded, has a slightly different sugar component (ribose) and contains the pyrimidine base *uracil* instead of thymine. This carrier molecule which is called "messenger-RNA" (mRNA) transports the information to the site of actual protein synthesis, the ribosome. This organelle also contains a specific RNA, *ribosomal* RNA (r-RNA) of which several species are known. Together with a number of proteins they constitute the work-bench for the protein assembly. Smaller RNAs, *transfer-RNAs* (t-RNA), which are charged with the respective amino acids, line up according to the prescribed sequence, and the amino acids are joined together by peptide bonds (*translation*). The correct fit is recognized by a specific site in the t-RNA, the anti-codon, which is base-paired to a triplet in the messenger-RNA. After completion the, now uncharged, t-RNA is released and the ribosome disassembles when the total protein molecule is formed.

Apart from the triplets coding for specific amino acids there are also a few which cause protein chain termination.

It was originally thought that the sequences of base triplets in the DNA and that of amino acids in the protein is strictly colinear. It came as a surprise when it was found that many genes contain stretches whose counterparts are not found in the protein, with other words, their information is not used. These "non-coding" regions are called "introns" while the others are named "exons". Both introns and exons are transcribed. The sorting out takes place at the level of m-RNA by a mechanism called "m-RNA-splicing".

Another finding has to be mentioned in this context: The sequence of genes cannot be considered as permanently fixed. An exchange may take place by transposition of genes or parts of them from one place to another. These movable elements are called "transposons". Whether they are a general phenomenon or restricted to certain specific genes is still an open question. It is clear, however, that some genes are more easily transposed than others. By this mechanism the arrangement of genetic information can be drastically changed with concomitant alterations in phenotypic expression. Whether radiation can act as inducing agent is not yet sufficiently investigated although it seems plausible.

Not all proteins for which the cell carries information are synthesized all the time. This is particularly obvious in mammalian systems where the high degree of specialization is often mediated only by very few enzymes. The red blood cell (which has however lost its nucleus in humans) and its predecessors produce for instance essentially only haemoglobin. Even in bacteria only parts of the available proteins are synthesized depending on the actual culture conditions. A famous example of regulation which was proven by experiments is the so-called "JACOB-MONOD"-scheme. *E. coli* is very versatile in making use of a number of compounds as energy sources. One compound is lactose which has, however, to be split into glucose and galactose by the enzyme β-galactosidase. If glucose is offered its action is unnecessary and its synthesis would be wasteful. The enzyme is therefore found only at a very

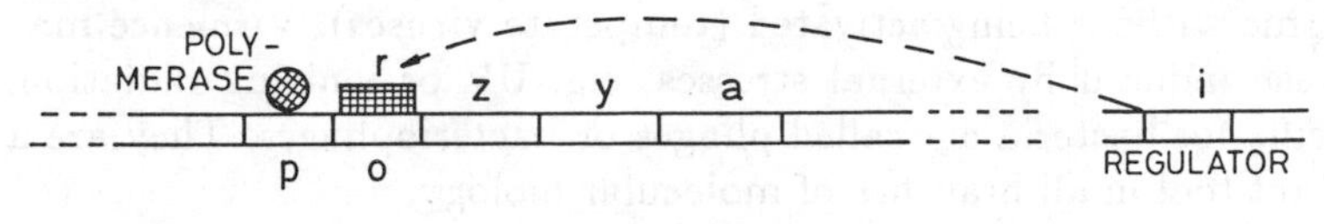

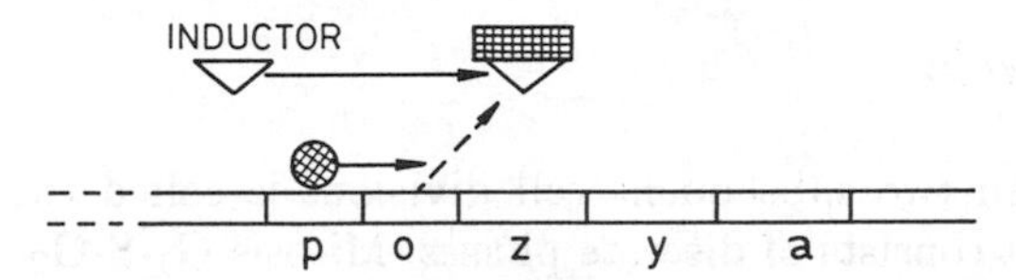

Figure II.3.2. Regulation of transcription according to the model of JACOB and MONOD: In the repressed state the polymerase which binds to the promotor site (p) is blocked by the repressor r at the operator o so that the structural genes z, a, and a cannot be transcribed. After induction the repressor is modified in such a way that it no longer binds to the operator, and transcription can proceed

low level. Only if lactose is present as the sole carbon source its concentration rises considerably. The scheme of regulation is depicted in Fig. II.3.2: The information for the enzyme β-galactosidase is encoded on the structural gene. Its transcription is controlled from a preceding section, *the operator*. It can be blocked by a specific protein, the *repressor*. The transcribing enzyme, DNA-dependent RNA-polymerase, binds to the DNA at the *promotor* from where synthesis starts. If the operator is blocked by the repressor transcription cannot proceed. The presence of lactose in the absence of glucose leads to an inducing chemical signal by which the repressor is altered in such a way that it no longer binds to the operator and β-galactosidase is synthesized. The combination of promotor, operator and structural gene is called an *operon*. The lac-operon consists in fact of three structural genes which are jointly regulated, but this is not important in this context. The information for the repressor protein is found at a different site, the *regulator*. It must be pointed out that the mechanism just described is only one example although the best known. Whether similar schemes operate in mammalian cells cannot be said at present. Not all bacterial enzymes are regulated: some are continuously synthesized. They are called "constitutive". Induction and repression play a role in connection with "SOS-repair" (Chap. 13).

Genetic information storage is also found below the level of cellular organisms. Viruses contain nucleic acids (DNA or RNA) coding for virus-specific information. They are encapsulated in a protein coat. Since they do not possess the enzymatic machinery for information-processing and replication they can reproduce only in cells. They attach to the surface of host cells (this interaction is specific) and inject the nucleic acid into them. The cellular machinery is then used to produce virus-specific protein and to replicate their nucleic acids. By this mechanism viruses are produced in great numbers. The infected cell finally dies and upon lysis the newly-formed particles are released into the environment. Sometimes their genetic information is only integrated into the cellular genome without being activated (temperate viruses). Virulence may be in this case induced by external stresses, e.g. UV or ionizing radiation. Viruses specific for bacteria are called phages or bacteriophages. They are a very important tool in all branches of molecular biology.

II.4 The Cell Division Cycle

The sequence of events between two subsequent cell divisions is called the "cell cycle". In all higher cells it consists of discrete phases: Mitosis-G_1-S-G_2-Mitosis (Fig. II.4.1). In G_1 the cell prepares for DNA-replication which takes

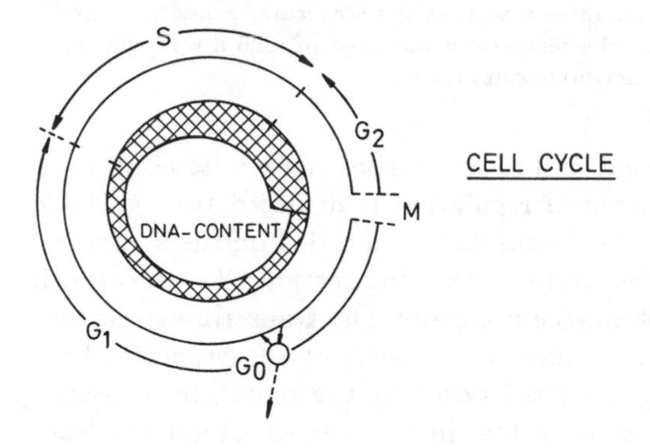

MITOSIS

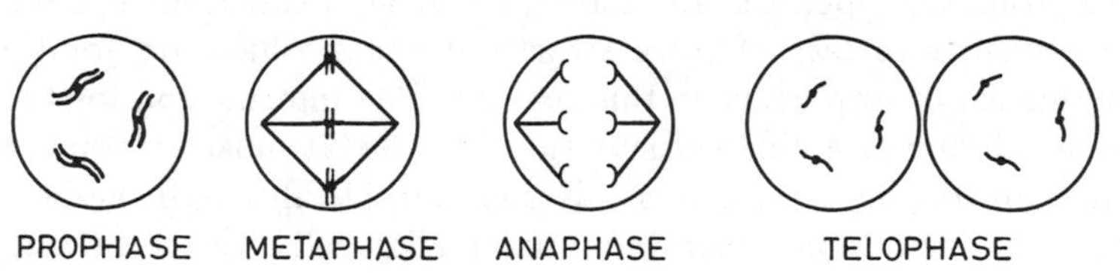

Figure II.4.1. Scheme of the cell cycle. Mitosis is depicted below

place in S; G_2 precedes mitosis. The time taken by S, G_2 and M (mitosis) is nearly constant for a given cell type under defined conditions whereas G_1 may vary substantially depending on the actual rate of division. In a normal undisturbed tissue, not all division competent cells are actively progressing towards division because the number of cells to be replaced is comparatively low. Only under severe stress is the full potential made use of. It seems as if the adjustment is made by shortening or lengthening the G_1-phase. It has therefore been suggested that non-cycling cells are not really in G_1, but in an "attention phase" called G_0 from which they may be triggered into active division. Whether this is really so is still a matter of debate. Mitosis consists of four stages: prophase when the chromosomes start to condense, metaphase when the longitudinally split chromosomes assemble at the equatorial plate, anaphase, when the two halves separate and are distributed to the daughter nuclei and the telophase when two new nuclear membranes are formed preceding cell division.

The minimum cycle time (i.e. $M + G_1 + S + G_2$) depends on the cell type. It is about 10–20 hours for mammalian cells, but only about two hours for yeast cells. Bacteria do not have a cell cycle structure as just described, but synthesize DNA continuously.

In normal untreated populations the cells progress through the cycle in an unrelated way, i.e. all stages may be found together at a certain instant. It is, however, possible by special techniques to produce synchronized populations where all cells progress in parallel. Since radiation sensitivity depends on cycle stage (Chap. 10) they are an important experimental tool in radiation biology.

II.5 Gene Cloning

As pointed out several times before the methods of modern molecular biology have opened up new avenues of research also in radiation biology, e.g. in the elucidation of mutation induction and of repair mechanisms. In many cases the important step is the "cloning" of the respective gene, i.e. its separation from the rest of the genome and its selective propagation *in vitro*. The key to these approaches lies in microbial "vectors", i.e. specially constructed plasmids which can be multiplied in bacteria, and whose information may be expressed in eucaryotes, especially in mammalian cells. The essential tools for their construction are the "restriction endonucleases". These are enzymes which induce double-strand breaks at certain well-defined base-sequences. They may be "blunt" or "frank" like the cutting by a sharp knife or they may possess overlapping "sticky" single-strand ends. The latter are more interesting in the present context since they allow easily the ligation of different pieces produced by the same cutting enzyme.

The procedure shall be sketched here by explaining the cloning of repair enzymes (Fig. II.5.1): A necessary condition is the availability of mutants which lack the property coded for by the respective gene. This makes selection possible.

The search for radiation-sensitive mammalian cell lines is partly motivated by this. In bacteria and yeast where a great number of repair-deficient mutants has been known for a long time the situation is much easier.

The starting point is a bacterial plasmid carrying one or more genes conferring resistance to certain compounds (so-called "resistance markers". They may serve as easy means of identification. Both – the plasmid as well as the cell's wild type genomic DNA – are cut by a suitable restriction nuclease

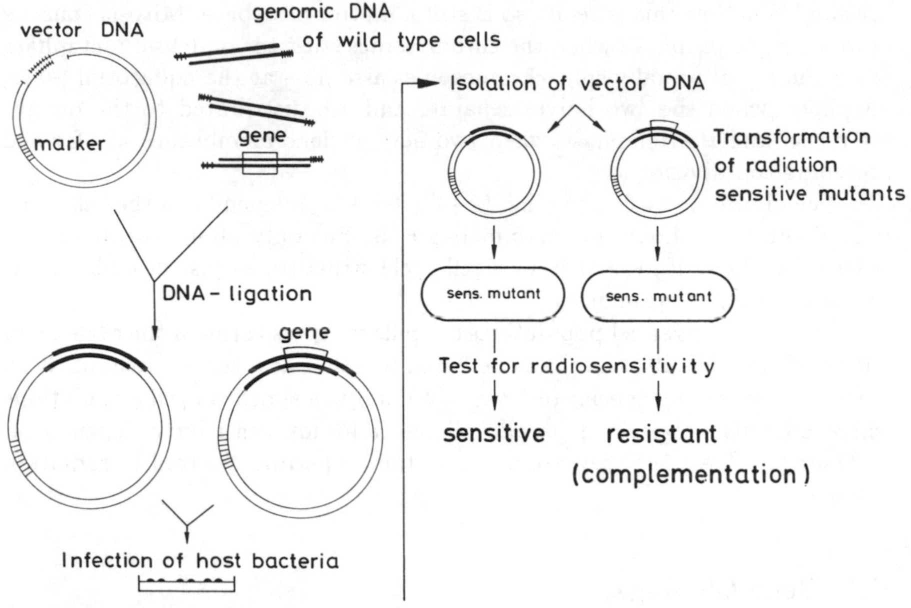

Figure II.5.1. Scheme of "gene cloning". Details see text

which produces double-strand breaks with "sticky" ends. By this procedure a great number of short DNA-pieces is obtained one of which harbours also the required gene. With the aid of a ligase plasmid and genomic DNA-stretches are joined in a completely stochastic manner. The "recombinant" plasmids thus obtained are propagated in bacteria which do not carry the resistance markers so that only those will grow to colonies in selective media which have taken up the plasmid. The DNA of a great number of bacteria is isolated and used to transform sensitive cell mutants. If one of them exhibits now radioresistance one may safely assume that the plasmid used for transformation contained the respective repair gene. By applying several types of restriction enzymes and multiple repetition of the general procedure this gene can be further and further delineated and in the end even be sequenced.

Further reading
BERGER and KIMMEL 1987

References

ABE, M., HIRAOKA, M. (1985): Localized hyperthermia and radiation in cancer therapy. Int. Radiat. Biol. *47*, 347-369.

ABRAHAMSON, S., BENDER, M.A., CONGER, A.D., S. WOLFF (1973): Uniformity of radiation-induced mutation rates among different species. Nature *245*, 460-462.

ADAMS, G.E., FLOCKHART, I.R., SMITHEN, C.E., STRETFORD, I.J., WARDMAN, P., WATTS, M.E. (1976): Electronic-affinic sensitization VII. A correlation between structures, one-electron reduction potentials and efficiencies of nitroimidazoles as hypoxic sensitizers. Radiat. Res. *67*, 9-20.

ADAMS, G.E., FOWLER, J.F., WARDMAN, P. (1978): Hypoxic cell sensitizers in radiobiology and radiotherapy. Brit. J. Cancer Suppl. *37*, Suppl. III.

ADAMS, G.E., JAMESON, D.G. (1980): Time effects in molecular radiation biology. Radiat. Environm. Biophysics *17*, 95-114.

AHNSTRÖM, G., EDVARDSSON, K.A. (1974): Radiation-induced single strand breaks in DNA determined by rate of alkaline strand separation and hydroxylapatite chromatography: an alternative to velocity sedimentation. Int. J. Radiat. Biol. *26*, 493-497.

ALMASSY, Z., KREPINSKY, A., BIANCO, A., KOETELES, G.J. (1987): The present state and perspectives of micronucleus assay in radiation protection. A review. Appl. Radiat. Isot. *38*, 241-249.

ALPER, T. (1979): Cellular radiobiology. Cambridge University Press: Cambridge.

ALPER, T. (ed.) (1975): Cell survival after low doses of radiation: theoretical and clinical implications. The Institute of Physics, J. Wiley and Sons, London.

ALPER, T., MOORE, J.L., BEWLEY, D.K. (1967): LET as determinant of bacterial radio-sensitivity and its modification by anoxia and glycerol. Radiat. Res. *32*, 277-293.

ANDREONI, A., CUBEDDU, R.(eds.) (1984): Porphyrins in tumor phototherapy. New York London: Plenum Press.

ATTIX, F.H., ROESCH, W.C., TOCHILIN, E (1968): Radiation dosimetry, 3 volumes. Academic Press: New York.

AUERBACH, C. (1976): Mutation research. Chapman and Hall: London.

AUXIER, J.A., SNYDER, W.S., JONES, T.D. (1968): Neutron interaction and penetration in tissue. In: ATTIX et al. (1968), Vol. I, p. 275-316

AVERY, O.T., MacLEOD, C.M., McCARTHY, M. (1944): Studies on the chemical nature of the substance inducing transformation of pneumococcal types. J. Exp. Med. *79*, 137-158.

BAIR, W.J. (1974): Toxicology of plutonium. Advances Rad. Biol. *4*, 255-316.

BAKER, M.L., DALRYMPLE, G.V., SANDERS, J.L., MOSS, A.J., Jr. (1970): Effects of radiation on asynchronous and synchronized L cells under energy deprivation. Radiat. Res. *42*, 320-330.

BARENDSEN, G.W. (1967): Mechanisms of action and different ionizing radiations on the proliferative capacity of mammalian cells. Theor. Exp. Biophysics I, 168-232.

BARENDSEN, G.W. (1968): Responses of cultured cells, tumours and normal tissues to radiations of different linear energy transfer. Curr. Top. Radiat. Res. *4*, 293-356.

BARENDSEN, G.W. (1970): Local energy requirements for biological radiation damage and their modification by environmental conditions. Proc. 2nd Symp. Microdosimetry EURATOM, p. 83-106.

BARENDSEN, G.W. (1978): Fundamental aspects of cancer induction in relation to the effectiveness of small doses of radiation. In: IAEA (1978), Vol. II, p. 263-275.

BARENDSEN, G.W. (1979): Influence of radiation quality on the effectiveness of small doses for induction of reproductive death and chromosome aberrations in mammalian cells. Int. J. Radiat. Res. *36*, 49-63.

BARENDSEN, G.W. (1982): Dose fractionation, dose rate and iso-effect relationships for normal tissue response. Int. J. Radiat. Oncol. Biol. Phys. *8*, 1981-1997.

BARKAS, W.H. (1963): Nuclear research emulsions. Academic Press: New York.

BASERGA, R. (1986): Molecular biology of the cell cycle. Int. J. Radiat. Biol. *49*, 219-226.

BAUM, J.W., VARMA, M.N., WINGATE, C.L., PARETZKE, H.G., KUEHNER, A.V. (1974): Nanometer dosimetry of heavy ion tracks. Proc. 4th Symp. Microdosimetry (BOOZ, J., EBERT, H.G., EICKEL, R., WAKER, A., eds.), EURATOM, p. 93-112.

BEAUREGARD, G., POTIER, M. (1984): Radiation inactivation of membrane proteins: Molecular weight estimates in situ and after Triton-100 solubilization. Anal. Biochem. *140*, 403-408.

BEDFORD, J.S. (1987) : The influence of proliferative status on responses to fractionated and low dose rate irradiation. In: FIELDEN, FOWLER, HENDRY and SCOTT (1987), p. 461-467.

BEIR, Committee on the Biological Effects of Ionizing Radiations (1972): The effects on populations of exposure to low levels of ionizing radiation. Natl. Academy of Science - Natl. Research Council: Washington.

BEIR, Committee on the Biological Effects of Ionizing Radiations (1980): The effects on populations of exposure to low levels of ionizing radiations. Washington: National Academy of Sciences.

BELLAMY, A.S., HILL, B.T. (1984): Interactions between clinically effective antitumor drugs and radiation in experimental systems. Biochim. Biophys. Acta *738*, 125-166.

BEN-HUR, E. (1984): Involvement of poly(ADP-ribose) in the radiation response of mammalian cells. Int. J. Radiat. Biol. *46*, 659-672.

BEN-HUR, E., ELKIND, M.M., BRONK, B.V. (1974): Thermally enhanced radioresponses of cultured Chinese hamster cells: Inhibition of repair of sublethal damage and enhancement of lethal damage. Radiat. Res. *58*, 38-51.

BERGER, M. (1973): Report NMSIR 73-107. US National Bureau of Standards.

BERGER, M. (1985): Energy loss straggling of protons in water. Radiat. Prot. Dosimetry *13*, 87-90.

BERGER, M. (1988): Microdosimetric event size distribution in small sites in water vapour irradiated by protons. Phys. Med. Biol. *33*, 583-595.

BERGER, S.L., KIMMEL, A.R. (eds.) (1987): Guide to molecular cloning techniques. Meth. Enzymol. 152. Academic Press, San Diego London, 812 pp.

BERGONIÉ, J., TRIBONDEAU, L. (1906): Interprétation de quelques résultats de la radiotherapie et essai de fixation d'une technique rationelle. Compt. Rend. Acad. Sci. *143*, 983-985.

BERNHARDT, J.H. (1988): The establishment of frequency dependent limits for electric and magnetic fields and evaluation of indirect effects. Rad. Env. Biophys. *27*, 1-28.

BERTALANFFY, F.D., LAU, C. (1962): Cell renewal. Int. Rev. Cytol. *13*, 359-366.

BJOERN, L.O., BORNMAN, J.F. (eds.) (1983): Effect of ultraviolet radiation on plants. Physiol. Plantarum *58*, 349-450.

BLAKELY, E.A., TOBIAS, C.A., YANS, T.C.H., SMITH, K.C., LYMAN, J.T. (1979): Inactivation of human kidney cells by high-energy monoenergetic heavy-ion beams. Radiat. Res. *80*, 122-160.

BLAKELY; E.A., NGO, F.Q., CURTIS, S.B., TOBIAS, C.A. (1984): Heavy ion radiobiology: cellular studies. Adv. Radiat. Biol. *11*, 295-390.

BLOK, J., LOHMAN, H. (1973): The effects of gamma radiation in DNA. Curr. Top. Radiat. Res. Quarterly *9*, 165-245.

BLOOM, E.T., AKIYAMA, M., KUSUNOKI, Y., MAKINODAN, T. (1987): Delayed effects of low dose radiation on cellular immunity in atomic bomb survivors residing in the United States. Health Phys. *52*, 585-594.

BOICE, J.D., FRAUMENI, J.F. (eds.) (1984): Radiation carcinogenesis – Epidemiology and biological significance. New York: Raven Press, 490 p.

BOND, V.P., FEINENDEGEN, L.E., BOOZ, J. (1988): What is a low dose of radiation? Int. J. Radiat. Biol. *53*, 1-12.

BOND, V.P., FLIEDNER, T.M., ARCHAMBEAU, J.O. (1965): Mammalian radiation lethality. Academic Press: New York London, 340 p.

BONKA, H. (1982): Strahlenexposition durch radioaktive Emissionen aus kerntechnischen Anlagen im Normalbetrieb. Köln, Verlag TÜV Rheinland, 495 p.

BOOTSMA, D. u.a. (1987): Homology of mammalian, Drosophila, yeast and E. coli repair genes. In: FIELDEN, FOWLER, HENDRY and SCOTT (1987), p. 412-418.

BOOZ, J. (1976): Microdosimetric spectra and parameters of low LET-radiations. 5th Symp. Microdosimetry, EURATOM EUR 5452 d-e-f, p. 311-344, Brussels.

BOOZ, J., SMIT, T. (1977): Local distribution of energy deposition in and around the follicles of a ^{125}J contaminated thyroid. Curr. Top. Radiat. Res. Quarterly *12*, 12-32.

BOREK, C. (1979): Malignant transformation in vitro: Criteria, biological markers and application in environmental screening of carcinogens. Radiat. Res. *79*, 209-232.

BOREK, C. (1980): X-ray-induced in vitro neoplastic transformation of human diploid cells. Nature *283*, 776-778.

BOREK, C., ONG, A. MASON, H. (1987): In vivo – in vitro systems in multistage carcinogenesis. Toxicol. Industrial Health *3*, 347-356.

BRASH, D.E. (1988): UV mutagenic photoproducts in Escherichia coli and human cells: a molecular genetics perspective in human skin cancer. Photochem. Photobiol. *48*, 59-66.

BRECCIA, A., GREENSTOCK, C.L., TAMBA, M. (eds.) (1984): Advances on oxygen radicals and radioprotectors. Bologna: Lo scarabeo, 190 p.

BREIMER, L.H. (1988): Ionizing radiation induced mutagenesis. Brit. J. Cancer *57*, 6-18.

BRENDEL, I., SCHÜTTMANN, W., ARNDT, D. (1978): Cataract of lens as late effect of ionizing radiation in occupationally exposed persons. In: IAEA (1978), Vol. I, p. 309-310.

BRIDGES, B.A., WOODGATE, R. (1985): Mutagenic repair in E.coli: Products of the recA-gene and of the umuD and umuC genes act at different steps in UV-induced mutagenesis. Proc. Nat. Acad. Sci. USA *82*, 4193-4197.

BRIDGES, R.A., LAW, J., MUNSON, R.J. (1968): Mutagenesis in Escherichia coli. II. Evidence for a common pathway for mutagenesis by ultraviolet light, ionizing radiation and thymine deprivation. Molec. Gen. Genetics *103*, 266-273.

BROERSE, J.J., HENNEN, L.A., van ZWIETEN, M.J. (1985): Radiation carcinogenesis in experimental animals and its implications for radiation protection. Int. J. Radiat. Biol. *48*, 167-188.

BROERSE, J.J., MacVITTIE, T.J. (eds.) (1984): Response of different species to total body irradiation. Boston Dordrecht Lancaster: Martinus Nijhoff Publ.

BROWN, H.D., CHATTOPADHYAY, S.K. (1988): Electromagnetic field exposure and cancer. Cancer Biochem. Biophys. *9*, 295-342.

BRYANT, P.E. (1984): Enzymatic restriction of mammalian cell DNA using Pvu II and Bam H1: evidence for the double strand break origin of chromosomal aberrations. Int. J. Radiat. Biol. *46*, 57-65.

BUDD, T., KWOK, C.S., MARSHALL, M., LYTHE, S. (1983): Microdosimetric properties of alpha particle tracks measured in a low pressure cloud chamber. Radiat. Res. *95*, 217-230.

BURKI, H.J. (1980): Ionizing radiation induced 6-thioguanune resistant clones in synchronous CHO cells. Radiat. Res. *81*, 76-84.

BURNS, F.J., ALBERT, R.E., HEIMBACH, R.D. (1968): RBE for skin tumours and hair follicle damage in the rat following irradiation with alpha particles and electrons. Radiat. Res. *36*, 225-241.

BUTTS, J.J., KATZ, R. (1967): Theory of RBE for heavy ion bombardment of dry enzymes and viruses. Radiat. Res. *30*, 855-871.

BUXTON, G.W. (1987): Radiation chemistry of the liquid state: (1) Water and homogeneous aqueous solutions. In: FARHATAZIZ and RODGERS (1987), p. 321-350.

CARLSSON, G.A. (1985): Theoretical basis for dosimetry. In: KASE, BJAERNGARD and ATTIX (1985), p. 2-77.

CASTRO, J.R. et al. (1987): Clinical results of charged particle therapy. In: FIELDEN, FOWLER, HENDRY and SCOTT (1987), p. 910-915.

CASTRO, J.R. und 13 weitere Autoren (1983): Treatment of cancer with heavy charged particles. Int. J. Radiat. Oncol. Biol. Phys. *8*, 2191-2198.

Cellular Radiation Biology (1965): The Williams and Wilkins Co., Baltimore.

CERUTTI, P.A. (1974): Effects of ionizing radiation on mammalian cells. Naturwissenschaften *61*, 51-59.

CHADWICK, K.H., LEENSHOUTS, H.P. (1981): The molecular theory of radiation biology. Springer: Berlin Heidelberg New York.

CHAPMAN, J.D. (1980): Biophysical models of mammalian cell inactivation by radiation. In: MEYN and WITHERS (1980), p. 21-32.

CHAPMAN, J.D., REUVERS, A.O., BORSA, J., GREENSTOCK, C.L. (1973): Chemical radioprotection and radiosensitization of mammalian cells growing in vitro. Radiat. Res. *56*, 291-306.

CHARLESBY, A. (ed.) (1964): Radiation Sources. Oxford: Pergamon.

CHARLTON, D.E., GOODHEAD, D.T., WILSON, W.E., PARETZKE, H.G. (1987): Energy deposition in cylindrical volumes. MRC Radiobiology Unit, Chilton, Didcot, UK, Monograph 85/1.

CHATTERJEE, A., MACCABEE, H.B., TOBIAS, C.A. (1973): Radial cut-off LET and radial cut-off dose calculations for heavy charged particles in water. Radiat. Res. *54*, 479-494.

CHRISTENSEN, R.C., TOBIAS, C.A., TAYLOR, W.D. (1972): Heavy-ion-induced single- and double-strand-breaks in ϕX-174 replicative form DNA. Int. J. Radiat. Biol. *22*, 457-577.

CHU, E.H.Y. (1965): Effects of ultraviolet radiation on mammalian cells. I. Induction of chromosome aberrations. Mutat. Res. *2*, 75-94.

CLEARY, S.F. (1977): Biological effects of microwave and radiofrequency radiation. CRC Crit. reviews on environmental control *8*, 121-166.

CLEAVER, J.E. (1974): Repair processes for photochemical damage in mammalian cells. Advances Rad. Biol. *4*, 1-76.

CLEAVER, J.E. (1978): Absence of interaction between X-rays and UV-light inducing Ouabain- and thioguanine-resistant mutants in Chinese hamster cells. Mutat. Res. *52*, 247-253.

CLEAVER, J.E., BOREK, C., MILAM, K., MORGAN, W.F. (1985): The role of poly(ADP-ribose) synthesis in toxicity and repair of DNA damage. Pharmac. Ther. *31*, 269-293.

CLEAVER, J.E., CORTES, F., KARENTZ, D., LUTZE, L.H., MORGAN, W.F., PLAYER, A.N., VUKSANOVIC, L., MITCHELL, D.L. (1988): The relative biological importance of cyclobutane and (6–4) pyrimidine-pyrimidone dimer photoproducts in human cells: evidence from a Xeroderma pigmentosum revertant. Photochem. Photobiol. *48*, 41-50.

CLEAVER, J.E., KARENTZ, D. (1987): DNA repair in man: regulation by a multigene family and association with human desease. BioEssays *6*, 122-127.

COLE, C.A., DAVIES, R.E., FORBES, P.D., d'ALOISIO, L.C. (1983): Comparison of action spectra for acute cutaneous responses to ultraviolet radiation: man and albino hairless mouse. Photochem. Photobiol. *37*, 623-631.

COLLINS, A., DOWNES, C.S., JOHNSON, R.T. (eds.) (1984): DNA repair and its inhibition. Oxford Washington: IRL Press, 371 p.

COLLINS, A., JOHNSON, R.T. (1987): DNA repair mutants in higher eucaryotes. J. Cell Sci. Supplement *6*, 61-82.
COLLINS, A., JOHNSON, R.T., BOYLE, J.M. (eds.) (1987): Molecular biology of DNA repair. J. Cell Science Supplement *6*,
CONKLIN, J.J., WALKER, R.I. (eds.) (1987): Military radiobiology. Orlando: Academic Press, 404 p.
COUCH, D.B., FORBES, N.L., HSIE, A.W. (1978): Comparative mutagenicity of alkylsulfate and alkanesulfonate derivatives in Chinese hamster ovary cells. Mutat. Res. *57*, 217-224.
COUNCIL ON SCIENTIFIC AFFAIRS (1987): Radon in homes. J. Amer. Med. Ass. (JAMA) *258*, 668-672.
COX, R., MASSON, W.K. (1979): Mutation and inactivation of cultured mammalian cells exposed to beams of accelerated heavy ions. III. Human diploid fibroblasts. Int. J. Radiat. Biol. *36*, 149-160.
COX, R., THACKER, J., GOODHEAD, D.T. (1977): Inactivation of cultured mammalian cells by aluminium characteristic ultrasoft X-rays. II. Dose-response of Chinese hamster and human diploid cells to aluminium X-rays and radiations of different LET. Int. J. Radiat. Biol. *31*, 561-576.
CROMPTON, N.E.A. (1987): Cytogenetic effects of low dose and low dose rate X- and gamma-radiation. Ph.D. thesis, Univ. of Giessen, Germany.
CRONKITE, E.P., FLIEDNER, T.M. (1972): The radiation syndromes. Handbuch der medizinischen Radiologie, Strahlenbiologie Teil 3. Springer: Berlin Heidelberg New York, p. 299-340.
CURTIS, S.B. (1970): The effects of track structure on OER at high LET. In: Charged particle tracks in solids and liquids. The Institute of Physics: London, p. 140-142.
CURTIS, S.B. (1979): The biological properties of high-energy heavy charged particles. In: OKADA (1979), p. 780-787.
CURTIS, S.B. (1986): Lethal and potential lethal lesions induced by radiation – a unified repair model. Radiat. Res. *106*, 252-2027.
CZEIZEL, A., SANKARANARAYANAN, K. (1984): The load of genetic and partially genetic disorders in man I. Mutat. Res. *128*, 73-103.
DALRYMPLE, G.V., GAULDEN, M.E., KOLLMORGEN, G.M., VOGEL, H.H. (eds.) (1973): Medical radiation biology. W.B. Saunders: Philadelphia London Toronto.
DE NETTANCOURT, D., SANKARANARAYANAN (eds.) (1979): Radiation-induced non-disjunction. Mutat. Res. *61*, 1-119.
DEFAIS, M.J., HANAWALT, P.C., SARASIN, A.R. (1983): Viral probes for DNA repair. Adv. Radiat. Biol. *10*, 1-38.
DeJONG, P.J., GROSOVSKY, A.J., GLICKMAN, B.W. (1988): Spectrum of spontaneous mutation at the APRT locus of Chinese hamster ovary cells: an analysis at the DNA sequence level. Proc. Nat. Acad. Sci. USA *85*, 3499-3503.
DENEKAMP, J. (1986): Cell kinetics and radiation biology. Int. J. Radiat. Biol. *49*, 357-380.
DERTINGER, H., JUNG, H. (1969): Molekulare Strahlenbiologie. Springer: Heidelberg Berlin New York.
Deutsche Risikostudie Kernkraftwerke (1979): Verlag TÜV Rheinland.
DEWEY, W.C. et al. (1980): Cell biology on hyperthermia and radiation. In: MEYN and WITHERS (1980), p. 589-621.
DEWEY, W.C., STONE, L.E., MILLER, H.H., GISLAK, R.E. (1971): Radiosensitization with 5-Bromodeoxyuridine of Chinese hamster cells X-irradiated during different phases of the cell cycle. Radiat. Res. *47*, 672-688.
DIETZEL, F. (1978): Thermoradiotherapie. Urban u. Schwarzenberg: München.
DOETSCH, G. (1961): Anleitung zum praktischen Gebrauch der La-Place-Transformation. Oldenbourg: München.
DOUDNEY, C.O. (1976): Mutation in ultraviolet light damaged microorganisms. In: WANG (1976), Vol. II, p. 309-374.

DOUDNEY, C.O., HAAS, F.L. (1959): Mutation induction and macromolecular synthesis in bacteria. Proc. Natl. Acad. Sci. US *45*, 709-722.

DOUGHERTY, T.J., POTTER, W.R., WEISHAUPT, K.R. (1984): The structure of the active component of hematoporphyrin derivative. In: ANDREONI and CUBEDDU (1984), p. 23-35.

DROBETSKY, E.A., GROSOVSKY, A.J., GLICKMAN, B.W. (1987): The specificity of UV-induced mutations at an endogeneous locus in mammalian cells. Proc. Nat. Acad. Sci. USA *84*, 9103-9107.

Du FRAIN, R.J., LITTLEFIELD, G., JOINER, E.E., FRAME, E.L. (1979): Human cytogenetic dosimetry: a dose-response relationship for α-particle radiation from ^{241}Am. Health Physics *37*, 279-289.

DuBRIDGE, R.B., CALOS, M.P. (1988): Recombinant shuttle vectors for the study of mutation in mammalian cells. Mutagenesis *3*, 1-9.

DUNCAN, W., NIAS, A.H.W. (1977): Clinical radiobiology. Churchill Livingstone: Edinburgh London New York.

EBERT, M., HOWARD, A. (eds.) (1972): "Radiation effects and the mitotic cycle" (Conference report). Curr. Top. Radiat. Res. Quarterly *7*, 244-391.

EISENBUD, M. (1987): Environmental radioactivity from natural, industrial and military sources. Orlando: Academic Press, 475 p.

ELESPURU, R.K. (1987): Inducible responses to DNA damage in bacteria and mammalian cells. Env. Mol. Mutagenesis *10*, 97-116.

ELKIND, M.M., SUTTON, H. (1960): Radiation response of mammalian cells grown in culture. I. Repair of X-ray damage in surviving Chinese hamster cells. Radiat. Res. *13*, 556-593.

ELKIND, M.M., WHITMORE, G.F. (1967): The radiobiology of cultured mammalian cells. Gordon and Breach: New York.

ELLIS, F. (1968): The relationship of biological effects to dose-time fractionation factors in radiotherapy. Curr. Top. Radiat. Res. IV, 357-397.

EPP, E.R., WEISS, H., LING, C.C. (1976): Irradiation of cells by single and double pulses of high intensity radiation: oxygen sensitization and diffusion kinetics. Curr. Top. Radiat. Res. Quarterly *11*, 201-250.

EVANS, H.J., BUCKTON, K.E., HAMILTON, G.E., CAROTHERS, H. (1979): Radiation-induced chromosome aberrations in nuclear-dockyard workers. Nature *277*, 531-534.

EVANS, R.D. (1968): X-ray and gamma-ray interactions. In: ATTIX and ROESCH (1968), Vol. I, p. 94-156.

FABIAN, P. (1980): Der gegenwärtige Stand des Ozonproblems. Naturwiss. *67*, 109-120.

FAIN, J., MONIN, M., MONTRET, M. (1974): Energy deposited by a heavy ion around its path. Proc. 4th. Symp. Microdosimetry (BOOZ, J., EBERT, H.G., EICKEL, R., WAKER, A., eds.), EURATOM, p. 169-188.

FARHATAZIZ, RODGERS, M.A.J. (eds.) (1987): Radiation chemistry – principles and applications. Weinheim: Verlag Chemie VCH, 642 p.

FAVRE, A., TYRELL, R., CADET, J. (eds.) (1987): From photophysics to photobiology. Elsevier: Amsterdam New York Oxford, 488 p.

FEINENDEGEN, L.E. (1978): Biological damage from radioactive nuclei incorporated into DNA of cell; implications for radiation biology and radiation protection. Proc. 6th Symp. Microdosimetry (EBERT, H.G., BOOZ, J., eds.). Harwood: London, p. 3-35.

FEINENDEGEN, L.E., TISLJAR-LENTULIS, G.T., EBERT, M. (eds.) (1977): Molecular and microdistribution of radioisotopes and biological consequences. Curr. Top. Radiat. Res. Quarterly *12*, 1-576.

FEINENDEGEN, L.E. (1985): Microdosimetric approach to the analysis of cell responses at low dose and low dose rate. Radiat. Protection Dosimetry *13*, 299-306.

FIELD, S.B. (1976): An historical survey of radiobiology and radiotherapy with fast neutrons. Curr. Top. Radiat. Res. Quarterly *11*, 1-86.

FIELDEN, E.M., FOWLER, J.F., HENDRY, J.H., SCOTT, D. (eds.) (1987): Radiation Research Vol. 2 (Proceedings of the 8th International Congress of Radiation Research Edinburgh). London: Taylor and Francis, 985 p.

FINNEY, D.J. (1962): Probit analysis. Cambridge University Press: Cambridge.

FISHBEIN, L., FLAMM, W.G., FALK, H.L. (1970): Chemical mutagens. Academic: New York und London.

FLIEDNER, T.M., NORTHDURFT, W. (1979): Structure and function of stem cell pools in mammalian cell renewal systems. In: OKADA u.a. (1979), p. 640-647.

FOWLER, J.F. (1984): Review: Total doses in fractionated radiotherapy – implications of new radiobiological data. Int. J. Radiat. Biol. *46*, 103-120.

FRANKENBERG, D., FRANKENBERG-SCHWAGER, M., HARBICH, R. (1984): Split-dose recovery is due to the repair of double-strand breaks. Int. J. Radiat. Biol. *46*, 541-584.

FRANKENBERG-SCHWAGER, M., FRANKENBERG, D., BLOECHER, D., ADAMCZYK, C. (1980): The linear relationship between DNA double strand breaks and radiation dose is converted into a quadratic function by repair. Int. J. Radiat. Biol. *37*, 207-212.

FREY, R., HAGEN, U. (1974): Oxygen effect on τ-irradiated DNA. Radiat. Env. Biophys. *11*, 125-133.

FRIEDBERG, E.C. (1985a): DNA repair. New York: W.H. Freeman.

FRIEDBERG, E.C. (1985b): Nucleotide excision repair of DNA in eucaryotes: comparisons between human cells and yeast. Cancer Survey *4*, 529-556.

FRIEDBERG, E.C. (1987): The molecular biology of nucleotide excision repair: recent progress. J. Cell Science Supplement *6*, 1-24.

FRITZ, T.E., NORRIS, W.P. u.a. (1978): Relationship of dose-rate and total dose to responses of continuously irradiated beagles. In: IAEA (1978), p. 71-82.

FRY, R.J.M., GRAHN, D., GRIEM, M.L., RUST, J.H. (eds.) (1970): Late effects of radiation. Taylor and Francis: London.

GALLO, U., SANTAMARIA, L (eds.). (1972): Research progress in organic, biological and medicinal chemistry. Vol. 3, I und II. North Holland: Amsterdam London.

GATES, F.L. (1930): A study of the bactericidal action of ultraviolet light. III. The absorption of ultraviolet light by bacteria. J. Gen. Physiol. *14*, 31-42.

GLICKMAN, B.W., DROBETSKY, E.A., deBOER, J., GROSOVSKY, A.J. (1987): Ionizing radiation induced point mutations in mammalian cells. In: FIELDEN, FOWLER, HENDRY and SCOTT (1987), p. 562-567.

GLOCKER, R., MACHERAUCH, E. (1965): Röntgen- und Kernphysik. G. Thieme: Stuttgart.

GOMER, C.J. (ed.) (1987): Photodynamic therapy. Photochem. Photobiol. *46*, 561-952.

GOODHEAD, D.T. (1980): Models of radiation inactivation and mutagenesis. In: MEYN and WITHERS (1980), p. 231-247.

GOODHEAD, D.T. (1985): Saturable repair models of radiation action in mammalian cells. Radiat. Res. *104*, Suppl. 8, 8-58.

GOODHEAD, D.T. (1987): Biophysical models of radiation action – introductory review. In: FIELDEN, FOWLER, HENDRY and SCOTT (1987), p. 306-311.

GOODMAN, G.B. et al. (1987): Clinical evaluation of Pi-meson radiotherapy at TRIUMF. In: FIELDEN, FOWLER, HENDRY and SCOTT (1987), p. 928-933.

GORDON, D., SILVERSTEIN, H. (1976): quoted from THORINGTON 1980.

GRAHN, D. (1970): Biological effects of protracted low dose radiation exposure of man and animals. in: FRY u.a. (1970), p. 101-136.

GRAHN, D., SACHER, S.A., LEA, R.A., FRY, R.J.M., RUST, J.H. (1978): Analytical approaches to and interpretations of data on time rate and cause of death of mice exposed to external gamma irradiation. In: IAEA (1978), IAEA, Vol. II, p. 49-58.

GUENTHER, K., SCHULZ, W. (1983): Biophysical theory of radiation action. Berlin: Akademie Verlag, 354 p.

HAGEN, U. (1985): Grundlagen der Strahlenbiochemie. Hb. Nuklearmedizin (H. KRIEGEL, Her.), Stuttgart: G. Fischer, p. 293-310.

HAHN, G.M. (1978): The use of microwaves for the hyperthermic treatment of cancer: advantages and disadvantages. Photochem. Photobiol. Reviews *3*, 277-302.

HAHN, G.M. (1982): Hyperthermia and cancer. New York London: Plenum Press, 285 p.

HAHN, G.M., LITTLE, J.B. (1972): Plateau-phase cultures of mammalian cells. Curr. Top. Radiat. Res. Quarterly *8*, 39-83.

HALL, E.J. (1978): Radiobiology for the radiologist. Harper and Row: Philadelphia, 460 p.

HALL, E.J., HEI, T.K. (1985): Oncogenic transformation with radiation and chemical. Int. J. Radiat. Biol. *48*, 1-18.

HALL, E.J., HEI, T.K. (1986): Oncogenic transformation of cells in culture: pragmatic comparisons of oncogenicity, cellular and molecular mechanisms. Int. J. Radiat. Oncol. *12*, 1909-1921.

HALL, E.J., ROSSI, H.H. (1974): Californium-252 in teaching and research. IAEA Vienna (Techn. Rep. Ser. 159).

HAMM, R.N., TURNER, J.E., RITCHIE, R.H., WRIGHT, H.A. (1985): Calculations of heavy ion tracks in liquid water. Radiat. Res. *104*, 20-26.

HAMM, R.N., TURNER, J.E., WRIGHT, H.A., RITCHIE, R.H. (1984): Calculated ionization distributions in small volumes in liquid water irradiated by protons. Radiat. Res. *97*, 16-24.

HAMM, R.N., WRIGHT, H.A., KATZ, R., TURNER, J.E., RITCHIE, R.H. (1978): Calculated yields and showing-down spectra for electrons in liquid water: Implications for electron and photon RBE. Phys. Med. Biol. *23*, 1149-1161.

HAN, A., ELKIND, M.M. (1978): Ultraviolet light and X-ray damage interaction in Chinese hamster cells. Radiat. Res. *74*, 88-100.

HAN, A., ELKIND, M.M. (1979): Transformation of mouse C3H/10T1/2 cells by single and fractionated doses of X-rays and fission-spectrum neutrons. Cancer Res. *39*, 123-130.

HAN, A., HILL, C.K., ELKIND, M.M. (1984): Repair processes and radiation quality in neoplastic transformation of mammalian cells. Radiat. Res. *99*, 249-261.

HANAWALT, P.C. (1987): Preferential DNA repair in expressed genes. Env. Health Perspectives *76*, 9-14.

HANAWALT, P.C., COOPER, P.K., GANESAN, A.K., SMITH, C.A. (1979): DNA repair in bacteria and mammalian cells. Ann. Res. Biochem. *48*, 783-836.

HANAWALT, P.C., HAYNES, R.H. (1967): quoted after SMITH (1972):.

HANAWALT, P.C., SARASIN, A. (1986): Cancer-prone hereditary deseases with DNA processing abnormalities. Trends in Genetics *2*, 124-129.

HANAWALT, P.C., SETLOW; R.B. (1975): Molecular mechanisms for repair of DNA. Plenum Press: New York London.

HANSEN, J.W., OLSEN, K.J. (1984): Experimental and calculated response of a radiochromic dye film dosimeter to high LET radiations. Radiat. Res. *97*, 1-5.

HAQQ, C.M., SMITH, C.A. (1987): DNA repair in tissue specific genes in cultured mouse cells. In: FIELDEN, FOWLER, HENDRY and SCOTT (1987), p. 418-423.

HAR-KEDAR, I., BLEEHEN, N.M. (1976): Experimental and clinical aspects of hyperthermia applied to the treatment of cancer with special reference to the role of ultrasonic and microwave heating. Adv. Radiat. Biol. *6*, 229-266.

HARM, H. (1976): Repair of UV-irradiated biological systems: Photoreactivation. In: WANG (1976), Vol. II, p. 219-265.

HARM, W. (1963): Repair of lethal ultraviolet damage in phage DNA. In: SOBELS, F.H. (ed.): Repair from genetic damage. Pergamon: Oxford, p. 107-124.

HARM, W. (1980): Biological effects of ultraviolet radiation. Cambridge: Cambridge University Press, 215 p.

HARMON, J.T., NIELSEN, T.B., KEMPNER, E.S. (1985): Molecular weight determination from radiation inactivation. Meth. Enzymol. *117*, 65-94.

HAXEL, O. (1966): Entstehung, Eigenschaften und Wirkung ionisierender Strahlen. Handbuch der medizinischen Radiologie Band I/1. Springer: Berlin Heidelberg, p. 1-107.

HAYNES, R.H., ECKARDT, F. (1979): Analysis of dose-response patterns in mutation research. Can. J. Genetics Cytol. *21*, 277-302.

HAYNES, R.H., KUNZ, B.A. (1981): DNA repair and mutagenesis in yeast. In: The molecular biology of the yeast Saccharomyces (STRATHERN, J.N., JONES, E.W., BROACH, J.R., eds.), Cold Spring Harbor, Vol. I, p. 371-414.

HEI, T.K., KOMATSU, K., HALL, E.J., ZAIDER, M. (1988): Oncogenic transformation by charged particles of defined LET. Carcinogen. *9*, 747-750.

HELENE, C. (1987): Excited states and photochemical reactions in DNA, DNA-photosensitizer, and DNA-protein complexes. A review. Photochem. Photobiophys. Suppl., p. 3-22.

HENDRY, J.H. (1985): Survival curves for normal tissue clonogens: a comparison of assessments using in vitro, transplantation and in situ techniques. Int. J. Radiat. Biol. *47*, 3-16.

HENRICHS, K., ELSASSER, U., SCHOTOLA, C., KAUL, A. (1985): Dosisfaktoren für Inhalation oder Ingestion von Radionuklidverbindungen (Altersklasse 1 Jahr). Neuherberg: Institut für Strahlenhygiene des BGA, ISH-Heft 78, 405 p.

HERMENS, A.F., BARENDSEN, G.W. (1969): Changes of proliferation characteristics in a rat rhabdomyosarcoma before and after X-irradiation. Eur. J. Cancer *5*, 173-189.

HETZEL, F.W., KRUUV, J., FREY, H.E. (1976): Repair of potentially lethal damage in X-irradiated V79 cells. Radiat. Res. *68*, 308-319.

HEUCK, F., SCHERER, E. (1985): Strahlengefährdung und Strahlenschutz (Handbuch der med. Radiologie XX). Berlin Heidelberg New York Tokyo: Springer Verlag.

HICKSON, D., HARRIS, A.L. (1988): Mammalian DNA repair – use of mutants hypersensitive to cytotoxic agents. Trends in Genetics *4*, 101-106.

HILL, C.K., HAN, A., ELKIND, M.M. (1984): Fission-spectrum neutrons at low dose rate enhance neoplastic transformation in the linear low-dose region (0–10 cGy). Int. J. Radiat. Biol. *46*, 11-16.

HINE, G.J., BROWNELL, G.L. (1956): Radiation dosimetry. Academic Press: New York.

HITTELMAN, W.N. (1984): Prematurely condensed chromosomes: a model system for visualizing effects of DNA damage, repair and inhibition. In: COLLINS, DOWNES and JOHNSON (1984), p. 341-371.

HOEIJMAKERS, J.H.J. (1987): Characterization of genes and proteins involved in excision repair of human cells. J. Cell Science Supplement *6*, 111-126.

HOEL, D.G. (1987): Cancer risk models for ionizing radiation. Env. Health Perspect. *76*, 121-124.

HOFER, K.G., KEOUGH, G., SMITH, J.M. (1977): Biological toxicity of Auger emitters: molecular fragmentation versus electron irradiation. In: FEINENDEGEN, TISLJAR-LENTULIS und EBERT, p. 335-354.

HOLICK, M.F., MacLAUGHLIN, J.A., PARRISH, J.A., ANDERSON, R.R. (1982): The photochemistry and photobiology of vitamin D. In: SMITH (1982), p. 147-194.

HOLLAENDER, A. (ed.) (1971): Chemical mutagens, 2 Volumes. Plenum: New York London.

HOLLAENDER, A., EMMONS, C.W. (1941): Wavelength dependence of mutation production in the ultraviolet with special emphasis on fungi. Cold Spring Harbor Symp. Quart. Biol. *9*, 179-186.

HUG, O., KELLERER, A. (1966): Stochastik der Strahlenwirkung. Springer: Heidelberg Berlin New York.

HUMPHREY, R.M., DEWEY, W.C., CORK, A. (1963): Effect of oxygen in mammalian cells sensitized to radiation by incorporation of 5-Bromodeoxyuridine into the DNA. Nature *198*, 268-269.

HUTCHINSON, F. (1961): Sulfhydryl groups and the oxygen effect on irradiated dilute solutions of enzymes and nucleic acids. Radiat. Res. *14*, 721-731.

HUTCHINSON, F. (1985): Chemical changes induced in DNA by ionizing radiation. Progr. Nucl. Acid Res. Mol. Biol. *32*, 115-154.
HUTCHINSON, F., TINDALL, K.R. (1987): Mechanisms of gamma ray mutagenesis inferred from changes in DNA base sequence. In: FIELDEN, FOWLER, HENDRY and SCOTT (1987), p. 557-561.
HÜTTERMANN, J., KÖHNLEIN, W., TÉOULE, R. (ed.) (1978): Effects of ionizing radiation on DNA. Springer: Berlin Heidelberg New York.
IAEA, Int. Atomic Energy Agency (1978): Late biological effects of ionizing radiation. Vol. I and II. International Atomic Energy Agency: Wien.
IAEA, Int. Atomic Energy Agency (1986): Biological dosimetry: chromosomal aberration analysis for dose assessment. Wien: IAEA, Technical Report Ser. No. 260, 68 pp.
ICRP 21, Int. Commission on Radiological Protection (1971): Data for protection against ionizing radiation from external sources, Oxford: Pergamon Press, 100 p.
ICRP 23, Int. Commission on Radiological Protection (1975): Reference man: anatomical, physiological and metabolic characteristics. Pergamon Press: Oxford New York Frankfurt.
ICRP 26, Int. Commission on Radiological Protection (1977): Recommendations of the International Commission on Radiological Protection. Ann. ICRP *1*, No. 3.
ICRP 29, Int. Commission on Radiological Protection (1979): Radionuclide release into the environment assessment of doses to man. Report 29. Pergamon Press: Oxford.
ICRP 30, Int. Commission on Radiological Protection (1979): Limits for intakes of radionuclides by workers. Annuals of the OCRP *2*, 3/4.
ICRU 16, Int. Commission on Radiation Units and Measurements (1970): Linear energy transfer. ICRU: Washington.
ICRU 30, Int. Commission on Radiation Units and Measurements (1979): Quantitative concepts and dosimetry in radiobiology. ICRU: Washington.
ICRU 36, Int. Commission on Radiation Units and Measurements (1983): Microdosimetry. Bethesda, Md.
ICRU 37, Int. Commission on Radiation Units and Measurements (1984): Stopping powers for electrons and positrons. Bethesda, Md.
ICRU 40, Int. Commission on Radiation Units and Measurements (1986). The quality factor in radiation protection, report 40. Bethesda: Int. Commission on Radiation Units and Measurements, 32 p.
ILIAKIS, G. (1988): Radiation induced potentially lethal damage: DNA lesions susceptible to fixation. Int. J. Radiat. Biol. *53*, 541-584.
ILIAKIS, G., PANTELIAS, G.E., SEANER, R. (1988): Effect of arabinofuranosyladenine on radiation-induced chromosome damage in plateau-phase CHO cells measured by premature chromosome condensation: implications for repair and fixation of α-PLD. Radiat. Res. *114*, 361-378.
ISHIHARA, T., SASAKI, M.S. (eds.) (1983): Chromosome damage in man. New York: Alan Liss.
ISHIMARU, T., OTAKE, M., ICHIMARU, M. (1979): Dose-response relationship of neutrons and τ-rays to leukemic incidence among atomic bomb survivors in Hiroshima and Nagasaki by type of leukemic, 1950-1971. Radiat. Res. *77*, 377-394.
JAGGER, J. (1967): Introduction to research in ultraviolet photobiology. Prentice Hall: Englewood Cliffs.
JAGGER, J. (1977): Phototechnology and biological experimentation. In: SMITH (1977), p. 1-26.
JENSH, R.P., BRENT, R.L. (1987): The effect of low-level prenatal X-irradiation on postnatal development in the Wistar rat. Proc. Soc. Exp. Biol. Med. *184*, 256-263.
JOHANSEN, I., GULBRANDSEN, R., PETTERSEN, R. (1974): Effectiveness of oxygen in promoting X-ray-induced single-strand breaks in circular lambda-phage DNA and killing of radiation-sensitive mutants of Escherichia coli. Radiat. Res. *58*, 384-397.
JOHNS, H.E., CORMACK, D.V., BENESUK, S.A., WHITMORE, G.F. (1952): quoted after JOHNS and LAUGHLIN (1956).

JOHNS, H.E., LAUGHLIN, J. S. (1956): Interaction of radiation with matter. In: HINE, G., BROWNELL, G.L. (1956): p. 50-125.
JORI, G. (1987): Photosensitizing properties of porphyrins and photodynamic therapy. Photobiochem. Photobiophys. Supplement, 373-384.
KANAI, T., KAWACHI, K. (1987): Radial dose distribution for 18.3 MeV/n alpha beams in tissue equivalent gas. Radiat. Res. *112*, 426-435.
KAPPOS, A., POHLIT, W. (1972): A cybernetic model for radiation action in living cells. Int. J. Radiat. Biol. *22*, 51-...
KASE, K.R., BJAERNGARD, B.F., ATTIX, F.H. (eds.) (1985): The dosimetry of ionizing radiation (Vol. 1). Orlando: Academic Press, 416 p.
KASE, K.R., NELSON, W.R. (1978): Concepts of radiation dosimetry. Pergamon Press: New York.
KATZ, R. (1988): Radiobiological modeling based on track structure. In: KIEFER (1988), p. 57-84.
KATZ, R., ACKERSON, B., HOMAYOONFAR, M., SHARMA, S.C. (1971): Inactivation of cells by heavy ion bombardment. Radiat. Res. *47*, 402-425.
KAYE, S.V., RHOVER, P.S. (1975): Radiobiological assessment of nuclear power stations. Advances Radiat. Biol. *5*, 47-82.
KEENE, J.P. (1963): Optical absorption in irradiated water. Nature *197*, 47-48.
KELLERER, A.M., CHMELEVSKY, D. (1975): Criteria for the applicability of LET. Radiat. Res. *63*, 226-234.
KELLERER, A.M., ROSSI, H.H. (1972): The theory of dual radiation action. Curr. Top. Radiat. Res. Quarterly *8*, 85-158.
KELLERER, A.M., ROSSI, H.H. (1978): A generalized formulation of dual radiation action. Radiat. Res. *75*, 471-488.
KEMPNER, E.S., SCHLEGEL, W. (1979): Size determination of enzymes by radiation inactivation. Analyt. Biochem. *92*, 2-10.
KENNEDY, A.R., LITTLE, J.B. (1984): Evidence that a second event in X-ray induced oncogenic transformation in vitro occurs during cellular proliferation. Radiat. Res. *99*, 228-248.
KESSEL, D. (1986): Photosensitization with derivatives of haemotoporphyrin. Int. J. Radiat. Biol. *49*, 901-908.
KESSEL, D., DOUGHERTY, T.J. (eds.) (1983): Porphyrin photosensitization. Adv. Exp. Med. Biol. *160*, 294 p.
KIEFER, H., KOELZER, W. (1986): Strahlen und Strahlenschutz. Berlin Heidelberg: Springer Verlag, 145 p.
KIEFER, J. (1974): On the interpretation of the oxygen effect. 4th Symp. Microdosimetry EURATOM: Brüssel, p. 441-462.
KIEFER, J. (1975): Theoretical aspects and implications of the oxygen effect. In: NYGAARD u.a. (1975), p. 1025-1039.
KIEFER, J. (ed.) (1977): Ultraviolette Strahlen. de Gruyter: Berlin New York.
KIEFER, J. (1982): On the interpretation of heavy ion survival data. In: Radiation Protection (BOOZ, J., EBERT, H.G., eds.). Commission of the Eur. Communities EUR 8395en, p. 729-742.
KIEFER, J. (1986): Cellular and subcellular effects of very heavy ions. Int. J. Radiat. Biol. *48*, 873-892.
KIEFER, J. (1987): Semi-empirical calculations of stopping power, effective charge and ranges of heavy ions (1–400 MeV/u) in water. 3rd Workshop on Heavy Charged Particles in Biology and Medicine, Darmstadt. GSI-Report 87-11.
KIEFER, J. (ed.) (1988): Quantitative mathematical models in radiation biology. Heidelberg: Springer Verlag.
KIEFER, J., STRAATEN, H. (1986): A model of ion track structure based on classical collision dynamics. Phys. Med. Biol. *31*, 1201-1209.
KIMBALL, R.F. (1978): The relation of repair phenomena to mutation induction in bacteria. Mutat. Res. *55*, 85-120.

KIMBALL, R.F. (1987): The development of ideas about the effect of DNA repair on the induction of gene mutations and chromosomal aberrations. Mutat. Res. *186*, 1-34.

KITTLER, L., LÖBER, G. (1977): Photochemistry of the nucleic acids. Photochem. Photobiol. Rev. *2*, 39-132.

KLIAUGA, P., DVORAK, R. (1978): Microdosimetric measurements of ionization by monoenergetic photons. Radiat. Res. *73*, 1-20.

KOENIG, F., KIEFER, J. (1988): Lack of dose rate effect for mutation induction by gamma rays in human TK6 cells. Int. J. Radiat. Biol. *54*, 891-898.

KOHN, K.W. (1986): Assessment of DNA damage by filter elution assays. In: SIMIC, GROSSMAN and UPTON (1986), p. 101-118.

KONINGS, A.W.T. (1987): Effects of heat and radiation on mammalian cells. Rad. Phys. Chem. *30*, 339-349.

KRAFT, G. (1987): Radiobiological effects of very heavy ions: inactivation induction of chromosome aberrations and strand breaks. Nucl. Sci. Applic. *3*, 1-28.

KRIEGEL, H., SCHMAHL, W., GERBER, G.B., STIEVE, F.E. (eds.) (1986): Radiation risks to the developing nervous system. Stuttgart New York: Fischer, 436 p.

LAFLEUR, M.V.M., LOMAN, H. (1986): Radiation damage to ϕX174 DNA and biological effects. Radiat. Env. Biophys. *25*, 159-174.

LAMOLA, A.A., TURRO, N.J. (1977): Spectroscopy. In: SMITH (1977), p. 27-62.

LASKOWSKI, W. (1981): Biologische Strahlenschäden und ihre Reparatur. Berlin New York: de Gruyter, 88 p.

LATARJET, R. (1972): Interaction of radiation energy with nucleic acids. Curr. Top. Radiat. Res. Quarterly *8*, 1-38.

LAWLEY, P.D. (1966): Effects of some chemical mutagens and carcinogens on nucleic acids. Progr. Nucl. Acid Res. Mol. Biol. *5*, 89-132.

LEA, D.L. (1956): Action of radiation on living cells. Cambridge University Press.

LENNARTZ, M., COQUERELLE, T., BOPP, A., HAGEN, U. (1975): Oxygen effect on strand breaks and specific endgroups in DNA of irradiated thymocytes. Int. J. Radiat. Biol. *27*, 577-587.

LETT, J.T., ALTMAN, K.L. (eds.) (1987): Relative radiation sensitivities of human organ systems. Adv. Radiat. Biol. *12*, 296 p.

LING, C.C., MICHAELS, H.B., EPP, E.R., PETERSON, E.C. (1978): Oxygen diffucion into mammalian cells following ultrahigh dose rate irradiation and lifetime estimates of oxygen-sensitive species. Radiat. Res. *76*, 522-532.

LITTLE, J.B. (1981): Influence of non-carcinogenic secondary factors on radiation carcinogenesis. Radiat. Res. *87*, 240-250.

LITTLE, J.B. (1986): Characteristics of radiation induced neoplastic transformation in vitro. Leukemia Res. *7*, 719-725.

LIU, S.Z., LIU, W.H., SUN, J.B. (1987): Radiation hormesis: its expression in the immune system. Health Phys. *52*, 579-584.

LLOYD, D.C. et al. (1987): A collaborative exercise on cytogenetic dosimetry for simulated whole and partial body accidental irradiation. Mutat. Res. *179*, 197-208.

LUCKEY, T.D. (1980): Hormesis with ionizing radiation. Boca Raton: CRC Press.

LUNING, K.G. (1975): Test of recessive lethals in the mouse. Mutat. Res. *27*, 357-366.

MALAISE, E.P., CHAVANDRA, N., TUBIANA, M. (1973): The relationship between growth labelling index and histological type of human soli tumours. Eur. J. Cancer *9*, 305-312.

MAXFIELD, W.S., HANKS, G.E., PIZZARELLO, D.J., BLACKWELL, L.H. (1973): Acute radiation syndrome. In: DALRYMPLE et al., p. 190-208.

MCGRAWTH, R.A., WILLIAMS, R.W. (1966): Reconstruction in vivo of irradiated Escherichia coli deoxyribonuclcic acid; the rejoining of broken pleces. Nature *212*, 534-535.

McNALLY, N.J. (1982): Cell survival. In: PIZZARELLO (1982), p. 27-68.

McNALLY, N.J., de RONDE, J., FOLKARD, M. (1988): Interaction between X-ray and alpha-particle damage in V79 cells. Int. J. Radiat. Biol. *53*, 917-920.

MENZEL, H.G., BOOZ, J. (1976): Measurement of radiation energy deposition spectra for protons and deuterons in tissue equivalent gas. Proc. 5th Symp. Microdosimetry (BOOZ, J., EBERT, H.G., SMITH, G.R., eds.), EURATOM, p. 61-74.

MESELSON, M., STAHL, F.W., VINOGRAD, J. (1957): Equilibrium sedimentation of macromolecules in density gradients. Proc. Natl. Acad. Sci. US *43*, 581-588.

MEYN, R.E., WITHERS, H.R.(eds.(1980): Radiation biology in cancer research. Raven Press: New York.

MICHAEL, B.D., HARROP, H.A., MAUGHAN, R.L. (1979): Fast response methods in the radiation chemistry of lethal damage in interaact cells. In: OKADA u.a., p. 288-297.

MICHAELS, H.B., EPP, E.R., LING, C.C., PETERSON, E.C. (1978): Oxygen sensitization of CHO cells at ultrahigh dose rates: prelude to oxygen diffusion studies. Radiat. Res. *76*, 510-521.

MICHAELSON, S.M. (1977): Microwave and radiofrequency radiation. World Health Organization Document ICP/CEP 803.

MICKLEM, H.S., LOUTIT, J.F., FORD, C.E. (1968): Tissue grafting and radiation. ABIS Monograph. Academic Press: New York and London.

MIDANDER, J., DESCHAVANNE, P.J., DEBIEU, D., MALAISE, E.P., REVESZ, L. (1986): Reduced repair of potentially lethal radiation damage in glutathion synthetase-deficient human fibroblasts after X-irradiation. Int. J. Radiat. Biol. *49*, 403-413.

MILLER, M.W., MILLER, W.M. (1987): Radiation hormesis in plants. Health Phys. *52*, 607-616.

MITCHELL, D. (1988): The relative cytotoxicity of (6–4)photoproducts and cyclobutane dimers in mammalian cells. Photochem. Photobiol. *48*, 51-58.

MITCHELL, J.B., BEDFORD, J.S., BAILY, S.M. (1979): Dose rate effects in mammalian cells in culture. III. Comparison of cell killing and cell proliferation during continuous irradiation for six different cell lines. Radiat. Res. *79*, 537-551.

MITZEL-LANDBECK, L., HAGEN, U. (1976): Strahlenwirkung auf Biopolymere. Chemie in unserer Zeit *10*, 65-74.

MODRICH, P. (1987): DNA mismatch correction. Ann. Rev. Biochem. *56*, 435-466.

MOGGACH, P.G., LEPOCK, J.R., KRUUV, J. (1979): Effect of salt solutions on the radiosensitivity of mammalian cells as a function of the state of adhesion and the water structure. Int. J. Radiat. Res. *36*, 435-452.

MORGAN, K.Z., TURNER, J.E. (1973): Principles of radiation protection. Krieger Publ. Co.: New York.

MORISON, W.L. (1985): Photoimmunology. Photochem. Photobiol. *40*, 781-878.

MOROSON, H.L., QUINTILIANI, M. (eds.) (1970): Radiation protection and sensitization. Taylor and Francis: London.

MOSSMANN, K.L., THOMAS, D.S., DRITSCHILO, A. (1986): Environmental radiation and cancer. J. Envir. Sci. Health *C4*, 119-161.

MOTT, N.F., MASSEY, H.S.W. (1965): The theory of atomic collisions. Clarendon: Oxford.

MOUSTACCHI, E. (1987): DNA repair in yeast: Genetic control and biological consequences. Adv. Radiat. Biol. *13*, 1-30.

MOZUMDAR, A., MAGEE, J.L. (1966): Models of tracks of ionizing radiation for radical reaction mechanisms. Radiat. Res. *28*, 203-214.

MUNZENRIDER, J.E. et al (1987): Clinical results of Proton beam therapy in Boston. In: FIELDEN, FOWLER, HENDRY and SCOTT (1987), p. 916-921.

MÜLLER, W.E.G., ZAHN, R.K. (1977): Bleomycin, an antibiotic that removes thymine from double stranded DNA. Prog. Nucl. Acid Res. Molec. Biol. *20*, 22-59.

NAIRN, T.D., HUMPHREY, R.M., ADAIR, G.M. (1988): Transformation of UV-hypersensitive Chinese hamster ovary cell mutants with UV-irradiated plasmids. Int. J. Radiat. Biol. *53*, 249-260.

NAMBI, K.S., SPOMAN, S.D. (1987): Environmental radiation and cancer in India. Health Phys. *52*, 653-658.

NCRP, National Council on Radiation Protection and Measurements (1987): Recommendations on limits for exposure to ionizing radiation. Report 91. Bethesda, 72 p.

NEARY, C.J., SAVAGE, J.R.K. (1966): Chromosome aberrations and the theory of RBE. II. Evidence from track segments experiments with protons and α-particles. Int. J. Radiat. Biol. *11*, 209-223.

NEARY, G.J. (1965): Chromosome aberrations and the theory of RBE. 1. General considerations. Int. J. Radiat. Biol. *9*, 477-503.

NEEL, J.V., MOHRENWEISER, H.M. (1984): Failure to demonstrate mutations affecting protein structure of function in children with congenital defects or born prematurely. Proc. Nat. Acad. Sci. USA *81*, 5499-5503.

NOSSKE, D., GERICH, B., LANGNER, S. (1985): Dosisfaktoren für Inhalation oder Ingestion von Radionuklidverbindungen (Erwachsene). Neuherberg, Institut für Strahlenhygiene des BGA, ISH-Heft 63, 405 p.

NOVICK, A., SZILARD, L. (1949): Experiments on light reactivation of ultraviolet-inactivated bacteria. Proc. Natl. Acad. Sci. US *35*, 591-600.

NYGAARD, O.F., ADLER, I.H., SINCLAIR, W.K. (eds.) (1975): Radiation Research. Academic Press: New York San Francisco London.

O'NEILL, J.P., BRIMER, P.A., MACHANOFF, R., HIRSCH, G.P., HSIE, A.W. (1977): A quantitative assay of mutation induction at the hypoxyxanthine-guanine phosphoribosyltransferase locus in Chinese hamster ovary cells (CHO/HPGRT system): Development and definition of the system. Mutat. Res. *45*, 91-101.

OBE, G., NATARAJAN, A.T., PALITTI, F. (1982): Role of DNA double strand breaks in the formation of radiation induced chromosomal aberrations. Progress in Mutation Res. *4*, 1-9.

OKADA, S., IMAMURA, M., TERASIMA, T., YAMAGCHI, H. (eds.) (1979): Radiation Research. Tokio.

ORR-WEAVER, T., SZOSTAK, J.W. (1985): Fungal recombination. Microbiol. Revs. *49*, 33-58.

ORTON, C.G. (1986a): Bioeffect dosimetry in radiation therapy. In: ORTON (1986b), 1-72.

ORTON, C.G. (ed.) (1986): Radiation dosimetry. Plenum Press: New York London, 328 p.

ORTON, C.G., COHEN, L. (1988): A unified approach to dose-affect relationships in radiotherapy. I. Modified TDF and linear quadratic equations. Int. J. Radiat. Oncol. Biol. Phys. *14*, 549-556.

OSSANNA, N., PETERSON, K.R., MOUNT, D.W. (1986): Genetics of DNA repair in bacteria. Trends in Genetics *2*, 55-58.

OVERGAARD, J. (1978): The effect of local hyperthermia alone, and in combination with radiation. on solid tumours. In: STREFFER et al., p. 49-61.

PAINTER, R.B. (1985): Inhibition of mammalian cell DNA synthesis by ionizing radiation. Int. J. Radiat. Biol. *49*, 771-782.

PAINTER, R.B., YOUNG, B.R. (1975): X-ray induced inhibition of DNA-synthesis in Chinese hamster ovary, human HeLa and mouse L cells. Radiat. Res. *64*, 648-656.

PAINTER, R.B., YOUNG, B.R. (1987): DNA synthesis in irradiated mammalian cells. J. Cell Science Supplement *6*, 207-214.

PARETZKE, H.G. (1980): Advances in energy deposition theory. Advances in radiation protection and dosimetry in medicin (THOMAS, R.H., PEREZ-MENDEZ, V., eds.), Plenum Publ., p. 51-73.

PARETZKE, H.G. (1987): Radiation track structure theory. In: Kinetics of nonhomogeneous processes (FREEMAN, G.R., ed.), Wiley-Interscience: New York, p. 89.

PARRISH, J.A., ANDERSON, R.R., URBACH, J., PITTS, D. (1978): UV-A. Biological effects of ultraviolet radiation with emphasis on human responses to longwave ultraviolet. Plenum Press: New York London, 262 p.

PARRISH, J.A., KRIPKE, M.L., MORISON, W.L. (eds.) (1983): Photoimmunology. New York London: Plenum Medical Book Co., 304 p.

PATERSON, M.C. (1979): Environmental carcinogenesis and imperfect repair of damage DNA in Homo sapiens: Casual relation revealed by rare hereditary disorders. In: GRIF-

FIN, A.C., SHAW, C.R. (eds.), Carcinogens: Identification and mechanism of action. Raven Press: New York 1979, p. 251-276.
PEAK, J.G., PEAK, M.J., SIKORSKI, R.S., JONES, C.A. (1985): Induction of DNA-protein crosslinks in human cells by ultraviolet and visible radiations: action spectrum. Photochem. Photobiol. *41*, 295-302.
PERRY, P., WOLFF, S. (1974): New Giemsa method for the differential staining of sister chromatids. Nature *251*, 156-158.
PHILIPPS, T.L. (1979): Current status, opportunities and problems in clinical combined chemoradiotherapy. In: OKADA, IMAMURA, TERASHIMA and YAMAGUCHI (1979), p. 822-831.
PIZZARELLO, D.J. (ed.) (1982): Radiation biology. Boca Raton: CRC Press, p. 298.
PLANEL, H. et al. (1987): Influence on cell proliferation of background radiation or exposure to very low chronic gamma radiation. Health Phys. *52*, 571-578.
POTTEN, C.S. (1985): Radiation and skin. London Philadelphia: Taylor and Francis, 225 p.
POTTIER, R., TRUSCOTT, T.G. (1986): The photochemistry of haematoporphyrin and related systems. Int. J. Radiat. Biol. *50*, 421-452.
POWERS, E.L. (1982): Responses of cell to radiation sensitizers: methods of analysis. Int. J. Radiat. Biol. *42*, 629-651.
PRESTON, D.L., KATO, H., KOPECKY, K.J., FUJITA, S. (1987): Studies of the mortality of A-bomb survivors. 8. Cancer mortality, 1950-1982. Radiat. Res. *111*, 151-178.
PRESTON, D.L., PIERCE, D.A. (1988): The effect of changes in dosimetry on cancer mortality estimates in the atomic bomb survivors. Radiat. Res. *114*, 437-466.
PRICHARD, H.M., GESELL, T.F. (1984): Radon in the environment. Adv. Radiat. Biol. *11*, 391-428.
PUCK, T.T., MARCUS, P.I. (1956): Action of X-rays on mammalian cells. J. Exptl. Med. *103*, 653-666.
PURDIE, J.W., INHABER, E.R., KLASSEN, N.V. (1978): Increased sensitivity of anoxic mammalian cells irradiated with single pulses at very high dose rates. 6th Symp. Microdosimetry (BOOZ, J., EBERT, H.G., eds.), p. 1023-1032. Harwood: London.
QUASTLER, H. (1945): Studies on Roentgen death in mice. I. Survival time and dosage. Am. J. Roentgenol. *54*, 449-456.
QUINTILIANI, M. (1986): The oxygen effect in radiation inactivation of DNA and enzymes. Int. J. Radiat. Biol. *50*, 573-594.
RADFORD, I.R. (1988): The dose response for low LET radiation induced DNA double strand breakage methods of measurement and implications for radiation action models. Int. J. Radiat. Biol. *54*, 1-12.
RADMAN, M., WAGNER, R. (1988): The high fidelity of DNA duplication. Sci. Amer. *259*, 24-31.
RAJU, M.R. (1980): Heavy particle radiotherapy. Academic Press: New York.
RAJU, M.R., RICHMAN, C. (1972): Negative pion radiotherapy: physical and radiobiological aspects. Curr. Top. Radiat. Res. Quarterly *8*, 159-233.
RAUTH, A.M., TAMMEMAGI, M., HUNTER, G. (1974): Nascent DNA synthesis in ultraviolet irradiated mouse, human and Chinese hamster cells. Biophys. J. *14*, 209-220.
REVELL, S.H. (1966): Evidence for a dose-squared term in the dose-response curve for red chromatid discontinuities induced by X-rays, and some theoretical consequences thereof. Mutat. Res. *3*, 34-53.
REVELL, S.H. (1974): The breakage and reunion theory and the exchange theory for chromosomal aberration in use by ionizing radiation: A short history. Adv. Radiat. Biol. *4*, 367-416.
REVESZ, L. (1985): The role of endogeneous thiols in intrinsic radiation protection. Int. J. Radiat. Biol. *47*, 361-368.
RIEGER, R., MICHAELIS, A. (1967): Chromosomenmutationen. VEB Gustav Fischer: Jena.

ROBERTS, J.J. (1978): The repair of DNA modified by cytotoxic mutagenic and carcinogenic chemicals. Adv. Radiat. Biol. 7, 212-436.

ROBERTS, Jr., H.J., MICHAELSON, S.M., LU, S.-T. (1986): The biological effects of radiofrequency radiation: a critical review and recommendations. Int. J. Radiat. Biol. *50*, 379-420.

ROBINSON, K.R. (1985): The response of cells to electric fields: a review. J. Cell Biol. *101*, 2023-2027.

RODGERS, R.C., DICELLO, J.F., GROSS, W. (1973): The biophysical properties of 3,9 GeV nitrogen ions. II. Microdosimetry. Radiat. Res. *54*, 12-23.

RODGERS, R.C., GROSS, W. (1974): Microdosimetry of monoenergetic neutrons. 4th Symp. Microdosimetry EURATOM EUR 5122 d-e-f, Brüssel, p. 1027-1042.

ROOTS, R., CHATTERJEE, A., BLAKELY, E., CHANG, P., SMITH, K., TOBIAS, C. (1982): Radiation responses in air-, nitrous oxide- and nitrogen-saturated mammalian cells. Radiat. Res. *92*, 245-254.

ROOTS, R., KRAFT, G., GOSSCHALK, E. (1985): The formation of radiation induced DNA breaks: the ratio of double strand breaks to single strand breaks. Int. J. Radiat. Oncol. Biol. Phys. *11*, 259-265.

ROSENSTEIN, B.S., DUCORE, J.M. (1983): Induction of DNA strand breaks in normal human fibroblasts exposed to monochromatic UV and visible wavelengths in the 240–546 nm range. Photochem. Photobiol. *38*, 51-55.

ROSENSTEIN, B.S., MITCHELL, D.L. (1987): Action spectra for the induction of pyrimidine(6–4)pyrimidone photoproducts and cyclobutane pyrimidine dimers in normal human skin fibroblasts. Photochem. Photobiol. *45*, 775-780.

ROSS-RIVEROS, P., LEITH, J.T. (1979): Response of 8 tumour cells to hyperthermia and X-irradiation. Radiat. Res. *78*, 296-311.

ROSSMAN, T.G., KLEIN, C.B. (1988): From DNA damage to mutation in mammalian cells: a review. Env. Molec. Mutagenesis *11*, 119-113.

RUBIN, P., CASARETT, G.W. (1973): Concepts of clinical radiation pathology. In: DALRYMPLE u.a., p. 160-189.

RUPERT, C.S. (1974): Dosimetric concepts in photobiology. Photochem. Photobiol. *20*, 203-212.

RUPERT, C.S., HARM, W., HARM, H. (1972): Photoenzymtic repair of DNA. In: BEERS, R.F, Jr., HERRIOTT, R.M., TILGMAN, R.C., p. 64-78.

RUPP, W.D., WILDE, C.E., RENO, D.L., HOWARD-FLANDERS, P. (1971): Exchanges between DNA strands in ultraviolet-irradiated Escherichia coli. J. Mol. Biol. *61*, 25-42.

SAGAN, L.S. (ed.) (1987): Radiation hormesis. Health Phys. (Special Issue) *52*, 517-618.

SANCAR, A. (1987): DNA repair in vitro. Photobiochem. Photobiophys. Supplement, 301-316.

SANKARANARAYANAN, K. (1974): Recent advances in the assessment of genetic hazards of ionizing radiation. Atomic Energy Rev. *12*, 47-74.

SARASIN, A., BOURRE, F., BENOIT, A., DAYA-GROSJEAN, L., GENTIL, A. (1985): Molecular analysis of mutagenesis in mammalian cells. Int. J. Radiat. Biol. *47*, 479-488.

SAUER, M.C., SCHMIDT, K.H., JONAH, C.D., NALEWAY, C.A., HART, E.J. (1978): High-LET pulse radiolysis: O_2- and oxygen production in tracks. Radiat. Res. *75*, 519-528.

SAUERBIER, W. (1976): UV damage at the trancriptional level. Adv. Radiat. Biol. *6*, 50-107.

SAUNDERS, W. et al (1986): Helium-ion radiation therapy at the Lawrence Berkely Laboratory: Recent results of a northern California Oncology Group clinical trial. Radiat. Res. *104*, 227-...

SAVAGE, J.R.K. (1978): Some thoughts on the nature of chromosomal aberrations and their use as a quantitative endpoint for radiobiological studies. 6th Symp. Microdosimetry (BOOZ, J., EBERT, H.G., eds.), Harwood Publ.: London, p. 39-54.

SCHÄFER, V., HEINRICH, G. (1977): Erzeugung von UV-Strahlen. In: KIEFER, (1977), p. 47-178.

SCHOLES, G. (1978): Primary events in the radiolysis of aqueous solutions of nucleic acids and related substances. In: HÜTTERMANN u.a. (1978), p. 158-170.
SCHOLZ, M., KRAFT-WEYRATHER, W., RITTER, S., KRAFT, G. (1989): Cell cycle delays induced by heavy ion irradiation of synchronous mammalian cells. Adv. Space Res., in the press.
SCHROY, C.B., TODD, P. (1979): The effects of caffeine on the expression of potentially lethal and sublethal damage in τ-irradiated cultured mammalian cells. Radiat. Res. *78*, 312-316.
SCHULL, W.J. (1987): The status of the assessment of radiation risk to humans. In: FIELDEN, FOWLER, HENDRY and SCOTT (1987), p. 627-633.
SCHULTE-FROHLINDE, D., von SONNTAG, C. (1985): Radiolysis of DNA and model systems in the presence of oxygen. In: Oxydative Stress (SIES, H., ed.), London: Academic Press, p. 11-40.
SCHULZE, R., KIEFER, J. (1977): UV-Strahlen: Allgemeine Einführung und Grundbegriffe. In: KIEFER (1977), p. 1-16.
SCOTT, D., LYONS, C.Y. (1979): Homogeneous sensitivity of human peripheral blood lymphocytes to radiation-induced chromosome damage. Nature *278*, 756-758.
SEARLE, A.G. (1974): Mutation induction in mice. Adv. Radiat. Biol. *4*, 131-203.
SEARLE, A.G. (1977): Use of doubling doses for the estimation of genetic risks. First European Symposium on rad-equivalence, EURATOM: Brüssel, EUR 5725e, p. 133-144.
SEARLE, T. (1987): Radiation – the genetic risk. Trends in Genetics *3*, 152-157.
SEDGWICK, S.G. (1986): Inducible DNA repair in microbes. Microbiol. Sciences *3*, 76-83.
SEDGWICK, S.G. (1987): Stability and change through DNA repair. In: Accuracy in Molecular Processes (KIRKWOOD, T.B.L., ROSENBERGER, R.F., GALAS, D.J., eds.), London: Chapman and Hall, p. 233-389.
SELIGER, H.H. (1977): Environmental Photobiology. In: SMITH, K.C. (1977), p. 143-174.
SETLOW, J.K. (1966): The molecular basis of biological effects of ultraviolet radiation and photoreactivation. Curr. Top. Radiat. Res. *2*, 195-248.
SETLOW, R.B., CARRIER, W.L. (1964): The disappearance of thymine dimers from DNA: an error correcting mechanism. Proc. Nat. Acad. Sci. US *51*, 226-231.
SETLOW, R.B., SETLOW, J.K. (1962): Evidence that ultraviolet-induced thymine dimers in DNA cause biological damage. Proc. Nat. Acad. Sci. US *48*, 1250-1257.
SHALL, S. (1984): ADP-ribose in DNA repair: a new component of DNA excision repair. Adv. Radiat. Biol. *11*, 2-71.
SHEPPARD, S.C., REGITNIG, P.J. (1987): Factors controlling the hormesis response in irradiated seeds. Health Phys. *52*, 599-606.
SHIGEMATSU, I., KAGAN, A. (eds.) (1986): Cancer in atomic bomb survivors. Tokio: Japan Scientific Societies Press, 196 p.
SILBERBERG, R., TSAO, C.H., ADAMS, J.H., LETAW, J.R. (1984): Radiation doses and LET distributions of cosmic rays. Radiat. Res. *98*, 209.
SIME, E.H., BEDSON, H.S. (1973): A comparison of ultraviolet action spectra for vaccinia virus and T2 bacteriophage. J. Gen. Virology *18*, 55-60.
SIMIC, M.G., GROSSMAN, L., UPTON, A.C. (eds.) (1986): Mechanisms of DNA damage and repair. New York London: Plenum Press, 578 p.
SIMPSON, J.A. (1983): Elemental and isotopic composition of the galactic cosmic rays. Ann. Rev. Nucl. Part. Sci. *33*, 323.
SINCLAIR, W.K. (1987): Risk, research and radiation protection. Radiat. Res. *112*, 191-216.
SINCLAIR, W.K., PRESTON, D.L. (1987): Revisions in the dosimetry of the A-bomb survivors at Hiroshima and Nagasaki and their consequences. In: FIELDEN, FOWLER, HENDRY and SCOTT (1987), p. 588-594.
SINGER, B. (1975): The chemical effects of nucleic acid alkylation and their relation to mutagenesis and carcinogenesis. Prog. nucleic acid res. molec. biol. *15*, 219-284.

SMITH, C.A. (1987): DNA repair in specific sequences in mammalian cells. J. Cell Science Supplement *6*, 225-244.
SMITH, K.C. (1972): Dark repair of DNA damage. In: GALLO und SANTAMARIA, p. 356-382.
SMITH, K.C. (ed.) (1977): The Science of Photobiology. Plenum Press: New York and London.
SMITH, K.C. (ed.) (1982): The science of photomedicine. Plenum Press: New York London, 658 p.
SMITH, K.C., HANAWALT, P.C. (1969): Molecular photobiology. Academic Press: New York and London.
SNYDER, M.H., KIMBLER, B.F., LEPPER, D.B. (1977): The effect of caffeine on radiation-induced division delay. Int. J. Radiat. Biol. *32*, 281-284.
SOELEN, G., EDGREN, M., SCOTT, O.C.A., REVESZ, L. (1987): Cellular glutathione content and k values. Int. J. Radiat. Biol. *51*, 39-44.
SONNTAG, C. von, ROSS, A.B. (1987): Bibliographies on radiation chemistry: pulse radiolysis of nucleic acids and their base constituents. Radiat. Phys. Chem, im Druck.
SOWBY, F.D. (1985): Statement from the Paris meeting of the International Commission on Radiological Protection. Phys. Med. Biol. *30*, 863-864.
STEEL, G.G. (1977): Growth kinetics of tumours. Clarendon Press: Oxford.
STRAHLENSCHUTZKOMMISSION (1985): Wirkungen nach pränataler Bestrahlung. Stuttgart New York: G. Fischer, 202 p.
STRANDQVIST, M. (1944): Studien über die kumulative Wirkung von Röntgenstrahlen bei Fraktionierung. Acta Radiol. (suppl.) *55*, 1-300.
STRAUSS, B.S. (1968): DNA repair mechanisms and their relation to mutation and recombination. Curr. Top. Microbiol. Immunol., Vol. 44, p. 1-89, Springer: Berlin Heidelberg New York.
STREFFER, C. (ed.) (1987): Effects after combined exposure to ionizing radiation and chemical substances. Int. J. Radiat. Biol. *51*, 959-1110.
STREFFER, C., van BEUNINGEN, D., DIETZEL, F., RÖTTINGER, E., ROBINSON, J.E., SCHERER, E., SEEBER, S., TROTT, K.-R. (eds.) (1978): Cancer therapy by hyperthermia and radiation. Urban u. Schwarzenberg: Baltimore München.
STUCHLY, M.A. (1979): Interaction of radiofrequency and microwave radiation with living systems. A review of mechanisms. Radiat. Env. Biophys. *16*, 1-14.
SUTHERLAND, R.M. (1988): Cell and environment interactions in tumor microregions: the multicell spheroid model. Science *240*, 177-184.
SUTHERLAND, R.M. (1988): Cell and environment interactions in tumor microregions: The multicell spheroid model. Science *240*, 177-184.
SUTHERLAND, R.M., DURAND, R.E. (1973): Hypoxic cells in an in-vitro tumour model. Int. J. Radiat. Biol. *23*, 235-245.
SUTHERLAND, R.M., DURAND, R.E. (1976): Radiation response of multicell spheroids – an in vitro tumour model. Curr. Top. Radiat. Res. Quarterly *11*, 87-139.
SWALLOW, A.J. (1973): Radiation Chemistry. Longman: London.
THACKER, J. (1973): The possibility of genetic hazard from ultrasonic radiation. Curr. Top. Radiat. Res. Quarterly *8*, 235-258.
THACKER, J. (1979): The involvement of repair processes in radiation-induced mutation of cultured mammalian cells. In: OKADA u.a. (1979), p. 612-620.
THACKER, J. (1986): The use of recombinant DNA techniques to study radiation induced damage, repair and genetic changes in mammalian cells. Int. J. Radiat. Biol. *50*, 1-30.
THACKER, J. (1987): Radiation mutagenesis in bacteria and mammalian cells. In: FIELDEN, FOWLER, HENDRY and SCOTT (1987), p. 544-549.
THACKER, J., STRETCH, A., STEPHENS, M.A. (1977): The induction of thioguanine-resistant mutants of Chinese hamster cells by τ-rays. Mutat. Res. *42*, 313-326.
THACKER, J., STRETCH, A., STEPHENS, M.A. (1979): Mutation and inactivation of cultured mammalian cells exposed to beams of accelerated heavy ions. II. Chinese hamster V79 cells. Int. J. Radiat. Biol. *36*, 137-148.

THOMPSON, L.H., HUMPHREY, R.M. (1970): Proliferation kinetics of mouse L-P59 cells irradiated with ultraviolet light: A time-lapse photographic study. Radiat. Res. *41*, 183-201.

THOMPSON, L.H., SUIT, H.D. (1969): Proliferation kinetics of X-irradiated mouse L cell studies with time lapse photography II. Int. J. Radiat. Biol. *15*, 347-362.

THORINGTON, L. (1980): Actinic effects of light and biological implications. Photochem. Photobiol. *32*, 117-129.

THRALL, D.E., GERWECK, L.E., GILLETTE, E.L., DEWEY, W.C. (1976): Response of cells in vitro and tissues in vivo to hyperthermia and X-irradiation. Adv. Radiat. Biol. 6, 211-228.

TILL, J.E., MCCULLOCH, E.A. (1961): A direct measurement of the radiation sensitivity of normal mouse bone marrow cells. Radiat. Res. *14*, 213-222.

TIMME, T.L., MOSES, R.E. (1988): Review: Deseases with DNA damage-processing defects. Am. J. Med. Sci. *295*, 40-48.

TOBIAS, C.A., BLAKELY, E.A., NGO, F.Q.H., YANG, T.C.H. (1980): The repair-misrepair model of cell survival. In: MEYN and WITHERS (1980): p. 195-230.

TOLMACH, L.J., JONES, R.W., BUSSE, P.M. (1977): The action of caffeine on X-irradiated HeLa cells. I. Delayed inhibition of DNA synthesis. Radiat. Res. *71*, 653-665.

TOLMACH, L.J., TERASIMA, T., PHILIPPS, R.A. (1965): X-ray sensitivity changes during the division cycle of HeLa 33 cells and anomalous survival kinetics of developing microcolonies. In: Cellular Radiation Biology (1965), p. 376-399.

TOWN, C.D., SMITH, K.C., KAPLAN, H.S. (1973): Repair of X-ray damage to bacterial DNA. Curr. Top. Radiat. Res. Quarterly *8*, 351-399.

TRIMBLE, B.K., DOUGHTY, J.H. (1974): The amount of hereditary desease in human populations. Ann. Human Genet. (London) *38*, 199-223.

TRONNIER, H. (1977): Medizinische Wirkungen ultravioletter Strahlen. In: KIEFER (1977), p. 567-597.

TSUNEMOTO, H. u.a. (1987): Clinical results of proton beam therapy in Japan. In: FIELDEN, FOWLER, HENDRY and SCOTT (1987), p. 922-927.

TUBIANA, M. (1987): Recent developments in combinations of radiotherapy and chemotherapy. In: FIELDEN, FOWLER, HENDRY and SCOTT (1987), p. 743-749.

TYRELL, R. (1980): Mutation induction by and mutational interaction between monochromatic wavelength radiations in the near ultraviolet and visible ranges. Photochem. Photobiol. *31*, 37-46.

TYRELL, R. (1984): Damage and repair from non-ionizing radiation. In: Repairable lesions in microorganisms (HURST, A., NASIM, A., eds.), Academic Press: London Orlando, p. 86-125.

TYRELL, R.M., PIDOUX, M. (1987): Action spectra for human skin cells: estimates of the relative cytotoxicity of the middle UV, near UV and violet regions of sunlight. Cancer Res. *47*, 1825-1829.

ULLRICH, R.L. (1982): Radiation carcinogenesis. In: PIZZARELLO (1982), p. 111-128.

ULMER, K.M., GOMEZ, R. F., SINSKEY, A. J. (1979): Ionizing radiation damage to the folded chromosome of Escherichia coli K12: Repair of double strand breaks in deoxyribonucleic acid. J. Bacteriol. *138*, 486-491

UNDERBRINK, A.G., POND, V. (1976): Cytological factors and their predictive role in comparative radiosensitivity: a general summary. Curr. Top. Radiat. Res. Quarterly *11*, 251-306.

UNDERBRINK, A.G., SPARROW, A.H., POND, V. (1968): Chromosomes and cellular radiosensitivity II: Use of interrelationship among chromosome volume, nucleotide content and D_0 of 120 diverse organisms in predicting radiosensitivity. Radiat. Bot. *8*, 205-238.

UNSCEAR, United Nations Scientific Committee on the Effects of Atomic Radiation (1966): Report of the United Nations Scientific Committee on the Effects of Atomic Radiations. United Nations: New York.

UNSCEAR, United Nations Scientific Committee on the Effects of Atomic Radiation (1972): Ionizing radiation: Levels and effects. United Nations: New York.
UNSCEAR, United Nations Scientific Committee on the Effects of Atomic Radiation (1977): Sources and effects of ionizing radiation. United Nations: New York.
UNSCEAR, United Nations Scientific Committee on the Effects of Atomic Radiation (1982): Ionizing Radiation: Sources and biological effects. New York: United Nations, 773 p.
UNSCEAR,, United Nations Scientific Committee on the Effects of Atomic Radiation (1986): Genetic and somatic effects of ionizing radiation. New York: United Nations, 336 p.
URBACH, F. (1987): Cutaneous photocarcinogenesis. Envir. Carcinog. Revs. *C5*, 211-234.
VAETH, J.M. (ed.) (1969): The interrelationship of chemotherapeutic agents and radiation therapy in the treatment of cancer. Frontiers of Radiation Therapy and Oncology 4, S. Karger: Basel New York.
Van der LEUN, J.C. (1984): UV-Photocarcinogenesis. Photochem. Photobiol. *39*, 861-868.
Van der LEUN, J.C. (1987): Animal experiments in photocarcinogenesis. Photobiochem. Photobiophys. Supplement, 353-360.
Van der LEUN, J.C., van WEELDEN, H. (1986): UV-B phototherapy: principles, radiation sources, regimes. Curr. Probl. Dermatol. *15*, 39-51.
VARGHESE, A.J. (1972): Photochemistry of nucleic acids and their constituents. Photophysiol. *7*, 208-266.
VARMA, M.N., BAUM, J.W. (1980): Energy deposition in nanometer regions by 377 MeV/ nucleon 20Ne-ions. Radiat. Res. *81*, 355-363.
VARMA, M.N., BAUM, J.W., KUEHNER, A.V. (1977): Radial dose, LET and W for oxygen ions in nitrogen and tissue equivalent gases. Radiat. Res. *70*, 511-518.
VARMA, M.N., BAUM, J.W., KUEHNER, A.V. (1980): Stopping power and radial dose distribution for 42 MeV bromine ions. Phys. Med. Biol. *25*, 651-656.
VARMA, M.N., PARETZKE, H.G., LYMAN, J.T., HOWARD, J. (1976): Dose as a function of radial distance from a 930 MeV helium ion beam. Proc. 5th Symp. Microdosimetry (BOOZ, J., EBERT, H.G., SMITH, R.G., eds.), EURATOM, p. 75-95.
VERLY, W.G. (1974): Monofunctional alkylating agents and apurinic sites in DNA. Biochem. Pharmacol. *23*, 3-8.
VOGLER, H.H., Jr. (1973): Radiation cataracts. In: DALRYMPLE u.a., p. 232-236.
WACHSMANN, F., DREXLER, G. (1976): Graphs and tables for use in radiology. Springer: Heidelberg New York.
WALBURG, H.E., Jr. (1974): Experimental radiation carcinogenesis. Adv. Radiat. Biol. *4*, 210-254.
WALBURG, H.E., Jr. (1975): Radiation induced live shortening and premature aging. Adv. Radiat. Biol. *5*, 145-181.
WALDREN, C.A., RASKO, I. (1978): Caffeine enhancement of X-ray killing in cultured human and rodent cells. Radiat. Res. *73*, 95-110.
WALIGORSKI, M.P.R., HAMM, R.N., KATZ, R. (1986): The radial distribution of dose around the path of a heavy ion in liquid water. Nucl. Tracks Radiat. Measurements *11*, 309-319.
WALKER, G.C. (1984): Mutagenesis and inducible responses to DNA damage in E.coli. Microbiol. Revs. *48*, 60-93.
WALTERS, R.A., ENGER, M.D. (1976): Effects of ionizing radiation on nucleic acid synthesis in mammalian cells. Adv. Radiat. Biol. *6*, 1-49.
WANG, S.Y. (ed.) (1976): Photochemistry and Photobiology of nucleic acids. Vol. I: Chemistry; Vol. II: Biology. Academic Press: New York San Francisco London.
WARD, J.F. (1986): Ionizing radiation induced DNA damage: identities and DNA repair. In: SIMIC, GROSSMAN and UPTON (1986), p. 135-138.
WARDMAN, P. (1976): The use of nitroaromatic compounds as hypoxic cell sensitizers. Curr. Top. Radiat. Res. Quarterly *11*, 347-398.

WARTERS, R.L., HOFER, K.G., HARRIS, C.R., SMITH, J.M. (1977): Radionuclids toxicity in cultured mammalian cells: elucidation of the primary site of radiation damage. Curr. Top. Radiat. Res. Quarterly *12*, 389-407.

WEBB, R.B. (1977): Lethal and mutagenic effects of near-ultraviolet radiation. Photochem. Photobiol. Rev. *2*, 169-262.

WEBER, K.J. (1988): Models of cellular radiation action – an overview. In: KIEFER (1988), p. 3-27.

WEISSMANN, M., SCHINDLER, H., FEHER, G. (1976): Determination of molecular weights by fluctuation spectroscopy: application to DNA. Proc. Nat. Acad. Sci. US *73*, 2776-2780.

WELLS, R.I., BEDFORD, J.S. (1983): Dose rate effects in mammalian cells. IV. Repairable and non repairable damage in non cycling C3H 10T1/2 cells. Radiat. Res. *94*, 105-134.

WHO, World Health Organization (1978): Health implications of nuclear power production. Kopenhagen.

WHO, World Health Organization (1982): Non ionizing radiation protection. Kopenhagen: WHO Regional Office for Europe, 267 p.

WILSON, W.E., PARETZKE, H.G. (1981): Calculation of distributions for energy imparted and ionization by fast protons in nanometer sites. Radiat. Res. *87*, 521-537.

WINACKER, E.L. (1985): Gene und Klone. Weinheim: Verlag Chemie VCH, 454 p.

WINGATE, C.L., BAUM, J.W. (1976): Measured radial distribution of dose and LET for alpha and proton beams in hydrogen and tissue equivalent gas. Radiat. Res. *65*, 1-19.

WITHERS, H.R. (1967): The dose-survival relationship for irradiation of epithelial cells of mouse skin. Brit. J. Radiol. *40*, 335-343.

WITHERS, H.R. (1975): The four R's of radiotherapy. Adv. Radiat. Biol. *5*, 241-272.

WITHERS, H.R. (1975a): Responses of some normal tissues to low doses of τ-radiation. In: ALPER (1975); p. 369-375.

WITHERS, H.R., THAMES, H.D., PETERS, L.J. (1982): Differences in the fractionated response of acute and late responding tissue. Progress in Radio-oncology II, p. 257-296.

WITHROW, T.J., LUGO, M.H., DEMPSEY, M.J. (1980): Transformation of balb 3T3 cells exposed to a germieidal UV lamp and a sunlamp. Photochem. Photobiol. *31*, 135-142.

WITKIN, E.M. (1966): Radiation induced mutation and their repair. Science *152*, 1345-1353.

WITKIN, E.M. (1976): Ultraviolet mutagenesis and inducible DNA repair in Escherichia coli. Bacteriol. Rev. *40*, 869-907.

WOLFF, S. (1972): Chromosome aberration induced by ultraviolet radiation. Photophysiol. 7, 189-207.

WOLSKY, A. (1982): The effect of radiations on developmental processes. In: PIZZARELLO (1982), p. 149-192.

WOOD, R.D., SEDGWICK, S.G. (1986): Molecular aspects of mutagenesis. Mutagenesis *1*, 399-405.

ZAIDER, M., ROSSI, H.H. (1986): Microdosimetry and its application to biological processes. In: ORTON (1986b), p. 171-242.

ZAIDER, M., ROSSI, H.H. (1988): On the application of microdosimetry to radiobiology. Radiat. Res. *113*, 15-24.

ZIMMER, K.G. (1961): Quantitative Radiation Biology. Oliver and Boyd: Edinburgh and London.

ZOELZER, F., KIEFER, J. (1984): Wavelength dependence of inactivation and mutation induction to 6-thioguanine resistance in V79 Chinese hamster fibroblasts. Photochem. Photobiol. *40*, 49-53.

Subject Index